ADVANCED AIRCRAFT DESIGN

Aerospace Series List

Advanced Aircraft Design: Conceptual Design, Analysis and Optimization of Subsonic Civil Airplanes	Torenbeek	June 2013
Design and Analysis of Composite Structures: With Applications to Aerospace Structures, Second Edition	Kassapoglou	April 2013
Aircraft Systems Integration of Air-Launched Weapons	Rigby	April 2013
Design and Development of Aircraft Systems, Second Edition	Moir and Seabridge	November 2012
Understanding Aerodynamics: Arguing from the Real Physics	McLean	November 2012
Aircraft Design: A Systems Engineering Approach	Sadraey	October 2012
Introduction to UAV Systems, Fourth Edition	Fahlstrom and Gleason	August 2012
Theory of Lift: Introductory Computational Aerodynamics with MATLAB and Octave	McBain	August 2012
Sense and Avoid in UAS: Research and Applications	Angelov	April 2012
Morphing Aerospace Vehicles and Structures	Valasek	April 2012
Gas Turbine Propulsion Systems	MacIsaac and Langton	July 2011
Basic Helicopter Aerodynamics, Third Edition	Seddon and Newman	July 2011
Advanced Control of Aircraft, Spacecraft and Rockets	Tewari	July 2011
Cooperative Path Planning of Unmanned Aerial Vehicles	Tsourdos et al	November 2010
Principles of Flight for Pilots	Swatton	October 2010
Air Travel and Health: A Systems Perspective	Seabridge et al	September 2010
Unmanned Aircraft Systems: UAVS Design, Development and Deployment	Austin	April 2010
Introduction to Antenna Placement and Installations	Macnamara	April 2010
Principles of Flight Simulation	Allerton	October 2009
Aircraft Fuel Systems	Langton et al	May 2009
The Global Airline Industry	Belobaba	April 2009
Computational Modelling and Simulation of Aircraft and the Environment: Volume 1 - Platform Kinematics and Synthetic Environment	Diston	April 2009
Handbook of Space Technology	Ley, Wittmann Hallmann	April 2009
Aircraft Performance Theory and Practice for Pilots	Swatton	August 2008
Aircraft Systems, Third Edition	Moir and Seabridge	March 2008
Introduction to Aircraft Aeroelasticity and Loads	Wright and Cooper	December 2007
Stability and Control of Aircraft Systems	Langton	September 2006
Military Avionics Systems	Moir and Seabridge	February 2006
Design and Development of Aircraft Systems	Moir and Seabridge	June 2004
Aircraft Loading and Structural Layout	Howe	May 2004
Aircraft Display Systems	Jukes	December 2003
Civil Avionics Systems	Moir and Seabridge	December 2002

ADVANCED AIRCRAFT DESIGN

CONCEPTUAL DESIGN, ANALYSIS AND OPTIMIZATION OF SUBSONIC CIVIL AIRPLANES

Egbert Torenbeek

Delft University of Technology, The Netherlands

A John Wiley & Sons, Ltd., Publication

Registered office
John Wiley & Sons Ltd, The Atrium, Southern Gate, Chichester, West Sussex, PO19 8SQ, United Kingdom

For details of our global editorial offices, for customer services and for information about how to apply for permission to reuse the copyright material in this book please see our website at www.wiley.com.

Library of Congress Cataloging-in-Publication Data

Torenbeek, Egbert.
  Advanced aircraft design : conceptual design, analysis, and optimization of subsonic civil airplanes / Egbert Torenbeek.
      pages cm
  Includes bibliographical references and index.
  ISBN 978-1-118-56811-8 (cloth)
 1. Transport planes–Design and construction.   2. Jet planes–Design and construction.
3. Airplanes–Performance.   I. Title.
  TL671.2.T668   2013
  629.133'34–dc23

                                                                            2013005449

A catalogue record for this book is available from the British Library.

ISBN: 9781119969303

Typeset in 10/12pt Times by Aptara Inc., New Delhi, India

Contents

Foreword   xv

Series Preface   xix

Preface   xxi

Acknowledgements   xxv

1   Design of the Well-Tempered Aircraft   1
1.1   How Aircraft Design Developed   1
        1.1.1   Evolution of Jetliners and Executive Aircraft   1
        1.1.2   A Framework for Advanced Design   4
        1.1.3   Analytical Design Optimization   4
        1.1.4   Computational Design Environment   5
1.2   Concept Finding   6
        1.2.1   Advanced Design   6
        1.2.2   Pre-conceptual Studies   7
1.3   Product Development   8
        1.3.1   Concept Definition   10
        1.3.2   Preliminary Design   11
        1.3.3   Detail Design   13
1.4   Baseline Design in a Nutshell   13
        1.4.1   Baseline Sizing   13
        1.4.2   Power Plant   15
        1.4.3   Weight and Balance   16
        1.4.4   Structure   16
        1.4.5   Performance Analysis   17
        1.4.6   Closing the Loop   18
1.5   Automated Design Synthesis   19
        1.5.1   Computational Systems Requirements   19
        1.5.2   Examples   20
        1.5.3   Parametric Surveys   21
1.6   Technology Assessment   22
1.7   Structure of the Optimization Problem   25
        1.7.1   Analysis Versus Synthesis   25
        1.7.2   Problem Classification   26
        Bibliography   27

2	**Early Conceptual Design**	**31**
2.1	Scenario and Requirements	31
	2.1.1 What Drives a Design?	31
	2.1.2 Civil Airplane Categories	33
	2.1.3 Top Level Requirements	35
2.2	Weight Terminology and Prediction	36
	2.2.1 Method Classification	36
	2.2.2 Basic Weight Components	37
	2.2.3 Weight Limits	39
	2.2.4 Transport Capability	39
2.3	The Unity Equation	41
	2.3.1 Mission Fuel	43
	2.3.2 Empty Weight	44
	2.3.3 Design Weights	45
2.4	Range Parameter	46
	2.4.1 Aerodynamic Efficiency	47
	2.4.2 Specific Fuel Consumption and Overall Efficiency	48
	2.4.3 Best Cruise Speed	49
2.5	Environmental Issues	51
	2.5.1 Energy and Payload Fuel Efficiency	51
	2.5.2 'Greener by Design'	54
	Bibliography	56
3	**Propulsion and Engine Technology**	**59**
3.1	Propulsion Leading the Way	59
3.2	Basic Concepts of Jet Propulsion	60
	3.2.1 Turbojet Thrust	60
	3.2.2 Turbofan Thrust	61
	3.2.3 Specific Fuel Consumption	62
	3.2.4 Overall Efficiency	63
	3.2.5 Thermal and Propulsive Efficiency	63
	3.2.6 Generalized Performance	65
	3.2.7 Mach Number and Altitude Effects	66
3.3	Turboprop Engines	67
	3.3.1 Power and Specific Fuel Consumption	67
	3.3.2 Generalized Performance	68
	3.3.3 High Speed Propellers	69
3.4	Turbofan Engine Layout	70
	3.4.1 Bypass Ratio Trends	70
	3.4.2 Rise and Fall of the Propfan	72
	3.4.3 Rebirth of the Open Rotor?	74
3.5	Power Plant Selection	74
	3.5.1 Power Plant Location	75
	3.5.2 Alternative Fuels	76
	3.5.3 Aircraft Noise	77
	Bibliography	78

4   Aerodynamic Drag and Its Reduction   81
4.1   Basic Concepts   81
         4.1.1   Lift, Drag and Aerodynamic Efficiency   82
         4.1.2   Drag Breakdown and Definitions   83
4.2   Decomposition Schemes and Terminology   84
         4.2.1   Pressure and Friction Drag   84
         4.2.2   Viscous Drag   85
         4.2.3   Vortex Drag   85
         4.2.4   Wave Drag   86
4.3   Subsonic Parasite and Induced Drag   87
         4.3.1   Parasite Drag   87
         4.3.2   Monoplane Induced Drag   90
         4.3.3   Biplane Induced Drag   91
         4.3.4   Multiplane and Boxplane Induced Drag   94
4.4   Drag Polar Representations   95
         4.4.1   Two-term Approximation   95
         4.4.2   Three-term Approximation   96
         4.4.3   Reynolds Number Effects   97
         4.4.4   Compressibility Correction   98
4.5   Drag Prediction   99
         4.5.1   Interference Drag   100
         4.5.2   Roughness and Excrescences   101
         4.5.3   Corrections Dependent on Operation   102
         4.5.4   Estimation of Maximum Subsonic L/D   102
         4.5.5   Low-Speed Configuration   104
4.6   Viscous Drag Reduction   106
         4.6.1   Wetted Area   107
         4.6.2   Turbulent Friction Drag   108
         4.6.3   Natural Laminar Flow   108
         4.6.4   Laminar Flow Control   110
         4.6.5   Hybrid Laminar Flow Control   111
         4.6.6   Gains, Challenges and Barriers of LFC   112
4.7   Induced Drag Reduction   114
         4.7.1   Wing Span   114
         4.7.2   Spanwise Camber   115
         4.7.3   Non-planar Wing Systems   115
         Bibliography   115

5   From Tube and Wing to Flying Wing   121
5.1   The Case for Flying Wings   121
         5.1.1   Northrop's All-Wing Aircraft   121
         5.1.2   Flying Wing Controversy   123
         5.1.3   Whither All-Wing Airliners?   124
         5.1.4   Fundamental Issues   126
5.2   Allocation of Useful Volume   127
         5.2.1   Integration of the Useful Load   128
         5.2.2   Study Ground Rules   128

5.2.3	*Volume Ratio*	129
5.2.4	*Zero-Lift Drag*	130
5.2.5	*Generalized Aerodynamic Efficiency*	131
5.2.6	*Partial Optima*	132
5.3	Survey of Aerodynamic Efficiency	134
5.3.1	*Altitude Variation*	134
5.3.2	*Aspect Ratio and Span*	135
5.3.3	*Engine-Airframe Matching*	136
5.4	Survey of the Parameter *ML/D*	138
5.4.1	*Optimum Flight Conditions*	138
5.4.2	*The Drag Parameter*	139
5.5	Integrated Configurations Compared	140
5.5.1	*Conventional Baseline*	141
5.5.2	*Is a Wing Alone Sufficient?*	143
5.5.3	*Blended Wing Body*	144
5.5.4	*Hybrid Flying Wing*	146
5.5.5	*Span Loader*	147
5.6	Flying Wing Design	149
5.6.1	*Hang-Ups or Showstopper?*	149
5.6.2	*Structural Design and Weight*	150
5.6.3	*The Flying Wing: Will It Fly?*	151
	Bibliography	152
6	**Clean Sheet Design**	**157**
6.1	Dominant and Radical Configurations	157
6.1.1	*Established Configurations*	157
6.1.2	*New Paradigms*	159
6.2	Morphology of Shapes	159
6.2.1	*Classification*	160
6.2.2	*Lifting Systems*	160
6.2.3	*Plan View Classification*	162
6.2.4	*Strut-Braced Wings*	163
6.2.5	*Propulsion and Concept Integration*	164
6.3	Wing and Tail Configurations	165
6.3.1	*Aerodynamic Limits*	165
6.3.2	*The Balanced Design*	167
6.3.3	*Evaluation*	168
6.3.4	*Relaxed Inherent Stability*	169
6.4	Aircraft Featuring a Foreplane	169
6.4.1	*Canard Configuration*	170
6.4.2	*Three-Surface Aircraft*	172
6.5	Non-Planar Lifting Systems	173
6.5.1	*Transonic Boxplane*	173
6.5.2	*C-Wing*	175
6.6	Joined Wing Aircraft	177
6.6.1	*Structural Principles and Weight*	178
6.6.2	*Aerodynamic Aspects*	179

	6.6.3	Stability and Control	180
	6.6.4	Design Integration	181
6.7	Twin-Fuselage Aircraft	182
	6.7.1	Design Integration	185
6.8	Hydrogen-Fuelled Commercial Transports	186
	6.8.1	Properties of LH2	187
	6.8.2	Fuel System	188
	6.8.3	Handling Safety, Economics and Logistics	189
6.9	Promising Concepts	189
	Bibliography	190

7	**Aircraft Design Optimization**	**197**
7.1	The Perfect Design: An Illusion?	197
7.2	Elements of Optimization	198
	7.2.1	Design Parameters	198
	7.2.2	Optimal Control and Discrete-Variable Optimization	199
	7.2.3	Basic Terminology	200
	7.2.4	Single-Objective Optimization	201
	7.2.5	Unconstrained Optimizer	202
	7.2.6	Constrained Optimizer	204
7.3	Analytical or Numerical Optimization?	206
	7.3.1	Analytical Approach	206
	7.3.2	Multivariate Optimization	207
	7.3.3	Unconstrained Optimization	209
	7.3.4	Constrained Optimization	210
	7.3.5	Response Surface Approximation	211
	7.3.6	Global Models	212
7.4	Large Optimization Problems	213
	7.4.1	Concept Sizing and Evaluation	213
	7.4.2	Multidisciplinary Optimization	214
	7.4.3	System Decomposition	215
	7.4.4	Multilevel Optimization	217
	7.4.5	Multi-Objective Optimization	218
7.5	Practical Optimization in Conceptual Design	219
	7.5.1	Arguments of the Sceptic	219
	7.5.2	Problem Structure	220
	7.5.3	Selecting Selection Variables	220
	7.5.4	Design Sensitivity	222
	7.5.5	The Objective Function	222
	Bibliography	223

8	**Theory of Optimum Weight**	**229**
8.1	Weight Engineering: Core of Aircraft Design	229
	8.1.1	Prediction Methods	230
	8.1.2	Use of Statistics	231

8.2   Design Sensitivity   232
        8.2.1   Problem Structure   232
        8.2.2   Selection Variables   233
8.3   Jet Transport Empty Weight   234
        8.3.1   Weight Breakdown   234
        8.3.2   Wing Structure (Item 10)   235
        8.3.3   Fuselage Structure (Item 11)   236
        8.3.4   Empennage Structure (Items 12 and 13)   237
        8.3.5   Landing Gear Structure (Item 14)   238
        8.3.6   Power Plant and Engine Pylons (Items 2 and 15) 238
        8.3.7   Systems, Furnishings and Operational Items (Items 3, 4 and 5) 238
        8.3.8   Operating Empty Weight: Example   239
8.4   Design Sensitivity of Airframe Drag   239
        8.4.1   Drag Decomposition   240
        8.4.2   Aerodynamic Efficiency   242
8.5   Thrust, Power Plant and Fuel Weight   243
        8.5.1   Installed Thrust and Power Plant Weight   243
        8.5.2   Mission Fuel   245
        8.5.3   Propulsion Weight Penalty   245
        8.5.4   Wing and Propulsion Weight Fraction   248
        8.5.5   Optimum Weight Fractions Compared   249
8.6   Take-Off Weight, Thrust and Fuel Efficiency   249
        8.6.1   Maximum Take-Off Weight   249
        8.6.2   Installed Thrust and Fuel Energy Efficiency   251
        8.6.3   Unconstrained Optima Compared   252
        8.6.4   Range for Given MTOW   253
        8.6.5   Extended Range Version   254
8.7   Summary and Reflection   254
        8.7.1   Which Figure of Merit?   254
        8.7.2   Conclusion   256
        8.7.3   Accuracy   257
        Bibliography   257

9   Matching Engines and Airframe   **261**
9.1   Requirements and Constraints   261
9.2   Cruise-Sized Engines   262
        9.2.1   Installed Take-Off Thrust   262
        9.2.2   The Thumbprint   263
9.3   Low Speed Requirements   265
        9.3.1   Stalling Speed   265
        9.3.2   Take-Off Climb   266
        9.3.3   Approach and Landing Climb   266
        9.3.4   Second Segment Climb Gradient   267
9.4   Schematic Take-Off Analysis   267
        9.4.1   Definitions of Take-Off Field Length   268
        9.4.2   Take-Off Run   269

	9.4.3	*Airborne Distance*	270
	9.4.4	*Take-Off Distance*	270
	9.4.5	*Generalized Thrust and Span Loading Constraint*	271
	9.4.6	*Minimum Thrust for Given TOFL*	273
9.5	Approach and Landing		273
	9.5.1	*Landing Distance Analysis*	273
	9.5.2	*Approach Speed and Wing Loading*	274
9.6	Engine Selection and Installation		275
	9.6.1	*Identifying the Best Match*	275
	9.6.2	*Initial Engine Assessment*	276
	9.6.3	*Engine Selection*	277
	Bibliography		278

10	**Elements of Aerodynamic Wing Design**		**281**
10.1	Introduction		281
	10.1.1	*Problem Structure*	282
	10.1.2	*Relation to Engine Selection*	283
10.2	Planform Geometry		283
	10.2.1	*Wing Area and Design Lift Coefficient*	285
	10.2.2	*Span and Aspect Ratio*	286
10.3	Design Sensitivity Information		286
	10.3.1	*Aerodynamic Efficiency*	287
	10.3.2	*Propulsion Weight Contribution*	288
	10.3.3	*Wing and Tail Structure Weight*	289
	10.3.4	*Wing Penalty Function and MTOW*	290
10.4	Subsonic Aircraft Wing		291
	10.4.1	*Problem Structure*	291
	10.4.2	*Unconstrained Optima*	292
	10.4.3	*Minimum Propulsion Weight Penalty*	294
	10.4.4	*Accuracy*	294
10.5	Constrained Optima		295
	10.5.1	*Take-Off Field Length*	296
	10.5.2	*Tank Volume*	296
	10.5.3	*Wing and Tail Weight Fraction*	297
	10.5.4	*Selection of the Design*	297
10.6	Transonic Aircraft Wing		298
	10.6.1	*Geometry*	298
	10.6.2	*Wing Drag in the Design Condition*	299
	10.6.3	*Modified Wing Penalty Function*	300
	10.6.4	*Thickness Ratio Limit*	301
	10.6.5	*WPF Affected by Sweep Angle and Thickness Ratio*	303
10.7	Lift Coefficient and Aspect Ratio		304
	10.7.1	*Partial Optima*	304
	10.7.2	*Constraints*	306
	10.7.3	*Refining the Optimization*	307

10.8	Detailed Design	309
	10.8.1 Taper and Lift Distribution	309
	10.8.2 Camber and Twist Distribution	310
	10.8.3 Forward Swept Wing (FSW)	311
	10.8.4 Wing-Tip Devices	312
10.9	High Lift Devices	313
	10.9.1 Aerodynamic Effects	313
	10.9.2 Design Aspects	314
	Bibliography	315
11	**The Wing Structure and Its Weight**	**319**
11.1	Introduction	319
	11.1.1 Statistics can be Useful	319
	11.1.2 Quasi-Analytical Weight Prediction	320
11.2	Methodology	321
	11.2.1 Weight Breakdown and Structural Concept	321
	11.2.2 Basic Approach	323
	11.2.3 Load Factors	324
11.3	Basic Wing Box	326
	11.3.1 Bending due to Lift	326
	11.3.2 Bending Material	331
	11.3.3 Shear Material	333
	11.3.4 In-Plane Loads and Torsion	334
	11.3.5 Ribs	334
11.4	Inertia Relief and Design Loads	335
	11.4.1 Relief due to Fixed Masses	336
	11.4.2 Weight-Critical UL and Design Weights	337
11.5	Non-Ideal Weight	338
	11.5.1 Non-Taper, Joints and Fasteners	339
	11.5.2 Fail Safety and Damage Tolerance	340
	11.5.3 Manholes and Access Hatches	340
	11.5.4 Reinforcements, Attachments and Support Structure	341
	11.5.5 Dynamic Over Swing	342
	11.5.6 Torsional Stiffness	342
11.6	Secondary Structures and Miscellaneous Items	344
	11.6.1 Fixed Leading Edge	345
	11.6.2 Leading Edge High-Lift Devices	345
	11.6.3 Fixed Trailing Edge	346
	11.6.4 Trailing Edge Flaps	346
	11.6.5 Flight Control Devices	348
	11.6.6 Tip Structures	348
	11.6.7 Miscellaneous Items	349
11.7	Stress Levels in Aluminium Alloys	349
	11.7.1 Lower Panels	350
	11.7.2 Upper Panels	350
	11.7.3 Shear Stress in Spar Webs	352

11.8   Refinements   352
        11.8.1   Tip Extensions   352
        11.8.2   Centre Section   353
        11.8.3   Compound Taper   354
        11.8.4   Exposed Wing Lift   355
        11.8.5   Advanced Materials   355
11.9   Application   357
        11.9.1   Basic Ideal Structure Weight   357
        11.9.2   Refined Ideal Structure Weight   358
        11.9.3   Wing Structure Weight   359
        11.9.4   Accuracy   359
        11.9.5   Conclusion   360
        Bibliography   361

12   Unified Cruise Performance   363
12.1   Introduction   363
        12.1.1   Classical Solutions   363
        12.1.2   Unified Cruise Performance   364
        12.1.3   Specific Range and the Range Parameter   365
12.2   Maximum Aerodynamic Efficiency   366
        12.2.1   Logarithmic Drag Derivatives   368
        12.2.2   Interpretation of Log-Derivatives   369
        12.2.3   Altitude Constraint   370
12.3   The Parameter *ML/D*   371
        12.3.1   Subsonic Flight Mach Number   371
        12.3.2   Transonic Flight Mach Number   372
12.4   The Range Parameter   374
        12.4.1   Unconstrained Optima   374
        12.4.2   Constrained Optima   376
        12.4.3   Interpretation of η_M   376
        12.4.4   Optimum Cruise Condition   378
12.5   Range in Cruising Flight   379
        12.5.1   Bréguet Range Equation   379
        12.5.2   Continuous Cruise/Climb   380
        12.5.3   Horizontal Cruise, Constant Speed   381
        12.5.4   Horizontal Cruise, Constant Lift Coefficient   381
12.6   Cruise Procedures and Mission Fuel   382
        12.6.1   Subsonic Flight   382
        12.6.2   Transonic Flight   383
        12.6.3   Cruise Fuel   384
        12.6.4   Mission Fuel   385
        12.6.5   Reserve Fuel   387
12.7   Reflection   388
        12.7.1   Summary of Results   388
        12.7.2   The Design Connection   389
        Bibliography   390

A	**Volumes, Surface and Wetted Areas**	**393**
A.1	Wing	393
A.2	Fuselage	394
A.3	Tail Surfaces	395
A.4	Engine Nacelles and Pylons	395
A.5	Airframe Wetted Area	395
	Bibliography	396

B	**International Standard Atmosphere**	**397**

C	**Abbreviations**	**399**

Index		**403**

Foreword

Aircraft design is a very fascinating and motivating topic for pupils, students and young researchers. They are interested in the engineering subject, knowing that this is a complex subject with the aerodynamics to make the aircraft fly, with the structural layout to accommodate some sort of payload and keep the integrity of the vehicle, and with the aspects of flight mechanics to stabilize and control the aircraft, just to mention the basic aspects. In the scientific world, the faculties of aerospace engineering follow this principle and consider the basic disciplines such as aerodynamics, lightweight structures, flight mechanics and space technologies as the fundamentals to provide the envelope for aeronautics and space for the engineering students. Aircraft design is normally not considered a specific discipline worthy of inaugurating a specific chair. Some exceptions, however, do exist. The Delft University of Technology was one of the first Technical Universities in Europe to inaugurate a specific chair for aircraft design, and with the nomination of Egbert Torenbeek in 1980 they found a very strong personality who has further developed the scientific approach and methodology for preliminary aircraft design. The Technical University of München (TUM) in 1995 established a new chair for aeronautical engineering with the specific focus on aircraft design and I was nominated for this chair. This shows that the focus of integrated aircraft design has only slowly found its role in the scientific world.

A similar view can also be seen in industry. During my time at Airbus, the Technical management was not fully convinced that the aircraft design had the same importance and role as the big engineering departments like aerodynamics, structures, systems, propulsion and cabin. On the other hand, Airbus suddenly discovered about some ten years ago with some urgency that they did not have enough engineers with sufficient global knowledge to understand the total aircraft as a complex system. A huge push was then started to develop within the company 'aircraft architects' and 'aircraft integrators', also highlighting, that the discipline 'aircraft design' with its specific knowledge and experience is of prime importance.

There is, however, a huge discrepancy between industry and research centres or universities with regard to integrated aircraft design. Industry claims and wishes that universities as well as research centres should not look too closely at aircraft integration; this is seen as the unique role of industry. Industry claims to be the only partner, who knows the market demand and who has to consider the right design approach with respect to time, cost, quality and risk before deciding on a new product and its introduction onto the market. Industry therefore would like to keep the universities out of the domain of aircraft design, and do not want to give too many details to the scientific community, on how to prepare an innovative aircraft design. On the other hand, students and young engineers have to be trained and have to learn and understand the basic features of aircraft design at university during their studies. Students are primarily not so much fascinated by details of low speed aerodynamics or the detailed design of a fuselage frame compared to designing an aircraft. They are motivated to develop aircraft models, sailplanes and want to know how to design this sort of flying vehicles and what is the approach to defining the size of the wing, tailplane and engines. The scientific approach to aircraft design is therefore a major topic for the universities and has to be part of the aeronautical engineering curriculum.

There are several good books on the market, one of the best in my view written by Egbert Torenbeek. But these books were written mainly in the years 1980 to 1990 and have established a lot of design data, collected from aircraft designs of the 1960s to the 1980s. Also at that time the focus was on the preliminary aircraft design, starting from the weight breakdown, defining wing and tailplane areas and checking stability and controllability.

Over the past twenty years, computer capabilities have improved considerably and a lot of aircraft design software programs are distributed on the market with some quite good success and good results as long as the aircraft design follows the classical design features. The new dimension which has been added to the aircraft design process is called multidisciplinary optimization (MDO) methodologies. The continuous increase in computer speed and capacity has first allowed FEM methods for all sort of structural layout and CFD methods for the aerodynamic design of aircraft components and the total aircraft to be developed. The next steps were then multidisciplinary tools, first, to integrate the different design boundaries such as high-speed and low-speed aerodynamics, and in a next step, today the multidisciplinary methods permit an aircraft to be designed by using the integration of aerodynamic, structural and flight mechanics design constraints and by using multidisciplinary optimization methodologies. MDO is the new design methodology for all aircraft design features and nearly all papers in aircraft design are now using some sort of multidisciplinary optimization approach.

I remember that some five years ago – sitting on the Wolga beach in Samara (Russia) during a seminar for aircraft design professors – we had some lively discussions on some aircraft optimization problems. We also learned that Egbert Torenbeek was working on a new book about advanced aircraft design. However, he had some doubts whether there were still enough people interested in learning about the complex aspects of advanced aircraft design, while all institutions are just working with big and complex software tools. He was not sure whether the aircraft community would like to see such a book. We encouraged him very much to continue. Egbert Torenbeek has a very high reputation among the aircraft design professors and I am very happy to see that he finally managed to finish his book. Having read several chapters, I really believe that his way of addressing a quasi-analytical approach to aircraft design is very valuable and an excellent complementary way to the common normal approach of computerized analysis.

In the next decades, the aeronautical industry will be faced with considerable new environmental challenges. The past success of air transport will be confronted with new questions like 'Which optimal flight altitude will have minimum impact on the atmosphere?' or 'How can new aircraft concepts with new engine options like Open Rotors improve fuel efficiency and also the environmental footprint for a given mission?' I am convinced that new aircraft concepts for the future will be required to cope better with the increasing environmental restrictions which air transport will have to face. This book will be of great help and interest for these sorts of questions where the impact of new boundary conditions will have to be analyzed and investigated and where the large industrial computer software is not yet properly validated and verified. The physics-based approach of this book will help to better qualify the dominant parameters for different new and unconventional aircraft concepts and also help the reader to understand the assessment of benefits and risks of these concepts.

I wish this book a lot of success and hope that my colleagues from industry and the scientific community and especially the young scientists will appreciate this book as well.

Prof.h.c. Dr.-Ing. Dr.h.c. Dieter Schmitt
Aeronautical consultant. Former Head of Future Projects at Airbus SAS
Former Professor at TU München, Institute of Aeronautical Engineering
Blagnac, 25th November 2012

Series Preface

The Aerospace Series covers a wide range of aerospace vehicles and their systems, comprehensively covering aspects of structural and system design in theoretical and practical terms. This book complements the others in the Series by looking at the concept phase of design of the aircraft.

Aircraft Design is an early stage of activity in the evolution of an aircraft project starting at the concept and enduring until the preliminary design. It is time for broad thinkers, for people prepared to take risks and to understand the big picture. At this stage of an aircraft project the important issues are the shape of the aircraft, its fuel and load carrying capability and its mass leading to an assessment of its suitability to perform a mission. From ideas generated during this process will gradually emerge a solution that can be committed to design and manufacture.

The author introduces the topic with an overview of the advanced design process, considering design requirements and methodologies, considerations driving a design, followed by an example of early design mass prediction. The next stage deals with the selection of the aircraft general arrangement, an essential but complex issue which concerns new technology applications and operational properties. Decisions made at this stage involve and affect many disciplines in a project – many of those dealt with in other books in the Series. This is the challenging stage of integration and the role of the Chief Designer. Then an approach to explicit optimization by means of quasi-analytic relations is developed and the book concludes with analytical examples that are essential to advanced design in general and optimization in particular.

This is performed in a clear and concise manner to make the book a comprehensive treatise on the subject of advanced design of subsonic civil aircraft from initial sizing through to final drag calculations. There are lessons to be learned here also for military aircraft designers. It will be of great use to undergraduate and postgraduate students as well as to practitioners in the field of aircraft design and scientists in aerospace research and development. The author has given his work authority by basing it on many years of research at the Delft University of Technology where this subject is taught under the auspices of a Chair in the subject.

Peter Belobaba, Jonathan Cooper and Allan Seabridge

Preface

I don't know why people are frightened by new ideas.
It's the old ones that frighten me.

—John Cage, American composer

Advanced Design (AD) is the name for the activity of a team of engineers and analysts during the early stages of an aircraft design and development process. The point of departure is a set of top level requirements specifying payload/range capabilities, cabin accommodation, flight performance, operational, and environmental characteristics. The first design activity generates a conceptual baseline configuration defined by (electronic) drawings of its layout, a database specifying the physical characteristics and the essential technological assumptions, and an assessment of the feasibility of complying with the requirements. Designers may propose one or several concepts which are subsequently refined and compared during the second advanced design stage called the preliminary design. Conceptual design and preliminary design are crucial phases in the development process during which creativity and ingenuity are of paramount importance to support the far-reaching decisions that can make or break the programme as a whole.

Since the 1970s, aircraft design has become the subject of academic education and research at an increasing number of academic institutions which have an aerospace curriculum. Many topics typical of aircraft design projects are nowadays covered in academic courses, and educational handbooks, and an abundance of software tools have become available to support students in their design exercises. Although many academic courses pay modest attention to aircraft design, a design-oriented approach to the traditional aeronautical disciplines can contribute to an improved understanding of aeronautical science as a whole. However, design handbooks are essentially based on existing or even obsolete technology and may produce unrealistic results when applied to future advanced aircraft design projects. And design technologies are becoming more complicated due to the introduction of integrated product design technology and multidisciplinary design optimization, subjects not covered in most handbooks.

In writing this book it has been the author's aim to contribute to the advancement of aircraft design (teaching) by emphasizing clear design thinking rather than sophisticated computation or using a huge collection of statistical information. Another orientation came from industrial design staff and academic teachers who indicated that they would be particularly interested in assessments of unusual aircraft concepts and examples of practical optimization in the early design stage. It was decided to focus on subsonic transports and executive (business) aircraft. The present text combines the author's academic teaching approach with numerous results from in-depth investigations on advanced technologies and innovative aircraft configurations reported since the 1970s. Particular attention is paid to research by staff of the aircraft design chair at Delft University of Technology between 1980 and 2000. Although some information about design methodologies and statistical data of recent airplane models are included, the result is not intended to be used as a handbook in the first place. Most of the material presented is readily understood by those who have previous experience with airplane design. The niche market for this book is formed by MSc and PhD students doing design-oriented research, academic staff teaching design, advanced airplane designers and applied scientists at aeronautical research laboratories.

The contents of this book can be subdivided into the following groups of chapters.

1. Chapters 1 and 2 offer an overview of the advanced design process, design requirements and methodologies, considerations driving a design, and an example of early design weight prediction by applying the unity equation. Chapter 3 is a summary of modern gas turbine engine technology and configurations, defining characteristics such as overall efficiency and thrust lapse rates to be used in subsequent chapters. Chapter 4 introduces the reader to different methods of decomposing and predicting aerodynamic drag and to technologies for drag reduction. Many of these topics are familiar to experienced designers; some of them may be eye-openers to students or researchers.

2. Chapters 5 and 6 focus on the choice of the aircraft's general arrangement. This is an essential but complex issue since numerous decisions with respect to (new) technology applications and operational properties are involved and many of these decisions have a highly interdisciplinary sphere of influence. Chapter 5 deals with the basic question of how to allocate the useful load inside a generic combination of a wing and a fuselage body. In the past, this question gave rise to a discussion between analysts, some in favour of and some against the flying wing. However, the optimum configuration is not necessarily an all-wing aircraft or a traditional tube and wing (TAW). For instance, the blended wing body could become a viable alternative. Chapter 6 deals with clean-sheet design of aircraft which do not have a payload inside the wing. A qualitative assessment is made of several unusual concepts such as canard and three-surface configurations, highly non-linear lifting systems, the joined wing, twin-fuselage and hydrogen-propelled aircraft. An unusual configuration may be the best solution in the case of a dominant performance requirement or geometric constraint.

3. Chapters 7 to 10 are intended to develop an approach to explicit optimization by means of quasi-analytic relations between figures of merit – such as the maximum take-off weight or energy efficiency – and primary selection variables. Chapter 7 offers an overview of the general optimization problem, terminology and strategies to identify a feasible solution. Chapter 8 is primarily devoted to weight engineering, an essential discipline of aircraft design. Design-sensitive expressions are derived for the gross weight and its components. These are intended to show how a baseline design may be modified to improve different figures of merit, disregarding design constraints. Chapter 9 deals with matching the engines to the airframe by incorporating constraints on the installed engine power or thrust derived from high- and low-speed performance requirements. Chapter 10 derives analytical criteria for optimum wing planform area, aspect ratio, sweep angle and thickness ratio. Results are illustrated for a subsonic freighter and a transonic jetliner, both with a classical general arrangement.

4. The last chapters deal with subjects with a predominantly analytical character that are essential to advanced design in general and optimization in particular. Chapter 11 presents the derivation of a wing structure weight prediction methodology which satisfies most of the requirements for application to conceptual optimization. Chapter 12 explains why traditional criteria for optimum cruising flight cannot be applied to high-speed airplanes. The theory is unified for (optimum) cruise performance analysis of propeller- as well as jet-powered aircraft and includes a simplified estimation of mission and reserve fuel.

The quasi-analytical character of the present approach to conceptual design optimization cannot replace rigorous numerical methods. Intended primarily to support advanced designers and researchers and help them to understand the complex relationships between the effects on airplane characteristics of varying design parameters, the results may also be useful to validate complex design sizing and optimization programs. Moreover, the simplicity of the analytical criteria is useful to quickly estimate the effects of introducing alternative technologies for propulsion and airframe design. If used judiciously, quasi-analytical relationships can be sufficiently accurate to successfully answer 'what-if' questions and make trade-off studies such as weight growth problems, specification changes and considering derivative aircraft. From this perspective, the present book can be seen as a tribute to prominent scientists and designers from the past – such as I.H. Ashkenas, R.T. Jones, D. Küchemann, and G.H. Lee – who pioneered this approach during the era when computer-based aircraft design technology did not yet exist. The author hopes that this effort will contribute to the way of thinking of those who consider conceptual design as an art rather than a science: the art of conceiving and building well-tempered aircraft.

Acknowledgements

I am indebted to chair holders Michel van Tooren and Theo van Holten who offered me the hospitality of their disciplinary group SEAD and to Michiel Haanschoten for his professional assistance with ICT problems. I am grateful to the staff of DAR – in particular, Arvind Gangoli Rao, Gianfranco la Rocca, Dries Visser, Roelof Vos and Mark Voskuijl – for frequent interesting communications on propulsion and aircraft design and for giving valuable feedback after reading draft versions of chapters. Thanks are also due to Evert Jesse of ADSE who has been my prime consultant on the subject of weight prediction.

This book would never have been realized without the support of my wife. Dear Nellie Volker, considering my weakness to find a proper balance between the dedication you deserve and my insatiable fascination for aeronautical engineering, I am eternally grateful that you tolerated my periods of distraction and continued to respect me during the more than 10 years of writing this book.

E. Torenbeek
Delft University of Technology, The Netherlands, April 2013

1

Design of the Well-Tempered Aircraft

Let no new improvement in flying and flying equipment pass us by.

—Bill Boeing (1928)

As our industry has matured ... we have become increasingly enslaved to our data bases of
past successful achievements. Increased competitive pressures and emphasis on control of rapidly
escalating costs have combined to preclude the level of bold risks taking in exploring possible new
configuration options that might offer some further increase in performance, etc., but for which
no adequate data exist to aid development.

—J.H. McMasters [57] (2005)

1.1 How Aircraft Design Developed

1.1.1 Evolution of Jetliners and Executive Aircraft

The second half of the twentieth century has been truly revolutionary. In particular, the period
1945–1960 produced some highly innovative projects which demonstrated that propulsion
of transport aircraft by means of jet engines had become feasible. In combination with the
appearance of the sweptback wing, this resulted in a jump in maximum cruising speeds
from about 550 to more than 850 km/h (Figure 2.1). Having pioneered the B-47 swept-wing
bomber, Boeing introduced its basic jet concept to the 367-80 tanker transport and later to
the 707 passenger transport; see Figure 1.1(a). This concept proved successful and has been
adopted for jetliners almost universally since the 1960s. When one realizes that in the early
1950s designers did not yet avail themselves of the advantage of electronic computers, it will
be appreciated that this revolution in design technology was a monumental achievement.

Modern jetliners are mostly low-wing designs with two or four engines installed in nacelles
mounted underneath and to the fore of the wing leading edge. It should not be concluded, however, that since the Boeing 707 little progress has been made in configuration design. An early
example of an unusual mutation was the Sud-Est Caravelle, see Figure 1.1(b), the airliner that

Advanced Aircraft Design: Conceptual Design, Analysis and Optimization of Subsonic Civil Airplanes, First Edition. Egbert Torenbeek.
© 2013 by Egbert Torenbeek. Published 2013 by John Wiley & Sons, Ltd.

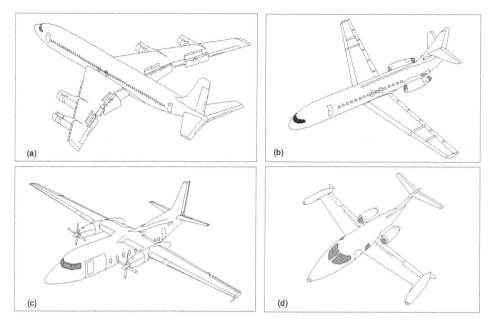

Figure 1.1   Prime examples of early post-WW II passenger aircraft. (a) Boeing 707 (1954): the first jet-powered airliner of US design. (b) Sud-Est Caravelle (1959): the first airliner with rear fuselage-mounted jet engines. (c) Fokker F 27 (1955): turboprop designed as a regional aircraft; still operational in 2012. (d) Gates Learjet 24B (1963): business jet designed in the early 1960s

pioneered jet engines attached to the rear fuselage. Even though this was a patented concept, several short-haul designs soon emerged with a similar layout and some of these were very successful. The introduction of bypass engines ($\sim$1960) and large turbofans ($\sim$1970) further improved the productivity and economy of jetliners. In combination with the strong worldwide economic expansion, this resulted in an unprecedented growth of air traffic and the almost complete extinction of competing modes of transportation over long distances, including the long-haul piston-powered and even the brand-new turboprop-powered propeller airliners.

Short-range jets initially suffered from poor low speed performances and high fuel expenditure. This market niche was filled by the four-engine Vickers Viscount and other turboprops designed in the 1950s. The twin-engine Fokker Friendship – see Figure 1.1(c) – had its Rolls-Royce Dart turboprop engines mounted to the high-set wing. This configuration was difficult to improve on and became the standard for similar propeller aircraft appearing later. Short-range turboprops have survived the twentieth century thanks to their excellent fuel economy and low operating costs. The idea of producing economy-size jets for large companies and wealthy individuals came around 1960. A prime example of a successful business jet was the Learjet depicted in Figure 1.1(d). Seating six in a slim fuselage ('no-one walks about in a Cadillac'), it outperformed jetliners of its time in maximum speed. Learjet's general arrangement, a low-wing design with engines attached to the rear fuselage and a high-set horizontal tail, has been adopted on most executive jets.

Since the introduction of the first jetliners, subsonic civil airplane technology development has advanced in an evolutionary way. During the time span between 1950 and 2000, considerable improvement has been accomplished in all technical areas, but none could be regarded as revolutionary. The basic properties of traditional designs – such as lift, drag, weight and flight performances – have become well understood. Computational methods supporting advanced design (AD) have steadily developed over a long period of time and a wealth of empirical evidence confirms their accuracy. Consequently, aircraft with a conventional layout can be developed with a high degree of confidence in the analysis. Though designing an innovative configuration will always be challenging from an engineering viewpoint, its application in an industrial project entails many challenges. This may lead to the situation that, after several years of costly configuration development, the project has to be terminated by a show stopper. It is also observed that airline management tends to avoid the uncertainties of an unusual general arrangement and prefers the purchase of a traditional configuration.

The conformity between modern airliners is not caused by the lack of conceptual creativity of designers; arguments supporting this statement can be found in publications such as [12] and [14]. In fact, several innovative designs proposed during the last decennia of the twentieth century have not been developed into a for-sale aircraft because airlines were reluctant to order them for non-technical reasons. The following projects serve as examples.

- The Boeing 7J7 project of the 1980s – Figure 1.2 (a) – was a 150-seat airliner in which new technologies were integrated: a fly-by-wire control system, unducted fan (UDF) engine technology, advanced system and flight deck technologies, and advanced aluminium alloys. The 7J7 did not find favour with the airlines mainly because the anticipated spike in fuel prices did not occur.
- Boeing's Sonic Cruiser – Figure 1.2 (b) – was designed to connect typical long-range city pairs at Mach 0.95 or above. In a business class layout for 100 seats it would attract passengers who would be willing to pay a fare premium to save several hours on long distance flights with increased comfort. The 300-seat version would be used for continental flights circumventing the large hubs. The Sonic Cruiser became the victim of the aftermath of the events following September 2001, when airlines began to re-evaluate their business models resulting in a

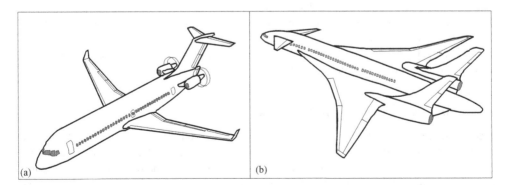

Figure 1.2 Boeing design projects which were not put into production. (a) 7J7 open rotor-powered narrow body airliner of the 1980s. (b) Sonic Cruiser long-range wide body Mach 0.95 airliner 1999–2002

preference for a more economical (slower) design which became the 787 [41]. The Sonic Cruiser was not developed into a for-sale product because potential customers would rather see its advanced technology developed for integration into an airplane optimized for lower Mach numbers.

1.1.2   A Framework for Advanced Design

The non-recurring costs of a commercial aircraft development programme are so enormous that even a relatively minor technical hiccup may be magnified into an unacceptable commercial risk. Consequently, a certain amount of conservatism is inherent in the development of civil aircraft design. In spite of this, conservatism in design is risky because it can lead to missed opportunities when maturing aerodynamic, structural and propulsive technologies are becoming available which find their best application in concepts different from the current dominant configuration.

In civil aircraft development programmes, far-reaching decisions concerning top level specifications, general arrangement, propulsion and enabling technologies are made before and during the concept finding and the conceptual design phases. The preliminary design phase is then entered during which the aircraft's characteristics are defined in more detail, initial assumptions are verified and the feasibility and risk level of the project are investigated. A year or more may elapse before management will decide to give the green light or withdraw from further development. The next phase consists of design verification (testing) and detail design during which major modifications of the basic configuration can be very labour-intensive and costly. Clearly, ESDU's trademark phrase, 'get it right the first time' is highly relevant for the initial aircraft system design process.

The observation has frequently been made that no more than a few percent of the pre-production costs are attributed by a few designers committing to a large fraction of total aircraft programme cost. In some cases this observation was made in favour of strengthening the advanced design capability of the aeronautical industry and/or the effort in academia to offer excellent aircraft design teaching. Although these arguments are fully justified, it is not always acknowledged that a large portion of aircraft programme costs is committed by merely specifying the need for the particular vehicle rather than by defining its technical and operational characteristics. If a new airplane has been developed for which no market exists, the project will be doomed to fail. The project design team cannot be blamed for a wrong go-ahead/exit decision and devoting more manpower to advanced design is not necessarily a panacea for avoiding misjudgement of the market. Although concept finding is not, in general, considered a part of the design project, it is at least as crucial to the success of a programme as the actual concept development phase.

1.1.3   Analytical Design Optimization

Since advanced design is highly relevant to the company's viability, one would expect that the discipline of design optimization has traditionally received a great deal of attention from the aeronautical community – in fact, this is not the case. Until the time of large-scale computer applications, only a few systematic efforts were made to develop a fundamental framework for non-intuitive decision-making. Most of these were small-scale programs initiated by individuals in research institutes and academia and their impact on the actual practice in design offices has not become entirely clear. Nevertheless, from the educational point of view, several approaches and trends from the past still deserve to be mentioned even though not all of them have received widespread recognition.

Early parametric surveys were made on a limited scale in the industry by experienced designers. Until the 1960s, efforts to include optimization in conceptual design were based on relatively simple methods with minimum take-off gross weight (TOGW) considered as the criterion for the figure of merit. The analytical approach to sizing and improving a design in the conceptual stage was discussed in 1948 by Cherry and Croshere Jr [19]. Though their methodology was based on experience with propeller airplanes, its systematic character appeared useful for jet aircraft as well. In 1958, G. Backhaus proposed a comprehensive (quasi-)analytical optimization of jet transports [20]. His article did not get the recognition it deserved, probably because it was published in German. Another pioneer of the analytical approach to concept optimization was D. Küchemann. During the 1960s, he and his co-workers at the Royal Aircraft Establishment in the UK developed analytical design methods of aircraft intended to fly over widely different ranges at different (subsonic, supersonic and hypersonic) speeds [21]. Part of this work was based on research in connection with the conception of Concorde and was compiled in a unique book [1]. The elegance and lucidity of Küchemann's analysis inspired the present author to initiate a systematic study of fundamental design considerations [27]; some of its results are included in the present book in a modified form. After the advent of computational design analysis and optimization technology in the 1970s, the (quasi-)analytical approach has appealed to only a few researchers; see, for example, W.H. Mason and B. Malone in [34, 35].

1.1.4 Computational Design Environment

During the first decennia after WWII, aircraft design was performed manually with the use of hand calculators and drawing boards. Despite the commercial success of several excellent airliners and business airplanes developed during this period, the 'paper method' is nowadays considered too labor-intensive and ineffective. Since the 1970s, the advancement of design technology changed fundamentally due to the availability of powerful computers and interactive graphics devices. Simultaneously, significant progress was made in the fields of computational engineering methods and numerical optimization, a trend set by early applications in astronautics and chemical engineering. The aeronautical community initially paid most attention to developing complex computer-based design synthesis programs such as those reported in [62] and [68]. Although automated design optimization has attracted much attention from research institutes such as NASA, reputed designers initially viewed these efforts with apprehension for reasons to be discussed in Chapter 2.

The penetration of ICT into all fields of aeronautics since 1980 has drastically changed the aircraft development scene. Whereas designs were traditionally almost exclusively produced by the aircraft company's design offices, reports presented at scientific conferences indicate that research institutions and universities have become new actors in the aircraft configuration design field. The remarkable expansion of multidisciplinary design optimization (MDO) and concurrent engineering methodologies have brought about a design climate change in aeronautics as well as in other engineering disciplines. Since the 1990s, the EU Framework Programs have stimulated the industry, research institutions and academia to cooperate in order to improve aircraft design technology. These efforts have resulted in improved possibilities for designers to gain insight into the impact of new technologies and concepts on the design quality in pre-competitive phases before excessive resources have to be committed.

With the intention of offering a fresh and practical approach, the present book emphasizes the fundamentals of aircraft conceptual design sizing and optimization. The treatment of advanced computational systems and the presentation of design data collections is considered to be outside its scope. Fortunately, those involved in design teaching, students and practising designers can avail themselves of an abundance of detailed guidelines for drawing up a conceptual aircraft design in excellent books quoted in the bibliography of this chapter. Several of these and other publications have been used to compile this overview. The author is also indebted to J. van Toor for his permission to quote freely from personal correspondence [43].

1.2   Concept Finding

> How an engineer generates good design concepts remains a mystery that researchers from engi-
> neering, computer and cognitive sciences are working together to unravel.
>
> —P. Raj [95]

1.2.1   Advanced Design

The essential transportation properties of a new aircraft type, its overall system concept, design data and detailed geometry are defined by the company's advanced design (AD) office which is responsible for the generation of aircraft concept proposals including the technical, technological, competitive and commercial aspects. Focussing on new product development, the AD team is active in the overall concept development and in defining its technical and operational properties. AD is a vital and essential part of product development and has a substantial influence on the company's competitiveness and effectiveness. Dependent on the internal organization, most of the AD tasks can be categorized into the following activities.

- Future projects. The prime task of a future projects team is carrying out pre-conceptual studies, conceptual design and proof of concept for a new ('clean sheet') design and making proposals for novel configurations. This complex activity has a highly multidisciplinary character which requires that individuals from functional groups such as flight physics, structures/materials and systems integration are involved in the AD process. The team must accomplish the projected task subject to boundary conditions such as top level requirements, certification rules, technical capabilities and economic environment of the company, customer operational aspects and other considerations.
- Tool development. Software tools for aircraft sizing, performance analysis, weight and cost prediction and optimization techniques are of vital importance for a successful design effort. Reflecting the expertise of the company, these tools are in general not available on the commercial market. The capability to investigate a wide variety of vehicles and alternative concepts requires the design tools to be continuously improved by making them more reliable and versatile and by incorporating and expanding design databases. Advanced designers will

also be active in merging new results from the (applied) research field with available methods and procedures. Chapter 11 illustrates how a design tool can be developed.

- Enabling technologies. Most of the company's R & D activities aim at applications with one of its (future) production programmes. AD identifies the required key technologies in accordance with the company's technology objectives and gives guidance in the development of new technologies enabling competitive products. Included activities are assessment of operational research and market analysis, and available manufacturing capabilities.
- Competition evaluation. The technical, technological and economical situation of the company's products is judged versus competing products and developments. This requires year-round exploration and modelling of competing airplanes under consideration by the same potential customers and creation of a well organized competition database.

In addition to these focussed activities, AD is responsible for highly constrained temporary tasks. These may entail, for instance, interaction with the company's sales department and with (potential) customers, external suppliers and partners. The engine selection process requires that regular contacts are made with engine manufacturers. During the validation and detailed design phases of an ongoing project, AD specifies and coordinates the peripheral activities carried out externally such as wind tunnel, structural and system testing. Another activity is developing proposals for upgrade programmes and future derivatives or modifications of the company's existing product line.

1.2.2   Pre-conceptual Studies

The starting point for any project development is an understanding of market requirements and answering the question why – rather than how – a new product will be developed. The underlying reason for any commercial aircraft programme is its ability to provide a profit to the company that designs and builds it as well as the customer that uses it. Reliable forecasts about the demand for new aircraft are obtained from continuously monitoring and assessing the advancements in aeronautical research and technology. The pre-conceptual study phase is intended to identify a product line within the company's capabilities that fits a potential market. This entails a complex process which ideally includes a dialogue between design, management, marketing and customer support. The pre-conceptual phase includes aircraft configuration trade studies identifying techniques and technology requirements suitable for integration into the new product. This is accomplished by initial aircraft sizing, engine matching, weight estimation and evolution of a family of aircraft with a given set of payload versus range combinations. At the end of the process, a management decision is expected for a selected configuration to be visualized in a provisional three view drawing.

Different terms exist for the pre-conceptual phase: companies may call it the pre-feasibility, concept finding or architectural phase. In fact, the notion of architecture refers to a transportation system – this could be an airline, a number of airlines or some other transportation service of which the future plane will be a constituent part – rather than to the characteristics of the aircraft itself. Pre-conceptual studies produce an agreed and binding set of definitions that will drive the design, generally known as top level requirements (TLRs). Together with airworthiness certification rules, these will form the principal framework of objectives and constraints for the following design phases, eventually leading to a new product development.

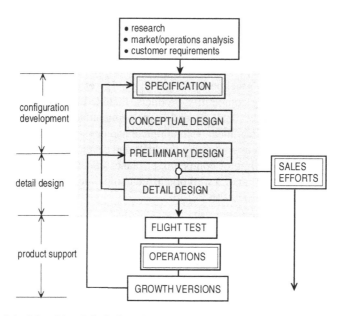

Figure 1.3   Schedule of the civil airplane development process [40]. Courtesy of J.H. McMasters

TLRs also include the criteria for a go-ahead or exit decision at the end of the product design phase. Although the concept finding phase may eventually lead to a new product, it is not usually considered as part of a design project; hence, concept finding entails a more continuous activity than project development.

Top level requirements identify characteristics that should not be subject to significant variations during project design since this could entail a violation of the transportation system architecture that has been identified as desirable for the new product. An illustration is the selection of the design cruise speed or Mach number. This parameter has a major impact on the aircraft geometry, propulsion, weight, operating costs, as well as on the way airline operations are carried out. Compared to a high cruise Mach number, a reduced speed is likely to result in a lighter aircraft structure, reduced installed engine power and less fuel consumption. But the low block speed may be detrimental to efficient and flexible operation, as well as commercial productivity.[1] Similar arguments apply to available field lengths for take-off and landing. Since these basic performance requirements are selected at the transportation system level, they should be considered as design constraints during conceptual sizing.

1.3   Product Development

The potential customer of a new airliner thinks that the life of a design begins with drawing up its requirements. However, advanced designers consider the conceptual design phase as the starting point of product development. The schedule of the development phases of a commercial or business airplane depicted in Figure 1.3 is helpful for understanding the design

---

[1]This rather complex selection problem must not be confused with mission performance analysis and optimization for an aircraft with given physical properties, a favoured, but not always properly treated topic of flight mechanics research (see Chapter 12).

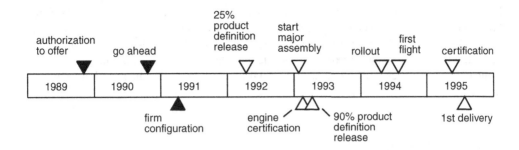

Figure 1.4   Boeing 777 project development milestones [36]

effort which is essential for a successful project. The complete process is subdivided into product design, manufacturing and testing phases. The product design process is broken down into conceptual design, preliminary design and detail design.

- Conceptual design encompasses sizing of the most promising overall aircraft concept and proof of its feasibility. Having a typical duration between 4 to 6 months for a business aircraft and 9 to 12 months for a mid-size airliner, conceptual design is characterized by cyclic design improvements and complexity increasing in time.
- The preliminary design phase aims at specifying the design concept at the main component level, sometimes including subsystem trades. Preliminary design typically lasts between 12 and 16 months.
- The detail design phase is entered when a management decision is taken to continue and give the project go-ahead.[2] This development phase is entered soon after the aircraft is committed to production and lasts between two and three years. The decision to freeze the configuration is taken early in the detail design phase when changes in the product definition are no longer appropriate.

As the aircraft goes through these phases, the level of detail and the confidence that the design will work are steadily increasing. For instance, during conceptual design, the interaction between major components such as fuselage frames, wing spars, fuel tanks, and landing gears is more important than their detailed geometry which materializes during preliminary design.

Aircraft manufacturers are active in different product lines in different markets and operate with different management methods; hence, the schedule and terminology discussed in this overview are far from universal. Dependent on the level of detail exerted in AD, the distinction between design phases is somewhat blurred, whereas some of the phases may overlap. Additional information on product development is given in the time schedule of well-defined events having the character of milestones. Figure 1.4 shows an example indicating major events such as programme go-ahead, configuration freeze, first flight, certification and first delivery. The

---

[2]The comparable term in system engineering terminology is full-scale development.

development time span between the first concept studies and certification is typically between three years for a light business aircraft and six years for a clean sheet wide body airliner. This illustrates why an aircraft must be conceived at least a decade ahead of the anticipated utilization period. Therefore, flexibility and extended duration have a strong impact on many application alternatives and the growth potential which a good design must have from its inception.

1.3.1 Concept Definition

The concept definition phase can be characterized as the highly creative and imaginative idea stage during which the component geometry, placement and connectivity of a future aircraft designed to fulfil the needs of a specific market are defined. Conceptual design also entails the development of a novel aircraft concept at an overall system level in competition with a more traditional layout. The objective is to explore a preferred configuration[3] to determine a layout which is technically superior and economically viable. This involves preliminary performance predictions and provision of three-dimensional electronic drawings with several cross-sections, an inboard profile showing the approximate placement and size of the major vehicle components. A weight and balance diagram provides another essential proof of concept.

There is an intimate relationship between the design objectives of an aircraft and the configuration concept capable of fulfilling these objectives. The overall concept describes a highly complex system which has to reach a compromise between contradicting requirements. Application of new technologies affecting all sub-systems is indispensable for economic success of the new product. The conceptual design phase is intended to generate a credible proposal of a feasible baseline design in order to convince management that it is worth the substantial resources required to develop and improve the design in further detail. Design tools are semi-empirical and low to medium fidelity methods used in trade-off studies and basic optimizations – most of the geometry is provisional. Validated design tools developed by the AD office are calibrated with statistical data bases, handbooks and historical trends, taking into account improvements expected from new technologies. The amount of data generated for a baseline design will be moderate and prediction errors are around 5%, typically.[4]

In an environment where designs are developed which fit into an existing product line, designers may investigate different fuselage cross-sections, wing positions and planforms, number and/or location of engines, empennage and undercarriage concepts. A few promising concepts are analyzed and selected for further study. Conceptual studies may also be carried out to investigate potential gains expected from new aerodynamic devices, structural concepts, materials and/or system technologies and/or an advanced engine concept. All of these features have a far-reaching effect when integrated into the design. Designers who are supposed to

---

[3]In the context of this chapter, the term configuration refers to the airplane's general arrangement, not to be confused with the same term used in flight mechanics defining operational parameters of a particular aircraft such as engine rating, flap and slat deflection angles, tailplane incidence, undercarriage position, etc.

[4]An error in the concept stage is defined as the difference between a predicted value and the established value of a parameter after completion and testing of the first aircraft. Since the certified aircraft will make its inauguration several years after concept definition, the detailed design has gone through many modifications. Hence, prediction errors are inaccuracies rather than mistakes.

explore a radical concept may consider an integrated configuration such as a blended wing body (BWB) or an all-wing aircraft (AWA) (Chapter 5). Less radical alternatives are a canard or three-surface aircraft, a twin-fuselage aircraft concept, a strut-braced or a nonplanar wing (Chapter 6). A detailed assessment of advantages and disadvantages must then be made by comparison with more conventional solutions. During all the stages of concept definition, the technical risks and costs of possible failure must be closely examined.[5]

Typical of concept design is its iterative character: primary components – wing, fuselage, nacelles, tailplane, landing gear, propulsion and other systems – are sized provisionally to result in a baseline design. Dependent on where improvements are desirable, the process may recycle to an earlier definition level at each point in time. A baseline design is not necessarily an optimized airplane and, for a traditional layout, the combined application of active constraints will normally give an adequate approximation to the best feasible design. If a novel solution is tried, the estimation of the aircraft effectiveness will be based in some areas on slender evidence and simple mathematical models. The best available model may then change rapidly with time and will probably be too crude to warrant rigorous treatment. In any case a comprehensive design optimization at the conceptual stage is of little value [66].

1.3.2   Preliminary Design

After selection of a baseline airplane concept, the design and analysis process will enter the preliminary design phase. As opposed to conceptual design which deals with the whole aircraft system, preliminary design aims at defining subsystems, making component trade-offs and optimization. Specialists from different functional groups contribute to this process of refining the initial vehicle concept – AD remains responsible for coordination. This team will (re)design the delivered baseline vehicle in sufficient detail to carry out supporting analysis and specify peripheral testing programmes but not with enough detail to specify each sub-assembly. Information to establish the programme feasibility is generated by means of sophisticated computational aerodynamical, mechanical and structural analysis, prediction of the economics and expected market penetration. Preliminary design can be characterized as setting goals for the extensive efforts to be made in the downstream detail design phase. Typical subjects are categorized as follows:

• Design definition. The baseline design team is committed to elaborate detailed analysis and sensitivity studies, with the aim of developing the best feasible configuration. This includes finding a balance between required volumes, main dimensions, weight distribution and engine performances. Details of the aircraft geometry, aerodynamic properties, structural loads and deformations and flying qualities have to be settled.
• Design validation. The predicted characteristics of the preferred configuration are verified by high-fidelity supporting analysis, simulations and test data. This becomes the final step of the preliminary design cycles.

---

[5]It is often said that 'In aircraft design one never gets a free lunch.' This statement applies in particular to conceptual design where it may be interpreted as follows: 'The greater the promise, the higher the risk of show stopper.'

A detailed analysis is usually made to determine the sensitivity of the configuration to technology inputs, performance objectives and design constraints. This makes sense since during preliminary design there still exists considerable freedom for refinements and improvements through optimization of variables such as detailed wing design, engine thrust, location of the power plant and empennage design. These trade studies have a widespread effect on most areas of the design and must therefore be carried out with scrutiny using high-fidelity analysis. Particular attention will be paid to the following issues:

- Detailed volumetric sizing and mass breakdown, centre of gravity (CG) location and loading restrictions, and moments of inertia, resulting in a considerable expansion of the design database.
- Definition of the aerodynamic shape of lifting surfaces, including high-lift devices, using computational fluid dynamics (CFD) methods and wind tunnel testing.
- Layout and sizing of main mechanical and structural concepts, aero-elastic analysis by means of finite element methods (FEM), and structural testing.
- Layout of the basic flight control system and control surfaces and devices, including prediction of flying qualities.
- Drawing up the specifications for buy-out components to be subcontracted to suppliers. This concerns the power plant (engines, propellers, nacelles) and other major aircraft systems such as the auxiliary power plant, environmental control, fuel system, hydraulic system, electrical system and avionics.
- Economic analysis in terms of operating costs. Commercial prospects are then predicted by means of economic analysis and a market penetration model.
- Analysis of environmental issues such as internal/external noise and engine emissions.

All activities are based on standards of the selected airworthiness codes and regulations as the primary measure for acceptance of design solutions. The scope and depth of the physics-based analysis and the fidelity of the computational models are increased to such a level that the impact on aircraft performance and cost of proposals to modify the baseline configuration can be quantified. Although CFD and FEM codes provide high-fidelity results, computer simulations cannot always be relied on for an accurate prediction of operational properties. Design validation will therefore require extensive wind tunnel testing[6] and testing of structural models to ensure that achieved performances will be no more than a few percent off the requirements.

Whereas the engineers involved in conceptual design generally belong to AD, specialists from various functional disciplines become involved when the preliminary design stage is entered. From then on the approach to be taken relies heavily on system engineering techniques. This phase may span a year or more with a team of dozens up to hundreds of engineers working in a multidisciplinary environment. The amount of detailed data generated is substantial, prediction errors should amount to no more than few percent. The end product is an optimized and verified airplane configuration resulting in the technical description of prototypes to be tested and the type specification of the aircraft. When the design project is

---

[6] A future addition to the presently available (high-cost) wind tunnel testing technology could be the use of free flying sub-scale models referred to as robot vehicle flight testing [56].

sufficiently mature, it may get an authorization to be offered for sale with written (contractually binding) guarantees on cost and performance. When market prospects appear to be promising, management authorizes the project go-ahead. In view of the high costs incurred by major configuration changes during the following design stage, this decision brings the iterative design cycle to an end by freezing the configuration.

1.3.3 Detail Design

The pieces of hardware that will actually be built and installed in the airframe are conceived during the detail design phase which is entered after the aircraft is committed to production. The objective of detail design is to specify the geometry of all components and plan their manufacturing processes. This encompasses drawing up instructions for the production department and the development of a careful plan for assembling the aircraft. Throughout this activity, drawings are progressively released for production. Results are complete production descriptions and specifications of all structures, systems and subsystems. Compared to preliminary design, detail design is much more labour-intensive and far more staff are involved. The participation of AD is restricted to coordinating measures to be taken when detail design appears to have an effect on the technical specifications. A typical example is when a significant empty weight growth has to be addressed by a weight reduction programme. After completion of the detail design phase, the manufacturing of components begins, followed by the final assembly of one or more flight-test vehicles, roll-out, first flight, flight testing, and certification. In the case of an all-new civil airplane development, the completion of the airworthiness certification process and first delivery to the customer will occur some five to eight years after the initial design efforts (Figure 1.4).

1.4   Baseline Design in a Nutshell

This book focusses on the conceptual and preliminary design definition rather than on the design validation and detail design. Although the described sizing procedure for conceiving a baseline design of a civil jet airplane is by no means a standard process, it contains a set of essential activity clusters. Conceptual design is a continuous iterative process that begins with an initialization of a few major characteristics to be discussed in Chapter 2. The following bird's eye view of a baseline design sizing process describes the logical order of activities visualized in Figure 1.5 leading to a converged design in as few iterations as possible. The result is a non-optimized but feasible jetliner design which complies with TLRs – the sizing process of a turboprop airplane is similar although sizing relationships are different. The diagram indicates an inner loop for satisfying payload/range and high speed performance requirements and an outer loop for making sure that low-speed properties comply with TLRs and other constraints. This approach is in the interests of an efficient iteration process but does not necessarily lead to the best feasible design. The next steps are optimization and trade-off studies as discussed in later sections.

1.4.1 Baseline Sizing

A baseline design is materialized in the form of a fairly comprehensive definition of its basic geometry, using a computer assisted design (CAD) system. Defining the geometry by

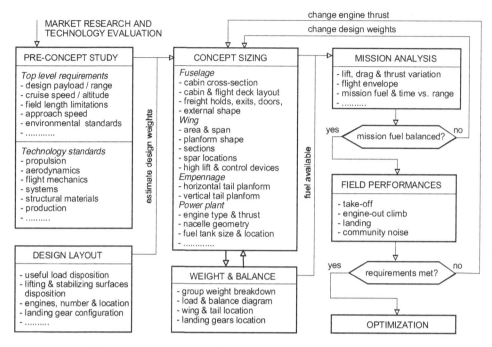

Figure 1.5 Schematic outline of a jetliner baseline design process

a parametric description helps the designer to predict the variation of mass distribution and aerodynamic properties due to variation of independent design variables such as wing area and aspect ratio, engine power or thrust, and empennage size. Figure 1.5 illustrates that initial 'guesstimates' are made of several important design characteristics.

• Design weights. Payload, fuel weight and take-off and landing gross weights corresponding to the design mission are estimated together with initial values of the empty weight and the zero fuel weight; see Chapter 2.
• Total take-off thrust and number of engines to be installed.
• Wing loading; that is, the maximum TOGW per unit of wing area.

These characteristics are revised as soon as more detailed design information becomes available. For designs with a conventional general arrangement, they are probably based on statistics, using input from the TLRs and data from similar existing or projected aircraft. For more innovative concepts, relevant publications on applied research should have been studied before starting the actual sizing.

For passenger aircraft and freighters with a traditional layout, the first geometric element to be conceived is typically a pressure cabin cross-section and floor planform, based on the specified transport function and comfort standards. Passenger cabin layout is based on selection of items such as number of passenger decks, number of seats abreast in a basic cross-section, seat pitch and number of seat rows for several accommodation standards, width and height of aisles and overhead bins, and allocation of restrooms and galleys. Pressure cabin dimensions are obtained by combining typical seating arrangements with (a) location and size of passenger and servicing entrance/exit doors and emergency exits; (b) flight deck geometry, entrance door and escape hatches; and (c) luggage/cargo hold geometry and loading/unloading hatches. Addition of a cabin wall with realistic thickness, fuselage nose and tail shapes results in a three-dimensional definition of the basic external fuselage shape. For airliners this is normally a streamlined hull with a cylindrical mid-section enclosing the contents arranged with good accessability inside the smallest possible skin. Cabin volume, floor area, and fuselage skin area are basic input for aerodynamic and weight prediction. In order to obtain an arrangement that is flexible enough to suit an airliner family concept, it may be necessary to repeat the fuselage sizing process for a future derivative or a freighter version.

The wing is the second main component to be sized. Wing geometry has to be matched to the performance requirements and is defined in terms of several parameters used in flight performance analysis. Prominent parameters of a lifting surface are the gross planform area, the span or the aspect ratio, the taper ratio, the thickness/chord ratio at the root and the tip, and the angle of sweep of the quarter-chord line. It is also desirable to specify a standard for aerodynamic wing design, such as supercritical or laminar flow technology. However, variation along the section shape and incidence along the span are the subject of downstream aerodynamic wing design (Chapter 10).

1.4.2   Power Plant

Figure 1.5 relates to the case that candidate engine types are under development. Each of the following options will lead to design processes that are different in the details.

- One option could be that a new aircraft design or a variant of an existing type is considered in response to the availability of a new engine type. In this case, installed engine performances and geometry are specified in detail by the engine manufacturer. In the interest of minimum project development costs, the basic properties of the existing aircraft are retained as much as possible, leading to a modest airframe modification. In this situation there is little or no opportunity for optimization.
- In order to match the engine to an existing airframe, its thrust is scaled up or down by means of a procedure known as engine rubberizing. A baseline engine is then characterized by its installed take-off thrust. Variations of engine performances with altitude and speed are described by generalized characteristics such as thrust as a fraction of take-off thrust – the thrust lapse rate – and specific fuel consumption (SFC) or overall efficiency of the engine. Scaling relationships for engine size and weight are derived from the engine manufacturer's information. Rubberizing includes the effects on the engine's installation characteristics which means that nacelles are scaled up or down so that performance losses due to nacelle drag and power off-takes are properly taken into account.
- A far more complicated situation occurs when different engine cycles are studied, for example, for investigating the effects of turbofan bypass ratio variation. This requires a thermodynamic cycle analysis for computing engine performances as well as prediction of engine plus nacelle geometry and weight, all of these in relation to the bypass ratio and the take-off thrust. It goes without saying that the various scaling rules should be drawn up in cooperation with the engine manufacturer's project design office.

The decision of where and how engines are attached to the airframe constitutes one of the most essential choices of layout design (Chapter 3).

1.4.3   Weight and Balance

The allowable variation of the useful load and its location determines the aircraft's loading flexibility. This is restricted by the allowable centre of gravity (CG) travel which in turn relates to the amount of (inherent or artificial) stability required for the configuration. The horizontal tail size is obtained from longitudinal stability and control requirements in combination with load and balance computations. For a given wing position relative to the fuselage and horizontal tail size/location, the following results are derived:

- Empty weights of major assemblies and their CG location, airplane CG location and moments of inertia about the three body axes. Some of these are be based on analysis, others are obtained from statistical information, using initial design weight estimates and design geometry.
- Envelopes of loaded aircraft fore and aft CG locations versus GW corresponding to likely combinations of payload and fuel.
- Allowable fore and aft CG locations for several important flight conditions derived from stability and control requirements, known as aerodynamic limits.

The required and allowable CG envelopes versus GW form the load and balance envelope which gives a tool for making sure that the operationally required CG range always fits within the aerodynamic limits. If that is not the case, measures have to be taken to obtain a balanced aircraft: (a) by relocating the wing relative to the fuselage; (b) by changing the wing sweep angle; (c) by relocating system components or operating items; or (d) by increasing the horizontal tail size. If these measures do not help, loading restrictions may be necessary – an undesirable situation. Vertical tail size follows from required directional stability, lateral control after engine failure and coping with cross-wind during the landing. When the airplane appears to be safely flyable in all practical conditions, the main landing gear elements are sized and arranged relative to the CG range so that the aircraft can be manoeuvred safely on the ground. Having located the landing gears, the designer may have to make another iteration of the CG envelope and/or horizontal tail sizing. The balancing process is thus highly iterative.

1.4.4   Structure

Part and parcel of the geometric development are the conception of a provisional structural configuration and distribution of the following system elements:

- Topology and major components of the cabin structure such as major fuselage frames where the wing, tail surfaces and engine pylons are attached, longerons, with pressure bulkheads, doors and hatches and pylons of fuselage-mounted engines.
- Principal structural elements of the wing: skin/stringer panels, spars and ribs, fixed leading and trailing edge structures, high lift devices, flight controls and their supports. Also specified are pylons of wing-mounted engines, landing gears and their attachments, and wheel wells.
- Basic structure of the empennage and its attachment to the fuselage body, taking into account provisions for flight controls and stabilizer incidence adjustment.

Size and location of major systems hardware such as the auxiliary power unit (APU) and the environmental control unit (ECU) are indicated to make sure that they can be fitted inside the available structure.

1.4.5 Performance Analysis

Aerodynamic and engine characteristics can be predicted as soon as the baseline design geometry is defined. First, lift and drag coefficients in the en-route configuration are computed as a function of the angle of attack and Mach number. Second, engine thrust and fuel consumption as a function of the altitude and Mach number are determined for several climb and cruise ratings. Before analyzing a flight profile, it is useful to check that the available thrust matches the drag at the desired cruise conditions. If that is not the case, the engines are scaled up or down in thrust and dimensions and part of the sizing loop is repeated, as indicated in Figure 1.5.

The objective of mission analysis is to compute the amount of fuel burned, distance travelled and time elapsed during each flight segment and during the complete flight from the airfield of departure to the destination. The flight profile of a civil aircraft (Figure 1.6) is composed of the segments take-off, climb and acceleration, cruise, descent, approach, aborted landing and diversion to an alternate airfield. The intention is to make sure that, when taking off, the required mission fuel is available to fly the design payload over the required distance, after which there remains sufficient reserve fuel to make a holding flight and divert to an alternate field. If a mission analysis is carried out after computation of the MTOW, it is mostly found that the range is not exactly equal to the design range specified in the TLRs. The next sizing iteration is then started as indicated in Figure 1.5. In principle, all design weights, many empty weight components and the amount of fuel have to be revised and the mission analysis is repeated until the weight distribution is balanced with the required range.

Aircraft handling and performances while taxying, taking off from and landing on the airfield, and in low-speed flight are primarily affected by take-off and landing weights, the number of engines and their forward and reverse thrust, wing area and planform shape, aerodynamic properties of high-lift and drag modulation devices, and landing gear braking capacity. The distance that a civil aircraft requires to make a safe continued or aborted take-off after failure of an engine during the take-off run is referred to as the balanced field length (BFL). This distance must not be longer than the available runway length. Moreover, it has to be verified that minimum climb-out performances can be attained in case the flight is continued

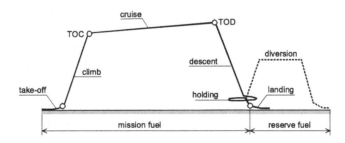

Figure 1.6 Typical flight profile

after engine failure. In commercial flights, a take-off is seldom complicated by engine failure – nevertheless, the pilot makes a flight schedule based on the case of an inoperative engine. Since in this situation low drag is most important, the high lift devices will not be deflected to their maximum extent. Climb performance lower limits apply during approach and landing as well, however, a higher lift coefficient is available compared to the take-off. Since for medium- and long-range (subsonic) aircraft the landing GW is significantly lower than the take-off GW, the approach and landing requirements are not often critical. The opposite is true for short-range aircraft for which – dependent on the deceleration due to aerodynamic drag devices, lift spoilers and wheel brakes – the approach speed and available landing distance often dictate the landing wing loading.

1.4.6   Closing the Loop

It is likely that at this stage one or more performances are inadequate and appropriate measures have to be taken. For example, if the BFL required is satisfactory but climb-out performance after engine failure is inadequate, one might consider increasing the wing span for the same area, that is, an increased aspect ratio. If both performances are unsatisfactory and cannot be improved by aerodynamic means such as more sophisticated high-lift devices, the take-off thrust may be inadequate. The final outcome of the iterations should be a certifiable baseline design which satisfies the requirements and constraints. Geometric details laid down in a three-view drawing and an inboard profile are augmented by more detailed information such as illustrated in Figure 1.7. In the case that all requirements in the TLRs are satisfied and the project can be continued, the baseline design becomes the subject of detailed analysis and verification during the preliminary design phase. If, even after reasonable modifications of the baseline aircraft, certain requirements cannot be met, the conclusion must be that the available technology for a successful project is lacking. This may lead to accepting (minor) changes in the TLRs or cancellation of the project.

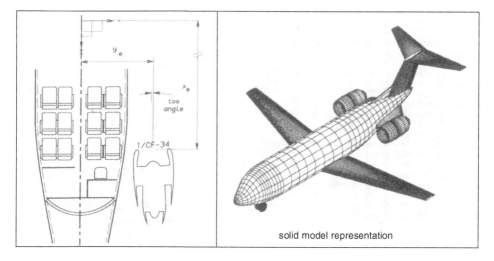

Figure 1.7   Airliner concept generated by a CAD and analysis system [84]

1.5 Automated Design Synthesis

Conceptual design tools used by aircraft manufacturers are not available on the software market since they contain sensitive information reflecting the expertise of the company. However, automated design synthesis (ADS) systems have been developed by consultancy firms, research institutes and academia. Automated design systems emphasize optimization rather than sizing. Although synthesis programs are used primarily for sizing a baseline design, they can be useful for other tasks as well. For example, figures of merit (FOMs) for evaluating the importance of design variables are operational/commercial performance and cost. Comparing the characteristics of aircraft optimized for different FOMs is a complex but revealing activity that can be carried out only with a ADS system. Investigations of unconventional concepts, technology trade-off studies between alternate designs and weight-related studies are almost invariably made by applications of ADS programs. And, last but not least, synthesis programs are useful for validating and improving drag and weight prediction tools.

1.5.1 Computational Systems Requirements

The aim of a high-fidelity design analysis is to obtain more credible predictions in the interest of reducing development risk. However, a comprehensive synthesis program is used in various stages of the design process and should therefore have the capability to provide computations with different levels of prediction accuracy. Low fidelity methods are used in the early conceptual phase to derive simple but statistically validated predictions based on only a small number of essential inputs. As the vehicle definition becomes more detailed, the fidelity of design tools increases. Accordingly, aerodynamic analysis ranges from (semi-)empirical handbook methods through several levels of computational complexity to interpretations of wind tunnel data. Other technical disciplines have similar levels. High fidelity analysis requires a detailed input which is not available until the preliminary design phase is entered. Internal and external geometry – in particular, the volumetric disposition and wetted areas – are essential inputs to the ADS program, and shape optimization requires that the geometry can be parameterized. This requires simplification by ignoring details such as small fairings between major components, wing leading and trailing edge extensions and kinks. The internal geometry is schematized by considering large structures to be built from generic components, such as the primary and secondary wing structure, the fuselage shell, floors, bulkheads and secondary structures.

ADS systems must be flexible enough to allow the design problem structure to be defined according to the designer's needs. They are usually highly modular so that the functions to be performed are clearly separated and can be assembled to perform various sequences of computations, depending on the problem. Each individual module provides the intermediate data from a particular technical discipline, thus forming the input of other modules as well as the geometric definitions needed for vehicle (component) definition. These modules should be verified prior to integration with other modules. Modularity leads to obtaining a more transparent design procedure since it simplifies data transfer and identification of errors during program development. A control program is used to exercise the individual modules and to coordinate data transfer to other modules. This set-up is required in order to ensure that in each stage of an iteration the definition of the vehicle as a system remains unambiguous and consistent. A correct problem structure is at least as important as the accuracy of the analysis methods used. Accordingly, a great deal of thinking and arranging of equations are required before the computations can be made. In contrast to many publications which emphasize the mathematical treatment of optimization, the designer's activity is a blend of interactive graphics manipulation and visualization, computation and organization of data flow. In an efficient integrated program the selection of the appropriate levels of computational fidelity and the preferred search strategy remain under the designer's control.

1.5.2 Examples

Synthesis programs contain simplified vehicle descriptions and merely mimic the process by humans to accomplish the design and analysis solutions provided by statistics and calculator-based methods from handbooks. The primary objective of the synthesis process is to carry out a trade-off between basic design parameters and characteristics that will warrant a more detailed study. In particular, wing size and engine power or thrust are primary candidates for trade-off. Computerized synthesis is a complex activity which integrates the contributions of various disciplines involved illustrated in Figure 7.8. A practical engineering approach requires a large computer program connecting the inputs and outputs of each discipline. The final result is a combined answer indicating the overall feasibility of the project.

Figure 1.8 shows schematically the principal lines of data flow of an early – but still representative – ADS program structure [62]. Analysis programs are modularized by discipline, the sequence of calculations is controlled by a separate module. Via the performance and economic evaluations block, design objectives are initially introduced as input into the program flow. Part of the concepts introduced into the modules is technology state-of-the-art information.

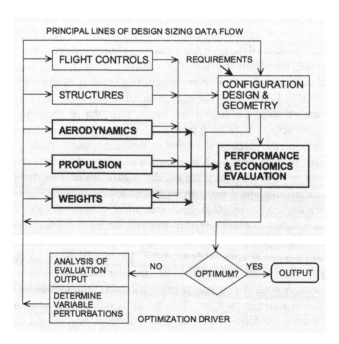

Figure 1.8 Schematic sizing and optimization data flow of a design synthesis program [62]

In aerodynamics this could be the sophistication of airfoil and high lift technology. Structural concepts' state of the art includes the degree of composite materials application in primary structures. Propulsion technology defines the gas generator compressor pressure and temperature levels and the bypass ratio. And in flight controls the degree of artificial stability is a typical concept for reducing tailplane size, drag and weight. Note that in Figure 1.8 the output from the disciplinary modules as well as the configuration design eventually goes to the performance and evaluation block. Some of the paths are direct, others are indirect via weights and configuration design. Flight systems, structures, aerodynamics and propulsion all feed weights as well as configuration design. However, since drag, thrust and aircraft weight are the only force vectors necessary for performance evaluation, the aerodynamics, propulsion and weight blocks are sufficient to deliver the data input. The design converges after several iterations and the outcome of the process is an evaluation of the design by the performance module. The user will then decide to accept the outcome or to continue the process through variations in the design. For this purpose, modern synthesis programs have a separate optimization driver containing a suite of algorithms to choose from.

Major design organizations are developing their own ADS programs. Commercially available programs developed by consulting companies are mostly based on handbooks [102] whereas several programs exist in the public domain using a similar approach as shown in Figure 1.8. Systems developed by NASA institutes are available in the public domain [75, 91, 96], systems developed in academia are reported in [80, 83, 89, 98]. Automated systems have proved effective for designing airplanes within the prescribed domain – some of them have been continuously improved and have become highly trusted. New approaches based on object-oriented and knowledge based engineering (KBE) may be particularly promising for developing sizing tools to study unusual air vehicle concepts [94, 100, 103].

1.5.3   Parametric Surveys

The traditional approach to conceptual sizing and optimization is to carry out parametric surveys by solving the system of equations for a few – typically, not more than five – selection variables defining the design space (Chapter 7). The most relevant properties of the evaluated designs are presented for inspection in tables and/or carpet plots [39] such as Figure 1.9. It should be noted that every combination of wing area and engine thrust represents a provisional but complete aircraft design for which the weight breakdown has been computed or interpolated. Since many of these designs do not comply with all TLRs, constraints are added to indicate the feasible region of the design space. The baseline design complies with all requirements, although a lower MTOW is obtained by selecting a larger wing area and a slightly smaller engine. Quasi-analytical methods for obtaining parametric surveys are derived and illustrated in Chapters 8 to 10.

Parametric surveys are rooted in the industrial approach, where they are considered valuable since the designer can make maximum use of experience and keeps complete control over decisions. These surveys accept relatively simple programming techniques such as spreadsheet methods adaptable to the situation [86]. This approach has the advantages that an *a priori* choice of a single FOM is not required and sensitivity to less-than-optimum design conditions remains visible; hence, maximum use is made of calculated results. The number of design variations and modifications can be selected in accordance with the time budget available.

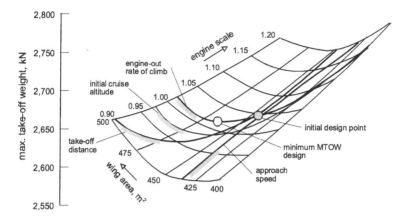

Figure 1.9 Carpet plot depicting results of a parametric survey. Wing loading and engine thrust are design variables, MTOW is the figure of merit

If the overall concept is within the designer's experience, parametric investigations may be helpful in optimization work although this pragmatic approach may not be entirely satisfactory for the following reasons.

- Due to the restricted number of selection variables, there is no guarantee that the solution represents the 'real optimum'.
- The method is practical for a small number of selection variables. The designer may resist increasing the design space since the addition of more variables will increase the amount of computation by an order of magnitude.
- The majority of the analyzed configurations are used for the purpose of the parametric study only and are discarded afterwards whereas small changes in the TLRs may make all previous results obsolete.

1.6 Technology Assessment

The challenge for conceptual designers active in an AD office is to strike a balance between the pros and cons of enabling technologies that are considered mature for application. They have to make an assessment of economic benefits, effects on safety, cabin comfort, reliability, maintainability, and environmental compatibility. The aim is to identify and select the most cost-effective breakthrough technical innovations and to find out whether they can be developed in time, and whether their integration into the aircraft system requires a departure from the conventional aircraft configuration. A long list of potential innovations can be made – the more radical of these tend to be achievable only with considerable development effort, costs and risk. However, using advanced technologies may drive up operating cost and degrade operational flexibility, maintainability or even passenger appeal. Assessment of emerging technologies must take into account possible program development risks. The enormous costs of developing jetliners are justified only when potential customers can be convinced that the new product will outperform existing types in its category both technically and economically.

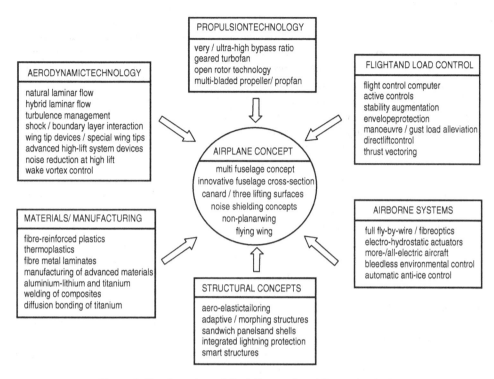

Figure 1.10 Overview of disciplines and enabling technologies

In general, this requires the introduction of an advanced but sufficiently mature technology standard in several design disciplines. Accordingly, conceptual design studies are frequently initiated to assess how the application of new technology will affect future designs rather than to develop a new aircraft development project. Disregarding the appearance of a new engine generation, the largest impact on operating economics is due to the accumulative effect of several relatively small advancements. Improvements in new airplane designs have therefore traditionally been evolutionary rather than revolutionary.

Conceptual design studies are based on choices from the most promising advancements in the state of the art; see Figure 1.10. For a particular project more alternatives can be added, other topics may already have been eliminated beforehand. Chapters 3 and 4 are dedicated to identifying improvements of aerodynamic and propulsive efficiencies. Specific technologies with a large impact on the overall airplane layout are treated extensively in Chapters 5 and 6. Factors driving the design of airliners are relevant to some extent for business aircraft design as well:

• Propulsion. Turbofans of the next generation of civil aircraft are expected to have an ultra-high bypass ratio and geared turbofans. A more radical concept is open rotor engine technology in the form of contra-rotating propfans or unducted fans. Integration into the airframe of these new engine concepts may require an innovative overall airplane concept. Turboprops for high speeds will drive multi-bladed crescent-shaped propellers.

- Airplane configurations. Evolutionary development of airliners may result in different concepts emphasizing low-cost transportation, improved cabin comfort, reduced atmospheric contamination and external noise. Radical concepts such as integrated configurations, non-planar wings, all-wing aircraft and propulsion by hydrogen fueleld engines require major investments in research and development.
- High-speed aerodynamics. Small airliners and business aircraft with a small wing leading edge sweep may achieve areas of natural laminar flow (NLF). Laminar flow control (LFC) by suction of the boundary layer through a porous skin is a potentially powerful fuel saver. Winglets and sheared tips reduce induced drag by up to 15%.
- Low-speed aerodynamics. Airplanes with stringent airfield performances require advanced high-lift devices. Options are variable camber, full-span flaps and powered lift by means of externally blown flaps or engines blowing over the wing. Reduction of aerodynamic flap/slat noise and manipulation of trailing vortices can have a favourable effect on air traffic management (ATM) and environmental compatibility.
- Flight control systems. Flight envelope protection and stability augmentation are accommodated by a flight control computer and active controls. Manoeuvre and gust load alleviation are considered additional provisions for saving structural weight and improving comfort. Direct lift control and/or thrust vectoring can be considered for unusual configurations to improve their controllability.
- Airborne systems. Except for very light aircraft, fly-by-wire has become the standard for airliners and business jets – fibre optics may be considered in the future. More-electric aircraft enable environmental and automatic icing control without making use of engine bleed air and simplify the auxiliary power unit.
- Materials and manufacturing. Composites are becoming increasingly dominant with applications of fibre-reinforced plastics in primary and secondary structures. Fibre metal laminates are applied in fatigue-sensitive areas and secondary structures. Aluminium-lithium alloys replace the traditional 5000 series, leading to about 10% weight reduction.
- Structural concepts. Affordable composite fuselages are becoming an option for many airliners and business jets. Lightning protection is required for all exposed composite structures. Aero-elastic tailoring is applied to avoid divergence or flutter and may be considered to enable a forward swept or very high aspect ratio wing. A morphing wing structure adapts the shape of high-lift devices off-design flight conditions.

Conceptual designers base their judgement concerning new technologies on experience and on how much credibility they attach to innovations, taking into account that airlines are often more conservative than airframe manufacturers. Neither of them do want to be involved in a high-risk technology leading to unexpected development, certification and maintenance costs. In this respect a (seemingly radical) technology such as active LFC is considered suspect even though its aerodynamic and operational feasibility has been demonstrated in R & T programs in the US and in Europe. Conceptual design technological standards are specified in relation to the operational profile and the time of introduction into service of the new type. For example, long-range airliners are usually large and cruise at high speeds – they require more advanced technology than regional aircraft spending more time at low altitudes and speeds. The preference for a particular technology will also be affected by the company's capabilities to incorporate them into their manufacturing processes. Although innovations generated by technological research are continuously reported worldwide, application in actual design practice can only be justified after a study of all aspects associated with their integration into the aircraft and its production process.

1.7   Structure of the Optimization Problem

The structure of an optimization problem is essential to make its results credible. This section is intended to stimulate the (inexperienced) advanced designer to avoid unrealistic results of a complex aircraft system optimization. Reference is made to Chapter 7 for the terminology used and more background information.

1.7.1   Analysis Versus Synthesis

It is useful to make a clear distinction between design analysis and synthesis. This is illustrated by showing the differences between performance analysis of a specified design and sizing a design for specified performance.

- Flight performance analysis applies to an aircraft (design) that is fully defined in terms of geometry, weight distribution, aerodynamic characteristics and propulsion system properties. The underlying physics – mainly flight mechanics and dynamics – is mono-disciplinary and the solution is fully determined. Validated programs based on substantial experience are available to assist the analyst. For instance, cruise performance analysis is a matter of routine although Chapter 12 demonstrates that for cruising at a transonic Mach number, this can be more complicated than is usually thought.
- Synthesizing an aircraft design that has to comply 'in an optimum fashion' with a specified set of TLRs. The designers' task is to find a satisfactory compromise between conflicting requirements which may lead to an indeterminate problem. The underlying physics originate from many interacting – but to a large extent fundamentally unrelated – aeronautical disciplines and the set of equations does not always have a unique solution. Although at first sight the freedom to select the independent design variables may seem overwhelming, constraints reduce the dimensionality of the design space. Guidelines based on statistics can be useful when choosing selection variables if a suitable objective function cannot be quantified, although these 'rules of a thumb' may be reliable only for configurations that are not too different from existing ones.

A practical example is selecting the preferable cruise altitude. The best initial cruise condition of an operational aircraft with given gross weight is readily obtained from performance analysis – it is usually defined as the altitude where the fuel consumption per distance travelled is minimal. For an aircraft taking off with its MTOW, this optimum is probably constrained by the available cruise thrust. On the other hand, the sizing problem is to find the definition of a design that yields the most economical performance by computing the effects on the cruise altitude of varying wing geometry and installed thrust. In the latter case, the designer has to cope with the following challenges:

- The sensitivity of dependent design variables (lift, drag, gross weight, initial cruise altitude) to variation in independent design variables (wing geometry and installed thrust) must be established.

- A criterion must be found defining 'the best aircraft' within the set of design constraints.
- A choice has to be made of representative conditions – in particular, payload and range – for which operational and economics characteristics are evaluated. This is often referred to as the design condition.

In principle, this problem definition determines the solution of the aircraft system optimization. However, even the basic task of defining the best cruise altitude must not be underestimated, whereas the problem of determining an optimum cruise Mach number is a challenging task. For both examples, the case of a given (operational) aircraft is fairly easy if not trivial, but computing the best design cruise Mach number of a clean-sheet design is extremely complicated, if not impossible. In both examples, the conceptual designer may decide that a maximum payload or a long-range condition is most important and that an all-out optimization does not have a high priority. Using the quasi-analytical approach discussed in Chapters 8 to 10 may then be preferable.

1.7.2   Problem Classification

In accordance with the structure of the aircraft system optimization problem and the previous considerations, the ground rules and the computational approach discussed in this book are clarified according to one of the following options.

1. Optimizing flight conditions for a given airframe-cum-engine combination. The problem is to derive a mission profile resulting in minimum fuel burn, flight duration or direct operating costs (DOC). This is basically an optimal control problem occurring in design studies as well as mission analysis of existing aircraft. Depending on the design phase (conceptual or preliminary), the solution may be simplified by accepting approximations, resulting in a closed-form equation for the mission fuel required to fly a given range or the range obtainable with a given mission fuel. Treated in Chapter 12, these subjects form essential elements of the aircraft sizing process.
2. Determining the optimum airframe properties for specified engines. This is a typical conceptual design exercise of matching the airframe to engines selected a priori in an optimum fashion. Wing geometry is then the primary candidate for a highly constrained optimization. A similar situation occurs if a new engine technology becomes available for installation in an existing aircraft (family). The aircraft's commercial value may then be enhanced by increasing the payload versus range capabilities, preferably with minor modifications of the airframe.
3. Finding the best combination of the airframe and a given number of engines. Selection of the engines and their position relative to the airframe has far-reaching consequences. In order to find the best match during the conceptual stage, basic airframe properties are varied and engine properties are scaled by rubberizing the engines. The final selection is likely to be made during preliminary or detailed design. Chapter 9 deals with the subject of engine and airframe matching and derives a FOM for initial engine selection.

This book focusses on how conclusions can be obtained and interpreted with regard to optimum conditions by means of a quasi-analytical approach. To this end, the problem is simplified to such an extent that the objective function is differentiable so that (closed form) analytical solutions are obtained. The emphasis is initially on unconstrained optima which are easily obtained by means of classical theory of local optima. These optima are interesting because they identify the ultimate design quality achievable for a given generic task – moving a payload over a given distance. Although practical considerations associated with technological limits may degrade this 'well-tempered aircraft design' into an unfeasible solution, the penalty to be paid for the best feasible solution becomes explicit by comparing it with the unconstrained optimum. The analytical approach may also give useful information on the sensitivity of a design to small changes in the TLRs and technological parameters.

Bibliography

[1] Küchemann, D., *The Aerodynamic Design of Aircraft*, Pergamon Press, Oxford, 1978.

[2] Torenbeek, E., *Synthesis of Subsonic Airplane Design*, Springer Verlag, Heidelberg, 1981.

[3] Roskam, J., *Airplane Design*, Roskam Aviation and Engineering Corporation, Kansas, 1985.

[4] Niu, M.C-Y., *Airframe Structural Design: Practical Design Information and Data on Aircraft Structures*, Conmilit Press Ltd., Hong Kong, and Technical Book Company, Los Angeles, CA, 1988.

[5] Niu, M.C-Y., *Composite Airframe Structures: Practical Design Information and Data*, Conmilit Press Ltd., Hong Kong, and Technical Book Company, Los Angeles, CA, 1992.

[6] Wilkinson, R., *Aircraft Structures and Systems*, Addison Wesley Longman Ltd., Harlow, 1996.

[7] Brandt, S.A., R.J. Stiles, J.J. Bertin, and R. Whitford, *Introduction to Aeronautics: A Design Perspective*, AIAA Education Series, AIAA, Inc., Washington, DC, 1997.

[8] Jenkinson, L.R., P. Simpkin, and D. Rhodes, *Civil Jet Aircraft Design*, Arnold, London, 1999,

[9] Schaufele, R., *The Elements of Aircraft Preliminary Design*, Aries Publications, Santa Anna, CA, 2000.

[10] Howe, D., *Aircraft Conceptual Design Synthesis*, Professional Engineering Publishing, London, 2000.

[11] Jenkinson, L.R., and J. Marchman, *Aircraft Design Projects*, Butterworth Heinemann, Oxford, 2003.

[12] Torenbeek, E., and H. Deconinck (eds), *Innovative Configurations and Advanced Concepts for Future Civil Aircraft*, von Kármán Institute for Fluid Dynamics, Lecture Series 2005-06, June 2005.

[13] Raymer, D.P., *Aircraft Design: A Conceptual Approach*, Fourth Edition, AIAA Education Series, AIAA, Inc., Reston, Virginia, 2006.

[14] Whitford, R., *Evolution of the Airliner*, The Crowood Press, Ramsbury, 2007.

[15] Gunston, B., *Airbus: The Complete Story*, Haynes Publishing, Sparkford, 2009.

[16] Obert, E., *Aerodynamic Design of Transport Aircraft*, IOS Press, Amsterdam, 2009.

[17] Kundu, A.K., *Aircraft Design*, Cambridge University Press, New York, 2010.

[18] Sadraey, M.H., *Aircraft Design: A Systems Engineering Approach*, John Wiley & Sons, Chichester, UK, 2013.

Design Development and Methodology

[19] Cherry, H.H., and A.B. Croshere Jr., "An Approach to the Analytical Design of Aircraft", *SAE Quarterly Transactions*, Vol. 2, No. 1, pp. 12–18, January 1948.

[20] Backhaus, G., "Grundbeziehungen für den Entwurf optimaler Verkehrsflugzeuge", *Jahrbuch der WGL*, pp. 201–213, 1958.

[21] Küchemann, D., and J. Weber, "An Analysis of Some Performance Aspects of Various Types of Aircraft Designed to Fly over Different Ranges at Different Speeds," *Progress in Aeronautical Sciences*, Vol. 9, pp. 324–456, 1968.

[22] Ladner, F.K., and A.J. Roch, "A Summary of the Design Synthesis Process", SAWE Paper No. 907, May 1972.

[23] Lange, R.H., "Parametric Analysis of ATT Configurations", AIAA Paper No. 72-757, 1972.

[24] Black, R.E., and J.A. Stern, "Creative Advanced Design: a Key to Reduced Life-Cycle Costs", AGARD CP-147-Vol.1, Paper No. 4, 1973.

[25] Fulton, R.E., "Overview of Integrated Programs for Aerospace-Vehicle Design", NASA TM-81874, September 1980.

[26] Harris, D.H.W., "Applying Computer Aided Design (CAD) to the 767", *Astronautics and Aeronautics*, pp. 44–49, September 1980.
[27] Torenbeek, E., "Fundamentals of Conceptual Design Optimization of Subsonic Transport Aircraft", TU Delft, Department of Aerospace Engineering, Report LR-292, August 1980.
[28] Steiner, J.E., "How Decisions are Made: Major Considerations for Aircraft Programs", AIAA 45th Wright Brothers Lecture, the Boeing Company, Seattle, August 24, 1982.
[29] Koen, B.V., "Definition of the Engineering Method", ASEE, Washington, DC, USA, 1985.
[30] Fetterman, D.E. Jr., "Preliminary Sizing and Performance of Aircraft", NASA TM 86357, 1985.
[31] Loftin, L.K., Jr., "Quest for Performance – the Evolution of Modern Aircraft", NASA SP-468, 1985.
[32] Daues, J.J., "Project Management Issues & Lessons Learned from Computer-Aided Design Applications", AIAA Paper No. 87-2912, September 1987.
[33] Michaut, C., D. Cavalli, H.T. Huynh, and H. Lethuy, "Preliminary Design of Aircraft", AIAA Paper No. 89-2152, July–August 1989.
[34] Mason, W.H., "Analytic Models for Technology Integration in Aircraft Design", AIAA Paper No. 90-3262, September 1990.
[35] Malone, B., and W.H. Mason, "Multidisciplinary Optimization in Aircraft Design Using Analytic Technology Methods", AIAA Paper No. 91-3187, September 1991.
[36] Petersen, T.J., and P.L. Sutcliffe, "Systems Engineering as Applied to the Boeing 777", AIAA Paper No. 92-1010, February 1992.
[37] Bushnell, D.M., "Far Term Visions in Aeronautics", *Transportation Beyond 2000: Technologies Needed for Engineering Design*, Part 1, NASA TP 10184, pp. 261–298, 1996.
[38] Mavris, D.N., and D.N. DeLaurentis, "A Probabilistic Approach for Examining Aircraft Concept Feasibility and Viability", *Aircraft Design*, Vol. 3, No. 2, pp. 79–101, 2000.
[39] ESDU, "Examples of Construction of Carpet Plots from Experimental Data", Data Item No. 04012, 2004.
[40] McMasters, J.H., and R.M. Cummings, "Rethinking the Airplane Design Process: An Early 21st Century Perspective", AIAA Paper No. 2004-0693, 2004.
[41] Friend, M.G., and C.P. Nelson, "From Concept to Market Reality: A Case Study of the Sonic Cruiser", *Innovative Configurations and Advanced Concepts for Future Civil Aircraft*, von Kármán Institute for Fluid Dynamics, Lecture Series 2005-06, June 2005.
[42] Kehayas, N., "Aeronautical Technology for Future Subsonic Civil Transport Aircraft", *Aircraft Engineering and Aerospace Technology*, Vol. 79, No. 6, pp. 600–610, 2007.
[43] Toor, J. van, and T. Ahn, "The Role of Advanced Design in Aircraft Development and Product Definition", private communication, 2010.

Transport Aircraft Technology

[44] Schneider, W., and M. Wittmann, "Cost Efficiency Considerations Concerning the Application of Advanced Technologies for Civil Transport Aircraft", SAWE Paper No. 1214, May 1978.
[45] Shevell, R.S., "Technological Development of Transport Aircraft – Past and Future", *Journal of Aircraft*, Vol. 17, No. 2, pp. 67–80, February 1980.
[46] Swihart, J.M., "The Next Generation of Commercial Aircraft: The Technology Imperative", ICAS Paper No. 80-0.2, October 1980.
[47] Maglieri, D.J., and S.M. Dollyhigh, "We Have Just Begun to Create Efficient Transport Aircraft", *Astronautics and Aeronautics*, Vol. 20, No. 1, pp. 26–38, February 1982.
[48] Williams, L.J., "Advanced Technology for Future Regional Transport Aircraft", SAE Paper No. 820731, 1982.
[49] Wright, H.T., "NASA Technology Program for Future Civil Transports", AIAA Paper No. 83-1603, 1983.
[50] Kayten, G.G., C. Driver, and D.J. Maglieri, "The Revolutionary Impact of Evolving Aeronautical Technologies", AIAA Paper No. 84-2445, 1984.
[51] Swihart, J.M., "The Progress of Aeronautics", AGARD Report No. R-782, Paper 1, June 1990.
[52] Flosdorff, H., J. Roeder, and J. Szodruch, "Technological Requirements Concerning Subsonic Transport Aircraft", paper presented at German Aerospace Congress, Bremen, 29 September-2 October 1992.
[53] Szodruch, J., and R. Hilbig, "Building the Future Aircraft Design for the Next Century", AIAA Paper 98-135, 1998.

[54] McMasters, J.H., and R.M. Cummings, "Airplane Design: Past, Present and Future", *Journal of Aircraft*, Vol. 39, No. 1, pp. 10–17, January–February 2002; also AIAA Paper No. 2001-0535.

[55] Jarry, P., "Market Drivers and Innovation Behind Airbus Products", ICAS Paper No. 02-0.5, September 2002.

[56] McMasters, J.H., and R.M. Cummings, "From Farther, Faster, Higher to Leaner, Meaner, Greener: Further Directions in Airplane Design", *Journal of Aircraft*, Vol. 41, No. 1, pp. 51–61, January–February 2004; also AIAA Paper No. 2003-0553.

[57] McMasters, J.H., "A U.S. Perspective on Future Commercial Airliner Design", *Innovative Configurations and Advanced Concepts for Future Civil Aircraft*, von Kármán Institute for Fluid Dynamics, Lecture Series 2005-06, June 2005.

[58] Frota, J., and Y. Vigneron, "Airbus Future Projects Approach to New Aircraft Concepts and Research and Technology", *Innovative Configurations and Advanced Concepts for Future Civil Aircraft*, Von Kármán Institute for Fluid Dynamics, Lecture Series 2005-06, June 2005.

[59] Kehayas, N., "Aeronautical Technology for Future Subsonic Civil Transport Aircraft", *Aircraft Engineering and Aerospace Technology*, Vol. 79, No. 6, pp. 600–610, 2007.

Computer-Assisted Synthesis and Design Integration

[60] Lee, V.A., H.G. Ball, E.A. Wadsworth, W.J. Moran, and J.D. McLeod, "Computerized Aircraft Synthesis", *Journal of Aircraft*, Vol. 4, No. 5, September–October, pp. 402–408, 1967.

[61] Boyles, R.Q., "Aircraft Design Augmented by a Man–Computer Graphic System", *Journal of Aircraft*, Vol. 5, No. 5, 1968.

[62] Wallace, R.E., "Parametric and Optimisation Techniques for Airplane Design Synthesis", AGARD LS-56, April 1972.

[63] Wallace, R.E., "A Computerized System for the Preliminary Design of Commercial Airplanes", AIAA Paper No. 72-793, August 1972.

[64] Ardema, M.D., and L.J. Williams, "Automated Synthesis of Transonic Transports", AIAA Paper No. 72-794, August 1972.

[65] Fulton, R.E., Sobieszczanski, J., and Landrum, E.J., "An Integrated Computer System for Preliminary Design of Advanced Aircraft", AIAA Paper No. 72-796, August 1972.

[66] Bishop, A.W., and A.N. Page, "An Approach to Design Integration", AGARD CP-147, Vol. 1, Paper No. 5, October 1973.

[67] Gregory, T.J., "Computerized Preliminary Design at the Early Stages of Vehicle Design", AGARD CP-147, Vol. 1, Paper No. 6, October 1973.

[68] Heldenfels, R.R., "Integrated Computer-Aided Design of Aircraft", AGARD CP-147, Vol. 1, Paper No. 16, October 1973; also AIAA Paper No. 73-410, March 1973.

[69] Galloway, T.L., and M.H. Waters, "Computer-Aided Parametric Analysis of General Aviation Aircraft", SAE Paper No. 730332, April 1973.

[70] English, C.H., "Interactive Computer Aided Technology: Evolution of the Design Manufacturing Process", McDonnell Aircraft Co., AIAA Meeting Paper, MC-AIR75-009, August 1975.

[71] Hitch, H.P.Y., "Computer Aided Aircraft Project Design", *Aeronautical Journal*, Vol. 81, No. 469, pp. 51–62, February 1977.

[72] Oman, B.H., "Vehicle Design Evaluation Program", NASA CR-145700, January 1977.

[73] Gregory, T.J., and L. Roberts, "An Acceptable Role for Computers in the Aircraft Design Process", AGARD CP-280, Paper No. 3, September 1979.

[74] Howe, D., "The Application of Computer Aided Techniques to Project Design", *Aeronautical Journal*, Vol. 83. No. 492, pp. 16–21, January 1979.

[75] Sliwa, S.M., and P.D. Arbuckle, "OPDOT: A Computer Program for the Optimum Preliminary Design of a Transport Airplane", NASA TM-81857, 1980.

[76] Smyth, S.J., "CADAM Data Handling from Conceptual Design through Product Support", *Journal of Aircraft*, Vol. 17, No. 10, pp. 753–760, October 1980.

[77] Fulton, R.E., "CAD/CAM Approach to Improving Industry Productivity Gathers Momentum", *Astronautics and Aeronautics*, pp. 64–70, February 1982.

[78] Stack, S.H., "A Computer-Aided Design System Geared Toward Conceptual Design in a Research Environment", AIAA Paper No. 81-0372, January 1981.

[79] Eckels, W.E., "Civil Transport Aircraft Design Technology", AIAA Paper No. 83-2463, October 1983.

[80] Haberland, C., J. Thorbeck and W. Fenske, "A Computer Augmented Procedure for Commercial Aircraft Preliminary Design and Optimization", ICAS Paper 84-4-8-1, 1984.

[81] Victor, K., and T. Weir, "Application of Computers to the Aircraft Design Process", AIAA, pp. 390–397. Paper No. 85-3044, October 1985.

[82] Haberland, C., and W. Fenske, "A Computer Augmented Procedure for Commercial Aircraft Configuration Development and Optimization", *Journal of Aircraft*, Vol. 23, No. 5, pp. 390–397, May 1986.

[83] Bil, C., "Applications of Computer-Aided Engineering to Subsonic Aircraft Design in a University Environment", ICAS Paper 86-3.1.1., 1986.

[84] Bil, C., *Development and Application of a Computer-Based System for Conceptual Aircraft Design*, Doctoral Dissertation, TU Delft, 1988.

[85] Michaut, C., C. Cavalli, H.T. Huynh, and H. Lethuy, "Preliminary Design of Civil Transport Aircraft" AIAA Paper No. 89-2152, August 1989.

[86] Hays, A.P., "Spreadsheet Methods for Aircraft Design", AIAA Paper No. 89-2059, August 1989.

[87] Michaut, C., D. Cavalli and H.T. Huynh, "Conceptual Design of Civil Transport Aircraft", ICAS Paper 90-2.1.2., 1990. September 1990.

[88] Denisov, V.E., "Application of Methods and Tools for Computer-Aided Design in Investigation of Prospects for Civil Aircraft Progress", ICAS Paper No. 90-2.1.4, September 1990.

[89] Haberland, C., W. Fenske, O. Kranz, and R. Stoer, "Computer-Aided Conceptual Aircraft Configuration Development by an Integrated Optimization Approach", ICAS Paper No. 90-2.6R, September 1990.

[90] Kroo, I.H., "An Interactive System for Aircraft Design and Optimization", AIAA Paper No. 92-1190, February 1992.

[91] ACSYNT Institute, "ACSYNT (Aircraft SYNThesis) V2.0", January 1993.

[92] Myklebust, A., and P. Gelhausen, "Putting the ACSYNT on Airplane Design", *Aerospace America*, Vol. 32, No. 9, pp. 26–30, September 1994.

[93] Bloor, J.M., "Efficient Parametrization of Generic Aircraft Geometry", *Journal of Aircraft*, Vol. 32, No. 6, pp. 1269-1275, 1995.

[94] Schneegans, A., O. Kranz, and Ch. Haberland, "Flying Objects: An Object-Oriented Toolbox for Multidisciplinary Design and Evaluation of Aircraft", ICAS Paper No. 98-6.4.4, September 1998.

[95] Raj, P., "Aircraft Design in the 21st Century: Implications for Design Methods insert paper presented at, 29th AIAA Fluid Dynamics Conference, Albuquerque, AIAA, 1998.

[96] McCullers, L.A., "FLOPS: FLight OPtimization System, Release 5.94 User's Guide", NASA Langley Research Center, December 1998.

[97] Xiaobin Xie, "A New Numerical Design Tool for Concept Evaluation of Propeller Aircraft", *Aircraft Design*, Vol. 2, No. 3, pp. 147–165, 1999.

[98] Österheld, C., W. Heinze, and P. Horst, "Preliminary Design of a Blended Wing Body Configuration Using the Design Tool PrADO", *in CEAS Conference on Multidisciplinary Aircraft Design and Optimization*, Maternushaus, Köln, pp. 119–128, June 2001.

[99] Pouchet, C., E. Forbes, D. DeLaurentis, and D. Mavris, "An Exploration of a Novel Approach to Aircraft Volumetric Sizing", AIAA Paper No. 2003-6710, 2003.

[100] Zhijie Lu, E-S Yang, D.A. DeLaurentis, and D.N. Mavris, "Formulation and Test of an Object-Oriented Approach to Aircraft Sizing", AIAA Paper No. 2004-4302, 2004.

[101] Rentema, D.W.E., "AIDA: Articifial Intelligence Supported Conceptual Design of Aircraft", Doctoral Dissertation, TU Delft, November 2004.

[102] Anemaat, W.A.J.,"Conceptual Airplane Design Systems", *in Encyclopaedia of Aerospace Engineering*, R. Blockly, and W. Shyy (eds), John Wiley & Sons, Chichester, October 2010.

[103] la Rocca, G., "Knowledge Based Engineering Techniques to Support Aircraft Design and Optimization ", Doctoral Dissertation, TU Delft, March 2011.

2

Early Conceptual Design

In today's world of high-speed computer programs, sophisticated analysis, and computer-aided design, the need still remains for quick, cursory methods of estimating weight, especially for early conceptual studies. One might say that there is still a need to take a quick look at the forest before examining a few of the trees.

—D.P. Marsh [13] (1982)

2.1 Scenario and Requirements

2.1.1 What Drives a Design?

Before starting the design of a clean-sheet aircraft, the first step is to gain an understanding of the dominant needs and conditions that will have a major effect on the design characteristics. Typically, this activity includes: (a) an assessment of the enabling technologies required to comply with the design and certification requirements; (b) comparative studies to evaluate the implications of choosing different conceptual general arrangements of the design; and (c) identification of the selection variables to be optimized in order to obtain an economically superior aircraft. Table 2.1 summarizes some of the dominant drivers for designing a transport aircraft. Many of these issues are incorporated in the design synthesis process, others have to be dealt with by specialists during the detail design phase. Most requirements are specified in the top level requirements (TLRs) in the form of performance limits, others are selected by the advanced design team. Finding a balance between all the relevant issues is a challenge that can best be countered by expertise and common sense rather than by a formalized (numerical) optimization process.

It is widely acknowledged that, for a given payload and mission performance, reduction of fuel weight and maximum take-off gross weight (MTOW) has a favourable effect on flight performance and on operating expenses. It is therefore not surprising that – during many decennia of commercial airplane development – designers have focussed on minimizing the MTOW. However, the sharp rise in oil prices since 2000, the forecasts regarding the depletion of fossil fuels and the worldwide concerns about CO_2 emissions have shifted the emphasis

Advanced Aircraft Design: Conceptual Design, Analysis and Optimization of Subsonic Civil Airplanes, First Edition. Egbert Torenbeek.
© 2013 by Egbert Torenbeek. Published 2013 by John Wiley & Sons, Ltd.

Table 2.1 Primary design drivers of commercial transport aircraft

Design driver	Relevant issues
Transport function	Payload/range, design speed
Passenger cabin	Accessibility, comfort, evacuation, aesthetics
Propulsion	Engine technology, power/thrust, fuel consumption
Structure	Strength, stiffness, fatigue life, manufacturing
Systems technology	Flight control system, avionics
Flight envelope	Dynamic pressure and altitude limits
Low-speed performance	Stalling and reference speeds, field length
Airfield	Runway conditions, aircraft handling
Flight safety	Flying qualities, engine-out performance
Environment	Community noise, emissions, wake vortex clearance
Economy	Productivity, direct operating costs
Operations	Load and balance, reliability, maintainability

to reducing fuel burn-off and using alternative fuels. Although the aims of reducing gross weight and fuel consumption are not necessarily conflicting, the second objective emphasizes aerodynamic and propulsive efficiency more than the first. In the interest of minimizing environmental damage, airplane manufacturers presently emphasize energy efficiency as an important element of merit (Section 2.5). A spin-off is the emergence of investigations on innovative or unusual airplane design concepts.

Designers have limited freedom to reduce empty weight by the appropriate selection of the airplane's layout and geometry – such as volumetric efficiency, surface areas and wing span – and through the use of advanced structural materials. Key items determining engine efficiency and installed thrust are take-off field length, climb-out performance and the (initial) cruise conditions. Design freedom is, however, constrained by safety and airworthiness requirements, operating costs and environmental issues. Even if the project is to deliver a clean sheet design, its initial concept will inevitably be influenced by company traditions and information on competing or announced designs.[1] The next step is then to consider improvements by introducing technological advances – or even embracing an unusual general arrangement – leading to adjustments of the initially estimated weight and drag components.

This chapter demonstrates that TLRs and statistical information on existing aircraft are sufficient in the early conceptual design phase to make a realistic estimation of empty weight, fuel weight at take-off and design weights. Key factors are data on the overall efficiency of the engines and the aerodynamic efficiency in cruising flight. The proposed elementary weight prediction method should be refined as soon as more information becomes available from sizing computations. Weight sensitivity information is essential input for optimization of independent design variables (Chapter 8).

---

[1]The use of statistical information on similar aircraft is useful to obtain a realistic initial baseline design. However, the use of statistics is not an asset to the creative design process and designers of a traditional configuration should not be surprised when their creation appears not to be superior to existing aircraft.

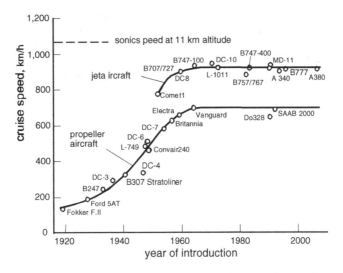

Figure 2.1 Historical development of maximum cruise speeds

2.1.2 Civil Airplane Categories

Essential parameters characterizing the general properties of a new design are the maximum cruise speed and the type of propulsion. The historical development of airliner maximum cruise speeds (Figure 2.1) shows that, between 1920 and 1960, propeller aircraft speed increased progressively from 150 to 650 km/h and levelled off thereafter. Modern turboprops do not exceed 700 km/h; further development of high-speed propellers may increase their cruise speed to 750 km/h. The introduction of jet engines during the 1950s expedited a jump in maximum speed to 900 km/h where it stayed during the rest of the twentieth century; some long-range business jets attain 950 km/h. Boeing's Sonic Cruiser project (Figure 1.2(b)) aimed at cruise speeds up to Mach 0.97 (1030 km/h) to be realized by means of a revolutionary general arrangement. Future long-range airliners optimized for environmentally friendly operation may cruise at no more than Mach 0.75 (800 km/h).

The required transport function of an airplane has a large impact on the best arrangement and location of its main components. The majority of presently operational airliners can be classified into one of the following categories:

- Regional propeller aircraft seating between 20 and 90 passengers in a single-aisle cabin.
- Regional jet aircraft seating between 50 and 110 passengers in a single-aisle cabin. The horizontal stabilizer is attached to the rear fuselage or to the fin.
- Narrow body jet aircraft seating between 110 and 220 passengers on a single-aisle single deck.
- Wide body jet aircraft seating at least 220 passengers on one or two twin-aisle passenger decks.

Propeller planes are powered by two turboprop engines in front of the low or high-set straight wing and the horizontal stabilizer is attached to the rear fuselage or the fin. Most jet transports

Table 2.2 Classification of business jets based on the manufacturer-defined role. Range applies to typical seating, NBAA reserves. Entry-level single-engine jets are excluded

Category	MTOW, ton	Number of seats	Cabin height, cm	Maximum cruise speed	Range, 1 000 km
Very light	< 5	2 + (3–4)	130–150	340–380 kt	2.0–2.5
Light	5–9	2 + (4–7)	135–170	400–460 kt	2.5–4.5
Mid–size	9–18	2 + (6–14)	140–185	430–465 kt	3.5–5.5
Large	18–38	(2–3) + (8–19)	165–190	Mach 0.80–0.86	5.0–8.5
Long distance	38–50	(3–4) + (8–19)	185–195	Mach 0.86–0.92	8.5–13

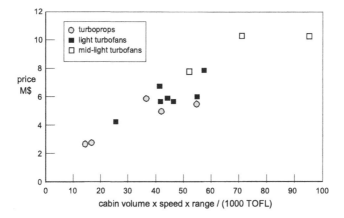

Figure 2.2 Productivity analysis of business aircraft [22]

are powered by two or four turbofan engines suspended below the leading edge of the low-set wing; they feature a horizontal tail attached to the rear fuselage. Several regional jets have two engines attached to the fuselage section behind the cabin in combination with a high-set horizontal stabilizer[2] – a T-tail. Most freighters are derived from passenger aircraft; dedicated freighters have a high-set wing.

Due to the availability of surplus aircraft, piston-powered propeller aircraft dedicated to the corporate and air taxi scene appeared soon after the Second World War. Nowadays, the high reliability of gas turbines for general aviation has enabled turboprop singles to make commercial flights during day- and nighttime. Turboprop twins have the added features of improved safety after engine failure and a higher cruise speed. High-end business aviation is associated with corporate jets.

Table 2.2 shows a proposed size classification of business jets which had their first flight between 1990 and 2010. Dedicated business jets have a multi-functional cabin, low-set wing, turbofans attached to the rear fuselage and a high-set horizontal tail. Market surveys show that the most significant parameters are cabin volume, speed, range and field length. Figure 2.2

---

[2]Jet transports with three engines were developed during the 1960s and 1970s: Hawker Siddeley Trident, Boeing 727, Lockheed 1011, Douglas DC-10 (modified into the MD-11), and the Tupolev 134 and 154.

demonstrates that the capability of satisfying the customers' need for productivity must be paid for by a proportional price increase, although design ingenuity will be useful in maintaining competitiveness.

2.1.3   Top Level Requirements

> It is not difficult to visualize a saving of hundreds of millions of dollars on the C-5A program if freedom had existed to increase the empty weight in the specification by, say, five percent (about 16,000 pounds).
>
> —F.A. Cleveland [9]

Apart from specifying the airworthiness rules [6] according to which the aircraft type has to be certified, TLRs define objectives and constraints in a degree of detail complying with the purpose of advanced design. The following is a typical set of requirements and technical objectives for jet transports:

- Payload, regional or continental (US domestic) flights: seating capacity in single- or two-class configurations.
  Payload, long-range (international) flights: seating capacity in three-class configuration with increased cabin baggage.
  Passenger accommodation: seat width and pitch, aisle width and height, cabin amenities, pressurization level, boarding/deboarding facilities.
  Baggage/freight holds: minimum volume for passenger baggage. Large airliners load under-floor baggage in standard containers.
- Range: the maximum distance over which the payload can be transported, for example, US continental 5 600 km (3 000 nm) or transpacific 14 000 km (7 600 nm). Policies concerning flight execution and fuel reserves are stated explicitly and a distinction can be made between economical and high-speed cruising conditions corresponding to different ranges. Regional airliners are mostly designed for multi-stage operations for which a number of flights are specified to be flown without intermediate refuelling.
- Cruise speed/altitude capability: minimum cruise speed or Mach number, often in combination with a minimum initial altitude. Requirements may pertain to standard and/or non-standard ambient conditions. The one-engine-inoperative (OEI) ceiling is important when flying over mountainous terrain or oceans.
- Low speed performance: constraints to the required take-off/field length (TOFL), engine-out climb gradients, approach speed and/or landing field length (LFL), in combination with a specified mission. Different limits may pertain to take-off and landing in standard sea level and non-standard ambient conditions. In particular, an airfield may be specified with a higher than standard temperature located at an elevated level ('hot and high').
- Airport compatibility: airfield classification, defining limitations to wing span and length, landing gear track and runway pavement loading.
- Environmental issues: maximum noise emission levels defined relative to certification requirements in FAR Chapter 36 and similar standards in ICAO Annex 16. For example, cumulative limit of 30 dB below FAA Stage 4. Engine exhaust emission targets during

take-off and landing are defined relative to internationally agreed criteria; in particular, CAEP/6 NO_x restrictions.

- Reliability and durability: intense airliner utilization emphasizes the need to achieve a specified lifetime in terms of a number of flight hours and/or flight cycles.

Many airliners are conceived in the framework of family planning rather than a single mission. Several versions with increased or decreased range and/or payload are generated during the life cycle of the project. This requires evaluation during the conceptual design of several derivative designs having their cabin cross-section in common with the basic version.

2.2 Weight Terminology and Prediction

Weight engineering is a principal discipline involved in aircraft design. The concepts of weight evaluation in the conceptual phase involve an extensive database which is transformed by regression techniques into statistical weight prediction methods for particular classes of aircraft. In the pre-conceptual phase, the only well-defined data are derived from the TLRs, previous project studies and statistical information. More design parameters become available as the design progresses and weight predictions become based more on analysis rather than statistics, leading to more accurate results.

2.2.1 Method Classification

Weight predictions of various types and levels of detail comply with the need for sizing information throughout the project development phases. The following classification is typical but should not be considered as standard:

- Class I – Pre-conceptual studies. Class 1 weight estimation uses a low-fidelity approach to get an impression of the complete vehicle weight without using design geometry. It is based on the TLRs and a database of existing aircraft. Input consists typically of the aircraft certification category, technology levels and one or more combinations of payload and range. A Class I weight estimation preferably uses this basic input in combination with carefully interpreted statistical information on weight fractions of similar aircraft.
- Class II – Conceptual design. At the stage when the airplane's configuration is selected and the basic geometry, performance and other general characteristics are determined, a more detailed weight prediction is made by adding weight components computed by means of medium-fidelity methods. Design weights obtained from a Class I weight estimation, a provisional three-view drawing and installed engine thrust or power are used as input. The weight components are collected in a group weight statement resulting in new values of the design weights. Prediction of structural weight components is based on quasi-analytical and analytical methods using geometric information augmented and calibrated by statistics. Systems and equipment weight components are based on statistics. Results are visible at group weight statement level, as well as the centre of gravity (CG) location of each group – both are required for the aircraft balancing process. There may also be a requirement for weight sensitivity to variation of wing geometry and installed power or thrust.

- Class III – Preliminary design. Further development and increasing information about design features make the weight prediction more complex and detailed. Computations are carried out in cooperation between AD engineers, weight engineers and specialists from functional groups. A Class III weight prediction involves a breakdown of structures, systems and equipment into elements at subsystem levels with information on weight augmented by CG locations and moments of inertia. Since refinement of the baseline design is the objective, weight analysis methods must accurately predict the effects of variations in relevant design variables. Information on design sensitivity is obtained from high-fidelity analysis methods, in particular computational fluid dynamics (CFD) and finite element methods (FEM) for lift and drag, loads analysis and sizing of structural components. A digital geometry data base is used as a frequently consulted source of input. Even at this AD stage, statistical calibrations cannot be completely avoided. Results are used for performance and flight dynamics analysis, design of flight control systems and as input for the detail design phase.
- Class IV – Detail design. Thousands of (electronic) drawings of hardware details are used by weight engineers to maintain an inventory of accurate masses and moments of inertia of all items designed in-house as well as bought-in products and subsystems. The weight engineering group monitors all recorded masses and makes comparisons with the airplane type specification. Results are visible at detail weight statement and assembly level. A small discrepancy between specification and actual weights is usually acceptable – however, a weight reduction programme is imposed on the organization in the case of a considerable excess weight.

This chapter demonstrates that a Class I weight estimation of a jetliner can be obtained with little engineering work – it is no great effort to derive a similar approach for a business jet or a propeller plane. The use of Class II methods to the initial sizing process and some optimization examples are discussed in Chapters 8 and 10. Derivation of a quasi-analytical method for wing structure weight prediction is the subject of Chapter 11. It is worth noting that – since the weight of an object varies slightly with its location on earth – weight engineers prefer to call their discipline mass engineering. In subsonic airplane design, this aspect is somewhat futile and it is assumed here that weight components are always based on the standard value of the gravitational acceleration: $g_0 = 9.80665$ m/s^2.

2.2.2   Basic Weight Components

The gross weight (GW) – that is, the all-up weight (AUW) of an operating aircraft – is subject to variations. The GW at take-off is the take-off weight (TOW), composed of the nominally constant[3] operating empty weight (OEW) and the useful load (UL),

$$W_{TO} = W_{OE} + W_{ul} \qquad (2.1)$$

The OEW consists of the empty weight of the aircraft delivered by the manufacturer, removable equipment and operator's items (OI). Due to different cabin layout standards, there can be

---

[3]The empty weight increases during the lifetime of an aircraft due to in-service modifications, structural repairs and contamination at airfields. Airlines pay due attention to operational weight growth since flight performance is sensitive to GW, especially in case of an engine failure during take-off.

Table 2.3 Airliner payload and cargo densities

Payload component	Mass, kg
Passenger, including carry-on baggage	75
Passenger luggage – domestic	20
Passenger luggage – international	25
Cubic metre of bulk cargo	160
Full cargo container (LD-3)	1 590

significant differences between the OEWs of aircraft of the same type. OIs are provisions for every flight: flight deck and cabin personnel, cabin amenities, safety equipment, system fluids, and unusable fuel. UL denotes the combination of payload (index pay) and fuel load (index f) in the take-off condition,

$$W_{ul} = W_{pay} + W_f \tag{2.2}$$

Airliner payload consists of the weight of passengers, including their luggage, and additional cargo – see Table 2.3.

The payload of a business aircraft depends on the category of operation, such as executive or company shuttle flying, and may or may not include the pilot(s). The payload of freighter aircraft is the total weight of cargo plus additional passengers, if any.

Figure 1.6 illustrates that fuel at take-off consists of mission fuel (index misf) – also known as block fuel or trip fuel – required to fly to the destination, and reserve fuel (index resf) for manoeuvring, holding, aborting the landing and making a diversion flight,

$$W_f = W_{misf} + W_{resf} \tag{2.3}$$

The GW decreases during the flight due to fuel burn-off. The zero fuel weight (ZFW) is the weight of the aircraft with payload on board and empty fuel tanks,

$$W_{ZF} = W_{OE} + W_{pay} \tag{2.4}$$

Equation (2.1) shows that the TOW is

$$W_{TO} = W_{ZF} + W_f = W_{OE} + W_{pay} + W_f \tag{2.5}$$

Landing weight (LW) is the GW when touching down after the flight,

$$W_L = W_{TO} - W_{misf} = W_{ZF} + W_{resf} \tag{2.6}$$

Fuel used during taxying after landing is counted as reserve fuel.

2.2.3   Weight Limits

Flight performance and loads on the aircraft structure are sensitive to variations in the GW and the distribution of weight components in the aircraft. In the interest of operational safety, well-defined weight limits are imposed which have to be adhered to in aircraft operation. Their definitions are found in relevant airworthiness requirements [5]. These weight limits determine the highest loads on the aircraft structure and, hence, they are also known as design weights.

Although many loading restrictions have to be considered in the preliminary design, the most pertinent of them are relevant to an early conceptual weight estimation.

- The maximum zero fuel weight (MZFW) is the maximum value of the aircraft GW with empty tanks. The MZFW is mostly based on the maximum loads which can be withstood by the wing structure and by the fuselage structure which absorbs stress at the introduction of the wing load.
- The structural payload (SPL) is the maximum payload that can be transported without exceeding the MZFW. Hence, SPL = MZFW– OEW.
- The total fuel loaded on the apron equals the pre-flight fuel for starting up the engines and taxying to the runway, mission fuel and reserve fuel. The maximum fuel load is determined by the fuel tank capacity.
- An aircraft may not take off at a gross weight in excess of the maximum take-off weight (MTOW). The MTOW is mostly determined by the capability of the structure to withstand loads exerted during taking off and by the maximum en-route GW. The MTOW limits the useful load to $UL \leq MTOW- OEW$.
- The maximum landing weight (MLW) must not be exceeded during the landing after a normal flight as well as in certain emergency conditions. The MLW depends on the strength of the landing gears and on the impact loads on the structure to which it is attached.

In order to increase their loading flexibility for short and medium-range flights, airliners can accommodate a payload which is (considerably) above the value for multi-class seating. The space limited or volumetric payload (VPL) is the maximum payload available in a cabin arrangement for single-class seating and/or volumetric limitations in cargo and baggage compartments. The VPL may not exceed the MPL nor the maximum seating capacity for which the aircraft is certificated. The maximum seating capacity corresponds to, for example, the number of emergency exits in the cabin. An aircraft with a large margin between structural and volumetric payload limits may suffer from either an empty weight penalty or inefficient use of available cabin space. The discrepancy between the MPL and the VPL can be one of the reasons for developing different versions of the basic aircraft. The volumetric payload capacity of a cargo aircraft is determined by the maximum number of containers, pallets and/or amount of bulk cargo that can be loaded in the freight hold.

2.2.4   Transport Capability

Commercial transport aircraft are conceived in such a way that maximum fuel cannot be loaded when the MPL is boarded. This is ascribed to three limitations on the UL which depend on the distance to be flown as depicted in Figure 2.3 (a). (1) The available PL on short flights is

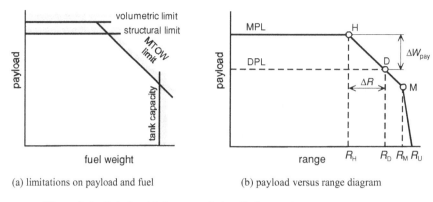

(a) limitations on payload and fuel   (b) payload versus range diagram

Figure 2.3   Relationship between design limits, payload and mission range

either the SPL or the VPL. (2) The UL for long flights is limited by the MTOW. Consequently, when the mission fuel is increased to fly a longer distance, the payload has to be reduced by the same amount. (3) An aircraft can fly its maximum distance when the fuel tanks are filled to their capacity. These limitations are translated into the payload versus range diagram in Figure 2.3 (b) which summarizes the transport capability. The following characteristic ranges are defined:

- The harmonic range R_H is the maximum range attainable when taking off with the MTOW and the MPL, corresponding to the MZFW and partly filled fuel tanks (point H). The MPL corresponds to a single-class economy seating arrangement.
- The design range R_D is the distance achievable when taking off with the MTOW and a design payload (DPL) less than the MPL and fuel (point D). The DPL mostly corresponds to a multi-class cabin of a medium or long range airliner.
- The maximum range R_M is attained when departing with the MFL and the MTOW (point M). The payload is (considerably) less than the MPL.
- The ultimate range R_U is attained when departing for a ferry flight with the MFL and zero payload.

A range increment $\Delta R = R_D - R_H$ requires a payload reduction ΔW_{pay} equal to the required fuel weight increment ΔW_f. The range increment is determined by ΔW_{pay} and the specific range $\Delta R / \Delta W_f$. Section 2.4 explains how the specific range is derived from the local slope of a payload vs. range diagram.

For an aircraft to be conceived, the TLRs generally specify either R_H or R_D. If two combinations of payload and range are specified, the sizing process results in two values of the MTOW. Both requirements are satisfied for the highest of the two take-off weights. A complication threatening the designer is hidden behind this simple statement since a complete sizing process is needed for both MTOW values, virtually resulting in two design cycles. The proposed prediction method can be of some assistance to find out as early as possible which payload/range combination is dominating.

2.3   The Unity Equation

An early Class I weight estimation can be based on little more than a specified mission and statistical information regarding weight fractions of existing or projected aircraft. Relevant weight fractions are the ratios of empty weight, payload and fuel weight to TOGW as obtained from Equation (2.5), resulting in the weight balance or unity equation,

$$\frac{W_{OE}}{W_{TO}} + \frac{W_{pay}}{W_{TO}} + \frac{W_f}{W_{TO}} = 1 \tag{2.7}$$

Although this equation applies to any aircraft loading condition and gross weight, it is used mostly for the aircraft when loaded to its MTOW. Weight fractions are seemingly simple but they are often misinterpreted. In order to apply Equation (2.7) correctly, the definitions of weight components and the associated fractions and the corresponding mission range must be unambiguous and in accordance with TLRs.

Figure 2.4 is a visualization of the unity equation showing the weight fraction breakdown for regional, standard body and wide body jetliners. The payload corresponds to the harmonic range, the fuel weight equals the MTOW minus the MZFW and the fuel tanks are partially filled. The left part of the figure applies to short/medium-range aircraft, with fuel fractions less than 0.2 and empty weight fractions between 0.55 and 0.65. Airliners with a fuel fraction around 0.10 are mostly short-range versions for which the unity equation prescribes a high empty weight fraction. Extended-range versions of a basic aircraft have a considerably higher MTOW to allow for more fuel whereas their (absolute) OEW shows a much smaller increment.

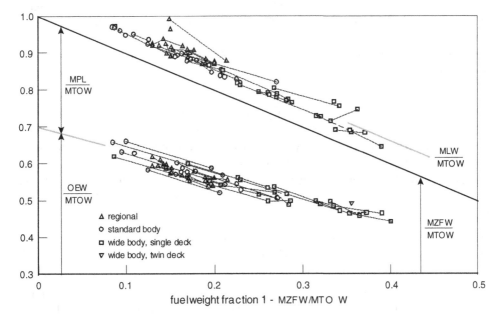

Figure 2.4   The unity equation based on data of commercial jet aircraft with EIS between 1980 and 2012. Members of an aircraft family are connected by dotted lines

Consequently, the line connecting different airliner versions has a downward slope. Standard body airliners have an OEW between 65 and 70% of the MZFW. Similar trends apply to wide body aircraft with fuel fractions generally between 0.25 and 0.40 and empty weight fractions between 0.45 and 0.55. Their OEW is mostly between 70 and 75% of the MZFW whereas the twin-deck A 380 forms a special category. Similar observations are made in [13] for business aircraft and freighters.

In spite of the large data collection in Figure 2.4, most OEW fractions show a fairly small deviation from the overall trend. Surprisingly, the same trend applies to the older generation of jetliners, which have been omitted for clarity. It seems that continuous attempts to reduce empty weights are hardly reflected in these statistics. For instance, the OEW of the B 787 with a predominantly composite structure is only a few percent below the average in Figure 2.4. The conclusion is justified that most improvements of hardware-oriented technology have been negated by measures to increase operational lifetime, reliability, economy and passenger comfort, leading to more durable but not necessarily lighter structures and systems. Another explanation could be that the increasing passenger capacities through the years, leading to wide body aircraft in particular, have had the consequence that attempts to defeat the square cube law [9] have not been completely successful. These considerations suggest that the OEW fraction of a new design is more closely related to the fuel weight fraction – hence, to range performance – than to any other characteristic. If an aircraft has a relatively low OEW fraction, this is not necessarily due to a very light structure. Instead, a combination of an extreme range requirement, relatively high aerodynamic drag and/or engine fuel consumption results in a large fuel load and a high MTOW; hence, the empty weight fraction will be low. On the other hand, when similar aircraft are compared for the same fuel weight fraction, the lighter empty aircraft is not necessarily a better design than the heavier. If structural heaviness is due to a modest wing loading (large wing) in combination with simple high-lift devices, the aircraft may have better aerodynamic and range performance and more growth potential than a highly wing-loaded design.

According to Equation (2.7), the required MTOW for a given payload/range capability is found from

$$W_{\text{TO}} = \frac{W_{\text{pay}}}{1 - (W_{\text{OE}}/W_{\text{TO}} + W_{\text{f}}/W_{\text{TO}})} \qquad (2.8)$$

This equation can be refined by decomposing the gross weight into (a) fixed weight (index fix) with values assigned or based on mission requirements, such as payload, crew, cabin furniture and equipment; and (b) variable weight (index var) components such as fuel, wing and landing gear structure and power plant which are estimated as a fraction of the MTOW. The unity equation is thus generalized as

$$W_{\text{TO}} = \frac{\Sigma_{i=1}^{n} W_{\text{fix}_i}}{1 - \Sigma_{j=1}^{m} W_{\text{var}_j}/W_{\text{TO}}} \qquad (2.9)$$

This first step of the weight estimation depends to a large extent on the availability of information about the category to which the airplane belongs. The unity equation is a useful point of departure for initial weight prediction based on the design mission fuel fraction as the most influential parameter. The following approach applies to airliners introduced into service between 1980 and 2010. A more accurate answer can be obtained by focussing on the specific category to which the airplane belongs, by using low-speed performance requirements as additional input and by calibration of the method for aircraft with similar state of the art and cruise Mach number. An example is found in [20].

2.3.1 Mission Fuel

Basically, the mission fuel is computed by adding the fuel burned during all phases of the flight (Chapter 12). Since this procedure requires information that is not yet available in the pre-conceptual phase, a quasi-analytical method based on simple flight physics can be used instead. Engine and aerodynamic efficiencies are derived from statistical data on turbofan-powered airliners. The same approach can be made for business jets and propeller airplanes by re-calibrating a few numbers quoted in the following derivation. The specific range is the distance that an aircraft travels while burning a given amount of fuel,

$$\frac{\mathrm{d}R}{\mathrm{d}W_\mathrm{f}} = \frac{\mathrm{d}R/\mathrm{d}t}{\mathrm{d}W_\mathrm{f}/\mathrm{d}t} = \frac{V}{F} \tag{2.10}$$

Introducing the aerodynamic efficiency L/D and the overall engine efficiency η_o the specific range in steady level flight is

$$\frac{V}{F} = \frac{\eta_\mathrm{o}H/g}{T} = \frac{\eta_\mathrm{o}H/g}{D} = \eta_\mathrm{o}\frac{L}{D}\frac{H/g}{W} \tag{2.11}$$

where H denotes the calorific value of jet fuel. The range in cruising flight (index cr) is obtained from integration of the specific range

$$R = \int \frac{V}{F}\,\mathrm{d}W_\mathrm{f} = -\int \frac{V}{F}\,\mathrm{d}W = -H/g \int_{W_0}^{W_1} \eta_\mathrm{o}\frac{L}{D}\frac{\mathrm{d}W}{W} \tag{2.12}$$

with W_0 and W_1 denoting the initial and final gross weight. For cruising at constant Mach number and angle of attack, η_o and L/D are constant and equal to the value for the initial condition. The integral in Equation (2.12) may then be solved analytically to yield the generalized Bréguet range equation

$$R_\mathrm{Br} = (H/g)\,\eta_\mathrm{o}\frac{L}{D}\ln\frac{W_0}{W_1} \tag{2.13}$$

The cruise fuel (index crf) weight equals $W_0 - W_1$ and the amount of fuel required for a given cruise range (index cr) is solved from Equation (2.13),

$$\frac{W_\mathrm{crf}}{W_1} = 1 - \exp\left\{\frac{-R_\mathrm{cr}}{(H/g)\,\eta_\mathrm{o}L/D}\right\} \tag{2.14}$$

which is accurately approximated by Taylor series expansion,

$$\frac{W_{\mathrm{crf}}}{W_1} = \frac{R_{\mathrm{cr}}}{(H/g)\,\eta_{\mathrm{o}}L/D + 0.5R_{\mathrm{cr}}} \tag{2.15}$$

In fact, this equation assumes that the average specific range equals the specific range when 50% of the fuel has been burned-off. A further simplification could be that the cruising flight starts with $W_1 = W_{\mathrm{TO}}$. In reality, the specific range during take-off, climbing, descending and manoeuvres is lower than during cruising. In order to correct for this, the mission range is increased by the lost range R_{lost}, or the fuel is increased by the lost fuel [10]. This leads to the following mission fuel fraction required to fly a given mission range

$$\frac{W_{\mathrm{misf}}}{W_{\mathrm{TO}}} = \frac{R_{\mathrm{mis}} + R_{\mathrm{lost}}}{(H/g)\eta_{\mathrm{o}}\,L/D + 0.5\,(R_{\mathrm{mis}} + R_{\mathrm{lost}})} \tag{2.16}$$

with $H/g = 4\,350$ km (2 350 nm) for jet fuel. For a lost range of 300 km (160 nm) this equation indicates that, depending on range, the lost fuel may vary between 2 and 20% of the cruise fuel. The concept of the lost range enables a simple computation of the mission fuel and forms an essential refinement for short-range aircraft.

2.3.2   Empty Weight

Although Figure 2.4 can be used to estimate the OEW fraction, a more appropriate prediction is made by decomposing the empty weight into several primary components, as follows:

$$W_{\mathrm{OE}} = C_{\mathrm{mpl}}W_{\mathrm{mpl}} + C_{\mathrm{MTO}}W_{\mathrm{MTO}} + W_{\mathrm{fix}} \tag{2.17}$$

• The first term summarizes the body group weight: items which are directly related to cabin dimensions determined by the MPL (index mpl). Fuselage and vertical tail structures, air conditioning, pressurization, electrical and electronic systems, passenger accommodation, cabin furnishing and equipment, and OIs are all classified in this category. The body group weight depends on the payload accommodation density (PAD) which can be defined as the ratio between the MPL and the cabin floor area or volume. When comparing airplanes with the same payload designed to the same standard of structural and systems technology, the design with the highest PAD will have the lowest body group weight. The PAD forms a useful criterion for comparing different airliner fuselage cross-sections and cabin layouts and for sizing the cargo holds of a freighter [13].
• The second term represents weight components that are primarily related to the MTOW, such as the wing and horizontal tail structure, the power plant and the landing gear weight.
• The third term represents the flight deck crew with their accommodation and documentation. For a given airplane category, this relatively small component is assumed to be independent of the payload and the MTOW.

The most influential term of Equation (2.17) is the factor C_{mpl} which can be calibrated by deriving the body group weight from the weight information of similar aircraft. To this end, the second and third terms are subtracted from the OEW and the resulting weight is divided

Table 2.4 Parameters for estimating the OEW of a basic jetliner

Number of decks	one	one	two
Number of aisles	one	two	four
C_{mpl}	1.25	1.50	1.75
C_{MTO}	0.20	0.21	0.22
W_{fix} (kN)	5	6	7

by the MPL. Approximate data for turbofan-powered airliners and business aircraft are shown in Table 2.4. The accuracy of the presented empty weight prediction is fair for basic airliners but yields too high values for stretched and too low values for shrunk versions. Business jets feature great variations in comfort level associated with a diversity of applications which should be accounted for in the empty weight estimation. For instance, very light jets (VLJs) are at the lower end of the gross weight spectrum with minimal cabin space, whereas long-range executive aircraft have luxury accommodation. Different PAD values lead to large differences in the body group weight. It is also emphasized that the method is exclusively intended for generating input to initialize early concept sizing.

2.3.3 Design Weights

The prediction of design weights is continuously refined during design development. The iterative character of this process requires the input of an educated guess, preferably using the concept of the unity equation. Substitution of Equations (2.16) and (2.17) into Equation (2.8) yields the MTOW,

$$W_{MTO} = \frac{W_{pay} + C_{mpl} W_{mpl} + W_{fix}}{1 - (C_{MTO} + W_{misf}/W_{MTO} + W_{resf}/W_{MTO})} \qquad (2.18)$$

The mission to be inserted in this equation is either the combination of the MPL and the harmonic range or the combination of the DPL and design range. The mission fuel fraction is obtained from Equation (2.16), the reserve fuel fraction is between 0.045 and 0.050. The OEW is obtained by substitution of the MTOW into Equation (2.17) and addition of the MPL yields the MZFW. Finally, Figure 2.4 suggests MLW = 1.10 MZFW.

The proposed method has the advantage that only information from TLRs and statistical data on the range parameterare needed. Assumptions regarding the geometry of the aircraft are not necessary unless a more accurate estimation of the aerodynamic efficiency is desired; in that case the engine's overall efficiency must be inserted separately into Equation (2.16). Although the method is calibrated for jet airplanes of traditional layout, the user will probably spend some time deriving more accurate statistical coefficients from an in-house data base. Application to a radical design will probably require modification of the method as a whole. This is not a trivial task, especially when a new technology such as mostly composite primary structures or an unusual configuration such as a blended wing body (BWB) concept is investigated.

2.4   Range Parameter

The primary factor determining the cruise fuel fraction is the range parameter $\eta_o\,L/D$ for the initial cruise condition. The challenge is to find a realistic value if the aircraft has not yet been conceptualized. The approach suggested in [16] is to derive statistical information for similar (competing) aircraft from their payload vs. range diagram. This allows computation of the specific range in cruising flight using the differential of Equation (2.16). The range parameter is then solved from the approximate solution

$$\eta_o\,\frac{L}{D} = \frac{\Delta R}{|\Delta W_{\text{pay}}|}\,\frac{W_{\text{MTO}}}{H/g} - \frac{R_{\text{H}} + R_{\text{lost}}}{H/g} \tag{2.19}$$

The slope of the payload vs. range diagram in Figure 2.3 (b) is measured or computed at or above the harmonic range R_{H}. This approach has been used to approximate the range parameter for present-day jetliners with the result depicted in Figure 2.5. The range parameter varies between roughly four for small regional aircraft and eight for wide body airliners. Large airliners have very efficient high-bypass ratio turbofans and fly faster at higher altitudes than regional aircraft. Altogether, their sophisticated design, advanced technology and the scale effect are manifest in high aerodynamic and overall efficiencies. The range parameter may also be derived directly from Equation (2.16),

$$\eta_o\,\frac{L}{D} = \frac{R_{\text{H}} + R_{\text{lost}}}{H/g}\left(\frac{W_{\text{MTO}}}{W_{\text{misf}}} - 0.5\right) \tag{2.20}$$

Application requires input of the lost range and the mission fuel to fly the harmonic range for existing aircraft, which equals MTOW-MZFW minus the reserve fuel.

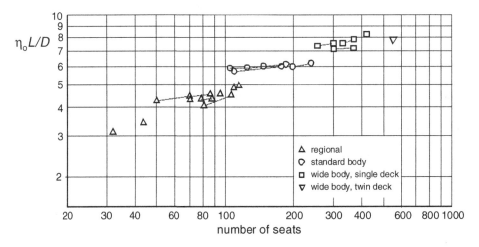

Figure 2.5   The range parameter for commercial jet aircraft derived from payload/range diagrams with $R_{\text{lost}} = 300$ km and $W_{\text{resf}} = 0.045\,W_{\text{MTO}}$

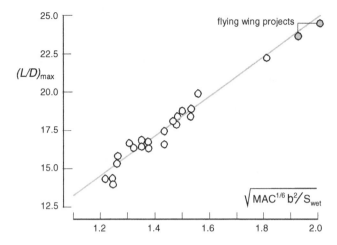

Figure 2.6   Maximum subsonic efficiency. MAC is the mean aerodynamic chord in metres. Adapted from [23]

2.4.1   Aerodynamic Efficiency

The ratio of lift to drag L/D is widely known as the aerodynamic efficiency.[4] This is determined by the airplane's drag polar (Chapter 4). For the usual two-term polar, Section 4.5 derives the maximum L/D in terms of four basic design parameters,

$$\left(\frac{L}{D}\right)_{\text{max}} = \frac{1}{2}\sqrt{\frac{\pi e}{C_{f_{eq}}}}\frac{b}{\sqrt{S_{\text{wet}}}} \tag{2.21}$$

The factor $C_{f_{eq}}$ is an equivalent skin friction drag coefficient based on the airplane's total wetted area S_{wet}. Oswald's efficiency factor e defines the reciprocal value of drag due to lift relative to the reference with $e = 1.0$. This depends mainly on the distribution of lift along the wing span. The factors S_{wet} and b are design variables which have a large influence on the drag. Since their determination is subject to many considerations, initial estimation of L/D has to be based on (measured) data of existing aircraft. Figure 2.6 depicts such a statistical correlation for transonic civil jet aircraft with a conventional layout, with the exception of two flying wing projects. Skin friction drag is a function of the Reynolds number; hence, the size of the aircraft has an effect on C_{fe}. The correction factor $\text{MAC}^{1/6}$ in Figure 2.6 accounts for this scale effect. An alternative statistical approach for conventional jet aircraft is

$$\frac{e}{C_{f_{eq}}} = 220\left(\frac{S_{\text{wet}}}{b\,l_{\text{ref}}}\right)^{1/6} \quad \text{with} \quad l_{\text{ref}} = 10\,\text{m (32.8 ft)} \tag{2.22}$$

---

[4]The name aerodynamic efficiency is strictly a misnomer since the efficiency of a technical process defines the ratio of (energy) output to input, which is always less than one.

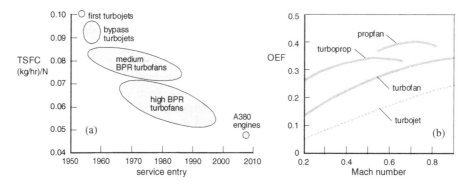

Figure 2.7 Fuel efficiency trends of gas turbine aero engines in cruising flight. (a) Historic development of turbojet and turbofan specific fuel consumption. (b) Variation of overall efficiency with speed

from which $(L/D)_{\max}$ is found by substitution into Equation (2.21). Winglets increase the Oswald factor by 5 to 10% but this is not clear in Figure 2.6 and Equation (2.22). It is also noted that jetliners and business jets cruise at transonic speeds. The next paragraph explains that the drag must then be corrected for wave drag due to compressibility of the air.

2.4.2   Specific Fuel Consumption and Overall Efficiency

- Since the fuel consumption of a gas turbine engine varies with the operating conditions, engine efficiencies cannot be compared on the basis of a single performance parameter. Depending on the category in which they are classified, a comparison between engines is usually based on their specific fuel consumption (SFC). Engine manufacturers define the thrust-based specific fuel consumption C_T (TSFC) of a jet engine as the fuel flow rate per unit thrust. In the SI system, TSFC is defined in (g/s)/N or in (kg/hr)/N, whereas the engine industry mostly uses the Imperial system, with TSFC in (lb/hr)/lbf. Figure 2.7 (a) depicts trends in the historic development of TSFC. Engine manufacturers often quote SFC for the take-off condition at sea level (SL) which is lower than in cruising flight. Turbofan TSFC increases with flight speed and decreases with the (tropospheric) altitude.[5]
- The SFC of a turboprop engine C_P is based on its power output. This is usually defined as the fuel flow per unit of equivalent engine power with dimensions in (g/s)/W, (kg/hr)/kW or (lb/hr)/hp. Based on the shaft power, the SFC is higher due to the jet thrust contribution. Turboprop SFC decreases with increasing altitude and flight speed.

Analysts derive the propulsive power and SFC from thrust and/or shaft power and fuel flow according to the engine manufacturer's data base, propeller thrust is derived from performance charts. Although the concepts C_T and C_P are straightforward and widely accepted, they do

---

[5]Many academic texts assume TSFC to be independent of flight speed and/or altitude, an acceptable simplification for (now obsolete) straight turbojets but erroneous for turbofans.

not mean the same thing to everybody and their magnitudes are entirely different. Confusion and mistakes can be avoided by the use of the overall engine efficiency (OEF),

$$\eta_o \stackrel{\text{def}}{=} \frac{\text{thrust power developed by the engine}}{\text{rate of fuel energy added to the engine}} = \frac{TV}{\dot{m}_f H} \tag{2.23}$$

The OEF is a dimensionless number which depends on engine technology and flight speed. For gas turbine engines it varies between 0.2 and 0.5 in cruising flight, see Figure 2.7 (b). The conversion of SFC into OEF is treated in Section 3.2. Another advantage of using OEF instead of SFC is the unification of flight performance analysis and optimization of aircraft with different propulsion systems. This obviates the separate treatment of jet and propeller aircraft performances which complies with the evolution of engine technology. Future generations of very-high bypass ratio turbofans and open rotor engines can be seen as a cross-breed of jet and propeller propulsion with a mix of genes inherited from both parents. The proposed unification leads to improved understanding of aircraft performance optimization as illustrated by the following example. Chapter 3 offers a more extensive summary of gas turbine engine properties which will be useful for the aircraft design engineer and analyst.

2.4.3 Best Cruise Speed

When dealing with optimum cruising flight, a distinction should be made between the optimum cruise condition of a given aircraft and the best aircraft design for a specified cruise condition.

1. The first problem is a classical subject of flight mechanics. It uses the specific range as the objective for the most fuel-efficient speed and has a straightforward solution. A more refined analysis is based on minimum direct operating costs (DOC), leading to a higher optimum cruise speed.
2. The second problem is far more complex since it involves designing the best combination of airplane and engine for a set of aircraft cruising at different Mach numbers. But even the solution of the first problem requires careful analysis.

These two cases illustrate the distinction between analysis and synthesis discussed in Section 1.7.

Equation (2.11) shows that the specific range in cruising flight is proportional to the range parameter $\eta_o L/D$. Since η_o and L/D are both sensitive to Mach number variation, the range parameter $\eta_o L/D$ is a suitable objective for optimizing the cruise condition and for comparing the fuel economy of aircraft propulsion systems. For a given combination of airframe and engines in steady level flight, the range parameter is determined by two control variables – typical combinations are altitude and speed or lift coefficient and Mach number. In the case of a constraint on altitude, speed, or engine rating, only one control variable remains to be optimized. For historical reasons, this problem first was treated analytically for propeller aircraft, with the classical solution that the cruise speed for maximum range equals the minimum drag speed V_{MD}. The maximum range of jet aircraft was first analyzed during the 1940s by assuming constant TSFC and ignoring compressibility drag. A well-known analytically derived result was the best flight speed at a given altitude with no limit on thrust

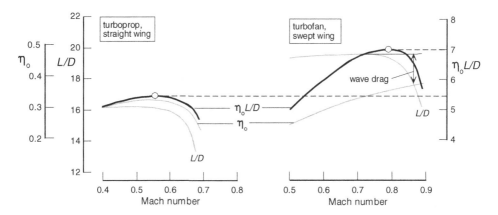

Figure 2.8   Effects of flight Mach number variation on the range parameter for turboprop- and turbofan-powered aircraft flying at optimum altitude

which was found to be $3^{1/4}$ times V_{MD}. Although this result soon proved to be useless for transonic (jet) aircraft, it has survived in academic teaching since 1950 (Chapter 12).

Figure 2.8 depicts how the range parameter is affected by speed for a short-haul propeller airplane and a long-range turbofan-powered airliner. Both airplanes are assumed to cruise at the minimum drag condition for each Mach number, dictating that the altitude is varied in order to match the varying Mach number. At subsonic Mach numbers, both planes have a constant $(L/D)_{max}$ and the range parameter is governed by the variation of the OEF with speed. For the propeller aircraft flying at $M > 0.55$, aerodynamic and propeller efficiencies deteriorate due to compressibility effects. The OEF of the swept wing turbofan airplane continues to increase with the Mach number but the drag increases at $M > 0.75$. As a result, the propeller plane achieves a flat maximum range parameter for Mach 0.55, the jetliner has a pronounced optimum at Mach 0.79. This example leads to the following observations.

- The turboprop performs best at low-subsonic speeds, the jetliner achieves a higher range factor above Mach 0.55 due to its higher L/D. If the comparison is made for regional aircraft with the same design payload and range, the turboprop will probably have a better specific range, lower fuel consumption and lower cruise speed; the turbofan will have a higher speed and productivity ([14], [21]).
- Transonic aircraft have their best range performance at a speed where drag due to compressibility determines the fuel-optimum cruise Mach number.

Between 1960 and 2000, design requirements dictated jetliner cruise speeds between Mach 0.75 and 0.85, dependent on range. The urgency of reducing engine emissions is the driving factor for recent investigations suggesting that future airliners may have a reduced design cruise Mach number. Notwithstanding an airliner productivity penalty due to the lower block speed, the economy may improve since the reduced aircraft acquisition costs and the cost of fuel could be the dominant compensating factors.

2.5 Environmental Issues

The target of a reduction in fuel burn of 50% between 2000 and 2020 is an extremely challenging one which is not achievable without important breakthroughs, both in technology and in concepts of operation.

—ACARE Strategic Research Agenda, 2002

In view of their effects on community noise and exhaust gas emissions into the atmosphere, engine selection has significant effects on the environment friendliness. The possibility of noise shielding by positioning engines favourably relative to aircraft components is likely to become an increasingly important element of conceptual design. On the other hand, the overview of what may be achieved by 'greener by design' categorizes the reduction of engine emissions as more demanding in the long term than the control of noise production around airfields [45].

With the exception of the calorific value of fuel, all the factors affecting engine overall efficiency have improved over the years. Although engine technology development is largely outwith the control of the advanced aircraft designer, far-reaching decisions have to be made when a new engine generation is introduced to a clean sheet design (Section 3.4). Reduction of engine noise and exhaust emissions is a primary driving factor during the engine selection process (Section 3.5). The bypass ratio of a turbofan is a primary selection variable and the application of very high bypass-ratio geared turbofans or open rotor engines will allow a major step forward in fuel economy but may lead to major complications when designing the airplane's general arrangement. Since energy efficiency is closely related to aerodynamic and overall engine efficiency, both figures are of crucial importance for environmental acceptance and economic competitiveness.

2.5.1 Energy and Payload Fuel Efficiency

The energy efficiency (EEF) of a passenger transport defines the available seat-kilometres per litre of fuel consumed. A similar characteristic is the payload fuel efficiency (PFE) defining the available payload-kilometre per unit fuel weight consumed. For a given payload and range, the accumulated cost of fuel consumed during commercial operations forms a large proportion of the operating expenses whereas the seat-km or ton-km production represents the income potential. Energy efficiency can be expressed in terms of parameters which are, to a considerable extent, under the control of the advanced designers – Section 8.6 treats payload fuel efficiency as a figure of merit in the design. Expressed as an instantaneous performance, the EEF is derived from the specific range V/F

$$E_{\mathrm{EN}} \overset{\text{def}}{=} \frac{N_{\mathrm{s}} \Delta R}{\Delta Q_{\mathrm{f}}} = \frac{N_{\mathrm{s}} V \Delta t}{(\dot{m}/\rho)_{\mathrm{f}} \Delta t} = \rho_{\mathrm{f}} g N_{\mathrm{s}} \frac{V}{F} \tag{2.24}$$

where ΔQ_{f} is the volume of fuel burn-off per distance flown ΔR and N_{s} denotes the number of cabin seats. Similarly, PFE can be written as

$$E_{\mathrm{PF}} \overset{\text{def}}{=} \frac{W_{\mathrm{pay}} \Delta R}{\Delta W_{\mathrm{f}}} = \frac{W_{\mathrm{pay}} V \Delta t}{F \Delta t} = W_{\mathrm{pay}} \frac{V}{F} \tag{2.25}$$

These expressions are clarified by substitution of V/F in horizontal cruising flight according to Equation (2.11),

$$E_{EN} = \rho_f H \frac{N_s}{W_{pay}} \eta_o \frac{L}{D} \frac{W_{pay}}{W} \quad \text{and} \quad E_{PF} = (H/g)\eta_o \frac{L}{D} \frac{W_{pay}}{W} \qquad (2.26)$$

Equation (2.26) shows that design drivers are the calorific value of fuel, engine overall efficiency, aerodynamic efficiency and the ratio of payload to gross weight, occasionally called the structural efficiency. The following examples illustrate that, different from what the name suggests, energy efficiency is not a dimensionless number between zero and one.

- For a typical payload mass of 95 kg per seat and a fuel energy density of 34.5 MJ per litre, $\rho_f H N_s / W_{pay} = 37$ seat-km per litre. A jetliner with $W_{pay}/W = 0.20$, $\eta_o = 0.35$ and $L/D = 18$ has a range parameter of 6.3 and a PFE of 46.6 seat-km per litre (95.1 seat-nm per gallon).
- The PFE of a cargo aircraft with the same range parameter and a payload fraction of 0.25 amounts to 6 850 ton-km per ton fuel (3 700 ton-nm per ton).

The average energy efficiency for a flight is obtained from the integration of the instantaneous value along the flight path; see Section 8.6.

Energy efficiency is often considered as measure of technical progress in commercial aviation. Since the introduction of long-range jetliners, it has increased each year with 0.6 seat-km per litre; that is, by a factor 2.5 between 1960 and 2000 (Figure 2.9).[6] Wide body passenger transports introduced into service after 2010 achieve E_{EN} values approaching 50 seat-km per litre. Since the actually realized efficiency depends on the cabin seating arrangement and distance flown, it varies between different airliner versions, cabin arrangements and operations. This makes the clear trend in Figure 2.9 rather coincidental. It is therefore preferable to compare energy efficiencies based on maximum payload rather than on seating capacity.

Figure 2.10 shows the effect on PFE of varying the harmonic range for a family of wide body airliners with equal MPL, aerodynamic and overall engine efficiency in cruising flight. Long-range aircraft have a disadvantage due to the high MTOW required to carry the large fuel load – indeed, very few existing long-range airliners have a harmonic range longer than 10 000 km. For ranges below 3 000 km, the fuel lost during climb and descent forms a significant fraction of the mission fuel which causes the efficiency to decrease and, in the present example, the highest PFE is achieved for approximately 3 500 km range. Most airline flights are made over shorter distances than the harmonic range, with a take-off weight less than the MTOW. Consider an aircraft designed for 10 000 km harmonic range (point D) with a PFE = 6 900 km. If this plane makes a 5 000 km flight with maximum payload (point E), its TOGW is 15% less than the MTOW, and PFE = 7 323 km. However, if the plane were designed for a harmonic range of 5 000 km (point F), its empty and fuel weights would be lower, resulting in a PFE of 7 900 km – 8% better than point E. Figure 2.10 also shows that,

---

[6]The fuel efficiency of a modern airliner is similar to that of a typical middle-class car on the highway. However, airliners travel ten times faster over long distances and achieve a higher load factor. Based on payload-kilometres produced, airliners achieve a (much) better energy efficiency than cars.

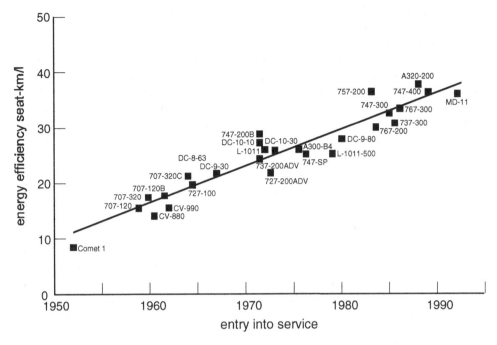

Figure 2.9   Historic energy efficiency trend for jetliners. Source: NASA

for ranges in excess of the harmonic range, the PFE decreases rapidly due to the replacement of payload by fuel.

Since the selection of a long-range design point entails a reduced energy efficiency, it has been observed that (a lot of) fuel can be saved if long-range flights were covered in two or more stages by medium-range aircraft [40]. This concept is verified in [30] for the Boeing 747 and the 777 and it is further elaborated in recent publications. According to [36], fuel savings by staging are possible for trips longer than about 6 000 km. The benefits increase almost linearly with the flight length, reaching about 13% for the maximum distances that are likely to be

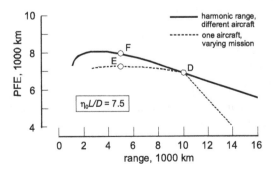

Figure 2.10   Payload fuel efficiency of jetliners with 50 ton payload

covered. Since 30% of the global fuel is burned on flights of 6 000 km and above, the staging of all long-haul flights would save 3% fuel if the same (long-range) aircraft were used, and 9% if (optimized) medium-range aircraft were used. However, a long-range transport aircraft is more flexible when an airline network is treated as a system. For instance, an airliner can make a return flight with two medium-range segments with no intermediate refueling. This may be one of the reasons why few customers opt for short/medium-range versions of a basic long-range airliner. Energy efficiency affects fuel economy as well as engine emissions and it is beneficial to get insight in its importance in the initial design stage. Much of the present book is devoted to finding out how selection variables affect these parameters and to what extent the designer's freedom of choice is limited. This is a complex matter since the (often opposing) effects of design technology are spread all over aircraft components and they are felt in all technical disciplines contributing to costs.

2.5.2   'Greener by Design'

Environmental issues are causing increasing concerns for the aviation community. The IPCC study on the global atmosphere [3] assesses radiative forcing[7] effects from aircraft in 1992, projects it forward to 2050 and predicts their overall effect to increase by a factor of five. Most effects are associated with engine combustion products. About one-third of the total radiative forcing stems from CO_2, for which fuel burn is the direct measure. Other forcing components are NO_x, H_2O, contrails, cirrus clouds, sulfate, and soot – the estimated ratings of their radiative forcing are mostly qualified as unreliable. Areas of uncertainty are the effects of the O_3 and CH_4 created by NO_x emissions and the effects of contrails and cirrus clouds. In particular, their dependence on conditions in the engine exhaust, flight altitude, latitude, climate and season are uncertain. Consequently, there are circumstances in which measures to reduce fuel burn may actually increase the overall contribution of engine emissions to climate change. This applies, for example, to the trade-off between increasing engine pressures and temperatures in favour of lower SFC and reducing them in order to reduce the generation of NO_x. In spite of these complications, a pertinent recommendation to the aircraft designer is to increase the energy efficiency of the vehicle by all affordable measures and to treat the cruise altitude explicitly as an independent design variable.

Primary issues driving civil aircraft development were summarized in 2001 by the Advisory Council for Aeronautics in Europe (ACARE). This organization aims at radical improvements of the air transport system in order to reduce its environmental impact. In terms of goals to be achieved in 2020–2025, specific measures are summarized as follows:

- Reduce fuel consumption and CO_2 emissions by 50%.
- Reduce NO_x emissions by 80%.
- Reduce perceived external noise by 50%.

These requirements are formulated for aviation as a whole and need to be broken down into specific goals for the individual players. The aircraft and engine industry, operators, air traffic management, certification and safety agencies, and the European Union – all of them take these goals seriously.

---

[7]Radiative forcing expresses the perturbation or change to the energy balance of the Earth's atmospheric system in watts per square metre.

In the UK, several organizations representing the manufacturing and air transport sectors as well as academia have established the Air Travel Greener by Design study group. Focussing on mitigating the environmental impact of aviation, three sub-groups were formed to specifically consider operation, market-based options and technology. The technology sub-group addressed airfield noise and local air quality and climate change, with the climate change considered to be the most important in the long term. The sub-group reported its findings in [40]; the following text is an extract from their recommendations.

> In the context of climate change, we consider the following to be key areas of research, study and technology validation. Some already benefit from substantial European funding, some are new proposals. All merit support.
>
> - Research into the atmospheric effects of aircraft emissions, including the dependence of the effects on flight altitude.
> - Research and demonstration of engine technology to achieve a substantial reduction in NO_x emissions, particularly at cruise conditions.
> - Extended in-service trials of hybrid laminar flow control, sufficient to assess the practicability of the system.
> - Design study of the blended wing-body concept, backed by research and technology demonstration to eliminate all significant technical doubt.
> - Feasibility study of the laminar flying wing.
> - Total system study of the provision of long distance air travel by medium range aircraft covering distances greater than 7 500 km (4 000 nm, eight hours flying time) in two or more stages.
>
> Study of the methodology for designing transport aircraft so as to minimize impact on climate change per unit thrust. The barrier to the introduction of the identified design and technology options that have the potential to reduce the greenhouse effect is the perceived imbalance between the significant technical and economical risks and the small potential economic benefit.

Increasing environmental acceptability of conventional airliner configurations will eventually become prohibitively expensive and some of the Greener by Design recommendations are not achievable without a radical change of the design concept. The blended wing body (BWB) (Figure 5.3) is an unorthodox but promising concept which has been investigated by many institutions. The less radical configuration depicted on Figure 2.11 features a very high aspect ratio laminar flow wing and ultra-high bypass ratio turbofans located above the empennage for noise shielding. Application of a new generation of airliners appears to promise that a major

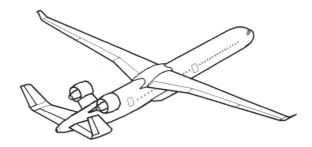

Figure 2.11   Will this be the ultimate green aircraft of the future?

step forward will be made during the first half of the twenty-first century to reduce the impact of aviation on the Earth's climate.

Bibliography

[1] Society of Allied Weight Engineers, *Introduction to Aircraft Weight Engineering*, SAWE Inc., Terminal Annex, Los Angeles, CA, 1996.

[2] Fielding, J.P., *Introduction to Aircraft Design*, Cambridge University Press, Cambridge, 1999.

[3] Intergovernmental Panel on Climate Change, *Aviation and the Global Atmosphere*, Cambridge University Press, Cambridge, 1999.

[4] Ferreri, D., *Marketing and Management in the High-Technology Sector: Strategies and Tactics in the Commercial Airplane Industry*, Praeger, Westport, CT, 2003.

[5] Roux, E., *Avions Civils à Reaction: Plan 3 Vues et Données Characteristiques*, Éditions Élodie Roux, Imprimé par lulu.com, 2007.

[6] De Florio, *Airworthiness: An Introduction to Aircraft Certification*, Butterworth-Heinemann, Oxford, 2011.

Initial Orientation

[7] Werner, R.A., and G.F. Wislicenus, 'Analysis of Airplane Design by Similarity Considerations", AIAA Paper No. 68-1017, October 1968.

[8] Caddell, W.E., "On the Use of Aircraft Density in Preliminary Design", SAWE Paper No. 813, May 1969.

[9] Cleveland, F.A., "Size Effects in Conventional Aircraft Design", *Journal of Aircraft*, Vol. 3, No. 6, pp. 483–512, November–December 1970 (also AIAA Paper No. 70-940).

[10] Engineering Sciences Data Unit, "Lost Range, Fuel and Time to Climb and Descent: Aircraft with Turbojet and Turbofan Engines", ESDU Data Sheet No. 74018, August 1974.

[11] Overend, W.J., "Design Criteria for Airline Operations", AIAA Paper No. 79-1849, August 1979.

[12] Smith, H.W., and R. Burnham, "The Outside Has to Be Bigger than the Inside", *Journal of Aircraft*, Vol. 18, No. 6, pp. 463–468, June 1981.

[13] Marsh, D.P., "Theory of Transport Aircraft Weight Fractions", SAWE Paper No. 1452, May 1982.

[14] Edlund, U., "Optimum Design Cruise Speed for an Efficient Short Haul Airliner", ICAS Paper No. 84-4.8.3, September 1984.

[15] Roskam, J., "Rapid Sizing Method for Airplanes", *Journal of Aircraft*, Vol. 23, No. 7, 1986, pp. 554–560.

[16] Torenbeek, E., "The Initial Calculation of Range and Mission Fuel During Conceptual Design", TU Delft, Department of Aerospace Engineering, Report LR 525, 1987.

[17] Association of European Airlines, "Requirements for Short/Medium Range Aircraft for the 1980s", Technical Affairs Committee, Report G.2110/R2, Third Edition, July 1981. Later edition: December 1989.

[18] Oelkers, W., "High Capacity Aircraft", ICAS Paper No. 92-1.10.3, September 1992.

[19] Martínez-Val, R., E. Pérez, T. Muñoz, and C. Cuerno, "Design Constraints in the Payload-Range Diagram of Ultra High Capacity Transport Airplanes", *Journal of Aircraft*, Vol. 31, No. 6, pp. 1268–1272, November–December 1994.

[20] Arjomandi, M., and N.K. Liseytsev, "A Simplified Method for Estimating the Take-Off Weight for Short-Haul Transports", *Aircraft Design*, Vol. 3, No 1, pp. 49–56, 2000.

[21] Richter, H., and S. Menczykalski, "Civil Propulsion of the Future", *Innovative Configurations and Advanced Concepts for Future Civil Aircraft*, Von Kármán Institute for Fluid Dynamics, Lecture Series 2005-06, June 2005.

[22] Sacco, G., and C. Lanari, "The Three Lifting Surface Configuration Concept and Lessons Learned from the Piaggio P180", *Innovative Configurations and Advanced Concepts for Future Civil Aircraft*, Von Kármán Institute for Fluid Dynamics, Lecture Series 2005-06, June 2005.

[23] Bolsunovsky, A.L., N.P. Buzoverya, B.I. Gurevich, V.E. Denisov, and O.V. Sonin, "Flying-Wing: Problems and Decisions", *Innovative Configurations and Advanced Concepts for Future Civil Aircraft*, von Kármán Institute for Fluid Dynamics, Lecture Series 2005-06, June, 2005.

[24] Torenbeek, E., and G. La Rocca, "Civil Transport Aircraft", *Encyclopedia of Aerospace Engineering*, R. Blockley and W. Shyy (eds), pp. 4043–4054, John Wiley & Sons, Chichester, 2010.

[25] Torenbeek, E., "Business Aviation", in *Encyclopedia of Aerospace Engineering*, R. Blockly, and W. Shyy (eds), John Wiley & Sons, Chichester, UK, October 2010.

[26] Poll, D.I.A., "A First Order Method for the Determination of the Leading Mass Characteristics of Civil Transport Aircraft", *Aeronautical Journal*, vol, 115, No. 1167, pp. 257–272, May 2011.

Energy Efficiency

[27] Poisson-Quinton, P., "Energy Conservation in Aircraft Design and Operational Procedures", AGARD Lecture Series LS-96, Paper No. 9, October 1978.

[28] Niedzballa, S.M., and D. Schmitt, "Comparison of the Specific Energy Demand of Aeroplanes and Other Vehicle Systems", *Aircraft Design*, Vol. 4., No. 4, pp. 163–178, December 2001.

[29] Nangia, R.K., "Efficiency Parameters for Modern Commercial Aircraft", *Aeronautical Journal*, Vol. 110, No. 1110, pp. 495–510, August 2006.

[30] Creemers, W.L.H., and R. Slingerland, "Impact of Intermediate Stops on Long-Range Jet Transport Design", AIAA Paper No. 2007–7849, September 2007.

[31] Hahn, A.S., "Staging Airliner Service", AIAA Paper No. 2007-7759, September 2007.

[32] Poll, D.I.A., "The Optimum Aeroplane and Beyond", *Aeronautical Journal*, vol, 113, No. 1141, pp. 151–164, March 2009.

[33] Kenway, G.K.W., R. Henderson, J.E. Hicken, N.B. Kuntawala, D.W. Zingg, J.R.R.A. Martins, and R.G. McKeand, "Reducing Aviation's Environmental Impact Through Large Aircraft for Short Ranges", AIAA Paper No. 2010-1015, January 2010.

[34] Lanfhans, S., F. Linke, P. Nolte, and H. Schnieder, "System Analysis for Future Long Range Operation Concepts", ICAS Paper No. 2010-11.3.4, September 2010.

[35] Martinez-Val, R., E. Perez, C. Cuerno, and J.L. Palacin, "Cost-Range Trade-Off in the Design and Operation of Long-Range Transport Airplanes", ICAS Paper No. 2010-1.5.1, September 2010.

[36] Poll, D.I.A., "On the Effect of Stage Length on the Efficiency of Air Transport", *Aeronautical Journal*, vol. 115, No. 1167, pp. 273–283, May 2011.

Emissions and Atmospheric Pollution

[37] Haberland, Ch., O. Kranz, and R. Stoer, "Impact of Operational and Environmental Aspects on Commercial Aircraft Design", ICAS-94-1.3.1, Anaheim, 1994.

[38] Deidewig, F., A. Döppelheuer, and M. Lecht, "Methods to Assess Aircraft Engine Emissions in Flight", ICAS Paper No. 96-4.1.4, September 1996.

[39] Szodruch, J., "The Environmental Challenge as Chance for the Next Century Aircraft Design", ICAS Paper No. 98-4.10.3., September 1998.

[40] Green, J.E., "Greener by Design – The Technology Challenge", *Aeronautical Journal*, Vol. 106, No. 1056, pp. 57–103, February 2002.

[41] Whellens, M.W., and R. Singh, "Propulsion System Optimisation for Minimum Global Warming Potential", Paper No. 7111, ICAS 2002 Congress, Toronto, September 2002.

[42] Rolls-Royce, "Powering A Better World", Environment Report, www.rolls-royce.com., 2003.

[43] Antoine, N.E., and I.M. Kroo, "Aircraft Optimization for Minimal Environmental Impact", *Journal of Aircraft*, Vol. 41, No. 4, pp. 2100–2109, 2004. (Also AIAA Paper No. 2002-5667, September 2002.)

[44] Green, J.E., "Greener by Design", *Innovative Configurations and Advanced Concepts for Future Civil Aircraft*, Von Kármán Institute for Fluid Dynamics, Lecture Series 2005–06, June 2005.

[45] Green, J.E., "Air Travel – Greener by Design. Mitigating the Environmental Impact of Aviation: Opportunities and Priorities", *Aeronautical Journal*, Vol. 109, pp. 361–416, September 2005.

[46] Green, J.E., "Civil Aviation and the Environment: The Next Frontier for the Aerodynamicist", *Aeronautical Journal*, Vol. 110, pp. 469–486, August 2006.

[47] Hahn, A.S., "Staging Airliner Service", AIAA Paper No. 2007-7759, September 2007.

[48] Nangia, R.K., "Highly Efficient and Greener Civil Aviation: Step Jump, Why & How", lecture presented at the Royal Aeronautical Society, Bristol Branch, RKN-SP-2007-100, October 2007.

[49] Sankrithi, M., "Environmentally Progressive Aircraft Design Leveraging Advanced Technologies", Boeing Commercial Airplanes, lecture presented to Department of Aerospace Engineering, TU Delft, April 2008.

[50] Shakariyants, S.A., "Generic Methods for Aero-Engine Exhaust Emission Prediction", doctoral thesis, TU Delft, September 2008.

[51] Gardner, R., "How Green In Your Contrail?", *Aerospace International*, pp. 14–17, March 2007.

[52] Kenway, G.K.W., R.P. Henderson, J.E. Hicken, N.W. Kuntawala, D.W. Zingg, J.R.R.A. Martins, and R.G. McKeand, "Reducing Aviation's Environmental Impact Through Large Aircraft for Short Ranges", AIAA Paper No. 2010-1015, January 2010.

[53] Rutherford, D., "The Role and Design of a CO2 Standard for New Aircraft", ICCT, 14 April 2010.

[54] Bradley, M.K., and C.K. Droney, "Subsonic Ultra Green Aircraft Research: Phase I Final Report". NASA/CR-2011-216847, April 2011.

[55] Henderson, R.P., J.R.R.A. Martens, and R.E. Perez, "Aircraft Conceptual Design for Optimal Environmental Performance", *Aeronautical Journal*, Vol. 116, pp. 1–22, January 2012.

3

Propulsion and Engine Technology

At 65 cents per gallon, the fuel price was too low to justify the UDF. If fuel were at a buck or so a gallon, they'd be clamoring.

— R. Welsh, GE Manager Commercial Operations (1988)

3.1 Propulsion Leading the Way

The history of civil aviation has shown that a prominent contribution to improved energy efficiency and economics stems from advancements in propulsion technology [28]. Since the end of the Second World War, gas turbine engines have evolved significantly through coordinated development in airplane and engine technology. For instance, the Boeing 707 used the evolution from the turboprop to the turbojet engine to fly faster and higher whereas the need for more range led to the development of bypass engines around 1960. Larger engines with high bypass ratios were developed during the 1960s to meet the high take-off thrust and reduced fuel consumption requirements of wide body airliners. Since 1970, the continuous development of turbofans into more efficient and reliable engines has been gradual but significant. Within half a century, their thermal efficiency has increased from less than 0.40 for the straight jet engine to about 0.50 for the modern turbofan engine. The take-off thrust of the largest turbofans has increased from 20 to 45 tons, specific fuel consumption (SFC) decreased by approximately 30%; see Figure 2.7 (a). The greatest improvement came from the propulsive efficiency improving from about 0.50 to 0.75. Open rotor turbofans may improve this further to 0.90. A new generation of very high bypass ratio engines is emerging and the next step to open rotor technology may mature sooner than expected. The installation of these advanced engines may necessitate a marked change in the general arrangement of new airplanes.

This chapter gives an overview of the terminology associated with gas turbine engine technology and performances, covering subjects specifically selected to be applicable to conceptual design and performance analysis. For this purpose, the most relevant properties are engine power and thrust, specific fuel consumption and engine efficiencies. Several generalized performance characteristics will be presented in non-dimensional form for use in cruise performance optimization and in engine scaling. Aircraft gas turbines operate on the Brayton

Advanced Aircraft Design: Conceptual Design, Analysis and Optimization of Subsonic Civil Airplanes, First Edition. Egbert Torenbeek.
© 2013 by Egbert Torenbeek. Published 2013 by John Wiley & Sons, Ltd.

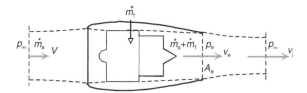

Figure 3.1   Using the momentum equation to derive turbojet and turbofan thrust

cycle; however, cycle analysis is outwith the scope of this text. Several excellent textbooks on gas turbine engine cycles, performance analysis and technology are mentioned in the bibliography.

3.2   Basic Concepts of Jet Propulsion

3.2.1   Turbojet Thrust

The thrust of a straight jet engine is derived from the momentum equation, using the control surface as indicated in Figure 3.1. At flight speed V, ambient air with mass flow rate $\dot{m}_a$ enters the inlet with momentum $\dot{m}_a V$, referred to as ram drag (or inlet momentum drag). The compressed air is heated by the injection of fuel with the flow rate $\dot{m}_f$ and then burning it at high pressure. The hot gas drives the turbine and exits the engine through the nozzle. Its total mass flow rate is $\dot{m}_a(1 + f)$, with f denoting the fuel/air mass ratio. Average properties in the nozzle exit with area A_e are velocity v_e and pressure p_e determinoing the gross thrust,

$$T_G = \dot{m}_a(1 + f)v_e + (p_e - p_\infty)A_e \tag{3.1}$$

where p_∞ denotes ambient pressure. The net thrust is equal to gross thrust minus the ram drag,

$$T_N = T_G - D_{ram} = \dot{m}_a\{v_e(1 + f) - V\} + (p_e - p_\infty)A_e \tag{3.2}$$

Engine manufacturers refer to this as the standard net thrust. Since its terms are well defined and accurately measurable, the standard thrust defines the performance of the (uninstalled) engine normally used by the aircraft manufacturer. While expanding to ambient pressure behind the nozzle exit, the jet efflux mixes with air. This leads to a (small) thrust contribution known as post-exit thrust. A simplified concept to include post-exit thrust is to assume a hypothetical uniform jet with velocity v_j after expansion. Another simplification is to ignore the fuel mass flow rate relative to the air mass flow ($f \ll 1$). This yields the ideal thrust,

$$T_{id} = \dot{m}_a(v_j - V) \tag{3.3}$$

The difference between standard net thrust and ideal thrust is generally small at subsonic speeds and is ignored in this text. We consider the thrust T as the net force exerted by the installed engine to propel the aircraft. A useful figure of merit for jet propulsion is the specific thrust which relates the thrust to the air throughput for generating it,

$$\frac{T}{\dot{m}_a} = v_j - V \tag{3.4}$$

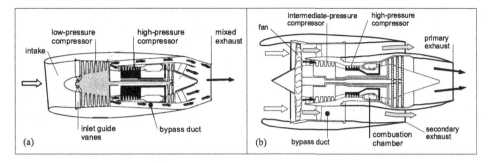

Figure 3.2 Schematic cross-sections of turbofan engines. (a) Low bypass ratio two-spool turbofan. (b) High bypass ratio three-spool turbofan. This engine has an intermediate pressure (IP) spool between the LP and HP spools, an exceptional configuration developed by Rolls-Royce

In order to generate a given thrust, a high specific thrust engine uses less air mass than an engine with a larger airflow. In general, it is lighter and can be installed in a smaller nacelle with lower drag. However, engines with a high specific thrust consume more fuel to deliver a given thrust and produce more jet noise. For many years, the only application of the straight jet engine has been in military applications.

3.2.2   Turbofan Thrust

Civil jet aircraft are powered by turbofan engines. In low-bypass turbofans the inlet airflow is first compressed by a low-pressure (LP) compressor as in Figure 3.2 (a) – in high-bypass engines, this is a single stage compressor known as the fan; Figure 3.2 (b). The inner (primary) part of a turbofan is called the core engine or gas generator. The core features a high pressure (HP) compressor, a combustion chamber and an HP turbine to develop high-energy gas by means of the same (Brayton) cycle as a turbojet. Part of its kinetic energy is extracted by a low-pressure (LP) turbine which drives the LP compressor or fan. The hot core flow leaves the engine through the primary nozzle. The fan compresses the cold airflow which bypasses the core through a duct and leaves the engine through the secondary nozzle. The total mass flow through a turbofan engine is

$$\dot{m}_a = \dot{m}_{a,h} + \dot{m}_{a,c} \tag{3.5}$$

with indices 'h' and 'c' referring to the hot primary and cool secondary engine flows, respectively. Turbofan thrust is derived by applying the momentum Equation (3.3) to both flows,

$$T = \dot{m}_{a,h}(v_{j,h} - V) + \dot{m}_{a,c}(v_{j,c} - V) \tag{3.6}$$

Some 20% of the energy transferred from the core engine to the fan is lost in the LP spool processes. With separate nozzle flows, the engine is most efficient when the cold jet velocity is the same percentage lower than flows. The bypass ratio (BPR) is the ratio of the secondary to the primary air mass flows,

$$B \stackrel{\text{def}}{=} \frac{\dot{m}_{a,c}}{\dot{m}_{a,h}} \tag{3.7}$$

Exhaust flow mixers improving propulsive efficiency are installed in many low BPR turbofans. If the two flows are mixed with no momentum loss into a uniform jet with velocity

$$v_j = \frac{v_{j,h} + B v_{j,c}}{1 + B} \tag{3.8}$$

it appears that Equation (3.3) applies to turbofan thrust as well. Since the BPR varies with operational conditions, it is commonly specified for the static condition at sea level. Turbofans can be characterized according to the following tentative scheme:

- Low bypass ratio turbofans ($B \leq 1.5$), formerly known as bypass engines, can be found in a few obsolete military aircraft and trainers.
- Medium bypass ratio ($1 < B \leq 4$) turbofans are installed in older jetliners and business jets, most of which do not comply with today's noise regulations.
- High bypass ratio ($4 < B \leq 10$) turbofans constitute the presently dominating category in jetliner and business jet applications.
- Very high bypass ratio ($B > 10$) turbofans will be installed in airliners developed for entry into service after 2010.

The BPR of a turbofan is closely related to its specific thrust. For a given thrust, increasing the BPR leads to a larger fan diameter, more airflow, reduced specific thrust, and lower jet noise. Many turbofans are offered with different values of the fan diameter, BPR and take-off thrust for a basically unchanged core engine. This has the advantage that turbofans with a single common core can be matched to the thrust required by different airplane designs. For an entire engine family with the same gas generator, development costs are only half as much as developing three separate engines. Commonality simplifies the manufacturing process and cuts parts inventories to a minimum. For airlines, core commonality translates into simplified inventory management, overall lower propulsion costs and more uniform maintenance procedures for a variety of engines spanning a wide range of thrusts.

3.2.3   Specific Fuel Consumption

The fuel weight flow per unit time burnt by a turbojet or turbofan engine varies approximately proportional to its thrust. In order to make a meaningful comparison between different engines, it is customary to use the specific fuel consumption (SFC). The thrust-specific TSFC denotes the fuel mass flow rate per unit thrust,

$$C_T = \frac{\dot{m}_f}{T} = \frac{F/g}{T} \tag{3.9}$$

where F denotes the fuel weight flow. A TSFC in SI units has the dimension g s^{-1} N^{-1} although it is customary to express TFC in kg h^{-1} N^{-1}. The TSFC in the SI system differs by a factor g from the value in Imperial units having the dimension lb hr^{-1} lbf^{-1}. The TSFC based on fuel weight flow is mostly used in aircraft performance analysis and differs by a factor g from the TSFC based on mass flow. Consequently, the weight-specific SFC in SI units is numerically equal to the mass-specific TSFC in Imperial units. Although TSFC is a widely used parameter, the inexperienced analyst may be confused about its definition since it does not mean the same thing to everybody.

3.2.4   Overall Efficiency

In case of doubt, it is recommended to convert SFC into overall efficiency (OEF), also known as total efficiency,

$$\eta_o \stackrel{\text{def}}{=} \frac{\text{thrust power developed by the engine}}{\text{rate of fuel energy added to the engine}} = \frac{TV}{\dot{m}_f H} = \frac{TM\,a}{FH/g} \tag{3.10}$$

The (lower) calorific value – or specific energy content – of jet fuel H is the chemical energy converted into thermal energy on complete combustion in air. The speed of sound is $a = a_{sl}\sqrt{\theta}$, with θ denoting the relative atmospheric temperature. Substitution of $a_{sl} = 340.29$ m/s and $H/g = 4\,350$ km for conventional (fossil) gas turbine engine fuel yields the conversion of TSFC in (kg/h)/N into OEF,

$$\eta_o = \frac{a_{sl}}{H/g}\frac{M\sqrt{\theta}}{C_T} = 0.0287\frac{M}{C_T/\sqrt{\theta}} \tag{3.11}$$

For TSFC in (lb/h)/lbf this becomes

$$\eta_o = \frac{a_{sl}}{H/g}\frac{M\sqrt{\theta}}{C_T} = 0.2816\frac{M}{C_T/\sqrt{\theta}} \tag{3.12}$$

Since overall efficiency is a dimensionless quantity between zero and one, it has the same value in both systems of units. For given flight conditions, the OEF can be derived from thrust and fuel flow (or SFC) stated in the engine specifications, corrected for installation losses. For cruising flight near the tropopause at a given Mach number, OEF is (nearly) independent of the altitude. The first generation of jetliners flew at $M \approx 0.80$ and were powered by straight jet engines with $C_T \approx 1.0$ (lb/h)/lbf. This complies with an OEF of only 20% which illustrates the jet's poor performance.[1] A typical $C_T = 0.56$ (lb/h)/lbf for current turbofans installed in long-range aircraft cruising at $M = 0.80$ corresponds with an OEF of 35%. It is worth noting that, even for a 100% efficient propulsion system, there exists a theoretical minimum TSFC; namely, $C_T = 0.2816\,M\sqrt{\theta}$. However, a more realistic lower limit is $C_T = 0.5\,M\sqrt{\theta}$ for $\eta_o = 0.55$. This would require present-day engines to be improved by some 50%.

3.2.5   Thermal and Propulsive Efficiency

The ideal turbofan engine converts the chemical energy contained by the fuel into the maximum possible kinetic energy which is then converted into propulsive thrust by the largest possible momentum increment across the engine. Accordingly, the OEF of this process can be decomposed into several process efficiencies.

---

[1]Concorde's propulsion system achieved an OEF of more than 40% in cruising flight at Mach 2. At the time of writing, this has not been exceeded by any operational subsonic airplane.

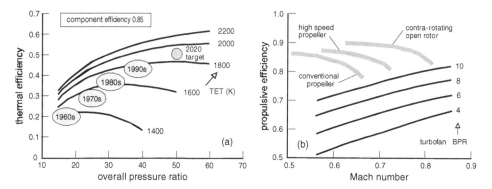

Figure 3.3   Efficiency trends for several generations of gas turbine engines used in commercial aircraft. (a) Turbofan thermal efficiency affected by cycle parameters. (b) Propulsive efficiency trends. The grey-scale curves for propellers and open rotors are performance envelopes

The thermal efficiency (TEF) is a measure of how efficiently the chemical energy contained by the fuel is converted into the kinetic energy increment of the air moving through the engine,

$$\eta_{th} = \frac{\text{rate of kinetic energy change across the engine}}{\text{rate of fuel energy added to the engine}} = \frac{\frac{1}{2}\dot{m}_a(v_j^2 - V^2)}{\dot{m}_f H} \tag{3.13}$$

The TEF can be seen as the product of combustion efficiency, thermodynamic efficiency of the gas generator's Brayton cycle, and efficiency of energy transfer from the gas generator to the engine air throughput. TEF depends primarily on the core engine's cycle parameters – it increases with the overall pressure ratio (OPR) and the turbine entry temperature (TET); see Figure 3.3 (a). Other important parameters are the efficiencies of the core compressors and turbines. Energy transfer suffers from total pressure losses in the power turbine and the fan and from mechanical losses due to gearing.[2] TEF increases with decreasing ambient temperature and increasing flight speed, one reason for airliners to have their best range performance at high altitudes and Mach numbers. It is expected that the TEF in the year 2020 will be approaching 55%.

The propulsive efficiency (PEF) of an engine is a measure of the thrust-producing momentum change for a given amount of kinetic energy increment of the air throughput,

$$\eta_p = \frac{\text{thrust power developed}}{\text{kinetic energy change across the engine}} = \frac{TV}{\frac{1}{2}\dot{m}_a(v_j^2 - V^2)} \tag{3.14}$$

Substitution of the thrust Equations (3.3) and (3.4) yields

$$\eta_p = \frac{\dot{m}_a(v_j - V)V}{\frac{1}{2}\dot{m}_a(v_j^2 - V^2)} = \frac{2}{1 + v_j/V} = \frac{2}{2 + T/(\dot{m}_a V)} \tag{3.15}$$

---

[2]Dependent on engine technology, polytropic compression and expansion efficiencies are approximately 90%, combustion and gear efficiencies are close to 99%.

Known as the Froude equation, this formula shows that the PEF is high for engines with a large air throughput producing a low specific thrust with a low-velocity jet. Figure 3.3 (b) illustrates that BPR and flight speed are the primary parameters affecting propulsive efficiency. Increasing the fan diameter – hence, air throughput – for a given core flow translates into higher BPR and increased PEF. For a $B = 5$ turbofan, a typical uninstalled engine PEF is 0.75 at Mach 0.80. In the year 2020 this value may have increased to 0.85.

The overall efficiency (OEF) defined by Equation (3.10) is the product of thermal and propulsive efficiencies,

$$\eta_0 = \eta_{th}\eta_p \tag{3.16}$$

Engine manufacturers foresee that the typical OEF of 0.35 for present-day turbofans will eventually be improved to 0.45 for open rotor engines. The question is not whether it will happen but when.

3.2.6 Generalized Performance

Propulsive thrust and fuel consumption of a turbofan engine are functions of its rating, flight speed, and ambient conditions. The engine rating is characterized by the high pressure (HP) spool rotational speed. Books on gas turbine engine performance derive generalized relationships which are useful for matching the engine size to the airframe, engine selection and airplane performance analysis. Thrust, engine air and fuel flow rates are generalized in the form of corrected performances. For example, corrected turbofan thrust is related to flight Mach number M and the HP rotor speed N,

$$\frac{T}{\delta} = f_T\left(M, N/\sqrt{\theta}\right) \tag{3.17}$$

The relative pressure δ and relative temperature θ denote the ambient pressure and temperature as fractions of sea level values in the standard atmosphere – see Appendix B – and $N/\sqrt{\theta}$ represents the corrected rotor speed. Fuel weight flow F is generalized as follows:

$$\frac{F}{\delta\sqrt{\theta}} = f_F\left(M, N/\sqrt{\theta}\right) \tag{3.18}$$

Elimination of the rotor speed yields

$$\frac{F}{\delta\sqrt{\theta}} = f_3\left(M, T/\delta\right) \tag{3.19}$$

whereas the TSFC is corrected for ambient temperature as

$$\frac{C_T}{\sqrt{\theta}} = \frac{f_F}{f_T} \tag{3.20}$$

Figure 3.4 depicts a typical generalized performance map of a turbofan engine. Within the range of typical cruise conditions, thrust is affected primarily by Mach number and, to a much

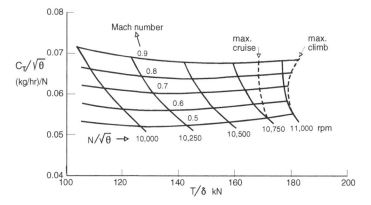

Figure 3.4 Generalized altitude performance of a high bypass turbofan

smaller extent, by the engine rating. Within the range of cruise altitudes and Mach numbers, the TSFC can be approximated by an empirical relationship, for instance,

$$C_T/\sqrt{\theta} = C_0(1 + C_M M) \tag{3.21}$$

The factors C_0 and C_M can be derived from the engine manufacturer's data. For a given engine, they are affected mainly by the BPR. The OEF is found by substitution into Equation (3.12),

$$\eta_0 = \frac{0.0287M}{C_0(1 + C_M M)} \tag{3.22}$$

As opposed to the TSFC, the OEF is independent of the altitude which demonstrates another advantage of its use.

3.2.7 Mach Number and Altitude Effects

In design studies where the engines are scaled up or down by the rubberizing method, the thrust is treated as a fraction of the maximum static take-off thrust (index TO) at sea level (SL) in the ICAO standard atmosphere (ISA). This fraction is known as the thrust lapse rate (TLR). If the corrected thrust is according to Equation (3.17), we obtain the thrust lapse parameter,

$$\tau \stackrel{\text{def}}{=} \frac{T/p}{T_{\text{TO}}/p_{\text{sl}}} = \frac{T}{\delta\,T_{\text{TO}}} \tag{3.23}$$

In a rubberizing process it is desirable to eliminate the rotor speed by combining Equations (3.10), (3.17) and (3.18), which yields the OEF in terms of two dimensionless quantities,

$$\eta_0 = f_\eta(\tau, M) \tag{3.24}$$

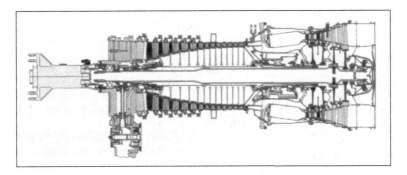

Figure 3.5 Rolls-Royce AE 1107: a member of a family of turboshaft and turboprop engines in the 6 000 shp (4 474 kW) class for military and commercial applications. *The Jet Engine*, Rolls Royce Technical Publications, Copyright 2013 Rolls-Royce plc

The carpet plot on log-log scale depicted in Figure 12.6 shows that Mach number variation has much more effect on OEF than the engine rating. For a cruising flight at a given Mach number, the thrust lapse rate with altitude depends on the engine cycle, in particular, the BPR. However, T/δ is constant in the stratosphere when the Reynolds number effects are disregarded. Consequently, if for a given engine rating and flight Mach number, a turbofan engine is rubberized, the thrust lapse rate stays practically constant.

3.3 Turboprop Engines

If the BPR efficiency of a turbofan engine is increased to more than about thirty and the fan duct is deleted, the resulting propulsion system is very similar to a turboprop engine-cum-propeller. In fact, a turboprop engine powers a shroudless propeller acting like a very low FPR. Compared to turbofan technology, the development of turboprop propulsion has been less spectacular [29]. An example of a modern turboprop engine used in commercial applications is shown in Figure 3.5.

3.3.1 Power and Specific Fuel Consumption

The output of a turboprop engine is the shaft horsepower (SHP) P_{sh} and jet thrust T_{jet}. Shaft power is converted into propulsive power by the propeller. Similar to the jet engine, the net jet thrust T_{jet} of a turboprop is equal to the gross jet thrust minus the ram drag of the inlet airflow. Compared to the propeller thrust, the jet thrust is small but not negligible. The total thrust horsepower (TTHP) available for propulsion is

$$P_{av} = \eta_{pr}P_{sh} + T_{jet}V \quad \text{with propeller efficiency} \quad \eta_{pr} \stackrel{\text{def}}{=} \frac{T_{pr}V}{P_{sh}} \qquad (3.25)$$

The TSHP is thus largely dependent on propeller efficiency which varies with the type of propeller, engine speed and airspeed. Airplane designers are primarily interested in the available power of the complete engine-cum-propeller combination. However, since the turboprop can be equipped with different propellers, the engine manufacturer expresses the available power output in terms of equivalent shaft horsepower (ESHP)

$$P_{eq} = P_{sh} + \frac{T_{jet} V}{\eta_{pr}} \tag{3.26}$$

where only the (relatively small) second term is affected by propeller efficiency. Note that, for static conditions, velocity and propeller efficiency are both zero and hence the jet thrust contribution is indefinite. To make allowance for jet thrust, it is assumed that 1 kW SHP gives 15 N jet thrust (2.5 lbf/hp).

The specific fuel consumption (SFC) of a turboprop engine is mostly referred to the power delivered. Power specific fuel consumption (PSFC) is then defined as fuel weight flow per unit ESHP,

$$C_P = \frac{F}{P_{eq}} = \frac{\dot{m}_f g}{P_{eq}} \tag{3.27}$$

In the SI system, PSFC has the dimension g s^{-1} W^{-1}, in British units it is in lb h^{-1} hp^{-1}. Obviously, the SFC based on shaft power will be higher than the SFC based on equivalent power. Due to the favourable effect of airspeed on engine thermal efficiency, SFC decreases slightly with flight Mach number. For engines with SFC based on ESHP, the OEF is computed from

$$\eta_0 = \frac{\eta_{pr} P_{eq}}{F H/g} = \frac{\eta_{pr}}{C_{P_{eq}} H/g} \tag{3.28}$$

with SFC based on SHP it amounts to

$$\eta_0 = \frac{\eta_{pr} + T_{jet} V/P_{sh}}{C_{P_{sh}} H/g} \tag{3.29}$$

These equations show that the OEF is determined primarily by the propeller efficiency.

3.3.2 Generalized Performance

Turboprop corrected performances are similar to those for jet and turbofan engines.

$$\text{Shaft power:} \qquad \frac{P_{sh}}{\delta \sqrt{\theta}} = f_P(N/\sqrt{\theta}, M) \tag{3.30}$$

$$\text{Air flow:} \qquad \frac{\dot{m}_a \sqrt{\theta}}{\delta} = f_{\dot{m}}(N/\sqrt{\theta}, M) \tag{3.31}$$

$$\text{Fuel flow:} \qquad \frac{F}{\delta \sqrt{\theta}} = f_F(N/\sqrt{\theta}, M) \tag{3.32}$$

$$\text{Gross jet thrust:} \qquad \frac{T_{jet}}{\delta} = f_T(N/\sqrt{\theta}, M) \tag{3.33}$$

If the jet is expelled opposite to the direction of flight the net total thrust amounts to

$$T = \frac{\eta_{pr} P_{sh}}{V} + T_{jet} \tag{3.34}$$

Turboprop performance is generalized by rewriting Equation (3.34) as follows:

$$\frac{T}{\delta} = \frac{\eta_{pr}}{M a_{sl}} f_P + f_T - M a_{sl} f_{\dot{m}} \tag{3.35}$$

The efficiency of a constant speed propeller is a function of the power coefficient C_P, the advance ratio J and the Mach number,

$$\eta_{pr} = f(C_P, J, M) \tag{3.36}$$

The power coefficient and the advance ratio are rewritten as

$$C_P \overset{def}{=} \frac{P_{sh}}{\rho N_{pr}^3 D_{pr}^5} = \frac{a_{sl}^2}{\gamma P_{sl} D_{pr}^5} \frac{P_{sh}/\delta\sqrt{\theta}}{(N_{pr}/\sqrt{\theta})^3} = f_{C_P}(N_{pr}/\sqrt{\theta}, M) \tag{3.37}$$

and

$$J \overset{def}{=} \frac{V}{N_{pr} D_{pr}} = \frac{a_{sl}}{D_{pr}} \frac{M}{N_{pr}/\sqrt{\theta}} = f_J(N_{pr}/\sqrt{\theta}, M) \tag{3.38}$$

where D_{pr} and N_{pr} denote the propeller diameter and RPM, respectively. Since the propeller rotates at a constant fraction of the engine RPM, the propeller efficiency is a function of the same parameters determining the engine performance,

$$\eta_{pr} = f(N/\sqrt{\theta}, M) \tag{3.39}$$

and combination with Equation (3.35) yields the propulsive thrust

$$\frac{T}{\delta} = f(N/\sqrt{\theta}, M) \tag{3.40}$$

This result proves that the generalized result for jet and turbofan engines applies to turboprop-powered aircraft as well. Hence, Equation (3.17) is valid for any gas turbine-based propulsion system. However, contrary to jet engines, the OEF of a turboprop system is sensitive to a variation in the Mach number as well as engine rating.

3.3.3 High Speed Propellers

During the 1990s, the development of high speed regional turboprops became the obvious step to counter the attack of regional jets on their traditional market. Since 2000, the orders for new and larger turboprops have been increasing again and it is anticipated that their maximum

Figure 3.6   The Europrop International TP 400 three-shaft turboprop engine with high-speed propeller.
Copyright Europrop International, reproduced with permission

speed will eventually increase to Mach 0.7. This is also the design speed of the Airbus A400M
military freighter which is powered by four of the West's most powerful Europrop TP400-D6
engines each producing 7 830 kW. Their propellers feature eight crescent-shaped composite
blades (Figure 3.6). High speed propellers were studied extensively resulting in the so-called
propfan. The combination of a gas turbine engine and a propfan is now categorized as an open
rotor engine.

3.4   Turbofan Engine Layout

3.4.1   Bypass Ratio Trends

The issue of how to get a step change in propulsion technology is all about (large) increases of
the BPR and the related key issues of drag, weight, and noise. Installing a bigger fan increases
the engine's air mass flow and its BPR with a corresponding reduction in fan pressure ratio
and specific thrust. This improves the OEF, leading to lower fuel consumption. Figure 3.7
shows the effects of BPR on SFC and OEF for a family of engines with a constant core OPR
and TET. Increasing the BPR from today's value of approximately five to 70 would reduce the
TSFC by more than 25% for the ideal cycle with 100% efficient components. With realistic
efficiencies, the TSFC is considerably degraded at all BPR levels but the favourable trend of

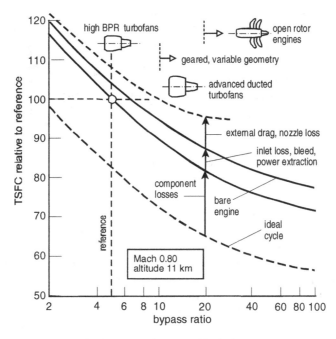

Figure 3.7   Effects of turbofan by-pass ratio on specific fuel consumption in cruise conditions

the increasing OEF remains. The gains are further diminished by inlet loss, bleed air/power extraction and nozzle losses. Moreover, the increased nacelle diameter and wetted area cause higher nacelle drag and weight, leading to increasing efficiency losses of the installed engine. For a given state of engine technology, an optimum BPR can be identified which leads to the best airplane efficiency, taking into account all installation losses and weight penalties. Since reduced specific thrust is the most powerful single measure to reduce engine noise, it follows that aggressive noise requirements may lead to configurations which are not optimal for fuel burn [45]. However, the characteristic buzz saw noise due to the supersonic fan tip speed is the dominant noise component becoming more annoying with higher by-pass ratio.

Increasing the BPR requires the optimum fan pressure ratio to be reduced. The upper limit of the fixed geometry high BPR turbofan segment is approximately $B = 10$ with a fan pressure ratio of 1.45, achieving a 10% reduced TSFC relative to $B = 5$. As the BPR further increases, the mismatch between fan and the low pressure turbine RPM for optimum fan and turbine efficiency becomes significant. A gearbox between the LP spool and the fan is needed for $B > 10$ to provide the optimum fan speed without an excessive number of turbine stages. The turbine is running at a more efficient higher speed, doing more work with fewer stages. Depending on its BPR, such a geared turbofan (GTF) requires a gear ratio between 2:1 and 4:1; see Figure 3.8 (a). The improved fan efficiency partly compensates for the 1% energy loss in the gearing which is transferred to a cooling system. Since the power developed by a GTF core exceeds the largest existing turboprop power, the planetary gear system forms the critical development issue. Adequate engine operation and stability for pressure ratios below 1.40 require variable pitch fan blades and/or a variable area cold air nozzle. Contra-rotating

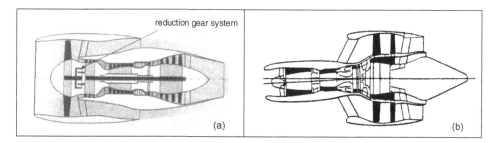

Figure 3.8 Very high bypass ratio turbofans. (a) Geared turbofan engine layout [46, 47]. (b) Rolls-Royce RB529 project with contra-rotating direct drive fans. Adapted from [31]

fans (CRF) may be favoured for the highest bypass ratios; see Figure 3.8 (b). Application of a BPR between 15 and 25 makes a mechanically complex engine inevitable and complicates its integration with the airframe.

A properly integrated engine installation forms the best balance between fuel efficiency, weight, noise production, maintainability, lifetime and cost. Different engine manufacturers cope with this matter in different ways. Pratt & Whitney have taken the lead in twenty-first-century engine technology by offering a GTF family of engines having the potential to cut fuel burn by 12%. Their production GTF is expected to cover the 100–180 kN (22 000–40 000 lbf) thrust range. The engine's relatively low fan tip speed reduces (accumulative) noise by 30 dB compared with Stage 3 requirements. In view of the increasing drag and weight of a shrouded turbofan, the radical step to open rotor engine technology is probably unavoidable when the BPR exceeds 30.

3.4.2   Rise and Fall of the Propfan

During the 1970s, NASA made an ambitious effort to stimulate the development of an advanced turboprop aircraft cruising at Mach 0.80 and altitudes up to 30 000 ft that could reduce fuel consumption by 30% compared to jetliners. This required the development of advanced high-speed propellers known as propfans. Proposed in 1975 by propeller manufacturer Hamilton Standard, propfans were introduced with multiple crescent-shaped highly loaded blades designed to maintain at least 80% propulsive efficiency. The company was awarded an advanced blade development contract and in 1981 began to design the composite blade set of a large single-stage demonstration propfan dubbed SR-7A which was tested in 1986. The complete engine with an eight-bladed unit flew on a modified Gulfstream II in 1987.

Propfan is a portmanteau word coined to describe a propulsion concept which combines some of the characteristics of a turboprop with those of a turbofan. Although early propfans had a much higher disk loadings than a conventional propeller, both were driven by a turboshaft engine via a gearbox. The essential difference with turbofans is the much higher propfan BPR between 25 and 40, variable-pitch blades and the absence of a rotor duct. The term propfan was originally applied to a multiple-bladed single rotor; however, since contra-rotation makes no fundamental difference, the term propfan still applies. Contra-rotation of the blades eliminates much of the swirl in the rotor slipstream, making the propulsive efficiency about 7% higher compared to the single-stage layout.

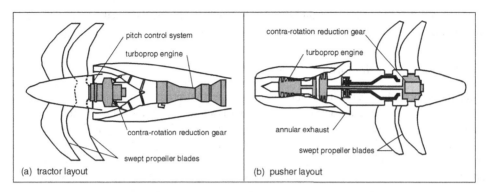

Figure 3.9 General arrangements of propfans with contra-rotating geared open rotors. The output is expressed as shaft horsepower (SHP) because there is a turboshaft engine and gearbox involved

In the 1980s, all the major airliner manufacturers considered adopting propfan technology for clean sheet designs. This required tractor engine arrangements which could be mounted to the wing leading edge or pusher arrangements mounted to the rear fuselage (Figure 3.9). Moving away from the geared propfan trend with the revolutionary unducted fan (UDF) concept, GE concentrated on the tail-mounted pusher configuration to limit cabin noise. Their UDF arrangement dispenses with the gearbox and features a gas generator to power a pair of CR statorless tree turbines carrying the rotor blades, as illustrated in Figure 3.10. American airframers were the most active investigators of the propfan. Boeing and McDonnell Douglas teamed with GE and P & W/Allison to evaluate the technology, culminating in the demonstrator engines GE 36 and PW 578-D mounted on 727-100 and MD-80 aircraft. In Europe, Rolls-Royce worked along the lines of a geared open rotor in pusher configuration but did not produce a full-scale demonstrator. The relatively low price of fuel at the time meant that potential concerns such as noise and reliability problems prevented the promising propfan technology from being adopted. The designs studied during the 1980s were at least three decennia ahead of their time, except for the Progress D-27 CR propfan which powers the

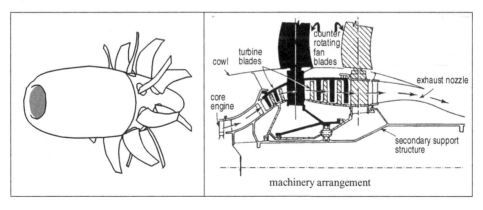

Figure 3.10 General Electric GE 36 unducted fan (UDF) of the 1980s

Antonov AN-70. This military freighter had its public debut in 1997 and is the only application of propfans in operational aircraft up to 2010.

3.4.3   Rebirth of the Open Rotor?

After the turn of the twentieth century, with soaring fuel prices and emphasis on reducing environmental emissions, the aeronautical industry is showing a renewed interest in the virtues of propfans. High-speed propellers (Figure 3.6) developed for speeds up to Mach 0.70 are becoming operational. Their diameter and detailed design are optimized for installation in a specific airplane. Different from turboprops, propfans are complete systems developed and produced by gas turbine engine manufacturers featuring variable pitch rotors with pressure ratios between 1.05 and 1.40, dependent on BPR. Application of contra-rotating open rotors lead to uninstalled cruise SFC reductions between 25 and 30% with similar noise levels compared to high BPR turbofans. Due to the varying blade pitch with speed, the gain in propulsive efficiency is even greater at low speeds; see Figure 2.7 (b). Hence, similar to turboprops, open rotor systems improve low speed performances which makes them especially fit for application in short-haul airplanes. However, complex airframe integration issues and acquisition costs will be high since open rotors are mechanically more complex than turboprops as well as turbofans. Major technical concerns exist regarding safety (blade failure), cabin noise, maintenance costs, reliability and fan efficiency at cruise speeds above Mach 0.75.

3.5   Power Plant Selection

Since 1980, aero engine manufacturers have proposed a large variety of advanced engine designs. Table 3.1 is a proposed classification of the propulsion concepts which appear to have a future of becoming realized. Apart from advanced turboprops, the main stream of engine development is devoted to direct drive as well as geared (very) high BPR turbofans – these are likely to be installed in new airliners becoming operational between 2010 and 2020. There is also a possibility that designing propfan-powered planes might be the right way to replace a substantial proportion of today's fleet of narrow body airliners. Engines that come as a logical choice might be based on turboprops such as those depicted in Figures 3.5 and 3.6.

Figure 3.11 is an overview of the effect on uninstalled and installed TSFC and OEF of the BPR and the specific thrust of a family of engines with a common core engine. Four candidates

Table 3.1   Classification of gas turbine engine systems

Power generator	Primary propulsive device	Abbreviation
Turboprop/turboshaft	conventional propeller	CTP
	high-speed propeller	HSTP
Turbofan	direct driven ducted fan	DDTF
	geared ducted fan	GTF
	contra-rotating ducted fans	CRDF
Propfan	single stage open rotor	SSPF
	contra-rotating open rotors	CRPF

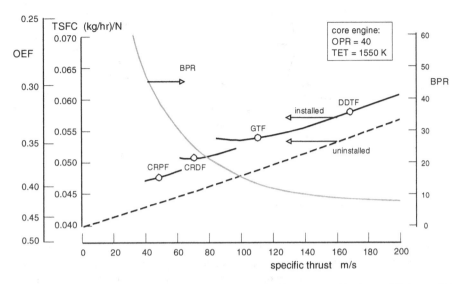

Figure 3.11 Spectrum of aero engine characteristics dependent on specific thrust. Flight conditions: Mach 0.76 at 35 000 ft

of advanced concepts are identified with the potential of reducing uninstalled TSFC between 10 and 30% relative to the high BPR turbofan. As indicated, installation effects – nacelle/cowl drag, flow interference drag for propfans, etc. – on system thrust are most important for very high BPR turbofans and open rotor engines (Figure 3.7).

3.5.1 Power Plant Location

The choice of a new propulsion concept has a far-reaching impact on the viability of a new airplane project. It is dictated by many aspects different from fuel economy, such as:

- costs and time to develop the engine to the operational status;
- fuel costs as a fraction of the total operating costs, taking into account the fuel price level in the future;
- the airplane's general arrangement, in particular the wing-mounted versus fuselage-mounted location;
- community noise noise footprint;
- cabin noise and effects on acoustic fatigue of the structure;
- containment requirements in case of propeller or rotor blade failures, and
- acceptance by the airlines, taking into account perceptions of the travelling public, expectations about safety, reliability, maintainability, etc.

The conceptual design stage is used for the process of matching the engine power or thrust to the airframe and for identifying the best location of the installed engines. High BPR turbofans are likely to be attached in the conventional way to the wing or the rear fuselage. Wing-mounted, very high BPR turbofans may be possible only with a high-set wing (Figure 3.12),

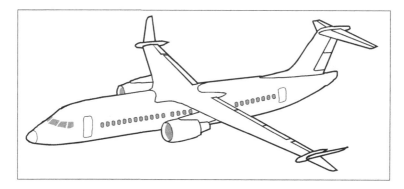

Figure 3.12   Regional jet powered by wing-mounted very high BPR geared turbofans

leaving the tail-mounted layout as a possible configuration for a low-set wing. Allocation of a propfan engine is complicated by the large diameter slipstream and its effect on aircraft stability and control. In order to eliminate or minimize the interaction between the slipstream and the flow over the wing and limit cabin noise, the preferred solution will probably be a pusher layout at the fuselage tail or a tractor layout suspended well below a high-set wing; see Figure 3.13. The final engine selection and power plant layout are probably made by interaction between the AD team, engine manufacturers and the airlines.

3.5.2   Alternative Fuels

The projected growth of air travel during the first decades of the twenty-first century will exceed the obtainable reduction of existing engine emissions. Moreover, the expected depletion of fossil fuel resources leads to worldwide concerns. Both developments have intensified the exploration and application of alternative fuels. Studies have narrowed the field to four principal candidates: biofuel, coal and gas-based synthetic fuel, liquid hydrogen and liquid methane. All four fuels are viable but biofuel and synthetic kerosene are judged to be the most attractive alternative, at least for some decades to come. The first generation of biofuels

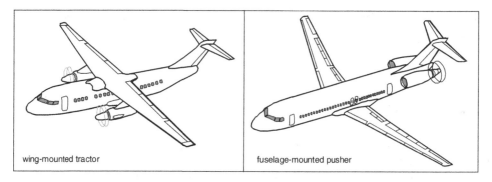

Figure 3.13   Preferred location of contra-rotating propfans

is becoming operational in commercial flights and promising developments of fuels obtained from algae feed stocks have been reported. The Fischer-Tropsch process is used to convert coal and natural gas into synthetic jet fuel. Since these fuels can be produced with physical properties essentially identical to those of jet-A fuel, no significant modifications will be required of aircraft or airports. The situation is entirely different if liquid hydrogen (LH_2) is used as jet fuel in the future. Since burning LH_2 offers extremely low emissions, it might become a suitable gas turbine fuel, on the provision that it is produced by renewable energy sources. However, LH_2 propulsion will entail a radical departure in the aircraft characteristics since they are characterized by very large tank volumes, a special fuel system and large differences in the weight distribution compared to conventional aircraft; see Section 6.8 for more information.

3.5.3   Aircraft Noise

With the advent of turbojet-powered civil aircraft and the steady growth of air travel which followed, noise was the first of the environmental impacts of civil aviation to cause public concern. Since the beginning of the jet age, the introduction of high BPR turbofans and lining materials have helped to reduce take-off sideline noise by about 25 EPNdB. With present BPRs slightly above the optimum for minimum fuel burn, engine manufacturers argue that there is still room for further noise reduction. The progressive reduction of aircraft noise at its source has been matched by tightening (Federal Aviation Regulations FAR) certification requirements and international (ICAO) noise regulations. In addition to these updated requirements, individual countries and airports are adopting their own stricter policies under pressure from local communities. The ACARE is pressing aircraft and engine manufacturers to achieve a considerable noise reduction between the years 2000 and 2020-2025 (Section 2.5). NASA has set the goal of reducing aircraft noise by an additional 20 dB during the same period of time. This is probably unlikely to be achievable without a radical change in aircraft design.

Throughout the history of commercial aviation, the responsibility for aviation noise and emission certification compliance fell primarily on the engine manufacturer – engine emissions were of secondary concern to airframe designers. In spite of the strong increase in air traffic over a long period of time, considerable progress has been made in reducing aviation noise even though noise certification requirements hardly have had any impact on the aircraft configuration. This situation is changing due to the fact that, for the latest commercial aircraft, airframe noise tends to dominate over engine noise. Three measurements points are used for noise certification. For commercial aircraft types, sideline, (take-off) climb and approach noise must remain below a limit based on the plane's MTOW and the number of engines. Jet noise typically dominates in sideline and climb for bypass ratios up to approximately five, fan noise dominates for higher bypass ratios. Aerodynamic noise is becoming increasingly relevant on the approach.

Prediction of the external noise footprint has become a standard activity of advanced aircraft design. The external noise produced by turbofan engines can be computed with NASA Langley's Aircraft Noise Prediction Program (ANOPP) [54]. However, the technology of open rotor engines is not yet sufficiently mature that noise prediction can be based on a standard routine.

Bibliography

[1] Torenbeek, E., *Synthesis of Subsonic Airplane Design*, Chapter 4 and Appendix H, Springer Verlag, Heidelberg, 1981.

[2] Kerrebrock, J.L., *Aircraft Engines and Gas Turbines*, Second Edition, MIT Press, Cambridge, MA, 1992.

[3] Hill, P.G., and C.R. Peterson, *Mechanics and Thermodynamics of Propulsion*, Second Edition, Addison-Wesley, Reading, MA, 1992.

[4] Mair, W.A., and D.L. Birdsall, *Aircraft Performance*, Chapter 5, Cambridge University Press, Cambridge, 1992.

[5] Smith, M.J.T., *Aircraft Noise*, Cambridge University Press, Cambridge, 1989.

[6] Ruijgrok, G.J.J., *Elements of Aviation Acoustics*, Delft University Press, Delft, 1993.

[7] Rolls-Royce plc., *The Jet Engine*, Fifth Edition, Renault Printing Company Ltd., 1996.

[8] Archer, R.D., and M. Saarlas, *An Introduction to Aerospace Propulsion*, Prentice-Hall, Inc., Englewood Cliffs, NJ, 1996.

[9] Hünecke, K., *Jet Engines: Fundamentals of Theory, Design and Operation*, Airlife Publishing Ltd., Shrewsbury, 1997.

[10] Walsh, P.P., and P. Fletcher, *Gas Turbine Performance*, Blackwell Science Ltd., Oxford, 1998.

[11] Saravanamutto, H.I.H., G.F.C. Rogers, and H. Cohen, *Gas Turbine Theory*, Fifth Edition, Pearson Education Ltd., Harlow, 2001.

[12] Mattingly, J.D., W.H. Heiser, and D.T. Pratt, *Aircraft Engine Design*, Second Edition, AIAA, New York, 2002.

[13] Cumpsty, N., *Jet Propulsion: A Simple Guide to the Aerodynamic and Thermodynamic Design and Performance of Jet Engines*, Second Edition, Cambridge University Press, Cambridge, 2003.

[14] Torenbeek, E., and H. Wittenberg, *Flight Physics: Essentials of Aeronautical Disciplines and Technology, with Historical Notes*, Chapter 5, Springer, Heidelberg, 2009.

Aero Engine Technology

[15] Jackson, A.J.B., "Some Future Trends in Aero Engine Design for Subsonic Transport", *Journal of Engineering for Power*, Transactions of the American Society of Mechanical Engineers, Vol. 98, pp. 281–289, April 1976.

[16] Wilde, G.L., "Future Large Civil Turbofans and Powerplants", *Aeronautical Journal*, Vol. 82, pp. 281–299, July 1978.

[17] Brewer, G.D., et al., "Study of Fuel Systems for LH_2 Fueled Subsonic Transport Aircraft", NASA CR 145369, Lockheed, California, July 1978.

[18] Fishbach, L.H., "Computerized Systems Analysis and Optimization of Aircraft Engine Performance, Weight, and Life Cycle Costs", AGARD CP 280, Paper No. 26, 1979.

[19] Gatzen, B.S., "Turboprop Design: Now and the Future", ICAS Paper No. 82-4.5.2., September 1982.

[20] Bennett, H.W., "Aero Engine Development for the Future", *Journal of Power and Energy*, Proceedings of the Institution of Mechanical Engineers, Part A, Vol. 197, pp. 149–157, July 1983.

[21] Williams, M.R., "Large Turbofans for the Year 2000", *Aeronautical Journal*, January 1984.

[22] Blythe, A.A., and P. Smith, "Prospects and problems of Advanced Open Rotors for Commercial Aircraft", AIAA Paper No. 85-1191, July 1985.

[23] Stuart, A.R., "The Unducted Fan Engine", AIAA Paper No. 85-1190, 1985.

[24] Lange, R.H., "A Review of Advanced Turboprop Transport Aircraft", *Progress in Aerospace Sciences*, Vol. 23, pp. 151–166, 1986.

[25] Grieb, H., and D. Eckardt, "Propfan and Turbofan as Basis for Future Economic Propulsion", 22nd Joint Propulsion Conference, AIAA Paper No. 86-1474, 1986.

[26] Blythe, A., "Potential Application of Advanced Propulsion Systems to Civil Aircraft", 15th ICAS Congress, London, Paper No. 86.3.8.3, September 1986.

[27] Miller, S.C., and H.W. Bennett, "Future Trends in Propulsion", ICAS Paper No. 86-86.04, September 1986.

[28] Wittenberg, H., "Aircraft Propulsion Leading the Way in Aviation", Report LR-532, TUDelft, Faculty of Aerospace Engineering, September 1987.

[29] Saravanamuttoo, H.I.H., "Modern Turboprop Engines", *Progress in Aerospace Sciences*, Vol. 24, pp. 225–248, 1987.

[30] Dunican, M.G., "Installation of Innovative Turbofan Engines on Current Transport Airplanes", AIAA Paper No. 87-2921, September 1987.

[31] Borradaile, J.A., "Towards the Optimum Ducted UHBR Engine", AIAA Paper No. 88-2954, July 988.

[32] Gordon, B.J., "The Development of the Unducted Fan", *Aerospace*, pp. 22–26, July 1988.

[33] Zimbrick, R.A., and J.L. Colehour, "An Investigation of Very High Bypass Ratio Engines for Subsonic Transports", AIAA Paper No. 88-2953, July 1988.

[34] Skavdahl, H., R.A. Zimbrick, J.L. Colehour, and G.P. Sallee, "Very High Bypass Ratio Engines for Commercial Transport Propulsion", ICAS Paper No. 88-4.11.2, September 1988.

[35] General Electric, "GE36/UDF, A New Thrust in Aviation", DGLR Jahrestagung, Darmstadt, Paper 88-011, 1988.

[36] Eckardt, D., and G. Brines, "Technology Readiness for Advanced Ducted Engines", 25th Joint Propulsion Conference, AIAA Paper No. 89-2479, 1989.

[37] Dupslaff, M., P. Wehlitz, and P. Schimming, "Propfan Technology", in European Forum: The Evolution of Regional Aircraft Technologies and Certification, organized by DGLR, AAF, RAeS, Friedrichshafen, April 6-7, 1989.

[38] Owens, R.E., K.L. Hassel, and D.E. Mapes, "Ultra High Bypass Turbofan Technologies for the Twenty-First Century", 26th Joint Propulsion Conference, AIAA Paper No. 90-2397, 1990.

[39] Zimbrick, R.A., and J.L. Colehour, "Investigation of Very High Bypass Ratio Engines for Subsonic Transports", *Journal of Propulsion and Power*, Vol. 6, pp. 490–496, July-August 1990. Also: AIAA Paper No. 88-2953.

[40] Sieber, J., and H. Jackwerth, "Betriebsverhalten Eines CRISP-Vortriebserzeugers", No. 91-193, DGLR Jahrestagung 1991.

[41] Stryker, H.Y., "21st Century Commercial Transport Engines", RAeS Sidney Branch Lecture, 1992.

[42] Saito, Y., M. Endoh, N. Sugahara, and K. Yamamoto, "Conceptual Study of Separated Core Ultra High Bypass Engine", AIAA Paper No. 92-3776, July 1992. Associated paper: AIAA Paper No. 92-3776, July 1992.

[43] Yaros, S.F., et al., "Synergistic Airframe-Propulsion Interactions and Integrations" White Paper prepared by the 1996-1997 Langley Aeronautics Technical Committee, NASA/TM-1998-207644, March 1998.

[44] Birch, N.T., "2020 Vision: the Prospects for Large Civil Airaft Propulsion", *Aeronautical Journal*, Vol. 104, No. 1038, pp. 281–289, August 2000.

[45] Richter, H., and S. Menczykalski, "Civil Propulsion of the Future", *Innovative Configurations und Advanced Concepts for Future Civil Aircraft*, Von Kármán Institute for Fluid Dynamics, Lecture Series 2005-06, June 6-10, 2005.

[46] Kurzke, J, "GasTurb 11 User's Manual, Design and Off-Design Performance of Gas Turbines", 2007.

[47] Dewanji, D., A.G. Rao, and J.P. van Buijtenen, "Conceptual Study of Future Aero-Engine Concepts", *International Journal of Turbo and Jet Engines*, Vol. 26, pp. 263–276, 2009.

Aircraft Noise

[48] Drell, H., "Impact of Noise on Subsonic Transport Design", SAE Paper No. 700806, October 1970.

[49] Kramer, J.J., and R.G. Dorsch, "NASA Aircraft Engine Noise Research", ICAS Paper no. 72-48, September 1972.

[50] Heidmann, M.F., "Interim Prediction Method for Fan and Compressor Source Noise", NASA TM X-71763, 1979.

[51] ICAO, "Aircraft Noise", *International Standards and Practices*, Annex 16 to the Convention of International Civil Aviation, 3rd Edition, 1985.

[52] Brentner, K.S., "Aerodynamic Impact on Noise and Emissions", Session IV of "Potential Impacts of Advanced Aerodynamic Technology on Air Transportation System Productivity", NASA Technical Memorandum 109154, September 1994.

[53] Kennepohl, F., R. Traub, R, Gumicio, and K. Heinig, "Influence of Bypass Ratio on Community Noise of Turbofans and Single Rotation Ducted Propfans", AIAA Paper No. 95-0135, June 1995.

[54] Kontos, K.B., B.A. Janardan, and P.R. Gliebe, "Improved NASA-ANOPP Noise Prediction Computer Code for Advanced Subsonic Propulsion Systems", Volume 1: ANOPP Evaluation and Fan Noise Model Improvement, 1996.

[55] Caves, R.E., L.R. Jenkinson, and D.P. Rhodes, "Development of an Integrated Conceptual Aircraft Design and Noise Model for Civil Transport Aircraft", ICAS Paper No. 98-6.4.3, September 1998.

[56] Kim, H.D., J.J. Berton, and S.M. Jones, "Low Noise Cruise Efficient Short Take-Off and Landing Transport Vehicle Study", AIAA Paper No. 2006-7738, September 2006.

[57] Hileman, J., "Airframe Design Considerations for a Silent Aircraft", presentation of work conducted by the Silent Engineering Design Team, Cambridge University Engineering Department, MIT, July 2007.

4

Aerodynamic Drag and Its Reduction

It has been noted that transportation is fundamentally 0% efficient as it involves moving mass from rest at one point to rest at another point, so that the energy of the system is unchanged. That it does take energy to accomplish this objective is due to the presence of drag, and the reduction of drag has been the primary focus of aircraft design over the last century.

—I.H. Kroo [72] (2001)

4.1 Basic Concepts

The drag estimation for a new design is not generally a single exercise, but a continuous process through its life from the early project study stage through preliminary design and development. Similar to weight prediction, the fidelity of the drag prediction methodology used during advanced design (AD) varies with the accuracy required, with the degree of airplane geometry definition, and with the amount of computed or experimental data available. Initially, predictions are mostly semi-empirical with a gradual shift to quasi-analytical and numerical methods. A wide variety of handbook methods, computations and wind tunnel data are used as the design proceeds and it is essential to be consistent with the definition of drag components. Although aircraft design over the past century has evolved into a process of increasing sophistication, the prediction of aerodynamic drag still poses a formidable challenge to the AD engineer. Even elemental flow physics driving drag can be quite complex. In addition, there are myriad ways in which flow fields around airplane components can interact to produce interference drag which is very difficult to predict accurately. But it is clear that full-scale aircraft drag prediction errors of 10 to 20% that have occurred in the past in certain development programs are outwith the range needed for success [38].

Advanced Aircraft Design: Conceptual Design, Analysis and Optimization of Subsonic Civil Airplanes, First Edition. Egbert Torenbeek.
© 2013 by Egbert Torenbeek. Published 2013 by John Wiley & Sons, Ltd.

In spite of the aircraft's complex shape, the external air force acting on it is traditionally resolved in the same way as for any other closed body moving through an airflow,[1] as follows:

- Drag is a positive force of magnitude D acting in the opposite direction to the airspeed vector $\vec{V}$.
- Lift with magnitude L acts normal to $\vec{V}$ in the airplane's plane of symmetry and is positive in the dorsal sense.

Most objects moving through air experience considerable drag and little or no lift; hence, $D/L \gg 1$. Airfoils form a special class of objects since they are designed to generate lift exceeding their drag by an order of magnitude: $L/D \gg 1$. Having the potential to generate a lift equal to several times the airplane's gross weight (GW), an airplane wing is classified as a structurally and aerodynamically extremely efficient lifting surface. Tail surfaces are airfoils designed primarily to provide lift for control in dynamic conditions and for stability – they are not intended for generating much lift in straight and level flight. Decisions taken during all stages of AD can have a major effect on the aerodynamic efficiency L/D; hence, it is essential to predict the sensitivity of drag to design variables as accurately as possible.

4.1.1   Lift, Drag and Aerodynamic Efficiency

Except for manoeuvres, the flight path angle γ of a passenger plane during climbing and descending flight normally amounts to less than ten degrees, whereas cruising flight is (nearly) steady and level throughout. Hence, it can be assumed that $\cos \gamma = 1$ during the complete flight. Accordingly, lift L equals the airplane's gross weight W and drag follows from

$$D = \frac{W}{L/D} \tag{4.1}$$

The L/D ratio enters the performance and design analysis directly since the thrust required is determined by the equilibrium of forces along the flight path,

$$T = D = \frac{W}{C_L/C_D} \tag{4.2}$$

Lift and drag coefficients are defined as:

$$C_L = \frac{L}{qS} \quad \text{and} \quad C_D = \frac{D}{qS} \tag{4.3}$$

where the dynamic pressure q is related to the altitude and the flight speed V or the Mach number M as follows:

$$q = \frac{1}{2}\rho V^2 = \frac{1}{2}\rho(Ma)^2 = \frac{1}{2}\rho M^2 \gamma RT = \frac{1}{2}\gamma p M^2 \tag{4.4}$$

The symbols a, R and γ are the sonic speed, the gas constant and the ratio of specific heats of air, respectively. Ambient air density, pressure and temperature are ρ, p and T, respectively.

---

[1] An airframe can be treated as a closed body by covering engine inlets and exhausts by streamlined nacelle fairings.

Drag components are mostly quoted in terms of counts – one drag count is defined as $\Delta C_D = 0.0001$. The standard reference area for lift and drag coefficients is the gross wing area S_w. This makes their numerical values dependent on the precise definition of the wing area for which no universally accepted standard exists. For example, Airbus and Boeing use different methods for including the wing area inside the fuselage whereas wings with a compound taper are sometimes approximated as an equivalent straight-tapered wing. Consequently, the zero-lift drag coefficient and the maximum lift coefficient quoted for different aircraft are not always comparable. However, since Equation (4.2) indicates that the drag in level flight is determined by the aerodynamic efficiency $C_L/C_D = L/D$, it does not depend on the reference area of both coefficients. At the time of writing, long-distance airliners achieve maximum values of L/D between 15 and 20, dependent mainly on their size. This parameter may go up to a value between 20 and 25 in the not too distant future.

The aerodynamic efficiency is important to the design engineer since it is a basic parameter in the engine/aircraft matching process and in flight performance analysis. Fuel consumption in cruising flight is proportional to the airplane GW and to engine thrust or power and a GW reduction contributes just as much to reducing fuel consumption as increasing the aerodynamic efficiency. Another important flight condition is the climb-out after taking off which is often a sizing criterion for the engine power or thrust to be installed.

4.1.2   Drag Breakdown and Definitions

Precise definitions of drag and its contributing components are essential for drag prediction and for an appreciation of drag reduction technology. However, the engineer is often confronted with (apparently) different terms for almost the same properties and the terminology found in the literature may create confusion due to a lack of international standardization.[2] Most of the confusion is associated with the complex task of analyzing and predicting the drag of a complete aircraft configuration. Dependent on the purpose of the work, the drag may be broken down according to one of the following basic schemes:

- The body surface exposed to the flow is divided into a large number of panels and the force on each panel is decomposed into a normal pressure component and a friction component tangential to it. Their components in the direction of the airflow are added to yield the total drag.
- The drag is decomposed in accordance with the physical mechanisms contributing to drag: the viscosity of air, vortical flow and shock waves.
- The drag is subdivided into parasite drag disregarding lift and induced drag caused by lift-generating mechanisms.

Figure 4.1 interrelates these schemes and indicates how flow phenomena, body shape and dimensions have an influence on drag components. The scheme is general since it applies to arbitrary closed bodies surrounded by incompressible as well as compressible (high-speed) flows. The physical mechanisms are fundamental in that they cannot be broken down into

---

[2]A comprehensive – but not widely used – set of definitions was drawn up by the British Aeronautical Research Council [17]; ESDU has produced a more recent overview [40]. It is also noticeable that the scheme in Figure 4.1 differs from schemes used in other publications such as [52] and [92].

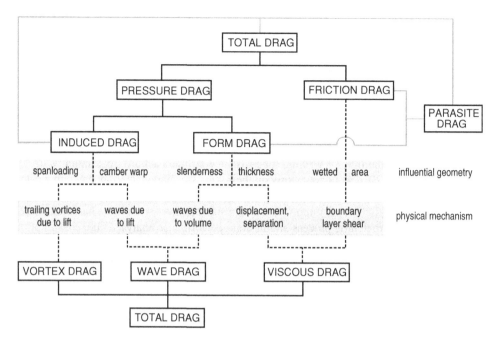

Figure 4.1   Schemes for decomposing the drag of a closed lifting body. The parasite drag of an airfoil is also known as profile drag

simpler elements. However, the simplicity of the scheme is deceiving since a number of drag sources are derived from interactions of fundamental mechanisms. For instance, viscous effects may affect the lateral variation of trailing vortex strength whereas three-dimensional non-viscous flow phenomena may contribute to viscous drag. And interactions between the (throttle-dependent) internal engine flow and the external flow may be the subject of debate between airframe and engine manufacturers about the drag accounting methodology.

Due to a lack of detailed information in the conceptual design process, the engineer may be forced to use a drag breakdown based on a simplified representation of the airplane's geometry. This leads to a somewhat heuristic approach known as a drag build-up technique, replacing the complex overall shape by individual components such as the wing, fuselage, tail surfaces and engine nacelles. If the result is a mixture of poor bookkeeping, incompatible drag analysis methods, inaccuracies caused by unforseen flow separations and interactions and surface imperfections, the reliability of the drag prediction appears to be unsatisfactory. An alternative drag breakdown may be needed during preliminary design by the analyst who uses measured aircraft (model) data to correct them for Reynolds number effects, trim drag and configuration effects.

4.2   Decomposition Schemes and Terminology

4.2.1   Pressure and Friction Drag

The most elemental drag breakdown distinguishes between the two ways in which the aerodynamic force is exerted on the exposed body surface. At each surface element the air pressure exerts a normal force whereas friction between the airflow and the surface due to shear causes a tangential force. Vectorial summation of these forces yields the resulting aerodynamic force R which is resolved into lift and drag: $\vec{R} = \vec{L} + \vec{D}$. Accordingly, drag is broken down into pressure drag and skin friction drag. In high speed level flight, pressure and friction drag are of similar magnitude, whereas pressure drag dominates at low speeds.

Compared to other breakdowns, the definitions of pressure and friction drag are straightforward and unambiguous. However, pressure and shear force distributions have to be obtained by computational fluid dynamics (CFD) methods or measured by experiments in a wind tunnel or in flight. Since these techniques are very time-consuming and require accurate shape information of the complete airframe, drag predictions for conceptual design are mostly based on a mixture of alternative schemes for decomposing drag [52]. For instance, insight into various forms of drag is gained from a breakdown based on schemes for modelling physical mechanisms as in the lower part of Figure 4.1. Different forms of energy are left behind the airplane in a trailing vortex sheet, a turbulent wake or a dead air region and, in high-speed flight, in the form of shock waves. The law of energy conservation requires the work produced in overcoming the drag to be equal to the energy increase of the surrounding atmosphere.

4.2.2   Viscous Drag

The viscosity of air leads to the formation of a boundary layer surrounding the body's surface. Except for a body surrounded by fully separated flow, the greater part of viscous drag is caused by surface friction due to shear stress in the boundary layer. Viscous drag is unique in that it is the only source of drag for which the force-generating mechanism acts tangentially to the external surface. For a given flow Reynolds number, friction drag is almost entirely determined by the state of the boundary layer: laminar or turbulent. Behind the body, the boundary layer is shed in the form of a turbulent wake causing a loss of momentum – the mechanical equivalence of viscous drag. Hence, skin friction drag can be derived from boundary layer analysis or by measuring the momentum loss in the wake of a wind tunnel model. As a secondary effect, the displacement effect of the boundary layer causes a modified pressure distribution. The result is a pressure imbalance between the front and rear parts of the body known as form drag, a relatively small component for a streamline body at a low incidence in subcritical flow.

Increasing the incidence of a transport or business aircraft leads to flow separation at the wing trailing edge which gradually spreads forwards. A large dead air region is eventually formed when the plane approaches the stall, causing a progressively increasing loss of lift and form drag. However, increasing the incidence of a thin airfoil causes separation at the (sharp) nose and a delta vortex is formed above a highly swept leading edge. Flow separation at the nose causes a loss of leading edge suction and a distinct drag inflation[3] when a critical angle of attack is exceeded. For transport aircraft this type of drag is avoided in normal flight conditions by proper aerodynamic design.

4.2.3   Vortex Drag

The finite span of a lifting surface causes the lower surface flow to move outboard toward the tip while the upper surface flow moves inward toward the centreline. The resulting cross-flow

---

[3]Drag inflation may occur at a modest incidence and must not be confused with the stall since it does not necessary lead to loss of lift: a delta vortex increases the lift of a low-aspect ratio wing.

velocities at the trailing edges combine with the free-stream flow, leading to the shedding of vortical flow that is particularly strong near the outboard regions and the tips. The vortex strength along the span is directly related to the lateral distribution of lift. The trailing vortex sheet contains rotational energy which must be paid for in the form of vortex drag. Since vortex formation is the unavoidable consequence of lift, vortex drag is also known as a component of (lift-)induced drag. Its coefficient C_{D_v} is derivable for high-aspect ratio straight wings by classical analysis such as L. Prandtl's lifting line theory which has been available since the early days of aeronautics. More general methods such as the vortex-lattice method and Trefftz plane analysis were derived for swept wings with low and high aspect ratios. Achieving minimal induced drag is a major concern of aircraft designers and remains an area of research activity [72].

Classical theory assumes the vortex sheet to be flat and located in the plane of the undisturbed flow. In reality, the vortex sheet is curved and, far behind the wing, the flow is deflected downward over the (average) downwash angle ϵ. In combination with the momentum equation, this flow model forms the basis for an alternative derivation of lift and drag due to lift [14]. It states that the apparent mass flow of air which is deflected downward by the wing generates lift and experiences a velocity decrement opposite to the direction of flight equal to $V(1 - \cos \epsilon) = 0.5\, V\epsilon^2$. The corresponding momentum loss is interpreted as the induced drag, which is identical to vortex drag when the flow is subcritical. In order to make sure that both concepts lead to the same drag, the apparent mass flow is assumed to be equal to the air mass flow through a circle with diameter equal to the wing span normal to the flow. The weakness of the flow model based on the momentum equation is that the sensitivity of the apparent mass flow to the wing shape is very difficult – if not impossible – to identify. The important exception is that, for given lift, the induced drag appears to be inversely proportional to the wing span squared.

4.2.4 Wave Drag

At high subsonic speeds (beyond Mach 0.7) flow expansions create pockets of supersonic flow embedded in the subsonic flow, and the aircraft enters the transonic flight regime. If a flow pocket gains sufficient supersonic speed, it is terminated by a shock wave. A shock wave is a very thin sheet of fluid where the state properties change almost instantaneously. The local velocity component approaching normal to the shock wave must be at least equal to the sonic velocity for a shock to exist; behind the shock it is below the sonic velocity. Through a shock wave, flow properties (pressure, velocity vector, density and temperature) change abruptly with an accompanying total pressure loss. The size of the supersonic pockets and the intensity of the shock waves increase progressively when the flight speed approaches Mach 1.

The loss of entropy in shock waves travelling with an aircraft causes wave drag, an important component of the total drag in high speed flight. Moreover, the severe and adverse surface pressure gradient at the foot of a strong shock causes thickening and even separation of the boundary layer, resulting in shock-induced form drag. Figure 4.1 makes a distinction between wave drag due to volume and wave drag induced by lift. The first of these depends on the progression of a body's cross-sectional area perpendicular to the flow, the latter depends mainly on the distribution of lift over the wing planform. Consequently, in the transonic regime, the aircraft drag coefficient depends on the Mach number and on the lift coefficient. Wave drag increases progressively when the critical Mach number M_{crit} is exceeded. Wave drag can only be accurately computed by CFD methods. Considering the mixed flow in the transonic regime, an attempt to derive a meaningful analytical expression for wave drag is elusive. One will normally resort to experimental data to determine the aerodynamic coefficients accurately in this operating regime [6].

4.3 Subsonic Parasite and Induced Drag

Figure 4.1 decomposes total airplane drag alternatively into parasite drag and induced drag. The dotted connections indicate their relationship with the three physical mechanisms of drag and how they are affected by the most influential geometrical properties. The primary aim of this breakdown is to make a clear distinction between non-lifting and lifting bodies and to identify how their drag varies when the altitude and speed change. Many academic texts on aircraft performance apply to flight Mach numbers where the flow surrounding the aircraft does not contain regions of supersonic flow. In this flight regime, the only contribution to induced drag is vortex drag. In other words: induced drag and vortex drag are identical and – apart from Reynolds number effects – lift and drag coefficients depend only on the angle of attack.

Equations (4.2) and (4.4) show that, in steady level subsonic flight, only the variation of drag with the dynamic pressure q is relevant. Parasite and induced drag are closely related – but not exactly equal – to zero-lift drag and drag due to lift, respectively (Section 4.4). Basically, airplane parasite drag increases proportional to q, whereas induced drag varies inversely proportional to it.[4] In the condition of minimum total drag, parasite and induced drag are (approximately) equal. The variation of drag with flight conditions changes markedly at transonic speeds where parasite drag and induced drag are both affected by wave drag. The dominant parameter in this speed domain is the flight Mach number rather than the dynamic pressure (Section 4.4).

4.3.1 Parasite Drag

Acting on lifting as well as non-lifting airplane components, parasite drag D_p is decomposed into skin friction drag and a pressure drag term known as form drag. Parasite drag was traditionally known as profile drag, a term still in use for airfoils. At small incidences to low-speed flow, a relatively thin airfoil or a slender body of revolution has predominantly attached flow and form drag is an order of magnitude smaller than friction drag. Flow separation at high incidences increases the form drag of an airfoil considerably, whereas a swept-up aft fuselage can be a source of flow separation and vortical flow, causing significant form drag. Improvement of airfoil characteristics has been the subject of continuous R & D up to the present time. A classical overview of theory, test results and design for a large number of wing sections designed is published in NASA TR 824 – most of this work is available in [1]. Together with an abundance of information in [3], this source has formed a basis for low-speed airplane wing design up to the 1960s. Aerodynamic research during the 1970s and 1980s resulted in design methods based on CFD technology which enable the AD engineer to

---

[4]This property differs fundamentally from the air drag of non-lifting objects, such as ground vehicles, since they experience only parasite drag.

design airfoils specifically adapted for the application. Airframe manufacturers nowadays use high-fidelity aerodynamic design tools developed in-house.

Although laminar boundary layer friction is much lower, in practical cases the boundary layer of the airplane components is almost entirely turbulent. The flat plate analogy is widely used in conceptual design to compute the parasite drag of aircraft components from the friction drag of a smooth flat plate at zero incidence in turbulent flow. Classical equations used are the Von Kármán-Schoenherr formula

$$0.242 = \sqrt{C_{f_i}} \log_{10} (C_{f_i} Re_1) \qquad (4.5)$$

or the Prandtl-Schlichting expression

$$C_{f_i} = \frac{0.455}{(\log_{10} Re_1)^{2.58}} \qquad (4.6)$$

with Reynolds numbers referred to the plate length. Since both equations apply to incompressible flow (index i) they must be corrected for reduced skin friction due to (subsonic) compressibility. Experimental data on the turbulent flat plate friction coefficient for Reynolds numbers between 10^6 and 10^8 adjusted to Mach 0.5 can be approximated as follows:

$$C_f = \frac{0.044}{Re_1^{1/6}} \qquad (4.7)$$

The flat plate analogy derives the parasite drag area from

$$F_p \stackrel{\text{def}}{=} C_{D_p} S = \Phi_f C_f S_{\text{wet}} \qquad (4.8)$$

where C_f is the skin friction coefficient of a flat plate with the same wetted area S_{wet} and Reynolds number as the component.[5] Accounting for flow supervelocities and pressure drag, the form factor Φ_f is the ratio between the body's parasite drag and the skin friction drag of the equivalent plate. The parasite drag area of a body can be interpreted as the area of a hypothetical plate normal to the free stream, having a drag coefficient of 1.0 and the same parasite drag as the body. This parameter can be seen as a quantitative dimensional value expressing the parasite drag of an aircraft configuration. For instance, approximate values are $F_p = 1.7$ m^2 for the Fokker 100 and $F_p = 7.1$ m^2 for the B 747-100.

Design handbooks might suggest that the prediction method for the friction drag coefficient and the form factor is highly accurate. In reality, the assumption of a fully turbulent boundary layer is not necessarily correct whereas the seemingly unambiguous methods for computing form factors are actually based on a specific class of wing sections. One reason is that the only form factors present in most handbook methods is the thickness ratio and sweep angle for wings, and the fineness ratio for fuselages. As an alternative, Figure 4.2 compares non-lifting wings and fuselages by relating Φ_f to the ratio of frontal to wetted area. Due to the higher supervelocities along the wing surface, Φ_f is 15 to 25% higher compared to a fuselage. And since wings mostly have a (much) lower average Reynolds number, their C_f is at least

[5]The wetted (or exposed) area of a body is the external surface which is in contact with the airflow. See Appendix A.

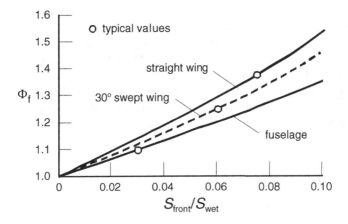

Figure 4.2 The form factor derived from experimental data

25% higher. Consequently, the wing's parasite drag area is significantly higher than that of a fuselage with the same wetted area. The discussion in Chapter 5 makes it clear that this observation is important when different airplane concepts are compared.

Aerodynamic design of a wing may require a more accurate drag prediction method than Equation (4.8). The parasite drag of a lifting surface is then computed by integrating the section profile drag for a number of airfoil sections along the span. Figure 4.3 shows how the section profile drag coefficient is affected by lift and camber. The drag variation with lift is related to flow supervelocities and boundary layer thickening at the trailing edge, rather than being induced by lift. Profile drag is minimal for the design lift coefficient c_{l_d} which is proportional to camber. For positive camber the minimum drag occurs for positive lift. The drag coefficient can be approximated for normal operating lift coefficients as follows:

$$c_{d_p} = (c_{d_p})_d\{1 + k_1(c_l - c_{l_d})^2\} \tag{4.9}$$

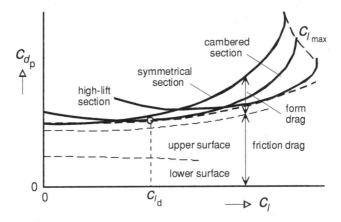

Figure 4.3 Section profile drag coefficient affected by camber and lift

The factor k_1 depends on the section shape, in particular its thickness ratio. Equation (4.9) underestimates the drag at high incidences and loses its validity at the stall, when the lift actually decreases. Using the flat plate analogy should be obviated for low-drag wing sections designed for natural laminar flow (NLF), for which [23] and [32] are recommended. Another application of the flat plate analogy is wing area optimization. Variation of the design lift coefficient is then combined with wing section variation by selecting an optimum camber for each wing area. For this purpose the (dotted) enveloping curve in Figure 4.3 is used by inserting a modified (reduced) value $\bar{k}_1$ into the wing profile drag coefficient,

$$C_{D_p} = (C_{D_p})_{L=0}(1 + \bar{k}_1 C_L^2) \tag{4.10}$$

4.3.2 Monoplane Induced Drag

The present section deals exclusively with flight at subcritical speed where induced drag is identical to vortex drag (Figure 4.1). Induced drag is generated predominantly by the wing, horizontal tail- and fore-plane surfaces. For an untwisted monoplane with elliptical chord and lift distribution, the classical result from the Lanchester/Prandtl lifting line theory is

$$D_i = \frac{L^2}{\pi q b_w^2} \quad \text{or} \quad C_{D_i} = \frac{C_L^2}{\pi A_w} \tag{4.11}$$

where b is the span measured in the lateral plane and $A_w = b_w^2/S_w$ denotes the aspect ratio. Equation (4.11) represents the theoretical minimum induced drag of a planar monoplane – it is considered as a reference for aerodynamic wing design. Since the lift distribution of the aircraft as a whole deviates from the elliptical, the induced drag coefficient is usually written as follows:

$$D_i = \frac{L^2}{\pi q b_w^2 e_v} \quad \text{or} \quad C_{D_i} = \frac{C_L^2}{\pi A_w e_v} \tag{4.12}$$

The span efficiency factor e_v depends on the distribution of lift along the span which is not necessarily independent of C_L. The theoretical span efficiency factor of a planar monoplane is less than or equal to 1.0. However, non-planar systems may achieve $e_v > 1.0$ and their induced drag can be significantly less than the reference given by Equation (4.11).

For a given lift distribution, the span efficiency factor of an symmetrical untwisted wing can be approximated in terms of the dimensionless lateral centre of lift located at η_c following [72]:

$$e_v = \{4.5(\pi \eta_c)^2 - 12\pi \eta_c + 9\}^{-1} \tag{4.13}$$

An elliptically loaded monoplane has the centre of lift at $\pi \eta_c = 4/3$ corresponding to $e_v = 1.0$. Dependent on the wing planform and the airplane's general arrangement, the span efficiency factor of a complete aircraft is (significantly) less than one. Figure 4.4 shows examples of an isolated wing as well as a wing in combination with a fuselage and/or engine nacelles. A well-designed trapezoidal wing achieves induced drag which is a few percentages above the theoretical minimum. The lift carry-over across the fuselage body is considered as reduced wing lift which increases induced drag in cruising flight by approximately 5%. Flow interaction

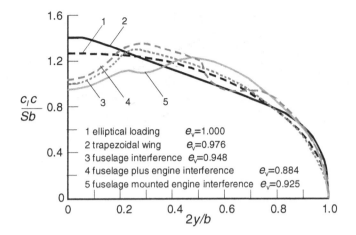

Figure 4.4   Lateral lift distribution and the span efficiency factor [38]

between wing and nacelles may lead to a significant modification of the lift distribution, with a drag penalty as indicated.

A twisted monoplane at zero lift carries positive lift over some parts of the span and negative lift over others – trailing vortices are then formed leading to induced drag. A wing with negative aerodynamic twist ('wash-out') at zero lift carries upward lift on the inboard wing and an equal downward lift on the outboard wing which reduces the induced drag for a selected range of lift coefficients. Compared to Equation (4.12), the following expression forms a more accurate representation:

$$C_{D_i} = \frac{(C_L - C_{L_x})^2}{\pi A_w e_x} + \left(C_{D_i}\right)_{\min} \tag{4.14}$$

where C_{L_x} and e_x are constants. Even though viscous drag is not included, Equation (4.14) describes an offset parabola which can be expanded into a three-term equation,

$$C_{D_i} = C_0 + C_1 C_L + C_2 C_L^2 \tag{4.15}$$

where C_0 and C_2 are positive and C_1 is (normally) negative. The ratio C_{D_i}/C_L is minimal for $C_L = \sqrt{C_0/C_2}$. Viscosity influences the lift distribution along the span and major interactions between viscous and vortex wake flows make an unambiguous decomposition of the associated drag components problematic. Incidentally, induced drag prediction for an airplane approaching the stall is not feasible – and unnecessary – in conceptual design.

4.3.3   Biplane Induced Drag

The induced drag analysis of a biplane system is complicated by the mutual interaction of the lifting surfaces. The aerodynamics of biplanes in combination with the general theorems of M. Munk [53] have been established in the form of Prandtl's classical biplane theory [54]. The

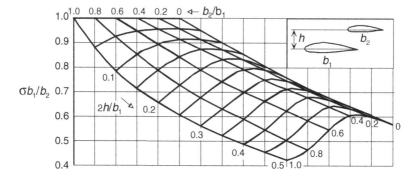

Figure 4.5   Prandtl's interference factor for biplanes

lift components of a biplane configuration are denoted L_1 acting on the forward wing with span b_1 and L_2 acting on the aft wing with span $b_2 < b_1$. Assuming that each wing has optimal (elliptical) loading, Prandtl obtained the total induced drag from the self-induced vortex drag of both wings in isolation according to Equation (4.11) with a drag increment caused by the mutual interaction between the two wings, resulting in

$$D_i = \frac{1}{\pi q} \left\{ \frac{L_1^2}{b_1^2} + 2\sigma \frac{L_1}{b_1} \frac{L_2}{b_2} + \frac{L_2^2}{b_2^2} \right\} \tag{4.16}$$

The interference factor σ, depicted in Figure 4.5, depends on the span ratio b_2/b_1 and on the gap h; that is, the vertical displacement between the wings. The equation approximating Prandtl's interference factor proposed by Laitone [59],

$$\sigma b_1/b_2 = 1 - \{1 + (b_1/2h)^2\}^{-1/2} \tag{4.17}$$

has an inaccuracy of less than 2% for $b_2/b_1 < 0.3$.

When Equation (4.16) is applied to a non-staggered biplane consisting of two wings with equal span b, it is readily found that minimum induced drag amounts to

$$(D_i)_{min} = \frac{L}{\pi q\, b^2} \frac{1+\sigma}{2} \quad \text{for} \quad L_1 = L_2 = \frac{1}{2} L \tag{4.18}$$

For instance, if the two wings have a vertical displacement of 20% span, Prandtl's interference factor amounts to $\sigma = 0.485$ which yields a span efficiency factor $e_v = 1.35$. Consequently, the biplane's induced drag is 26% less compared to a monoplane with the same lift and span. Munk's stagger theorem states that the total induced drag of a biplane system is unaltered if any of the lifting surfaces are moved in the direction of motion, provided the lateral lift distribution on each surface is constant. This principle is applied to a tandem wing configuration which features two highly staggered wings of equal span at the airplane's front and rear end. The condition of elliptic loading on both lifting surfaces is only satisfied for $b_1 > b_2$ which makes the biplane theory applicable to a wing-and-tail configuration with arbitrary stagger.

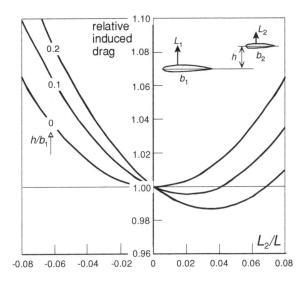

Figure 4.6   Induced drag of a wing-and-tail combination with $b_2/b_1 = 0.3$

The nose-plane of a canard configuration has a smaller span than the wing and its trailing vortex field causes the wing lift distribution to be heavily distorted relative to the elliptic loading. For this configuration Munk's stagger theorem is invalid, especially for a small vertical displacement between the two planes. Butler [64] and Kroo [63] demonstrated that the interference between the nose-plane and the wing reduces the induced drag of this configuration relative to classical biplane theory. They assumed an elliptical lift distribution on the nose-plane and optimized the main wing lift distribution immersed in the nose-plane's vortex field. The minimum total induced drag is summarized in a modified biplane equation,

$$D_i = \frac{1}{\pi q} \left\{ \sigma_c \frac{L_1^2}{b_1^2} + 2\sigma \frac{L_1}{b_1} \frac{L_2}{b_2} + \frac{L_2^2}{b_2^2} \right\} \tag{4.19}$$

with values of σ_c obtained from [63]. Equation (4.19) is applicable to a wing-and-tail combination ($b_1 > b_2$) with $\sigma_c = 1$ as well as to a canard airplane ($b_1 < b_2$) with $0 < \sigma_c < 1$.

As an example, Figure 4.6 depicts the induced drag of a wing-and-tail combination compared to the case when all the lift is carried by the wing. For a tail located in the wing plane ($h = 0$), there is a drag penalty if the tail load is upward or downward.[6] A high-set stabilizer may generate a modest induced drag reduction if it has an upward lift of about 3% of the total lift; however, even a small tail download causes a considerable drag penalty. Due to the high nose-plane lift inherent to a typical canard configuration its induced drag penalty is relatively high, especially if the classical biplane theory ($\sigma_c = 1$) were used. Figure 4.7 illustrates that applying a realistic $\sigma_c = 0.65$ reduces the drag penalty by more than 50%. It can safely be concluded that the lowest vortex drag of a canard-and-wing combination is obtained when the wing, with its large span, carries the highest possible fraction of the total lift.

---

[6]For a given flight condition, the tail load is determined by the airplane's centre of gravity (CG) location. The tailplane of a transport aircraft mostly carries negative lift.

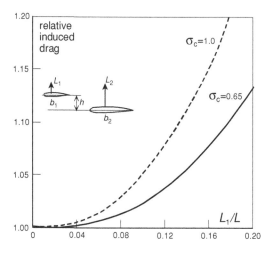

Figure 4.7 Induced drag of a canard airplane with $b_1/b_2 = 0.4$

4.3.4 Multiplane and Boxplane Induced Drag

The induced drag of an optimally loaded triplane is smaller than the minimum induced drag of a biplane with equal span and height. Induced drag approaches a lower limit when the number of multiplanes tends to infinity. However, L. Prandtl has proved that the minimum induced drag of a system of identical parallel lifting surfaces is equal to that of a closed lifting system consisting of a biplane interconnected by tip planes. This principle is materialized in the boxplane concept consisting of a biplane with wings of equal span connected at their tips by vertical planes. The front view of a boxplane in Figure 4.8 (a) shows that the horizontal surfaces carry the same lift, with an optimal lift distribution consisting of a constant part and an elliptical one. The circulation on the vertical planes is zero at their midpoint with equal

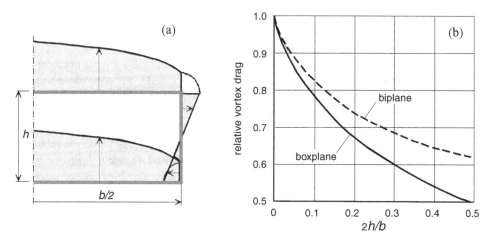

Figure 4.8 Obtainable vortex drag reduction of biplanes and boxplanes. (a) Optimum lift distribution of a boxplane. (b) Minimum vortex drag compared to a monoplane

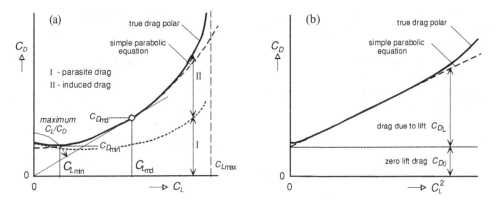

Figure 4.9   Drag polar for subsonic flight. (a) Breakdown into parasite and induced drag and condition for $(C_L/C_D)_{max}$. (b) Approximation by a two-term parabola

side force components acting outwards on the upper part and inwards on the lower part. The vertical planes are effective in reducing the trailing vortices which would appear if the tips were not jointed. Figure 4.8 (b) depicts the significant reduction in induced drag obtainable with the boxplane arrangement. For instance, for a height/span ratio of 0.3, the minimum induced drag is 60% of that for an equivalent monoplane – it reduces to 50% for a height equal to the semi-span. I.M. Kroo has shown in [72] that the boxplane does not have a particular aerodynamic advantage because it is a closed system. Its optimal lift distribution is not unique because one may superimpose a vortex loop with constant circulation on any closed geometry by varying the incidence between the lifting wings. This changes the load distribution but the lift and induced drag are unchanged.

4.4   Drag Polar Representations

Many aircraft are designed to fly at speeds where the flow is subcritical everywhere. If the airplane configuration and the Reynolds number are specified, lift and drag coefficients are unique functions of the angle of attack: $C_L = C_L(\alpha)$ and $C_D = C_D(\alpha)$. In performance analysis α is of little significance and can be eliminated. This yields a unique relationship between C_L and C_D universally known as the drag polar – a typical example is depicted in Figure 4.9 (a). For the ease of performance analysis it is long-established practice to approximate the true drag polar by a polynomial equation. This section demonstrates how a drag polar obtained from computational prediction, wind tunnel or flight tests can be represented by an analytical equation.

4.4.1   Two-term Approximation

The classical representation of the true drag polar consists of two terms,

$$C_D = C_{D_0} + C_{D_L} = C_{D_0} + K_L\, C_L^2 = C_{D_0} + \frac{C_L^2}{\pi\, A_w\, e} \qquad (4.20)$$

defining a parabola which is symmetrical about the $C_L = 0$ axis. This two-term approximation represents an artificial breakdown of drag into zero-lift drag and drag due to lift, with coefficients C_{D_0} and C_{D_L}, respectively. Such a subdivision does not have a fundamental justification since the quadratic term describes the variation of drag with lift in a simplified manner. This uses the airplane efficiency factor e, also called the Oswald factor after W.B. Oswald who first used it [16].

The usual approach to obtain C_{D_0} and K_L from a true polar is to make a plot of C_D versus C_L^2 and approximate it by a linear function as in Figure 4.9 (b). The true polar has a non-symmetrical relation with C_L and – since it is obtained by extrapolation of the straight line in this figure – C_{D_0} is (usually) lower than the true C_D at zero lift. In spite of this objection, the two-term approximation is mostly accurate in the range of normal operating conditions and has proven to be of considerable practical value. The Oswald factor accounts for the vortex-induced drag and for the lift-dependent part of the parasite drag. Some insight into its value is obtained by applying Equation (4.10) to the complete airplane,

$$C_{D_p} = C_{D_0}(1 + k_L C_L^2) \tag{4.21}$$

In combination with Equation (4.12) this results in

$$e = \frac{e_v}{1 + k_L C_{D_0} \pi A_w e_v} \tag{4.22}$$

The span efficiency factor can be assumed to be equal to $e_v = 0.95$. According to [9] k_L varies between 0.38 for straight wings, 0.40 for 20° and 0.45 for 35° sweptback wings. An alternative approximation according to [13] is: $k_L C_{D_0} = 0.007$. It is noticeable that Equation (4.22) indicates the Oswald factor to be lower for a high aspect ratio than for a low aspect ratio wing. Trimmed airliners flying in the en-route configuration have an Oswald factor typically between 0.75 for turboprops and 0.85 for jetliners.

The two terms of Equation (4.20) may alternatively be obtained from the graphical construction shown in Figure 4.9 (a). The condition for minimum drag (index md) is obtained by drawing a tangent line from the origin touching the polar curve and reading $C_{L_{MD}}$ and $(C_L/C_D)_{max}$. The coefficients C_{D_0} and C_{D_L} are then obtained from

$$C_{D_0} = \frac{C_{L_{MD}}}{2(C_L/C_D)_{max}} \quad \text{and} \quad \pi A_w e = 2C_{L_{MD}} \left(\frac{C_L}{C_D}\right)_{max} \tag{4.23}$$

This procedure uses just one point of the true drag polar namely, the minimum drag condition. Equation (4.23) is then very accurate and near this condition, a desirable characteristic for cruise performance analysis.

4.4.2 Three-term Approximation

Figure 4.9 (a) shows that the true polar has its minimum drag coefficient $C_{D_{min}}$ at the (positive) lift coefficient $C_{L_{min}}$. The straight line approximation according to Equation (4.20) ignores this asymmetry relative to the $C_L = 0$ axis. Instead, a three-term curve fit increases the C_L range with a close approximation to the true drag polar. For instance, Equation (4.20) can be replaced by

$$C_D = C_{D_{min}} + K_L(C_L - C_{L_{min}})^2 \tag{4.24}$$

or

$$C_D = C_{D_0} - 2K_L C_{L_{min}} C_L + K_L C_L^2 \quad \text{with} \quad C_{D_0} = C_{D_{min}} + K_L C_{L_{min}}^2 \tag{4.25}$$

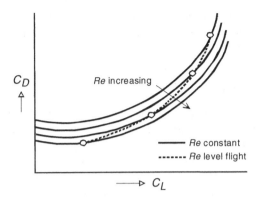

Figure 4.10   Drag polar curves at constant Reynolds number and interpolation for level flight

This equation is especially suitable to represent drag at transonic Mach numbers, for highly cambered wings and for airplanes with deflected high-lift devices. It should, however, be noted that the parameters C_{D_0} and K_L in Equation (4.25) are numerically different from those in Equation (4.20) since the concept of a constant Oswald factor in the two-term approximation does not apply to the three-term approximation. An even closer fit might be obtained by replacing a continuous drag polar approximation by two straight-line segments, each representing the angle of attack regions with different flow characteristics such as attached and separated flows [40].

4.4.3   Reynolds Number Effects

There exists a large gap between design handbook methods used to rough out a new configuration and the initial model testing results used to generate the first true performance prediction. In particular, variations in flight speed and altitude lead to variations in the Reynolds number affecting the lift and drag coefficients. For early performance estimates these variations are usually sufficiently small enough to be ignored. During preliminary design this may no longer be the case, in particular when drag polar curves for different Reynolds numbers measured in the wind tunnel have to be extrapolated to real flight conditions. The approximation according to Equation (4.20) has often been poorly understood as explained in the following comments on Figure 4.10 made by J.H. McMasters [47]:

> Unlike the theoretically derivable wing span efficiency factor, the Oswald factor has only quasi-physical meaning since it is one of two parameters in a simple two-term polynomial curve fit to experimental data. In steady level flight, the Reynolds number (of a wing) varies proportional to $1/\sqrt{C_L}$. This is not the case in a wind tunnel test, where drag data are usually measured at a constant Reynolds number. If the drag data show significant variation with Reynolds number, care must be taken in constructing the correct 'equivalent flight polar' from wind tunnel tests.

An airplane may be optimized for a narrow range of lift coefficients, for example, by selecting cambered wing sections and a wing setting angle so that the wing operates in a drag bucket. The simple two-term polynomial curve-fit is symmetrical with respect to the zero-lift condition and may deviate significantly from the true drag polar. For a point-optimized airplane, this may lead to a surprising value of the Oswald factor very close to or even exceeding $e = 1.0$ whereas an aircraft with a very high aspect ratio wing may be found to have a lower than expected Oswald factor.

4.4.4 Compressibility Correction

Compressibility effects are of minor importance when an aircraft flies at a low-subsonic speed. Compressibility drag must be taken into account for speeds in excess of Mach 0.5. Figure 4.1 decomposes wave drag into drag due to (wing) lift and drag due to (body) volume. The figure does not show the complication that interaction between a strong shock wave and the boundary layer causes boundary layer separation and a wake. Since compressibility drag is due to a combination of physical mechanisms, its computation is a complex problem. The drag coefficient at high-subsonic and transonic Mach speeds are not uniquely determined by the angle of attack but also by the Mach number,

$$C_L = C_L(\alpha, M) \quad \text{and} \quad C_D = C_D(\alpha, M) \quad \rightarrow \quad C_D = C_D(C_L, M) \qquad (4.26)$$

Figure 4.11 (a) depicts a set of drag polars for several Mach numbers whereas Figure 4.11 (b) is an alternative rendering of the same data. The figure shows that at high-subsonic speeds there is a noticeable Mach-effect on drag. Dependent on C_L, a subdivision can be made in three regions.

1. The increased stagnation temperature and Reynolds number cause the skin friction drag in compressible flow to decrease with increasing Mach number by up to 5% at Mach 1.0. This minor effect is visible for low values of C_L in Figure 4.11 (b).
2. When the (C_L-dependent) critical Mach number is exceeded, the drag rises slowly due to the increasing strength of the forward shocks and the gradual thickening of the boundary layer. Known as drag creep, this process starts at Mach 0.6 for $C_L = 0.6$.

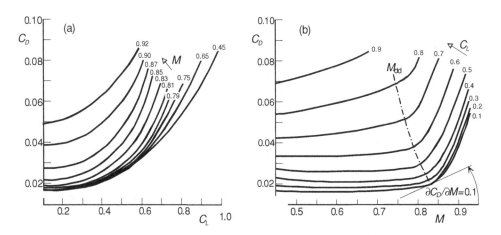

Figure 4.11 Drag coefficients of a transonic airliner. (a) C_D versus C_L for different Mach numbers. (b) C_D versus M for different lift coefficients

3. The drag rises sharply due to strong shocks causing flow separation and a wake when the speed exceeds the drag-divergence Mach number M_{dd}, an important design parameter for high-speed aircraft. The wing is usually responsible for the major part of the drag rise. Several criteria are in use to indicate when drag divergence begins. The preferable definition is based on the slope of the drag curve: $\partial C_D / \partial M = 0.10$. Figure 4.11 (b) shows that M_{dd} decreases with increasing C_L.

For a given transonic Mach number, the variation of lift is accompanied by a variation in the shock-wave strength and wave drag is interpreted as a component of induced drag. Compressibility drag is a function of C_L and M and is mostly accounted for as a correction ΔC_{D_c} of the low-speed drag. The compressibility drag penalty in fuel-optimal cruising flight is typically between five and ten counts; in high speed cruising it may increase to twenty counts. The aerodynamic design of a transonic airplane wing is dominated by the selection of M_{dd} since it is intimately related to the design point for long-range cruising. As shown in Section 10.6, this leads to different combinations of lift coefficient, sweep angle and average thickness ratio enabling optimum cruise performance. Transonic effects form a considerable complication of the design process and analytical computation of the compressibility correction is beyond the scope of the conceptual design stage.

4.5 Drag Prediction

Reducing the drag by aerodynamic design requires insight into its sensitivity to variations of the airplane geometry. The prediction of drag can be approached in several ways: quasi-analytically, numerically (via CFD) and by wind tunnel testing. Since not all of these strategies adequately recognize the design sensitivity, the selection of a method depends on the development phase and is influenced by the information available. Examples of drag prediction methods can be found in [7, 8, 9, 19, 21, 31, 39]. Application of most methods is time-consuming and rather than recommending a particular methodology, we discuss some approaches and simplifications useful for quasi-analytical design optimization in the conceptual design phase.

At the time of writing, no single method is capable of simultaneously predicting all airplane drag components and no comprehensive method is capable of treating a full aircraft configuration. Drag prediction in the conceptual design stage is usually based on a number of methods with results combined using the classical component build-up technique. This term refers to the approach of dividing the airframe into basic components and computing the zero-lift drag of each component exposed to the free flow in isolation. It is assumed that these components have streamline shapes with aerodynamically smooth surfaces. The summation of their contributions to drag is corrected to account for interacting flow fields around components placed in each other's vicinity and for the fact that practical airframe surfaces are not perfectly smooth. Notwithstanding the availability of modern computational drag analysis methods, these do not necessarily produce a better prediction than the component build-up technique. Moreover, high-fidelity CFD methods for drag prediction are very computer-intensive and require detailed geometric input which is not always available.

4.5.1   Interference Drag

Corrections for the assembly of isolated components into a full configuration allow for wetted area and other geometric modifications of adjacent components such as fairings to be added to reduce interference drag (Appendix A). The physical origins of interference drag are not different from the flow phenomena mentioned in Figure 4.1. Significant interference drag mostly arises from the intersection of a lifting surface and a body or the intersection of two lifting surfaces or bodies. Although interference may take myriad forms, in all cases a flow disturbance is responsible for the drag.

- Interacting body flows. A common form of interference drag can be found with interfering multiple body or wing-and-body components. A flow disturbance at the wing-to-fuselage junction may lead to separation and vortex shedding where the weakened viscous layer increases the probability of flow separation. The problem is aggravated when this form of interference occurs at high speeds where the wing shock pattern is changed. These effects are suppressed by selecting an appropriate wing setting angle relative to the fuselage and by modifying the junction by means of fairings or fillets. For a low-wing high-speed airplane this is effectively a local modification of the external fuselage cross-section shape. However, the use of these provisions leads to a penalty caused by additional wetted area. Flow interaction between the wing and the fuselage body also affects the lift distribution along the span, leading to a disturbance of the far-field wing wake, with an effect on induced drag. Similarly, interference drag is often caused by wing-mounted engine nacelles which is taken into account by modifying the span efficiency factor e_v in the design condition (Figure 4.4).
- Trim drag. The basic drag polar is determined for a neutral elevator position. Dependent on the centre of gravity (CG) location, a tail lift is necessary for trimming the airplane. The associated drag correction is primarily affected by the stability margin and by downwash at the tailplane. Known as trim drag, this correction can be estimated with the help of biplane theory (Section 4.3). A complication is that, for a given total lift, a download on the tail causes the wing lift to increase and in the high Mach number regime this leads to a significant drag penalty as demonstrated by Figure 4.12. Trim drag in cruising flight should amount to not more than a few drag counts.
- Propeller wake effects. Aircraft with tractor propellers have their wing immersed in their slipstreams. This increases dynamic pressure and changes the lift distribution. The primary effects are increased parasite drag and induced drag due to downwash behind the propellers. On the other hand, the swirling wake flow generates an upwash on one side of the propeller axis and a downwash on the other side and the combined effect on the wing load distribution may, in theory, reduce the induced drag. In practice, the Oswald factor of propeller transports may be reduced by 4%, typically [9].
- Throttle-dependent drag. The nacelle of a turbofan engine is generally designed so that spillage drag due to reduced intake mass flow is negligible in the cruise condition. Corrections may be necessary to account for interference between the flow on which the engines are operating and the flow around the airframe.

For conventional take-off and landing airplanes, propulsion system interference effects are mostly small. In order to improve low-speed performances, short take-off and landing (STOL)

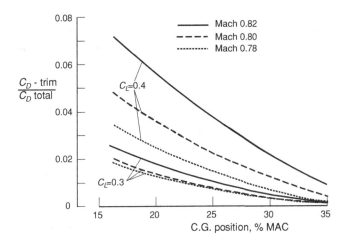

Figure 4.12   Percentage change in total drag due to trim for a DC-8-54. Adapted from [60]

airplanes have a large powerplant/airframe interference effect on maximum lift at the cost of a significant drag penalty. An in-depth analysis can be found in [4].

4.5.2   Roughness and Excrescences

Practical aircraft surfaces are not perfectly smooth. Roughness and excrescence drag may increase the zero-lift drag by 15% for large jetliners and up to 30% or more for small propeller transports.[7] Information in [48] and [50] can be used to make an estimation of the following penalties:

- Surface imperfections such as (protruding) rivets, skin joints, control surface gaps, removable panels, steps and gaps around windows and doors – alternatively classified as distributed surface roughness. For a certain type of construction, surface imperfection drag is often taken into account by an equivalent sand grain roughness defined as the size of sand grains, distributed uniformly on the surface, which would produce the same skin friction as the original surface with an arbitrary roughness. According to [9], jet transports have an average equivalent sand grain roughness of 0.0016 inch (0.04 mm) – this is too much for surfaces designed for laminar flow.
- Excrescences and protuberances cause a drag penalty that is predictable for each item individually. All transport aircraft are required to have the provisions associated with this category, namely, air data systems, drains, antennas, vents, lights and beacons, auxiliary power unit, and features required for rain dispersal, for ventilation/cooling, for air conditioning, and for fuel systems.

---

[7]Some airframe manufacturers characterize this non-ideal drag penalty as the only form of parasite drag. This can be confusing since parasite drag is widely interpreted as the drag of a nonlifting closed body (Figure 4.1).

Additional drag sources may have to be considered such as fuselage upsweep drag or unaccounted drag based on previous experience. Since these drag components are sensitive to detailed design, their estimation during conceptual design is based on statistical data [13].

4.5.3 Corrections Dependent on Operation

Drag is affected by the airplane's configuration, flight conditions, engine rating and the distribution of useful load. In the en-route configuration, drag is subdivided into uncorrected drag and corrections for the trim condition, compressibility and aero-elastic effects. The uncorrected drag is represented by a single drag polar for subsonic flight and a representative (variation of) Reynolds number. Engine power effects are often counted as a reduction of the overall efficiency. A compressibility correction accounts for Mach numbers in excess of 0.6 to 0.7. Aero-elastic effects are significant for large airliners flying at a high dynamic pressure. Deflection of high-lift devices, landing gear extension and engine failure also lead to considerable drag increments.

4.5.4 Estimation of Maximum Subsonic L/D

When different aircraft concepts are compared, the maximum aerodynamic efficiency $(L/D)_{\mathrm{max}}$ for the en-route configuration is a useful measure of the aerodynamic design quality. However, it is emphasized that selection of the aircraft shape is the outcome of many compromises with non-aerodynamic aspects such as structural weight, operational constraints and cost considerations. Moreover, turbofan-powered aircraft cruise optimally at a Mach number higher than M_{MD} where compressibility effects increase the drag by up to 5% (Section 2.4). Consequently, when aircraft designs are compared, it cannot be concluded that the one achieving the highest L/D is the best. In spite of these reservations, achieving a high aerodynamic efficiency is an important goal for designing transport as well as business airplanes. In the early design phase it is therefore safe to assume that the aerodynamic design will be optimized during the downstream development process. For example, parasite drag is minimized by selecting suitably cambered wing sections, application of smooth skin surfaces and by avoiding unnecessary excrescence drag. Major drag components will then approach the lower limit of statistical values demonstrated by existing aircraft. The purpose of this section is to present an approach to a quasi-analytical prediction of the maximum aerodynamic efficiency based on the classical two-term approximation of the drag polar.

For straight and level subsonic flight Equation (4.20) can be written as

$$\frac{C_L}{C_D} = \left(\frac{C_{D_0}}{C_L} + \frac{C_L}{\pi \, A_{\mathrm{w}} \, e} \right)^{-1} \tag{4.27}$$

which has its maximum value

$$\left(\frac{C_L}{C_D} \right)_{\mathrm{max}} = \frac{1}{2} \sqrt{\frac{\pi \, A_{\mathrm{w}} \, e}{C_{D_0}}} \quad \text{for} \quad C_{L_{\mathrm{MD}}} = \sqrt{C_{D_0} \pi \, A_{\mathrm{w}} \, e} \tag{4.28}$$

Since skin friction is the main contribution to lift-independent efficiency, a simple method for estimating C_{D_0} is the equivalent skin friction method. This considers C_{D_0} as the skin friction drag coefficient $C_{f_{eq}}$ based on the total airplane's wetted area $C_{D_0} = C_{f_{eq}} S_{wet}/S_w$. In combination with Equation (4.28) this yields

$$\left(\frac{C_L}{C_D}\right)_{max} = \frac{b_w}{2}\sqrt{\frac{\pi e}{C_{D_0} S_w}} = \frac{b_w}{2}\sqrt{\frac{\pi e}{C_{f_{eq}} S_{wet}}} \tag{4.29}$$

As explained in the previous sections, the drag coefficient of an aircraft component depends on the Reynolds number, the pressure distribution, the state of the boundary layer, the surface smoothness and many other details. The proposed quasi-analytical estimation method is based on the flat plate analogy. Equation (4.7) is used for the friction drag coefficient of a plate with a fully turbulent boundary layer with the mean full-configuration Reynolds number based on an average wetted plate length,

$$\overline{Re} \stackrel{def}{=} \frac{V}{\nu} \frac{S_{wet}}{b_w} \tag{4.30}$$

The uncorrected parasite drag is obtained from Equation (4.8) by multiplying the flat plate drag by the form factor accounting for supervelocities. It is approximated as

$$\Phi_f = 1 + r_\Phi \frac{S_{front}}{S_{wet}} \tag{4.31}$$

Figure 4.2 shows that typical values of r_Φ are 4.8 for a straight wing, 4.1 for a swept wing and 3.5 for a fuselage. The resulting ideal friction drag area (index id) of an aerodynamically smooth airframe surface is

$$C_{f_{id}} S = 0.044 \overline{Re}^{-1/6}(S_{wet} + r_\Phi S_{front}) \tag{4.32}$$

where S_{front} denotes the frontal area of the airplane excluding the engine air intakes. If detailed information is lacking, one may use $r_\Phi = 4.0$ for propeller aircraft and $r_\Phi = 3.5$ for jet aircraft. Although the airplane's frontal area is much smaller than its surface area, the factor r_Φ magnifies its effect on drag. This makes the values of the wing thickness and the fuselage fineness ratio very influential, an effect that is not visible in prediction methods based on only wetted areas. The drag increment due to surface roughness, excrescences and other non-ideal (index nid) sources of drag is made by multiplying the parasite drag of the aerodynamically smooth airplane by a correction factor $k_{nid} = C_{f_{eq}}/C_{f_{id}}$ resulting in the lift-independent drag area

$$C_{D_0} S = C_{f_{eq}} S_{wet} = 0.044 \, k_{nid} \, \overline{Re}^{-1/6}(S_{wet} + r_\Phi S_{front}) \tag{4.33}$$

In combination with Equation (4.28) this yields

$$\left(\frac{C_L}{C_D}\right)_{max} = 4.2 \left(\frac{e\,\overline{Re}^{1/6}}{k_{nid}}\right)^{0.5} \frac{b_w}{(S_{wet} + r_\Phi S_{front})^{0.5}} \tag{4.34}$$

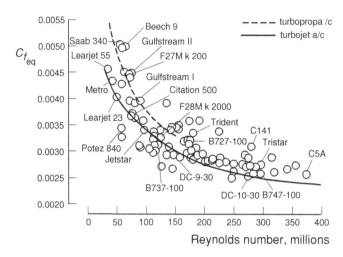

Figure 4.13 Statistical data for the equivalent friction drag coefficient with Reynolds number based on S_{wet}/b. Adapted from [44]

with Equation (4.22) to be inserted for the Oswald factor. Statistical data in Figure 4.13 have been used to derive $k_{\text{nid}} = 1 + 5\ 100\ \overline{Re}^{-0.50}$ for propeller aircraft and $k_{\text{nid}} = 1 + 255\ \overline{Re}^{-0.35}$ for commercial jet aircraft. Since k_{nid} varies typically between 1.25 and 1.50 there seems to be considerable potential for reduction of interference, roughness and excrescence drag and other non-ideal drag sources. It is worth noting that the significant effect of airplane size is visible in $\overline{Re}^{1/6}$ and even more pronounced in k_{nid}.

The previous derivation suggests that the subsonic performance potential of a civil airplane can be expressed in terms of the (aerodynamic) design efficiency

$$\eta_{\text{des}} \overset{\text{def}}{=} \sqrt{e/k_{\text{nid}}} \tag{4.35}$$

This figure of merit represents the ratio of $(C_L/C_D)_{\text{max}}$ to its ideal value with $e = 1.0$ and the skin friction drag coefficient $C_{f_{\text{id}}}$ of a perfectly smooth and streamlined aircraft. The highest achievable value at the time of writing is $\eta_{\text{des}} \approx 0.80$. The Oswald factor can be slightly increased by reducing the horizontal tail download and/or by installing winglets or sheared wing tips. The conceptual designer's leverage to significantly reduce the skin friction coefficient is limited unless a radical technology such as laminar flow control is incorporated (Section 4.6).

4.5.5 Low-Speed Configuration

The aircraft in the low-speed configuration has a drag polar curve for each setting of the high-lift devices with undercarriage up and down. Relative to the clean configuration, the drag with undercarriage retracted is increased due to (a) the increased wetted area and upper wing surface friction; (b) the modified wing lift distribution; and (c) the tail download required for trimmed flight. Figure 4.14 depicts drag polar curves of an airliner for several flap settings, with the undercarriage up and down. For $C_L > 1.0$ these curves can be approximated fairly accurately by straight lines – a three-term approximation would improve the accuracy. It is

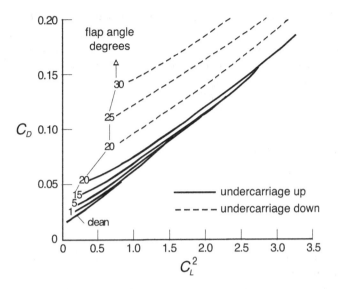

Figure 4.14   Drag polars of a wide body airliner for the low speed configuration

also noticeable that, with flaps down, the drag polar curves have a slightly smaller slope than with flaps up. Hence, the Oswald factor in a two-term approximation increases slightly relative to the value for the clean configuration.

A dominant flight condition for sizing the engines is the climb-out after the take-off with flaps down, undercarriage up, flying at the take-off safety speed V_2. For a given flap setting, the lift coefficient is determined by the requirement that the speed must have 20% reserve relative to the stalling speed: $V_2 \geq 1.2 \, V_S$. Experimental evidence suggests that the corresponding drag-to-lift ratio can be linearized for $1.2 < C_{L2} < 1.8$ as in Figure 4.15:

$$\left(\frac{C_D}{C_L}\right)_{V_2} = C_0 + \frac{C_{L2}}{\pi A_w E} \quad \text{with} \quad 1.2 \leq C_{L2} \leq 1.8 \tag{4.36}$$

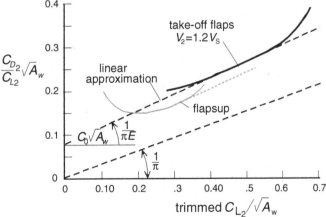

Figure 4.15   Corrected C_D/C_L at $1.2V_S$ for a jetliner with $C_{D_0} = 0.015$ flaps up

where $C_0 = 0.025$ accounts for the parasite drag with extended flaps and $E = 0.85$ is called the modified Oswald factor. The drag and lift coefficients are generalized by the factor $\sqrt{A_w}$ which makes Figure 4.15 applicable for different aspect ratios. It is emphasized that Equation (4.36) differs from the usual drag polar since it defines conditions at the take-off safety speed for different flap angles instead of a drag polar for a given flap configuration. This is clarified by comparison with the C_D/C_L ratio for the clean configuration.

4.6 Viscous Drag Reduction

Successful analysis, prediction, and control of the boundary-layer transition process for improved aerodynamic efficiency has been the ultimate goal – the Holy Grail – of aerodynamicists since the earliest days of aviation.

—J.R. Chambers [115] (2005)

Figure 4.16 depicts the cruise drag breakdown of a typical wide body transport aircraft giving an alternative view the relative importance of reducing parasite and induced drag. Major components are skin friction drag and vortex drag induced by lift accounting for about 50% and 40%, respectively. Form drag is built up from many small components – typically between five and ten counts – due to flow separations and vortices around the aft fuselage, flow interference near junctions, surface imperfections, excrescences, protuberances, and miscellaneous drag. Together they form a significant penalty which should be minimized by careful

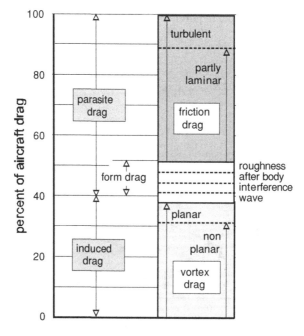

Figure 4.16 Example of transport aircraft cruise drag breakdown and drag reduction potential

design,[8] a subject beyond the scope of the present text. Wave drag at high speed has components due to body volume and lift and is sensitive to Mach number variation. The design condition in cruising flight is based on a trade-off between the desirability of block fuel conservation and block time reduction (Chapter 12). Compressibility drag in fuel economical flight is typically between five and ten counts.

The minimum drag in straight and level flight is obtained from Equation (4.29),

$$D_{\min} = 2\frac{W}{b_{\mathrm{w}}}\sqrt{\frac{C_{f_{\mathrm{eq}}}S_{\mathrm{wet}}}{\pi e}} \quad \text{for} \quad L = W \tag{4.37}$$

Although cruising flight is not exactly carried out in the minimum-drag condition, this equation clearly shows the relative importance of the design parameters which have a major influence on drag in cruising flight and, hence, on thrust and fuel consumption. The airplane GW and geometry dominate through the term W/b_{w} and the wetted area S_{wet}, aerodynamic design quality is apparent in the equivalent skin friction coefficient $C_{f_{\mathrm{eq}}}$ and the Oswald factor.

4.6.1   Wetted Area

The skin friction drag of each aircraft component exposed to the flow is proportional its wetted area – it is not affected by the aircraft's gross weight. Since reduction of the wetted area not only decreases drag but structural weight as well, minimizing the wetted area constantly needs the designer's attention. However, due to practical geometric constraints, feasible wetted area reductions are normally not spectacular unless major airplane configuration changes are adopted such as foreseen for the blended wing body (BWB, see Chapter 5).

- Fuselage skin area forms a major contributor which is, to a large extent, determined by the general arrangement of the airplane. It is constrained by cabin comfort and cargo hold volume requirements and can be minimized by carefully laying out the passenger cabin, baggage and cargo holds and by avoiding unused spaces.
- The wing is the other major contributor to wetted area. Aerodynamic design of a transonic aircraft wing is dominated by the Mach number and the lift coefficient for optimum cruising. Certain combinations of wing area and span, angle of sweep and wing thickness ratio lead to maximum aerodynamic efficiency, minimum trip fuel or minimum GW. Reduction of wing area is subject to low-speed performance requirements and a fuel tank volume constraint.
- Empennage wetted area depends on the size of the horizontal and vertical tail surfaces. Both are normally smaller when wing and/or fuselage wetted areas are reduced in size (Appendix A).

---

[8]It may appear easier to obtain a much greater percentage reduction in small drag sources than in the much larger contributions. For example, a 50% aft body drag reduction may be feasible, representing a useful 5% total drag reduction.

The traditional airliner configuration with a discrete fuselage – nicknamed tube and wing (TAW) – has a total wetted area between five and six times the wing planform area. Equation (4.34) shows that, for a given span, a 10% smaller wing area leads to roughly 4% wetted area and Reynolds number reduction, a 0.7% higher equivalent friction drag coefficient and a modest 1.6% improvement of L/D, on the provision that C_L is optimized by reducing the altitude.

The wetted area can be reduced significantly by adopting an unorthodox airplane configuration. The most radical mutation is the all-wing aircraft (AWA) which has a wetted area of about 2.5 times the gross wing area. For the same aspect ratio and equivalent skin friction coefficient, Equation (2.21) predicts the AWA's aerodynamic efficiency to be at least $\sqrt{2}$ times that of a TAW design, a value confirmed by two TsAGI projects depicted in Figure 2.6. Since the AWA planform area is typically twice the TAW gross wing area, its span is a factor $\sqrt{2}$ higher – unless the aspect ratio is reduced. In view of the restrictions by the ICAO airfield classification, such a large span is probably impractical. A dilemma for the designer of an AWA is therefore how to compare its properties with those of a TAW design. Bearing in mind that 'the outside has to be bigger than the inside' [75], a realistic approach might be to optimize an AWA with (1) a constraint on the internal volume required to contain the UL, systems and structure, and (2) an upper limit to the wing span. Such a design should be compared with an optimized conventional airplane with the same useful internal volume and wing span constraint, the subject of Chapter 5.

4.6.2 Turbulent Friction Drag

Friction drag can be reduced by manipulation of the turbulence structure within the sub-layer of a turbulent boundary layer using micro-structures on the airplane skin. This involves the introduction of riblets which suppress vortex formation near the wall. Riblets are formed by applying very fine grooves – a few microns high, about 0.05 mm wide – in the direction of flight, with sharp ridges in between. Riblets can be applied to the aircraft skin in the form of a self-adhesive film. A potential of 7 to 8% drag reduction has been demonstrated with experiments on flat plates in wind tunnel conditions and research has shown that riblets can deliver up to 2% aircraft drag reduction. This figure has been verified by flight tests with an Airbus A320 fuselage on which some 75% of the wetted surface was covered with 3M riblet film, resulting in an airplane drag reduction of the order of 1.5%. Application of riblets to the wing upper surface, horizontal stabilizer and the upper part of the fuselage (above the windows) would lead to approximately 1% in fuel burn on a long-haul airliner. The riblet film has shown to be durable over a test period of 18 months [85].

4.6.3 Natural Laminar Flow

The friction drag of surfaces with a laminar boundary layer is an order of magnitude less compared to a turbulent boundary layer. Delaying the onset of turbulent flow over 50% of the exposed airplane surface leads to a considerable reduction in skin friction drag (Figure 4.17). Natural laminar flow (NLF) is a passive technique that uses optimized pressure gradients to maintain a laminar boundary layer over large parts of a lifting surface. Transition to turbulence is a function of several types of instabilities and contamination:

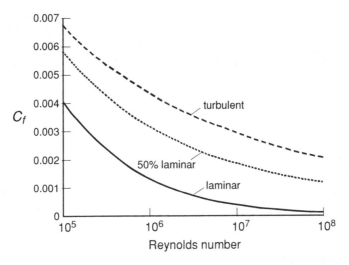

Figure 4.17   Flat plate skin friction coefficient

- Tollmien-Schlichting instability is caused by amplification of streamwise waves which is responsible for transition when the sweep angle is less than 25°.
- Cross-flow instability is the first cause of transition at high Reynolds numbers when the sweep angle is more than 25°.
- Attachment line instability is due to propagation along the wing leading edge of the disturbance caused by the fuselage boundary layer.

Obtaining NLF is closely connected to the airfoil shape, the leading edge sweep and the Reynolds number. The maximum airfoil thickness should be as far aft as possible; however, in high-speed flight this might lead to strong shock waves and high wave drag. Figure 4.18 is derived from results obtained in flight tests where transition locations over surfaces with a

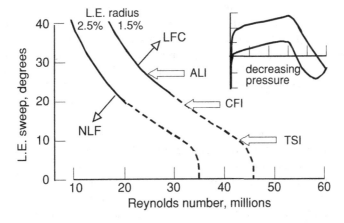

Figure 4.18   Extent of natural laminar flow over swept wings [108]

favourable pressure gradient were measured for varying leading-edge sweep. The shaded area indicates the approximate chordwise extent of laminar flow attainable on a wing with initially decreasing surface pressures towards the trailing edge. At relatively low chord Reynolds numbers and small leading edge sweep angles, a laminar boundary layer is possible over an appreciable stretch from the leading edge.

Present-day construction techniques result in the production of smooth and accurate aerodynamic surfaces over which long runs of NLF can be obtained. Stretches of laminar flow up to some 50% of the wing chord have been observed for airplanes featuring lifting surfaces with less than 18° of leading-edge sweepback. The wings of several business aircraft have been designed for NLF leading to drag reductions of 5 to 10% – application to high-subsonic transport aircraft leads to savings of a few percent. Consequently, NLF can be appreciated as a proven technique which finds application in business and commuter airplanes flying at subsonic speed. These successful applications also justify studies of applying NLF over fuselage surfaces [80].

4.6.4   Laminar Flow Control

Mid-size and large transport aircraft flying at high subsonic speed attain chord Reynolds numbers between 40 and 100 million at the root and between 8 and 20 million at the tip. Accordingly, the only chance of achieving laminar flow over most of the wing is by active control of the boundary layer. Laminar flow control (LFC) concepts using artificial mechanisms to maintain laminar flow over a large region of the wing have to be used. The use of wall suction is the favourite method of active LFC. This technology uses the continuous removal of a small amount of the laminar boundary layer by applying suction over the exposed surface through porous materials, multiple narrow surface slots or small skin perforations. The effect of suction over the entire airfoil as far back as the eventual transition point is that the laminar boundary layer is stabilized and transition to turbulence is delayed.

Active LFC is widely acknowledged to offer the potential for improvements in fuel usage that far exceed any known single aerodynamic technology. Theoretically, the fuel burned by transport aircraft might be decreased by up to 30% by applying boundary layer suction to both sides of the wing and the tail surfaces. A practical fall-back position which still provides significant savings is to apply suction control only to the upper wing surface which has the lowest pressures and hence the highest flow velocities and friction drag. The reduction in lower surface smoothness requirements enables the application of retractable leading edge devices for high lift. It also complies with the presence of inspection hatches and the higher susceptibility of damage by ground vehicles. Other cases where successful application of LFC must be considered elusive include most (metal) aircraft fuselages with roughnesses such as pitot probes, windshield wipers, doors and windows, and waviness due to pressurization.

LFC of the wing's upper surface, engines nacelles and the empennage is a well-researched aerodynamic technology with demonstrated drag benefits. Experience thus far indicates that laminar flow can be obtained on modern airfoil sections to mid-chord and beyond. However, a wing featuring LFC over the entire exposed primary structure is mechanically complex – see Figure 4.19 (a) – and costly to produce. The suction concept involves extensive ducting throughout the internal wing structure, auxiliary power sources or engine bleed air and an

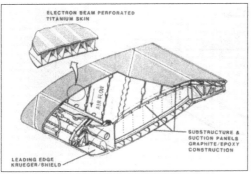 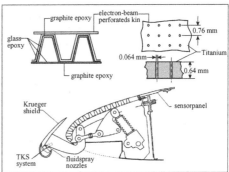

(a) structural wing concept   (b) perforated leading edge box with Krueger shield

Figure 4.19   Douglas LFC wing structural concept and leading edge flight test article. Copyright NATO STO - CSO

additional system to accurately control the suction airflow distribution. There are weight and maintenance penalties and some uncertainty in structural integrity must be accepted. Since results obtained in airline operation depend on the smoothness of the surface and on eventual contamination by insects and dust, a system for cleaning the leading edge is a prerequisite. The extension during take-off and landing of a Krueger flap or a shield containing a special cleaning fluid as depicted in Figure 4.19 (b) has proven to be effective, however, at the cost of reduced maximum lift.

4.6.5   Hybrid Laminar Flow Control

The hybrid laminar flow control (HLFC) concept integrates natural and active laminar flow technologies and avoids most of the objectionable characteristics of both. The leading-edge sweep limitation of NLF is overcome through suction in the leading edge box to control cross-flow and attachment line instabilities of swept wings. Wing shaping for favourable pressure gradients to suppress Tollmien-Schlichting instabilities allowing NLF over the wing box region removes the need for in-spar LFC suction and greatly reduces system complexity and cost [108]. Dependent on the wing size, HLFC may be used for long-chord inboard surfaces in combination with NLF for short-chord outboard surfaces. In 1990, Boeing flight tested the effectiveness of HLFC on a B 757 airplane using a large HLFC glove installed on a large section of the left wing (Figure 4.20). A titanium skin was used for suction to the front spar with an ability to reverse flow for purging. Existing manufacturing technology permitted construction of the glove to laminar-flow surface quality requirements. A Krueger shield as in Figure 4.19 (b) was installed for insect protection and high lift. All necessary systems required for practical HLFC were successfully installed. The flight test results located transition between 30 and 50% chord past the end of suction with less suction flow required than anticipated. The primary goals of this experiment – achieving HLFC at Reynolds numbers concurrent with medium-size transport airplanes and reducing industry risks to acceptable levels – were successfully accomplished [115].

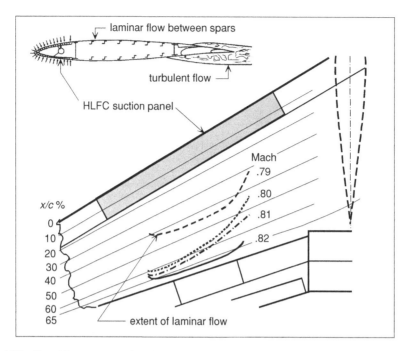

Figure 4.20 Transition contours showing the extent of laminar flow obtained on a B 757 test vehicle with HLFC [115]

4.6.6   Gains, Challenges and Barriers of LFC

Questions regarding long-term operational and reliability characteristics of current concepts still remain which must be resolved before a sustained performance of LFC over large areas of the aircraft can be guaranteed. The technologies of each of the three laminar-flow concepts have different drag reduction potential. LFC allows the aircraft to be sized for a considerably reduced fuel load and design gross weights, smaller engines and other benefits and this has the largest impact for long-range airliners. However, LFC is not a technology to be retrofitted and its application has to be envisaged and investigated in the early conceptual design stage.

Figure 4.21 plots the percentage improvement of L/D versus the wing area characterizing the airplane size. The performance gain due to NLF is now a fact that should not be ignored for business aircraft[9] and regional airliners. For a given transition Reynolds number, the percentage chord where laminar flow is obtained and the gain in L/D decrease with wing size. Figure 4.21 also shows a large improvement for HLFC compared to NLF, although the decreasing gain in L/D with size is manifest as well. For the largest transports, appreciably larger benefits are obtained with LFC suction position farther aft of the leading edge. Conceptually, even larger levels of drag reduction may be feasible via synergistic application if active laminarization is applied to non-conventional aircraft configurations that are especially fit for it.

---

[9]Piaggio Aero Industries reports that NLF degradation of their P-180 Avanti wing increases power required for level flight by 8 to 10%.

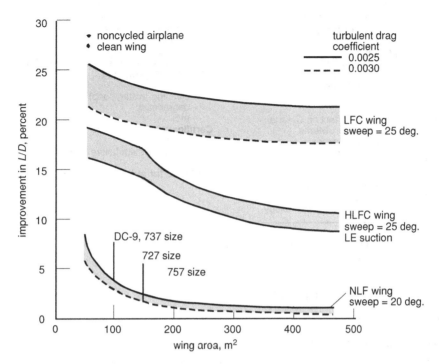

Figure 4.21 Improvement in aerodynamic efficiency due to laminar concepts as compared with a turbulent airplane [108]

Aerodynamic performance predictions by researchers indicate that, for a transport aircraft with laminar flow up to 60% chord on both sides of wing and tail, the fuel burned on long-range flights might decrease by 30% from that of fully turbulent airfoils. The performance improvements of HLFC are not as great as LFC, the potential gains are nevertheless substantial. Typical benefits of applying HLFC to a 300-passenger long-range twin engine subsonic transport are reported in [76]. Assuming 50% chord laminar flow over the upper wing surface and both surfaces of the empennage, HLFC provides about 15% reduction in mission fuel. Application to the engine nacelles with laminar flow to 40% of the nacelle length has the potential of at least an additional 1% reduction in fuel burned.

Real-world influences upon the operational functionality of LFC surfaces (Figure 4.22) include surface roughness/waviness, joints and steps, flight through ice clouds and ice protection, acoustic fields caused by the engines and localized vortical flow generation causing clogging effects. Maintaining laminar flow on an airplane wing is possible only with an extremely smooth surface requiring special manufacturing techniques and/or the use of composite materials. Research on building porous wing skins for LFC at an acceptable cost and keeping them free of dirt and insects has shown that composite materials such as graphite-epoxy or fibreglass may provide very smooth surfaces and freedom of waviness required for laminar flow. The major concern is not whether laminar flow can be obtained but whether it can be maintained reliably, in an economic fashion.

- **Validation of tools**
  - transition = ?
- **Application**
  - upper surface
  - both surfaces
- **Leading edge devices**
  - krueger
  - slat
  - suction effect on C_L - max
- **Technology problems**
  - suction system
  - leading edge contamination
  - cleaning
  - insects contamination
  - anti-icing / de-icing
  - surface quality
  - manufacturing technology
- **Suction power installation**
  - engine
  - auxiliary engine or APU
  - engine location: wing or fuselage?
- **Maintenance problems**
  - dirt
  - cleaning (water, soap)
  - degradation
  - etc.
- **Certification**
  - fuel reserves
  - suction system failure in flight
  - etc.
- **Real DOC for airlines?**

Figure 4.22 Questions and problems regarding design, certification, operation and reliability of LFC surfaces. Copyright NATO STO - CSO

4.7 Induced Drag Reduction

4.7.1 Wing Span

The induced drag of a long-range transport at cruise conditions (Figure 4.16) amounts to about 40% of the total drag, of which wave drag constitutes a small percentage. Reducing induced drag is therefore of paramount importance. Since induced drag at a given dynamic pressure is proportional to the span loading (W/b_w) squared, a span increase is an effective way of reducing it. For a given wing area, a 10% larger span increases the aspect ratio by 21%, whereas the Oswald factor and the Reynolds number are reduced by 3.3% and 10%, respectively. Equation (4.37) demonstrates that, dependent on the type of aircraft, the resulting improvement of the minimum drag for a given gross weight is roughly 7.5%. However, induced drag can have a much greater significance to advanced aircraft design and performance than might be inferred from cruise aerodynamics. In particular, induced drag is high at low speeds where it may account for 80 to 90% of the aircraft's total drag at a critical take-off condition with a failed engine. A mere 1% reduction of induced drag allows the aircraft to take off with almost the same percentage increased GW, which can be translated into a useful gain in payload or range. Increasing the span may also be a panacea when downstream of the conceptual phase the design appears to have performance shortcomings. And a span increment copes effectively with the heavier take-off weight of a derivative version in case the engines are sized to the one-engine-out take-off climb gradient.

The designer's freedom to increase the wing span is limited since a larger span requires more structural material to withstand the increased bending moments. In fact, wing structure weight is even more sensitive to span than induced drag and the problem of improving the aerodynamic efficiency through a span increase is clearly the subject of multidisciplinary design optimization (MDO). And a practical upper limit to the wing span – such as the '80 metre box' limit – may be a key constraint for large commercial transport aircraft.[10] If

---

[10]The use of folding wing tips was proposed in several design studies of high-capacity aircraft. However, folding tips as foreseen for the Boeing 777 during its design stage have not found favour with airlines.

increasing the wing span is not an option, the advanced designer can aim to increase the span efficiency factor e_v by using a non-planar wing or by embracing a general arrangement based on multiple lifting surfaces.

4.7.2   Spanwise Camber

A wing may exhibit a modest induced drag reduction if it produces a non-planar trailing vortex sheet. Dihedral/anhedral and a swept or curved trailing edge are all associated with wake curvature. As a corollary it can be stated that upward wing bending due to lift does not necessarily increase induced drag. The non-linearities associated with vortex sheet curvature requires a more refined analysis than the standard lifting line theory and these effects increase with incidence and wake roll-up. Relaxing the assumption of classical linear theory that the wing sheds a planar vortex sheet provides interesting possibilities to reduce drag. Early theoretical analysis was carried out by C.D. Cone Jr. and J.S. Letcher [55, 56]. The greatest increases in span efficiency occur for configurations which tend to release the major portion of the vortex wake over an appreciable vertical area near the tip. This leads to a geometry known as spanwise camber making it possible to get span efficiencies appreciably greater than one [68].

4.7.3   Non-planar Wing Systems

The high span efficiency of non-planar wings appears to accrue from the movement of vortical flow away from the mid-span line. Multiplane systems distribute the lift vertically as well as horizontally and each plane should be optimally twisted for minimum induced drag. Some aerodynamic principles of biplanes and boxplanes – see Section 4.3 and Figure 4.8 – were established during the 1920s for low speed aircraft. During the 1970s, the introduction of transonic jetliners stimulated investigations into the transonic biplane and the joined wing. These configurations may also have desirable non-aerodynamic effects when properly integrated into the overall airplane general arrangement. For instance, the joined wing geometry is explored primarily to improve structural depth when the bending moment on the wing system is high. Radical aircraft configurations based on the multiplane plane concept offer the possibility of significant reductions in induced drag for airplanes with a highly span-constrained wing and may lead to wetted area reductions or even complete elimination of tail surfaces. In these applications, induced drag reduction is just one of many design objectives that have to be investigated to obtain a balanced and realistic alternative for the conventional high-aspect ratio cantilever monoplane. A few of these design considerations are surveyed in the next chapters.

Bibliography

[1] Abbott, I.H., and A.E. Von Doenhoff, *Theory of Wing Sections: Including a Summary of Airfoil Data*, Dover Publications, Inc., New York, 1959.

[2] Lachmann, G.V. (ed.), *Boundary Layer and Flow Control*, Pergamon Press, Oxford, 1961.

[3] Hoerner, S.F., *Fluid Dynamic Drag*, Hoerner Fluid Dynamics, Bakersfield, CA, 1965.

[4] McCormick Jr., B.W., *Aerodynamics of V/STOL Flight*, Academic Press, New York, 1967.

[5] Hoak, D.E., et al., *USAF Stability and Control Datcom*, Flight Control Division, AFFDL, WPAFB, Ohio, USA, 1978.

[6] McCormick, B.W., *Aerodynamics, Aeronautics, and Flight Mechanics*, John Wiley & Sons, New York, 1979.

[7] Torenbeek, E., *Synthesis of Subsonic Airplane Design*, Appendices B, F and G, Delft University Press, Delft, 1981.

[8] Roskam, J., *Airplane Design*, Part VI: "Preliminary Calculation of Aerodynamic, Thrust and Power Characteristics", Roskam Aviation and Engineering Corporation, Lawrence, KA, USA, 1987.

[9] Shevell, R.S., *Fundamentals of Flight*, Chapters 11 and 12, Second Edition, Prentice-Hall, Englewood Cliffs, NJ, 1989.

[10] Eppler, R., *Airfoil Design and Data*, Springer Verlag, New York, 1990.

[11] Mair, W.A., and D.L. Birdsall, *Aircraft Performance*, Chapters 3 and 10, Cambridge University Press, Cambridge, 1992.

[12] Torenbeek, E., and H. Deconinck (eds), *Innovative Configurations and Advanced Concepts for Future Civil Aircraft*, von Kármán Institute for Fluid Dynamics, Lecture Series 2005-06, June 2005.

[13] Obert, E., *Aerodynamic Design of Transport Aircraft*, Chapter 40, IOS Press, Amsterdam, 2009.

[14] Torenbeek, E., and H. Wittenberg, *Flight Physics: Essentials of Aeronautical Disciplines and Technology, with Historical Notes*, Chapter 4, Springer Verlag, Heidelberg, 2009.

Theory and Prediction of Drag

[15] Von Kármán, T., and J.M. Burgers, "General Aerodynamic Theory: Perfect Fluids", in Vol. II of *Aerodynamic Theory*, Edited by W.F. Durand, 1935.

[16] Oswald, W.B., "General Formulas and Charts for the Calculation of Airplane Performance", NACA Report N0. 408, 1932. TR R-139, 1962.

[17] Aeronautical Research Council, "Definitions to be Used in the Description and Analysis of Drag", ARC Current Paper No. 369, 1958. Also *Journal of the Royal Aeronautical Society*, Vol. 62, pp. 796–801, 1958.

[18] Hoak, D.E., et al., "The USAF Stability and Control DATCOM", Air Force Wright Aeronautical Laboratories, TR-83-3048, October 1960 (Revised 1978).

[19] Morrison, W.D., "Advanced Airfoil Design Empirically Based Transonic Aircraft-Drag Buildup Technique", NASA Contractors Report 137928, January 1976.

[20] Feagin, R.C., and D.W. Morrison Jr., "Delta Method: An Empirical Drag Buildup Technique", NASA Contractors Report 151971, December 1978.

[21] ESDU, "Profile Drag of Axisymmetric Bodies at Zero Incidence for Subcritical Mach Numbers", Data Item 78019, London, 1978.

[22] Shevell, R.S., and F.P. Bayan, "Development of a Method for Predicting the Drag Divergence Mach Number and the Drag Due to Compressibility for Conventional and Supercritical Wings", Stanford University, Department of Aeronautics and Astronautics, Report SUDAAR 522, July 1980.

[23] Eppler, R., and D.M. Somers, "A Computer Program for the Design and Analysis of Low-Speed Airfoils", NASA TM 80210, August 1980.

[24] Young, A.D., J.H. Patterson, and L. Lloyd Jones, "Aircraft Excrescence Drag", AGARDograph No. 264, July 1981.

[25] Jacobs, P.F., "Experimental Trim Drag Values and Flow-Field Measurements for a Wide-Body Transport Model with Conventional and Supercritical Wings", NASA TP-2071, 1982.

[26] Kroo, I.M., and T. McGeer, "Optimization of Canard Configurations: An Integrated Approach and Practical Drag Estimation Method", ICAS Paper No. 82-6.8.1, September 1982.

[27] McGeer, T., and R.S. Shevell, "A Method for Estimating the Compressibility Drag of an Airplane", Stanford University, Report SUDAAR 535, 1983.

[28] Knaus, A., "A Technique to Determine Lift and Drag Polars in Flight", *Journal of Aircraft*, Vol. 20, pp. 587–593, July 1983.

[29] Hallstaff, T.H., and G.W. Brune, "An Investigation of Civil Transport Aft Body Drag Using a Three-Dimensional Wake Survey Method", AIAA Paper No. 84-0614, 1984.

[30] Covert, E.E. (ed.), "Thrust and Drag: its Prediction and Verification", *Progress in Astronautics and Aeronautics*, Volume 98, AIAA Education Series, New York, 1985.

[31] Jobe, C.E., "Prediction and Verification of Aerodynamic Drag, Part I: Prediction", *Progress in Astronautics and Aeronautics*, Volume 98, AIAA Education Series, New York, 1985.

[32] Drela, M., and M.B. Giles, "ISES:, A Two-Dimensional Viscous Aerodynamic Design Code", AIAA Paper No. 87-0424, January 1987.

[33] Slooff, J.W., "Technical Status Review on Drag Prediction and Analysis from Computational Fluid Dynamics: State of the Art", Foreword and Conclusions, AGARD Advisory Report No. 256, 1988.

[34] Naik, D.A., and C. Ostowari, "An Aerodynamic Comparison of Planar and Non-Planar Outboard Wing Planforms", ICAS Paper No. 88-4.3.2, September 1988.

[35] Haftmann, B., F-J. Debbeler, and H. Gielen, "Takeoff Drag Prediction for Airbus A300-600 and A-310 Compared with Flight Test Results", *Journal of Aircraft*, Vol. 25, pp. 1088–1096, December 1988.

[36] Boppe, C.W., "CFD Drag Predictions for Aerodynamic Design", AGARD Advisory Report No. 256, Lisbon, 1988.

[37] Vooren, J. van der, and J.W. Slooff, "CFD-Based Drag Prediction; State-of-the-Art, Theory, Prospects", lecture notes prepared for AIAA Professional Development Series on Drag Prediction and Measurement, NLR TP-9047L, August 1990.

[38] Boppe, C.W., "Aircraft Drag Analysis Methods", AGARD-FDP Special Course on Engineering Methods in Aerodynamic Analysis and Design of Aircraft, May 1991.

[39] ESDU, "Estimation of Airframe Drag by Summation of Components: Principles and Examples", Data Item 97016, London, 1997.

[40] ESDU, "Representation of Drag in Aircraft Performance Calculations (with Addenda A to G)", Data Item 81026, London, 1997.

[41] Chao, D.D., and C.P. van Dam, "Airfoil Drag Prediction and Decomposition", *Journal of Aircraft*, Vol. 36, pp. 675–681, July–August 1999.

[42] Kriszler, T., "Conceptual Design Methodology to Predict the Wave Drag of a Transonic Wing", RTO/AVT Conference, Ottawa, Canada, 18–21 October 1999.

[43] Dam, C.P. van, "Recent Experience with Different Methods of Drag Prediction", *Progress in Aerospace Sciences*, Vol. 35, No. 8, pp. 751–798, November 1999.

[44] Es, G.W.H. van, "Rapid Estimation of the Zero-Lift Drag Coefficient of Transport Aircraft", *Journal of Aircraft*, Vol. 39, pp. 597–599, July–August 2002.

[45] Vos, J.B., A. Rizzi, D. Darracq, and E.H. Hirschel, "Navier-Stokes Solvers in European Aircraft Design", *Progress in Aerospace Sciences*, Vol. 38, No. 8, pp. 601–697, November 2002.

[46] Laflin, K., S.M. Klausmeyer, T. Zickuhr, J.C. Vassberg, R.A. Wahls, J.H. Morrison, et al., "Data Summary from Second AIAA Computational Fluid Dynamics Drag Prediction Workshop", *Journal of Aircraft*, Vol. 42, No. 5, pp. 1165–1178, 2005.

[47] McMasters, J.H., private communication, June 2005.

[48] ESDU, "Excrescence Drag Levels on Aircraft", ESDU Data Item No. 94044, November 2007.

[49] Filippone, A., "Comprehensive Analysis of Transport Aircraft Flight Performance", *Progress in Aerospace Sciences*, Vol. 44, No. 3, pp. 192–236, April 2008.

[50] ESDU, "Examples of Excrescence Drag Prediction for Typical Wing Components of a Subsonic Transport Aircraft at the Cruise Condition", ESDU Data Item No. 93032 A, September 2009.

[51] Analytical Methods, I., "VSAERO – Integral Methods for Potential and Boundary Layer Flows", Redmond, WA, 2009.

[52] Gur, O., W.H. Mason, and J.H. Schetz, "Full-Configuration Drag Estimation", *Journal of Aircraft*, Vol. 47, No. 4, pp. 1356–1367, July–August 2010. (Also: AIAA Paper No. 2009-4109.)

Induced Drag

[53] Munk, M.M., "The Minimum Induced Drag of Airfoils", NACA Report No. 121, 1921.

[54] Prandtl, L., "Induced Drag of Multiplanes", NACA TN-182, March 1924.

[55] Cone Jr., C.D., "The Teory of Induced Lift and Minimum Induced Drag of Nonplanar Lifting Systems", NASA TR R-139, February 1962.

[56] Letcher Jr., J.S., "V-Wings and Diamond Ring Wings of Minimum Induced Drag", *Journal of Aircraft*, Vol. 9, No. 8, pp. 605–607, August 1972.

[57] Blackwell, J., "Numerical Method to Calculate the Induced Drag or Optimal Span Loading for Arbitrary Non-Planar Aircraft", NASA SP-405, May 1976.

[58] Sachs, G., "Minimum Trimmed Drag and Optimum C.G. Position", *Journal of Aircraft*, Vol. 15, pp. 456–459, August 1978.

[59] Laitone, E.V., "Positive Tail Loads for Minimum Drag of Subsonic Aircraft", *Journal of Aircraft*, Vol. 15, pp. 837–842, December 1978.

[60] Shevell, R.S., "Comment on "Ideal Tail Load for Minimum Drag", *Journal of Aircraft*, Vol. 15, pp. 639–640, September 1978.

[61] DeYoung, J, "Minimization Theory of Induced Drag Subject to Constraint Conditions", NASA CR-3140, June 1979.

[62] Laitone, E., "Prandtl's Biplane Theory Applied to Canard and Tandem Aircraft", *Journal of Aircraft*, Vol. 17, pp. 233–237, April 1980.

[63] Kroo, I.H., "Minimum Induced Drag of Canard Configurations", *Journal of Aircraft*, Vol. 19, pp. 792–794, September 1982.

[64] Butler, G.F., "Effect of Downwash on the Induced Drag of Canard-Wing Combinations", *Journal of Aircraft*, Vol. 19, p. 410, 1982.

[65] Yates, J.E., and C. duP. Donaldson, "A Fundamental Study of Drag and an Assessmant of Conventional Drag-Due-To-Lift Reduction Devices ", NASA Contractor Report 4004, 1986.

[66] Henderson, W.O., and B.J. Holmes, "Induced Drag: Historical Perspective", SAE Paper No. 892341, 1989.

[67] Naik, D.A., and C. Ostowari, "Effects of Nonplanar Outboard Wing Forms on a Wing", *Journal of Aircraft*, Vol. 27, No. 2, pp. 117–122, February 1990.

[68] Lowson, M.V., "Minimum Induced Drag for Wings with Spanwise Camber", *Journal of Aircraft*, Vol. 27, No. 7, pp. 627–631, July 1990.

[69] Kroo, I.M., and S.C. Smith, "Computation of Induced Drag with Nonplanar and Deformed Wakes", SAE Paper No. 901933, 1990.

[70] Smith, S.C., "A Computational and Experimental Study of Nonlinear Aspects of Induced Drag", NASA TP-3598, 1996.

[71] Smith, S.C., "Trefftz Plane Drag Minimization at Transonic Speeds", SAE Paper No. 971478, 1997.

[72] Kroo, I.H., "Drag Due to Lift: Concepts for Prediction and Reduction", *Annual Review of Fluid Mechanics*, Vol. 33, pp. 587–617, January 2001.

Drag Reduction Technology

[73] Hefner, J.N., and D.M. Bushnell, "An Overview of Concepts for Aircraft Drag Reduction", *Special Course on Concepts for Drag Reduction*, Von Kármán Institute for Fluid Dynamics, AGARD Report No. 654, Paper No. 1, June 1977.

[74] Whitcomb, R.T., "Methods for Reducing Subsonic Drag-Due-To-Lift", *Special Course on Concepts for Drag Reduction*, Von Kármán Institute for Fluid Dynamics, AGARD Report No. 654, Paper No. 2, June 1977.

[75] Smith, H.W., and R. Burnham, "The Outside Has to be Bigger than the Inside", *Journal of Aircraft*, Vol. 17, pp. 463–468, June 1981.

[76] Petersen, R.H., and D.V. Maddalon, "NASA Research on Viscous Drag Reduction", NASA TM-84518, 1982.

[77] Thomas, A.S.W., "Aircraft Drag Reduction Technology", *Improvement of Aerodynamic Performance Through Boundary Layer Control and High Lift Systems*, AGARD Conference Proceedings No. 365, Paper No. 11, 1984.

[78] Poisson-Quinton, P, "Parasite and Interference Drag Prediction and Reduction", *Special Course on Aircraft Drag Prediction and Reduction*, Paper No. 6, Von Kármán Institute for Fluid Dynamics, AGARD Report No. 23, 1985.

[79] Bushnell, et al., "Turbulent Drag Reduction Research", *Special Course on Aircraft Drag Prediction and Reduction*, Paper No. 17, Von Kármán Institute for Fluid Dynamics, AGARD Report No. 723, 1985.

[80] Dodbele, S.S., C.P. van Dam, P.M.H.W. Vijgen, and B.J. Holmes, "Shaping of Airplane Fuselages for Minimum Drag", AIAA Paper No. 86-0316, January 1986. Also NASA Contractors Report 3970, 1986.

[81] Wagner, R.D., D.W. Bartlett, and F.S. Collier Jr., "Laminar Flow: The Past, Present and Prospects", AIAA Paper No. 89-0989, 1989.

[82] Greff, E., "Aerodynamic Design for a New Regional Aircraft", ICAS Paper No. 90-2.7.1, September 1990.

[83] Schmitt, D., "Advanced Technology: Constant Challenge and Evolutionary Process", *Aeronautical Journal*, Vol. 94, pp. 335–340, 1990.

[84] Smith Jr., L., "Wake Ingestion Propulsion Benefit", AIAA Paper No. 91-2007, June 1991.

[85] Szodruch, J., "Viscous Drag Reduction on Transport Aircraft", AIAA Paper No. 91-0685, 1991.

[86] Robert, J.P., "Drag Reduction: An Industrial Challenge", *Special Course on Skin Friction Drag Reduction*, AGARD Report 786, Paper No. 2, March 1992.

[87] Chioccia, G., and S. Pignataro, "On the Induced Drag Reduction Due to Propeller-Wing Interaction", *Aeronautical Journal*, Vol. 32, pp. 328–336, October 1995.

[88] Thiede, P., "Drag Reduction Technologies", Proceedings of the CEAS/DRAGNET European Drag Reduction Conference, Potsdam, Germany, 19–21 June, 2000.

[89] Bushnell, D.M., "Aircraft Drag Reduction: A Review", *Journal of Aerospace Engineering*, Vol. 217, Part G, pp. 1–18, 2003.

[90] Young, T.M., "Fuel Sensitivity Analysis for Active Drag Reduction Systems", *Aeronautical Journal*, Vol. 108, pp. 215–221, April 2004.

[91] Reneaux, J., "Overview on Drag Reduction Technologies for Civil Transport Aircraft", paper presented at European Congress on Computational Methods in Applied Sciences and Engineering (ECCOMAS), July 2004.

[92] Jahanmiri, M., "Aircraft Drag Reduction: An Overview," Research Report 2011:02, Chalmers University of Technology, Department of Applied Mechanics, 2011.

Laminar Flow Control

[93] Sturgeon, R.F., et al., "Study of the Application of Advanced Technologies to Laminar Flow Control Systems for Subsonic Transport Aircraft", NASA CR 144975 and 133449, May 1976.

[94] Srokowski, A.J., and S.A. Orszag, "Mass Flow Requirements for LFC Wing Design", AIAA Paper No. 77-1222, 1977.

[95] Pfenninger, W., "Laminar Flow Control, Laminarization", *Special Course on Concepts for Drag Reduction*, AGARD Report 654, Paper No. 3, June 1977.

[96] Lange, R.H., "Design Integration of Laminar Flow Control for Transport Aircraft", *Journal of Aircraft*, Vol. 21, pp. 612–617, August 1984.

[97] Wagner, R.D., D.V. Maddalon, and M.C. Fischer, "Technology Developments for Laminar Boundary Layer Control on Subsonic Transport Aircraft", *Improvement of Aerodynamic Performance Through Boundary Layer Control and High Lift Sysytems*, AGARD Conference Proceedings No. 365, Paper No. 16, 1984.

[98] Braslow, A.L., and M.C. Fischer, "Design Considerations for Application of Laminar Flow Control Systems to Transport Aircraft", *Aircraft Drag Prediction and Reduction*, AGARD Report R-723, von Kármán Institute for Fluid Dynamics, 1985.

[99] Saric, W.S., "Laminar Flow Control with Suction: Theory and Experiment", *Aircraft Drag Prediction and Reduction*, von Kármán Institute for Fluid Dynamics, AGARD Report R-723, 1985.

[100] Holmes, B.J., C.J.Obara, G.L. Martin, and C.S. Domack, "Manufacturing Tolerances for Natural Laminar Flow Airframe Surfaces", SAE Paper No. 850863, 1985.

[101] Wagner, R.D., D.V. Maddalon, and D.F. Fischer, "Laminar Flow Control Leading Edge Systems in Simulated Airline Service", ICAS Paper No. 88.3.7.4, 1988.

[102] Redeker, G., et al., "Design of Natural Laminar Flow Glove for Transport Aircraft", AIAA Paper No. 90-3043, 1990.

[103] Arnal, D., "Boundary Layer Transition: Prediction, Application to Drag Reduction", *Special Course on Skin Friction Drag Reduction*, AGARD Report 786, Paper No. 5, March 1992.

[104] Bulgubure, C., and D. Arnal, "Dassault Falcon 50 Laminar Flow Flight Demonstrator", First European Forum on Laminar Flow Technology, Hamburg, March 1992.

[105] Arnal, D., and C. Bulgubure, "Drag Reduction by Boundary Layer Laminarization", *La Recherche Aerospatiale*, No. 3, pp. 157–165, 1996.

[106] Sawyers, D.M., and R.A.L. Wilson, "Assessment of the Impact of Hybrid Laminar Flow on a Large Subsonic Aircraft", paper presented at Second European Laminar Flow Forum, Bordeaux, June 1996.

[107] Joslin, R.D., "Overview of Laminar Flow Control", NASA TP-1998-208705, Langley Research Center, October 1998.

[108] Braslow, A.L., "A History of Suction-Type Laminar-Flow Control with Emphasis on Flight Research", Monographs in Aerospace History Number 13, NASA History Division, 1999.

[109] Boeing, "High Reynolds Number Hybrid Laminar Flow Control (HLFC) Flight Experiment", NASA CR-1999-209325 and CR-1999-209326, Boeing Commercial Airplanes Group, April 1999.

[110] Wong, P.W.C., and M. Maina, "Studies of Methods and Philosophies for Designing Hybrid Laminar Flow Wings", ICAS Paper 2.8.2, September 2000.

[111] Schrauf, G., and W. Kuhn, "Future Needs and Laminar Flow Technology", Fourth Community Aeronautical Days Conference, Hamburg, 29–31 January 2001, *Air and Space Europe*, Vol. 3, pp. 98–100, 2001.

[112] Young, T.M., and J.P. Fielding, "Potential Fuel Savings Due to Hybrid Laminar Flow Control under Operational Conditions", *Aeronautical Journal*, Vol. 105, pp. 581–588, 2001.

[113] Saric, W.S., and H.L. Reed, "Toward Practical Laminar Flow Control: Remaining Challenges", AIAA Paper No. 2004-2311, June 2004.

[114] Wong, P.W.C., and M. Maina, "Flow Control Studies for Military Aircraft Operations", AIAA Paper No. 2004-2313, 2004.

[115] Chambers, J.R., "Laminar-Flow Control: The Holy Grail of Aerodynamics", *Innovation in Flight*, NASA Special Paper 2005-4539, 2005.

5

From Tube and Wing to Flying Wing

It is logical that the first step in any preliminary study should be the definition of the minimum size vehicle that can conceivably enclose the required items.

—W.E. Caddell [17] (1969)

5.1 The Case for Flying Wings

Throughout the twentieth century, tailless aircraft and flying wings (FWs) have been continuously studied, but only a few have become operational. Between 1925 and 1950, a dedicated class of respected designers advocated radical concepts of the FW as the ultimate ideal of powered flight. The achievements of I. Cheranovski in the Soviet Union, F. Handley Page and A.V. Roe in the UK have become widely known. In Germany, A. Lippisch worked specifically on delta-wing research whereas the Horten brothers developed mainly tailless sailplanes [3]. However, the most prominent contribution to FW development was made in the USA by the legendary Jack Northrop.

5.1.1 Northrop's All-Wing Aircraft

J.K. Northrop began his research on flying wings as early as 1928. In 1940 he convinced the Army Air Corps that a long-range bomber airplane with an FW configuration would be superior to its more traditionally configured competitor, the Convair B-36. He acquired an order for the development of the XB-35 bomber with FW configuration and the development got underway in 1941. Northrop completely eliminated fuselage and empennage from his design and buried the engines inside the XB-35 and he called his design an all-wing aircraft (AWA). The XB-35 was powered by four piston engines, each driving two contra-rotating four-blade pusher propellers through a long shaft and gearbox. Although the production contract for this aircraft was cancelled in 1944, work was continued on testing a single production aircraft which flew for the first time in the summer of 1946.

Advanced Aircraft Design: Conceptual Design, Analysis and Optimization of Subsonic Civil Airplanes, First Edition. Egbert Torenbeek.
© 2013 by Egbert Torenbeek. Published 2013 by John Wiley & Sons, Ltd.

Figure 5.1   Prototype of the legendary YB-49 strategic bomber (1947). Courtesy of the U.S. Air Force

Towards the end of the Second World War, the turbojet engine had appeared on the scene and all ten XB-35s that had been built were to be converted to B-49 bombers powered by eight jet engines. The first YB-49 prototype (Figure 5.1) made its maiden flight in October 1947 and immediately proved to be more promising than its propeller-powered counterpart. Air Force officials testified that the aircraft showed considerable promise in speed and altitude, but had inadequate range. In 1949 the production contract of the B-49 was cancelled in favour of the competing piston-powered Convair B-36. The photo reconnaissance version YRB-49A, with six jet engines, made its first flight in 1950 (Figure 5.1). However, the definite production order was eventually granted to Boeing's revolutionary B-47 which flew considerably faster than the B-49. Shortly after cancellation of the B-49 program in 1950, Jack Northrop withdrew from airplane development as a disappointed man.

For three decades, Northrop refused to discuss why his promising airplane was scrapped so suddenly. In 1980, he finally admitted that the B-49 order was cancelled because the Air Force wanted the production line set up at Convair in Texas. Northrop was ordered to merge with the more established and competitive firm under conditions that he considered unfair to his company. During several decennia the aeronautical mainstream ignored the FW option until the Northrop company revealed in November 1988 that the B-2 (Figure 5.2) would be a flying wing like the B-49. This complex and costly bomber has a very unorthodox layout since its external shape was chosen primarily because of its excellent stealth properties rather than aerodynamic efficiency – its radar cross-section is only about 0.01 m^2 compared to 0.5 m^2 for the B-1 [25]. The B-2 became operational in the 1990s and demonstrated the feasibility of a military high-speed all-wing aircraft.

Figure 5.2 Northrop B-2 strategic bomber (1988). Courtesy of the U.S. Air Force

5.1.2 Flying Wing Controversy

Under certain assumptions, for wings greater than 12 per cent in thickness, the all-wing version will be the optimum envelope aerodynamically, whereas for wings thinner than 12 per cent, the wing-body configuration will be an optimum.

—I.H. Ashkenas, 1948

The flying wing is almost indistinguishable from the worst possible selection of all the wing volume to fuselage volume ratios that were presented.

—J.V. Foa, 1949

The aerodynamic superiority of appropriate wing-body configurations over the flying wing is generally a good deal more impressive than suggested by earlier analyses.

—J.V. Foa, 1983

Other advantages to the all-wing design have been established in recent years that should make up for the aerodynamic penalty on range expounded by Foa.

—W.R. Sears, 1990

The question whether the FW is the ultimate ideal vehicle for long-range flight initiated a long-lasting technical debate. It can be traced back to the emergence of Northrop's FW jet-propelled bomber when I.H. Ashkenas – who was involved in the development of the XB-35 and the YB-49 – authored a paper on maximum range performance of jet aircraft [6]. Ashkenas analyzed the maximum range of various jet propelled aircraft configurations and concluded that under certain assumptions the all-wing configuration would be an aerodynamically optimum concept. The paper was criticized by J.V. Foa [7] who interpreted Ashkenas' results differently and came to the conclusion that the flying wing, when jet propelled, is a poor choice for an aircraft configuration intended to achieve long range. Ashkenas replied '... with chagrin that the ideas I tempted to convey in the paper were so poorly expressed as to result in (Foa's) conclusions exactly opposite to my own'. Northrop's YB-49 contract was cancelled a month after Foa's critique was sent in to the Air Force. During the thirty years to come, aviation historians have argued about the technical and political aspects of the decision to cancel the promising B-49 [8]. In his 1983 reconsideration of the flying wing, Foa reconfirmed his earlier critique [9] which was supported by T.S. Schreiber [10].

The flying wing controversy was re-animated following the announcement in 1988 of the Northrop B-2 bomber (Figure 5.2) when Foa claimed that a jet powered flying wing such as the B-2 would suffer from an inferior range [12]. He repeated his critique on the early paper by Ashkenas [6] by emphasizing a mathematical error he found in it. W.R. Sears, Northrop's chief aerodynamicist during the B-49 development, replied in [11] and admitted the mathematical flaw. However, he put it in perspective and repeated that he had never agreed with Foa about his conclusions and contended that other advantages of the flying wing had been established in favour of the flying wing. Sears revealed that the bending and shear load reduction due to the favourable lateral mass distribution had proved to be very beneficial to the B-49. In 1992 P.J. Torvik published an independent examination of the maximum range problem for aircraft employing high-bypass turbofans [14]. He found a broader range of design parameters for which the FW produces the best range performance than is the case when a pure jet system is used.

5.1.3 Whither All-Wing Airliners?

Perhaps complete dependence on electronics for stability, control, load, and flutter suppression will allow the true flying wing as conceived by Jack Northrop over 50 years ago to become a practical alternative to the transport aircraft we know today.

—J.M. Swihart, Wright Brothers Lecture, 1987

The Northrop program of the 1940s demonstrated that the basic stability and control problems of a FW could be solved even before the days of readily available irreversible controls and artificial stability. The B-2 came as a surprise and proved the soundness of Northrop's approach for applying the concept to a bomber design. The appearance of the B-2 stimulated investigations into the feasibility of an all-wing airliner at aero-industries, research institutes and universities all over the world. Since the mid-1980s, the Russian institute TsAGI has investigated the suitability of the FW layout to very large airliners and freighters and critical technologies with the intention of making them viable [28, 43]. In the USA the investigations started in 1988 and focussed on the revolutionary blended wing body (BWB) concept originally developed by McDonnell Douglas under contract with NASA [39]. They were soon followed by pre-competitive research programs conducted in Europe. Although various proposed concepts have quite different characteristics, they consistently promise a 20 to 25% improvement

Figure 5.3   X-48B BWB flight test vehicle. Courtesy of NASA

in aerodynamic efficiency, less empty weight and lower operating costs compared to the conventional concept [32, 49]. The BWB concept is the first flying wing development in which major airframe manufacturers have taken an interest. For instance, Boeing's Phantom Works division operates a 8.5%-scale UAV remotely controlled test vehicle (Figure 5.3) which is based on its 451L study configuration. This technology demonstrator explores the flight-control properties and (post-)stalling behaviour of a BWB configuration. So far, no insurmountable flight control problems have been reported.

Advocates of the FW claim its potential of avoiding most of the parasite drag of major components other than the essential wing, thereby obtaining at least 20% gain in the aerodynamic efficiency in one airliner generation. Since the improvement of the range parameter during the jetliner era continues to be very modest (Figure 5.4), the FW should be regarded as an alternative to the traditional general arrangement nicknamed the tube and wing (TAW) configuration that must be considered as a serious challenge. However, in spite of the in-depth research programs between 1990 and 2010, civil airframe manufacturers have not yet embraced the FW layout for their clean sheet designs. Dominant issues that have not yet been solved are

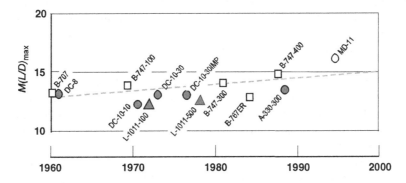

Figure 5.4   Historical improvement of long-range jetliner aerodynamic efficiency [33]

(1) how to design a suitable structure for a highly non-cylindrical pressure cabin; (2) how to settle the stability and controllability which have plagued the class of all-wing airplanes; and (3) how to prove that the radical FW concept can pass the certification process successfully.

It cannot be excluded that arguments in favour of or against the FW are often based on the superficial suspicion against the introduction of new technology. Instead, a satisfactory decision must be based on a rational investigation of a class of vehicles with various degrees of integration of the payload inside the wing. The aim of this chapter is to embark on a generic approach to the fundamental question of how the best distribution of the useful load can be established in the early conceptual design stage.

5.1.4   Fundamental Issues

The following issues have to be addressed during the early development stage of a new civil aircraft design concept:

- Allocation of useful load (UL) inside one or more fuselage bodies and/or lifting surfaces.
- The arrangement and shape of lifting surfaces so that the total lift is generated efficiently and the aircraft can be trimmed and stabilized in all anticipated operational conditions with little trim drag in normal flight.
- How to obtain a well-balanced design by suitably allocating the fuselage (if any), the wing, the power plant, the landing gears, and the horizontal tail or the foreplane (if any) relative to each other.

Although these issues are far from trivial even for the conventional TAW layout, its designers can rely on feasible solutions familiar from extensive research and satisfactory applications that have emerged from many decennia of experience. It is therefore unlikely that variations on the traditional arrangement will bring about a step forward unless radically new concepts and technologies are introduced. However, increasing the energy efficiency significantly by adapting the conventional layout may eventually become prohibitively expensive. The promise of an unorthodox configuration is primarily associated with increased functionality of major aircraft components. Anticipated gains can be substantial when this leads to reduced size or even deletion of some of these components. The main driving force behind the development of flying wings has always been the aspiration of designers to conceive an ideal aeronautical vehicle by removing all components not involved in the generation of lift. Since a wing is needed for lift – the argument went – the ideal airplane would be that wing and nothing else. The FW is impressive as a simple and elegant concept which, in its most radical form, is characterized by the following characteristics:

- The voluminous and pressurized passenger cabin, cargo hold and flight deck, as well as the power plant, fuel and on-board systems are all contained by the wing hull. Such a configuration lacks a fuselage, a horizontal tail and engine nacelles. In spite of the much bigger wing compared to that of a conventional airplane, the plane's total wetted area is significantly smaller.
- It appears obvious that elimination of the horizontal tail results in a useful parasite drag and empty weight reduction. Moreover, the FW has its mass distributed along the wing span in such a way that bending and shear loads on the structure due to lift are greatly reduced by the

opposing gravity and inertia loading. In spite of the increased wing size, this span loading effect yields a significant empty weight saving.

• The inherently low wing loading obviates the need for complicated high-lift devices required for aircraft of the traditional configuration having the same low-speed performances. This will further reduce wing weight and acquisition cost.

The anticipated overall effect should be a reduced empty weight and up to 30% savings in fuel required to fly long distances and, for given design payload/range capability, the take-off gross weight (TOGW) is significantly reduced. Moreover, engines with less thrust or (perhaps) a smaller number of engines can be installed. It is likely that these characteristics will result in substantially reduced operating costs.

The logic behind the FW applies to a strategic bomber such as the B-2 with its compact military load and equipment. However, a more detailed examination reveals that it may be flawed if applied to a jet airliner where other considerations are essential to achieve overall design superiority. In particular, the large cabin volume required and the low thickness ratio of a transonic wing may contribute to the adoption of a wing area that is too large to be aerodynamically efficient. In addition, the wing span of the FW is a design variable which is subject to the same limitations as for a TAW aircraft and this results in a lower wing aspect ratio. Although the arguments in its favour are difficult to disprove by fundamental considerations, many experienced designers are sceptical of the FW or even dislike it. It is also fair to point out that the FW gives the deceptive impression of simplicity whereas in reality the integration of all the major functionalities inside a single lifting hull has far-reaching consequences. Consequently, the design process of the FW appears to be extremely complex and viable only after committing to a comprehensive R & D program and a highly integrated design process.

If the strict definition of an FW is adhered to, the designer's freedom to investigate alternatives which may be (more) attractive from the overall system point of view is undesirably narrowed down. The primary objective of this chapter is therefore to compare the basic geometric, aerodynamic and performance aspects of a conventional layout with those of a 'pure' FW and with hybrid configurations retaining some of the FW's characteristics. The most promising FW applications might be an airliner with a 500+ passenger capacity and a very large freighter aircraft – the concept is not feasible for airliners carrying less than 200 passengers. Mature technological advances in aerodynamic and flight control technology, advanced materials and structures will be particularly beneficial in materializing this unorthodox concept.

5.2 Allocation of Useful Volume

The flying wing controversy has never been convincingly resolved. The parametric survey presented in the following sections may serve to clarify the relevance of the FW debate by providing quantitative guidelines useful in the early conceptual design stage.

> The actual size of the required volume changes the answer. Sometimes it is better to put the volume in a distinct fuselage, sometimes not. Thus this has become an interesting configuration issue. As I see it none of the papers to date have provided the answer in a simple, easy to understand analysis.
>
> —W. H. Mason, www.dept.aoe.vt.edu/mason

5.2.1   Integration of the Useful Load

The standard jetliner configuration has the pressurized cabin with payload allocated inside the fuselage, fuel tanks are mostly inside the wing. We refer to this concept as the discrete wing-cum-body (DWB). Although the wetted areas of fuselage and wing are roughly equal, the wing volume available for useful load is only a fraction of the fuselage volume. On the other hand, if the wing contains all the UL (and perhaps more) it will have a very large planform area and the optimum flight conditions will be quite different from a conventional design. In particular, it can be expected that an all-wing airliner (AWA) will cruise at a higher altitude and/or lower speed than usual and will therefore need a different wing shape and installed thrust.

It is anticipated that, for a large airliner and dependent on the design requirements, between the DWB and the AWA an even better configuration may be conceivable with a sizable part of the cabin and/or freight hold in the wing and the remaining part in a fuselage of reduced size. The potential to optimally distribute the UL inside such an integrated wing body (IWB) configuration adds another dimension to the design space: the degree of integration, characterized by the fraction of the UL volume assigned to the wing. This idea will be revealed in the following parametric survey based on [13], followed by practical consequences of the design integration applied to some generic hypothetical integrated concepts.

5.2.2   Study Ground Rules

The UL is assumed to be allocated inside the combination of a wing and a streamlined body of revolution with an equality constraint on the total useful volume containing the UL. It is assumed that a free exchange is possible of pressure cabin and fuel tank volume between wing and body. A matrix of generic transonic airplane configurations complying with this condition is depicted in Figure 5.5. The bodies are representative of a conventional airliner fuselage with a cylindrical passenger cabin, the straight-tapered wings are designed for the same cruise Mach number. The body and the wing are scaled up and down geometrically similar, although the wing aspect ratio is treated as a design variable. The present family of airplane configurations differs markedly from the blended wing body concept where the fuselage is moulded into a lifting body in the form of a very low-aspect ratio airfoil blended into the high aspect ratio outboard wings. It is interesting for the following reasons:

- Only a (high-aspect ratio) wing generates lift for low induced drag.
- A streamlined body of revolution has a better ratio of surface and frontal area to volume and a lower friction drag coefficient than a wing. Hence, the body has less parasite drag compared to a wing with the same volume.
- The internal useful volume of every configuration is utilized to the maximum extent – there is no unused space left. This yields the smallest surface area and empty weight required to contain the UL.

It may not seem appropriate to fill the available wing volume completely with UL since a bigger wing than required for a given useful volume will arguably reduce the induced drag. However, any unused wing volume can be filled by transferring UL from the body to the wing, thereby reducing body size and its parasite drag. The assumption that all the available useful volume is filled with payload and fuel is therefore appropriate. Although complete freedom

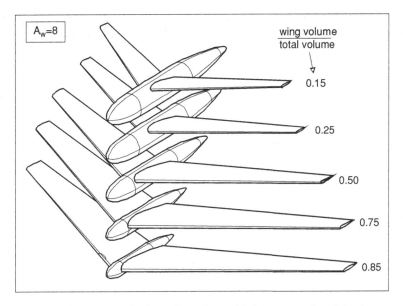

Figure 5.5   Wing and body configurations with the same total useful volume

in load allocation might be a hypothetical situation, several IWB concepts have been reported with part of the cabin and freight hold inside the inboard wing and the remaining part inside a body of reduced size [36]. The present study ground rules are therefore realistic even though in the real world they may lead to an objective function response surface which is not perfectly smooth.

5.2.3   Volume Ratio

The configurations in Figure 5.5 are characterized by their volume ratio, treated as the primary variable,

$$X = \frac{Q_w}{Q_{tot}} = \frac{\text{useful wing volume}}{\text{total useful volume}} \tag{5.1}$$

The useful wing (index w) and body (index b) volumes are

$$Q_w = X Q_{tot} \quad \text{and} \quad Q_b = (1 - X) Q_{tot} \tag{5.2}$$

Figure 5.5 shows a progression between a conventional design with $X = 0.15$ and an integrated layout with $X = 0.85$. The discussion is simplified by relating the volume ratio to the following classes of flight vehicles:

- Discrete wing and body (DWB) – $0 < X \leq 0.2$.
- Integrated wing body (IWB) – $0.2 < X \leq 0.9$.
- All-wing aircraft (AWA) – $0.9 < X \leq 1.0$.

In view of the limitations for the aerodynamic loading, a DWB with volume ratio $X < 0.1$ is not considered a practical vehicle. An AWA with $X = 1.0$ is also known as a pure flying wing.

5.2.4 Zero-Lift Drag

The drag area of the wing and body combination (index bw) is composed of contributions of exposed body and wing parasite drag,

$$(C_{D_0}S)_{\text{bw}} = (C_{D_\text{p}}S)_\text{b} + (C_{D_\text{p}}S)_\text{w} \tag{5.3}$$

Since the allocation of useful volumes inside the airplane forms the subject of this study, parasite drag coefficients are referred to body and wing (volume)$^{2/3}$ and denoted $\hat{C}_{D_\text{p}}$. Relationships between surface areas and volumes are found in Appendix A. In order to consider useful volumes rather than gross volumes, a volumetric efficiency factor is introduced,

$$\eta_\text{Q} = \frac{\text{useful volume}}{\text{gross volume}} \tag{5.4}$$

The zero-lift drag coefficient is referred to the gross wing area as follows:

$$(C_{D_0})_{\text{bw}} = (C_{D_\text{p}})_\text{w} \left\{ k_{\text{bw}}(X^{-1} - 1)^{2/3} + 1 \right\} \tag{5.5}$$

where the drag parameter[1] is defined as

$$k_{\text{bw}} \overset{\text{def}}{=} \left\{ \frac{(\eta_\text{Q})_\text{w}}{(\eta_\text{Q})_\text{b}} \right\}^{2/3} \frac{(\hat{C}_{D_\text{p}})_\text{b}}{(\hat{C}_{D_\text{p}})_\text{w}} \tag{5.6}$$

The significance of this parameter is clarified by substitution of $X = 0.5$ into Equation (5.5). It then appears that k_{bw} equals the ratio of body to wing parasite drag for equal volumes and volumetric efficiencies. A typical value $k_{\text{bw}} = 0.30$ indicates that the tube-like fuselage has a clear parasite drag advantage over the wing. The drag penalty needed to provide stability and controllability by means of the empennage is incorporated by factoring the body and wing parasite drag. The size of conventional airplane tail surfaces can be estimated using Appendix A which is based on statistics indicating that a DBW has horizontal and vertical tail areas of typically 25% and respectively 15% of the wing area. An AWA has no horizontal tail; it may have a single vertical tail or winglets with a total area of approximately 10% of the wing area. Between those two configurations the empennage size is accounted for by increasing the body parasite drag by 30% and the wing profile drag by 10%. It goes without saying that these numbers may have to be adapted to a specific design case. The drag parameter is assumed to be independent of the volume ratio X although the following reservations are made:

- Parasite drag coefficients are affected by Reynolds number variation. For a variation of X between 0.15 and 1.0 the mean wing chord increases by at least 50% leading to more than 7% reduction of its C_{D_p}. However, the body's C_{D_p} increases by 12% when its length is reduced by 50%. The net result is slightly in favour of the conventional DWB.

---

[1] A similar parameter was introduced by Ashkenas [6] who called it the geometric shape parameter k_n/k_w and proposed numerical values. Other authors [9, 10, 14] use similar parameters without information about numerical values.

- Merging body and wing reduces the net exposed and parasite drag areas. On the other hand, the merging causes unfavourable flow interferences which have to be suppressed by fairings and local cross-section adaptations resulting in parasite and vortex drag penalties. Although analysis of these drag components is complicated, it is likely that the net effect works out in favour of the AWA.
- It is not immediately obvious whether the volumetric efficiencies of the body and wing are independent of the volume ratio. The wing of a conventional DWB has a lower volumetric efficiency than the fuselage due to the volume required to install complicated high-lift systems and controls. However, since a flying wing needs less complicated trailing edge flaps, it is likely that its volumetric efficiency is higher than that of a conventional wing.

Accounting for these effects leads to considerable analytical complications. Since several trends are contradictory, the conclusions from the present simplified analysis probably remain valid.

5.2.5 Generalized Aerodynamic Efficiency

The power plant installation is not assumed to be integrated into the wing and body airframe and hence the total airplane drag can be decomposed into the airframe drag and engine installation drag. The nacelle parasite drag area is proportional (or nearly proportional) to the installed thrust. Hence, the nacelle drag has no effect on the optimization since it is equivalent to a percentage reduction of installed thrust and total drag. Consequently, the aerodynamic efficiency derived hereafter refers to the airframe excluding nacelles.

The airframe drag coefficient is represented by a two-term parabolic polar,

$$C_D = C_{D_0} + C_{D_L} = C_{D_0} + K_{\mathrm{L}}C_L^2 \tag{5.7}$$

The drag due to lift parameter K_{L} is inversely proportional to the wing aspect ratio; see Equation (4.20). A correction for compressibility is made by increasing the wing profile drag by a few drag counts and the induced drag factor K_{L} by a small percentage. The airframe aerodynamic efficiency is obtained from Equations (5.5) and (5.7),

$$\frac{C_L}{C_D} = \left(\frac{C_D}{C_L}\right)^{-1} = \left[\left\{1 + k_{\mathrm{bw}}\left(X^{-1} - 1\right)^{2/3}\frac{(C_{D_\mathrm{p}})_\mathrm{w}}{C_L}\right\} + K_{\mathrm{L}}C_L\right]^{-1} \tag{5.8}$$

A special case is an AWA in straight and level flight with

$$C_{L_{X=1}} \stackrel{\mathrm{def}}{=} \frac{W}{q\,Q_{\mathrm{tot}}^{2/3}}\frac{Q_\mathrm{w}^{2/3}}{S_\mathrm{w}} = X^{2/3}C_L \tag{5.9}$$

The term $W/(q\,Q_{\mathrm{tot}}^{2/3})$ is the lift coefficient based on the reference area $Q_{\mathrm{tot}}^{2/3}$ and is treated as an independent design variable representing the flight conditions as determined by the dynamic pressure $q = 0.5\,\gamma p M^2$. The term $Q_\mathrm{w}^{2/3}/S_\mathrm{w}$ depends on the wing taper, thickness and aspect ratios and can be computed using Appendix A – a typical value for a high-speed wing is between 0.07 and 0.08. Equation (5.8) is rewritten after substitution of (5.9) as follows:

$$\frac{C_L}{C_D} = \left[\{X^{2/3} + k_{bw}(1-X)^{2/3}\} \frac{(C_{D_p})_w}{C_{L_{X=1}}} + \frac{K_L C_{L_{X=1}}}{X^{2/3}} \right]^{-1} \tag{5.10}$$

A reference value for L/D is the maximum aerodynamic efficiency achievable for an AWA,

$$\left(\frac{C_L}{C_D}\right)_{ref} \overset{def}{=} \left\{ 2\sqrt{(C_{D_p})_w K_L} \right\}^{-1} \quad \text{for} \quad C_{L_{ref}} = \sqrt{\frac{(C_{D_p})_w}{K_L}} \tag{5.11}$$

This reference configuration is used to define the reduced lift coefficient

$$Y \overset{def}{=} \frac{C_{L_{X=1}}}{C_{L_{ref}}} \quad \rightarrow \quad \frac{C_L}{C_{L_{ref}}} = X^{-2/3} Y \tag{5.12}$$

This parameter is substituted into Equation (5.10) to yield the aerodynamic efficiency of a general configuration with a given useful volume as a fraction of the maximum achievable value for an AWA configuration with the same volume,

$$\frac{C_L/C_D}{(C_L/C_D)_{ref}} = 2 \left\{ \frac{X^{2/3} + k_{bw}(1-X)^{2/3}}{Y} + \frac{Y}{X^{2/3}} \right\}^{-1} \tag{5.13}$$

This expression is treated as a figure of merit of the wing and body configurations considered. Figure 5.6 depicts a contour plot of this FOM as a function of the volume ratio X and the reduced lift coefficient Y.

5.2.6 Partial Optima

The aerodynamic design efficiency as affected by the independent design variables X and Y has the following partial optima depicted in Figure 5.6 as dotted curves:

For given gross weight and flight conditions, the volume ratio resulting in a local optimizer is obtained from

$$\frac{\partial(C_L/C_D)}{\partial X} = 0 \quad \rightarrow \quad Y = X^{2/3} \left\{ 1 - k_{bw} \left(\frac{X}{1-X} \right)^{1/3} \right\}^{1/2} \tag{5.14}$$

For $k_{bw} = 0.3$ and $Y < 0.624$, this condition yields two solutions for X. The lowest of these defines an IBW with the highest obtainable L/D for given gross weight and flight conditions, the highest defines a local minimum aerodynamic efficiency.[2] For $Y \geq 0.624$ there exists no

---

[2]This is probably the condition referred to by J.V. Foa as the worst possible selection of the volume ratio (Section 5.1).

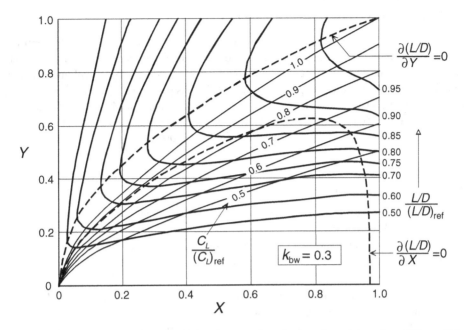

Figure 5.6 Generalized aerodynamic efficiency versus the volume ratio and the reduced lift coefficient

partial optimum volume ratio and the highest L/D is always obtained for an AWA. For given wing-body geometry, the flight condition for minimum drag is obtained from

$$\frac{\partial(C_L/C_D)}{\partial Y} = 0 \quad \rightarrow \quad Y = X^{2/3}\left\{k_{\text{bw}}(X^{-1}-1)^{2/3}+1\right\}^{1/2} \tag{5.15}$$

This defines the condition for the maximum aerodynamic efficiency of a wing and body configuration with given volume ratio,

$$\left(\frac{C_L}{C_D}\right)_{\text{max}} = \left(\frac{C_L}{C_D}\right)_{\text{ref}}\left\{k_{\text{bw}}(X^{-1}-1)^{2/3}+1\right\}^{-1/2} \tag{5.16}$$

for

$$C_L = C_{L_{\text{md}}} = \sqrt{\frac{C_{D_0}}{K_{\text{L}}}} \tag{5.17}$$

Since Equations (5.14) and (5.16) have no intersection, the aerodynamic efficiency does not exhibit a global maximum. In other words, if there is no constraint on altitude and speed, no configuration can be found which performs better than the AWA flying at its optimum flight condition.

5.3 Survey of Aerodynamic Efficiency

The variation of C_L/C_D illustrated in Figure 5.6 shows that there exists no global optimum volume ratio for maximum aerodynamic efficiency. The aim of this section is to investigate the effects of primary design parameter variation on the best obtainable useful volume allocation for configurations such as those depicted in Figure 5.5. This parametric survey is performed for a hypothetical airliner with 450 metric tons gross weight cruising at Mach 0.85 and altitudes between 30 000 and 50 000 ft. It is assumed that the volume ratio can be varied continuously with an equality constraint on the total useful volume of 2 000 m³.

5.3.1 Altitude Variation

Equation (5.13) is used to compute the effect of varying the flight altitude for the constant Mach number on L/D. The lift coefficient for each volume ratio is obtained from the combination of Equations (5.9), (5.12) and (5.10) which yields the result depicted in Figure 5.7. The following observations are made:

- A baseline DWB model with $X = 0.18$ flying at 35 000 ft altitude attains $L/D = 20.5$. This performance is close to $(L/D)_{max}$ and is representative of the year 2000 performance potential of the largest airliner airframes. The highest aerodynamic efficiency at 35 000 ft altitude is achieved by an IWB with $X = 0.35$. Its performance ($L/D = 22$) is not an impressive improvement over the DWB. Due to its low lift coefficient, the AWA does not perform any better than the DWB.
- Significantly higher aerodynamic efficiencies are obtainable at higher altitudes than today's airliner practice. At 40 000 ft the aerodynamic efficiency is almost constant for $X > 0.4$. In other words: any IWB flying at this altitude will be as good as the AWA and significantly

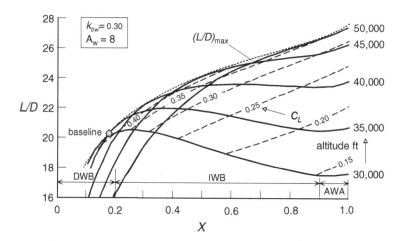

Figure 5.7 Aerodynamic efficiency versus volume ratio and cruise altitude for a large airliner. Data: $Q = 2\,000$ m³; $W = 4\,500$ kN; $M = 0.85$

better than the DWB baseline flying at 35 000 ft. For instance, a design with $X = 0.5$ flying at a modest $C_L = 0.35$ will improve L/D by 12.5% relative to the baseline DWB.

• The AWA's performance is not impressive at the usual cruise altitudes and achieves its best performance at unusually high altitudes. A pure flying wing reaches its full potential in excess of 50 000 ft.[3] The maximum aerodynamic efficiency $L/D = 27.5$, an improvement of 34% over the baseline DWB, is achieved for $C_L = 0.39$.

Integrated configurations with a substantial fraction of the useful volume in the wing achieve significantly better aerodynamic performance than the conventional wing body configuration provided they are designed to fly at a (much) higher altitude than usual.

5.3.2   Aspect Ratio and Span

Variation of the wing aspect ratio is very influential with the primary effect felt in the drag due to lift parameter $K_L = (\pi A_w e)^{-1}$. Another effect is manifest in Equation (A.2) on p. 437 indicating that the wing area required to obtain a specified wing volume is proportional to $A_w^{1/3}$. The following relationships apply if the wing profile drag coefficient is assumed to be independent of the aspect ratio:[4] $K_L \propto A_w^{-1}$, $C_{L_{X=1}} \propto A_w^{-1/3}$ and $k_{bw} \propto A_w^{-1/3}$. This leads to the observation that the aerodynamic efficiency of a wing with volume Q carrying a gross weight W at dynamic pressure q has a maximum value for

$$A_w = (2W/q)^{6/5}\{(t/c)/Q\}^{4/5}(C_{D_p}\pi e)^{-3/5}(1+\lambda)^{-2/5} \tag{5.18}$$

This is perhaps an unexpected result since the usual observation is that induced drag continues to decrease with increasing aspect ratio. The explanation is that increasing the aspect ratio reduces the wing volume unless the wing gets a larger planform area, leading to more wetted area and higher profile drag. Application of Equation (5.18) to the example wing results in a typical optimum $A_w = 8$ with the implication that an aerodynamically efficient AWA with a specified volume does not necessarily have a very high aspect ratio.

The parametric survey is now extended with a variation of aspect ratio and wing span for a constant cruise altitude of 35 000 ft. Equation (5.10) is used to derive the results shown in Figure 5.8 leading to the following observations:

• In spite of its potentially high performance, an AWA offers a disappointingly low L/D with little variation for aspect ratios between five and ten.
• Very high L/D values are obtainable for a DWB configuration if there is no constraint on the wing span. Increasing the baseline aspect ratio from eight to ten requires a fairly modest span increment and increases L/D by 12.5%, whereas an IWB needs a considerably larger span for the same improvement.

---

[3] Notably, this altitude has been quoted as the best cruising altitude of the Northrop B-49 flying wing bomber.
[4] Aspect ratio variation affects the chord Reynolds number and, hence, the skin friction and profile drag coefficients. This effect is not taken into account in the present analytical approach.

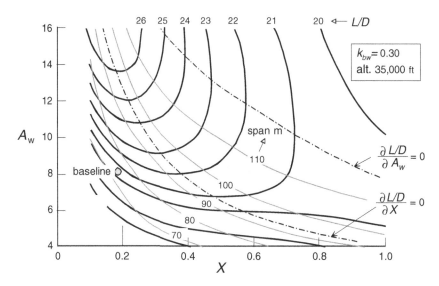

Figure 5.8 Aerodynamic efficiency versus volume ratio, aspect ratio and span. Data as in Figure 5.7

- For each volume ratio there exists an aspect ratio which maximizes L/D. It is possible to derive this partial optimum analytically but the result is unrealistic and for a span in excess of 110 m there is practically no gain in L/D. Irrespective of the aspect ratio, the optimum volume ratio occurs at approximately 95 m wing span.
- The partial optima with respect to volume ratio and span are incompatible; hence, there exists no unconstrained optimum combination.

The present result points to a DWB with the highest practical wing span to be the configuration yielding the best aerodynamic efficiency at 35 000 ft altitude. Due to the large effect of span on wing weight, the aspect ratio complying with an overall optimum system is significantly lower than the value for the maximum L/D (Chapter 10).

5.3.3   Engine-Airframe Matching

It has been observed that an airplane with a highly integrated wing body combination has a low wing loading and, for 35 000 ft flight altitude, its C_L is far below the value for maximum L/D. To improve the situation, the altitude of an IWB design should be higher than usual. However, selection of the altitude is closely associated with engine sizing and should not be based only on the aim of achieving the highest possible aerodynamic efficiency. The survey will therefore be augmented with a computation of the required engine size to match the airframe drag in level flight.

According to Equation (3.17) the corrected thrust T/p of a turbofan operating in the stratosphere at a fixed rating and Mach number is constant – for instance, equal to the value at the tropopause. The available thrust is therefore proportional to the ambient pressure and the

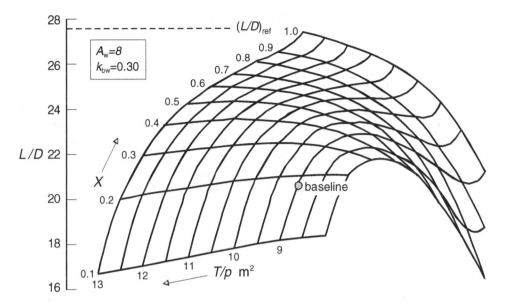

Figure 5.9 Aerodynamic efficiency versus volume ratio and corrected thrust. Data as in Figure 5.7

drag that can be balanced by a given engine thrust in straight and level flight is determined by $C_D = T/(q S_w)$. Accordingly, Equation (5.7) defines the achievable altitude and Mach number via a constraint on the lift coefficient,

$$C_L = \left(\frac{C_D - C_{D_0}}{K_L}\right)^{1/2} = \left\{K_L^{-1}\left(\frac{T/p}{\frac{1}{2}\gamma M^2 S_w} - C_{D_0}\right)\right\}^{1/2} \tag{5.19}$$

Substitution of C_{D_0} according to Equation (5.5) and S_w from Equation (5.9) yields the relationship between C_L/C_D, T/q and X satisfying the condition $T = D$. The result is depicted for the example airplane with $A_w = 8$ in Figure 5.9 and the following observations are made from this carpet plot:

- The obtainable L/D for a DWB with volume ratio between 0.15 and 0.20 is insensitive to engine size for corrected thrust values between 8.0 and 9.5 m². The highest aerodynamic efficiency for the DWB baseline flying at 35 000 ft is obtained for $T/p = 9$ m².
- If engines are installed with $T/p = 9$ m², the AWA does not perform any better than the baseline design, whereas L/D deteriorates rapidly for lower thrust. Compared with the baseline, the AWA needs engines with 45% more thrust to reach its full potential of $L/D = 27.5$.
- Configurations with $X > 0.30$ are sensitive to the installed thrust. An IWB with 50% of the useful volume in the wing achieves a respectable $L/D = 24.2$ but requires engines with 22% more thrust than the baseline DWB.

In conclusion, it is essential that the engine-airframe matching condition is included in any optimization of IWB configurations. This applies in particular to the pure flying wing.

5.4 Survey of the Parameter *ML/D*

The previous section has shown that different cruise altitudes can lead to significantly different appreciations of flying wing compared to conventional configurations if the Mach number is specified. Arguably, a constraint on the cruise Mach number, although practical for a particular design study, might be too restrictive when investigating a radically new concept. This section deals with a simplified analysis of the Mach number variation on the specific range and how it is affected by the volume fraction and the drag parameter.

5.4.1 Optimum Flight Conditions

Chapter 12 shows that, for constant engine TSFC, the specific range of a jet aircraft is proportional to the parameter ML/D. Although in real life this leads to an optimum cruise condition in the drag!rise, this complication is avoided when the following simplifications are accepted:

1. For each altitude the achievable Mach number in straight and level flight is determined by a constraint on the available thrust.
2. Engines are scaled up and down for constant cruise rating and the corrected thrust is assumed independent of the Mach number and ambient conditions.
3. The Mach number is below the drag rise so that a single parabolic drag polar can be used.[5]

Due to these assumptions the accuracy of the analysis is not very high, but the trends are thought to be correct. For a specified T/p the achievable flight Mach number is determined by the horizontal equilibrium,

$$M = \sqrt{\frac{T/p}{\frac{1}{2}\gamma C_D S}} \qquad (5.20)$$

Hence, the parameter ML/D is proportional to $C_L C_D^{-3/2} S^{-1/2}$, whereas the combination of C_L and S determines altitude and speed. The classical partial optimum condition for maximum ML/D is

$$C_L = \sqrt{\frac{C_{D_0}}{2K_L}} = \frac{C_{L_{md}}}{\sqrt{2}} \qquad \rightarrow \qquad M = 2^{1/4} M_{md} \qquad (5.21)$$

---

[5]It is worth noting that all publications in the bibliography on the present subject take the absence of compressibility drag for granted.

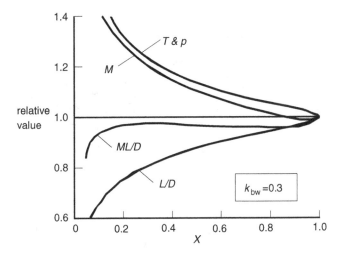

Figure 5.10   Flight conditions for maximum ML/D

Substitution into Equation (5.20) yields the flight condition

$$M = \sqrt{\frac{4\,T/\mu}{3\gamma\,C_{D_0}S}} \quad \text{and} \quad \frac{C_L}{C_D} = \frac{1}{3}\sqrt{\frac{2}{C_{D_0}K_L}} = 0.9428\left(\frac{C_L}{C_D}\right)_{max} \tag{5.22}$$

The highest value of ML/D appears to be proportional to $(C_{D_0}\sqrt{S})^{-1}$ and substitution of C_{D_0} according to Equation (5.5) yields

$$ML/D \propto \left[X^{1/3}\left\{1 + k_{bw}(X^{-1} - 1)^{2/3}\right\}\right]^{-1} \tag{5.23}$$

The example in Figure 5.10 shows that, when more useful volume is allotted to the wing, the optimum Mach number decreases whereas the altitude and L/D increase. The AWA achieves the highest aerodynamic efficiency for all values of the volume ratio. However, the parameter ML/D is practically invariable at a level of a few percentage below the AWA for all IWB configurations and DWB airplanes with $X > 0.15$.

5.4.2   The Drag Parameter

The drag parameter k_{bw} is a crucial factor when an FW is compared with the traditional TAW configuration. Its value depends on geometrical data such as fuselage fineness ratio, wing taper, thickness and aspect ratios, and volumetric efficiencies. The following approximation is based on Appendix A and on the assumption of equal volumetric efficiencies for wing and body:

$$k_{bw} = 1.20\left\{\frac{t/c}{A(1+\lambda)}\right\}^{1/3} \tag{5.24}$$

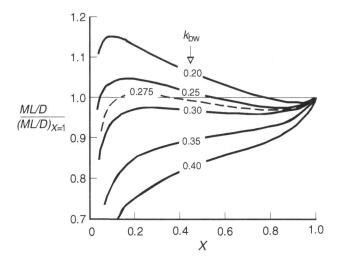

Figure 5.11   Maximum ML/D affected by the drag parameter

where A, λ and t/c denote the wing's aspect ratio, taper ratio and thickness ratio, respectively. Equation (5.24) also takes into account that body parasite drag depends on its fineness ratio. Typical values of k_{bw} are between 0.25 and 0.35 leading to the observation that the parasite drag of a fuselage is between 25 and 35% of the parasite drag of a wing with the same volume. The following conclusions are drawn from Figures 5.10 and 5.11.

- In accordance with [6], a pure flying wing has better specific range than a wing and body combination for $k_{bw} \geq 0.275$, a wing and body configuration is better for $k_{bw} < 0.275$.
- Since the best flight Mach number of a pure flying wing is lower than for a wing and body configuration, a wing section with higher thickness ratio can be selected in the interest of a more efficient passenger accommodation and lower structure weight.

The case for integrated configurations becomes stronger with the (future) introduction of open rotor engines which have their maximum efficiency at Mach numbers between 0.70 and 0.80 (Section 3.4). The aerodynamic efficiency increases considerably compared to the traditional TAW configuration if this speed range were selected as the design condition. It is obvious that such a radical deviation from the present cruise condition would require a far more comprehensive investigation than the present analysis.

5.5   Integrated Configurations Compared

The elementary derivation of the achievable aerodynamic efficiency has indicated that two main roads can be followed to improve the performance of large transport aircraft by means of a radical mutation.

1. Increasing the wing span is very effective from the aerodynamic point of view but considerations such as keeping structural weight within limits tend to dominate. For small to medium

size airplanes, aspect ratios up to 15 may become an option which requires an increased thickness ratio, advanced (composite) materials and/or selection of a general arrangement leading to a considerably reduced wing bending load. Examples are strut-braced wings and multi-body configurations (Chapter 6).

2. Compared to the traditional TAW layout – with constraints on cabin, cargo hold and fuel tank volumes and a large empennage – a substantial reduction of wetted area is only feasible for a tailless, highly integrated wing body aircraft configuration.

The previous analysis has indicated that it is not necessarily an advantage to adopt an AWA configuration for a transonic airliner because its wing area tends to be large, leading to a lift coefficient (far) below the flight condition for minimum drag. A possible answer is to decrease the dynamic pressure by increasing the altitude and/or reducing the flight speed. The first option leads to higher installed engine thrust, the second increases block time. Both options are not in the interest of good economics and in particular the second is likely to be objectionable to the airlines. This consideration does not necessarily apply to a large (dedicated) freighter for which the allocation of cargo in the wing reduces its en route bending moment considerably, allowing a span increase with less wing structure weight penalty than for a conventional design. The integrated configuration may form a viable candidate for this application.

The parametric survey in the previous sections is based on combinations of bodies and straight tapered wings as depicted in Figure 5.5. Although this simplification allows a generalized analytical approach, a practical integrated design may have a non-cylindrical body and/or an unorthodox wing planform. The survey is therefore not sufficiently flexible to conclude whether an integrated wing body can offer acceptable payload accommodation, operational and flight characteristics. In this section we compare a large long-range airplane of conventional layout with several concepts which have an integrated layout. The airliner designs have the same cruise speed and total useful volume, the freighter flies at a lower Mach number and has a larger cargo hold volume. Different from the parametric survey, the engine nacelles are counted as airframe components. Wetted area and drag predictions are based on simple methods such as those presented in Chapter 4 and Appendix A. Compressibility drag has not been taken into account.

5.5.1 Conventional Baseline

The conventional design is a long-range twin-deck Mach 0.80 airliner in the category Airbus A 380 and Boeing 747. Figure 5.12 (a) and Table 5.1 define its geometry and general arrangement. The wing span is limited to 80 m – the maximum value for ICAO field classification F. Application of Equations (A.5) and (A.2) shows that the fuselage contains 430% of the wing volume with 12% more wetted area. Empennage area is 35% of the wing area and engine nacelles and pylons surface areas are 10% of the airframe wetted area. Maximum subsonic L/D amounts to 20.6 which is at the upper limit of what is achievable for the largest of the year 2000 long-range airliners, although a wing span increase to 90 m would increase L/D by about 9%. Alternatively, a wing area reduction for the same span might be possible if the area is not constrained by the fuel tank volume required, resulting in reduced wetted area and parasite drag. However, wing area reduction for the same span entails more complicated high-lift devices and a higher structural aspect ratio, both leading to a higher empty weight. It

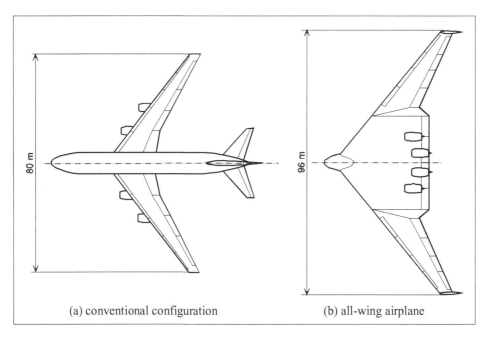

(a) conventional configuration   (b) all-wing airplane

Figure 5.12 Geometries of a large Mach 0.85 airliner of conventional layout and an all-wing aircraft, both accommodating a useful load volume of 2 000 m^3

Table 5.1   Design data for a twin-deck airliner with conventional layout

fuselage	equivalent diameter	7.5 m
	length	75 m
	gross volume	2 651 m^3
	gross wetted area	1 570 m^2
	cabin plus cargo hold volume	1 590 m^3
	cabin floor area	625 m^2
wing	span	80 m
	gross planform area	800 m^2
	exposed planform area	686 m^2
	thickness/chord ratios	0.14/0.10
	exposed gross volume	611 m^3
	wetted area	1 400 m^2
horizontal tail	exposed planform area	180 m^2
vertical tail	exposed planform area	100 m^2
engine nacelles	wetted area	350 m^2
airframe	gross volume	3 262 m^3
	total wetted area	3 852 m^2

turns out that there exists an optimum wing size resulting in a practical span and aerodynamic efficiency.

5.5.2   Is a Wing Alone Sufficient?

The AWA is a radical integrated concept which is not simply obtained from a conventional design by removing the empennage. A useful comparison can only be made when both designs carry the same useful load at the same cruising speed with acceptable stability and control properties. The AWA depicted in Figure 5.12 (b) has the following characteristics.

- The passengers are accommodated in the central section on a single deck with underfloor baggage. This section has the same gross volume as the conventional fuselage whereas the outboard wings together have the same volume as the conventional (exposed) wing.
- The gross planform area of 1 582 m² is almost twice the conventional wing area. For the same gross weight and flight conditions, the lift coefficient is reduced by 50% which allows the wing to be a few percentage chord thicker than the conventional wing.
- The AWA enables the designer to spread the UL laterally along the wing span rather than longitudinally along the fuselage. This reduces the inboard bending and increases the structural height, allowing a span increase for the same structure weight. In spite of the large span, the AWA aspect ratio of slightly less than six is unusually low for an airliner.[6]
- An inherent problem of tailless airplanes is how to achieve high lift and moment equilibrium simultaneously during take-off and landing. The low wing loading obviates the need for very efficient trailing edge flaps. Consequently, the AWA flies at a high angle of attack in low speed flight. Slats are applied to the outboard wing to increase the stalling angle of attack and to avoid uncontrollable pitch up during gusts or manoeuvres.
- Several tailless (military) airplanes developed in the past had formidable stability and control problems. The principles applied to the present AWA basically comply with the approach used on Northrop's B-49. Sweepback helps to place outboard control surfaces behind the centre of gravity (CG) but their lever arm is shorter than that of the conventional tail; hence, more effective controls are required. Pitch and roll control are provided by four elevators at the centre wing trailing edge and four elevons on each outboard wing. Vertical winglets with double-hinged rudders and drag producing split outboard elevons, known as drag ruddervons, provide directional stability and control.
- The allocation of engines close together above the centre wing trailing edge is an attractive feature of the AWA. This arrangement enhances take-off noise shielding, reduces noise in the cabin, increases passenger safety, and simplifies directional control after engine failure. However, the arrangement is vulnerable to non-contained engine failure since disintegration may damage other engines as well. This can make airworthiness certification a showstopper.

The present layout causes the empty CG and the useful load to be approximately midway in the primary structure whereas the power plant installation shifts it backward by at least 5% to

---

[6]Arguably, a flying wing with 96 m span can be parked inside an 80 × 80 m² box, although this argument may not convince airlines and airfield operators.

approximately 45% MAC. Since the aerodynamic centre is at approximately 30% MAC, the fully loaded airplane has a negative stability margin of 15% MAC, which is unacceptable even for an artificially stabilized aircraft. The overall concept of the AWA design in Figure 5.12 is therefore likely to be unfeasible which makes the estimation of L/D unrealistic. It is also worth noting that in level flight at 35 000 ft altitude the AWA achieves only 90% of its maximum L/D. The proposed AWA must be modified drastically to attain its full aerodynamic potential and it remains questionable whether the final result will be a useful design [32].

5.5.3   Blended Wing Body

The passenger cabin and/or cargo holds of a blended wing body (BWB) layout are integrated inside a lifting body shaped with a high thickness and low aspect ratio wing. The outboard wing sections together resemble a moderately loaded conventional wing. The high aerodynamic efficiency of the BWB is primarily due to the favourable ratio of total wetted area to cabin volume and floor area obtained by eliminating the horizontal tail. In order to put the BWB in the perspective of this chapter, a generic BWB design has been conceived according to the principles discussed in [32, 49] and other publications. The same values of useful load (volume), gross weight and cruise Mach number as used for the DWB and the AWA airplanes apply to the BWB. The top view of the layout drawing (Figure 5.13) depicts the arrangement of the passenger cabin and cargo hold. The bottom view shows how each half wing is approximated by four straight-tapered sections.

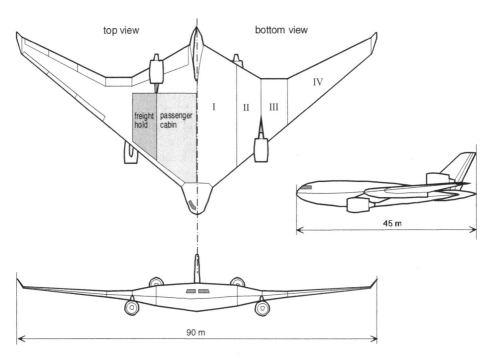

Figure 5.13   Sketch of a large airliner with blended wing body configuration

- The sections I together form the highly tapered central body containing the pressurized compartment, the landing gear and an optional underfloor fuel tank between the main gear bays. The aft spar constitutes the rear pressure bulkhead which is available for locating cabin evacuation doors. For a near-elliptical lift distribution along the span, the large central wing chord results in a low aerodynamic loading and the three-dimensional character of the flow helps to reduce the supervelocities on top of the body. This allows a thickness ratio to exceed that of a conventional wing by at least 2%. The available internal height allows the accommodation of a (partly) twin-deck cabin, leading to a favourable ratio of wetted to total cabin floor area. The inversely cambered central wing section is front loaded in high speed flight and has 17% thickness ratio.[7]
- Each section II contains a cargo hold and a main landing gear bay. The airfoil sections are basically symmetrical, tapering in thickness from 17% inboard to 15% outboard. The combined sections I and II of both sides have practically the same gross volume and planform area as the conventional baseline fuselage.
- The combination of both outboard sections III and IV forms an aspect ratio 6.70 wing containing 70% of the conventional baseline's tank volume. Thickness ratios are 15% at the root, 13% at the kink, and 11% at the tip. Rear-loaded supercritical sections are used for section IV.

The situation that the aircraft cannot be certificated in view of the risk that disintegration of one engine might damage other engines is avoided by using an unorthodox power plant arrangement which also leads to very small effects of thrust variation on longitudinal control. In the interest of avoiding significant trim drag, the plane is inherently unstable in pitch. Twist and camber distribution along the span aims at obtaining a near-elliptic lift distribution in high speed flight and simultaneously satisfies the condition of self-balancing – that is, trailing-edge controls are not deflected in the trimmed condition. The aim is to make trim drag in the cruise condition negligible which requires the balancing process to have meticulous control over the wing planform geometry and the allocation of systems and equipment. Variations in stability at high angles of attack must be carefully assessed to ensure that phenomena such as pitch up or tumbling will not occur [60]. In order to ensure recovery from high angle of attack regimes, the leading edge has a constant sweep angle and features outboard slats [36]. Moreover, a fuel tank in the nose for CG control in combination with artificial stabilization and flight envelope protection is provided to ensure safe low-speed flight.

Table 5.2 compares geometric data and aerodynamic performance of the BWB[8] with the DWB and AWA designs. Compared to the AWA, the span has been reduced from 96 m to a more realistic 90 m, the wing area is 16% smaller. The body-mounted vertical tail with 100 m^2 area features a double-hinged rudder to compensate for the small tail arm. Total wetted area is 21% less than for the conventional design. In spite of its modest span, the BWB has the highest maximum aerodynamic efficiency and it achieves 93.5% of this value cruising at the same altitude and speed as the conventional design. When cruising at a higher altitude, the BWB has the potential to achieve 24% gain in L/D relative to the conventional DWB.

---

[7]These characteristics were derived from published design projects. Since they are critical to the feasibility of a twin-deck BWB they must be verified by a CFD method in the conceptual design stage.
[8]Since the BWB differs considerably from the family of wing and body combinations in Figure 5.5, the parametric survey in Section 5.3 is not applicable to this configuration without essential modifications.

Table 5.2 Geometry and aerodynamic performance for five design concepts. Conditions: Mach 0.85 @ 35 000 ft except span loader: Mach 0.80 @ 40 000 ft

Design	Q_{tot} m^3	S m^2	S_{wet} m^2	$Q_{tot}^{2/3}/S_{wet}$	maximum L/D	C_L	L/D
Discrete wing & body	3 262	800	3 852	0.0571	20.6	0.47	19.7
All-wing aircraft	3 319	1 580	3 562	0.0625	25.7	0.24	23.2
Blended wing body	3 025	1 320	3 040	0.0688	26.1	0.28	24.4
Hybrid flying wing	3 230	1 050	3 445	0.0634	24.5	0.36	22.7
Span loader	6 700	1 995	4 800	0.0740	24.0	0.26	22.1

Its low wing loading makes the BWB more sensitive to gusts but gives it a better high-altitude buffet margin. Manoeuvre and gust load control are therefore desirable features of a BWB. From the aerodynamic performance point of view the BWB forms a promising alternative for the presently dominating general arrangement. However, due to its integrated character, the design process of a BWB is considerably more complex than usual.

5.5.4   Hybrid Flying Wing

The parametric survey in Section 5.3 shows that allocating more than 30% of the useful load volume in the wing can lead to significantly improved aerodynamic performance compared to both the conventional TAW and the AWA. An integrated wing body (IWB), named by TsAGI a hybrid flying wing (HFW), has a non-lifting body of reduced size (Figure 5.14). Inspired by this design, a generic HFW was conceived based on the same useful load volume, gross weight and cruise Mach number as used for the conventional DWB. The top view in Figure 5.15 depicts the general arrangement of the single deck passenger cabin. The bottom view shows that the wing geometry can be approximated by three straight tapered sections, giving the HFW the following characteristics.

- Sections I are both 20 m wide and 15% thick and form the inhabited parts of the wing. Compared with the BWB, section I has a smaller mean chord and its limited thickness does not allow a twin deck configuration.
- Baggage is carried in the belly under the fuselage cabin, main landing gear bays are in the fuselage as well as in wing section I. The same engine installation layout has been adopted as for the BWB.

Figure 5.14   TsAGI HFW airliner for 750 passengers with twin-deck fuselage. Reproduced from [32]. Copyright Elsevier 2001

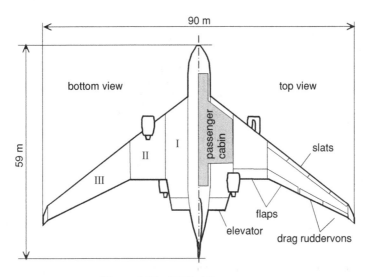

Figure 5.15   HFW airliner concept

- The combination of sections II and III is similar to an aspect ratio 7.20 conventional wing. It has thickness ratios 15% at the root, 12% at the kink and 10% at the tip.
- Since its wing loading is halfway between the DWB and AWA designs it is likely that the HFW needs trailing edge flaps for high lift and means to balance the pitching moment due to their deflection. Elevators at the trailing edge of section I and outboard elevons serve this purpose. An 80 m^2 vertical tail with a double-hinged rudder is mounted to the rear fuselage; directional control is augmented by ruddervons.

Compared to the BWB, the HFW has the same gross volume and wing span. Its gross wing area is 20% smaller, wetted area is 13% larger and maximum L/D is 6% lower. The HFW has a 19% higher maximum L/D than the DWB baseline but this gain would largely disappear if the DWB were given the same 90 m wing span. The HFW does not offer a higher L/D than the BWB, but the accuracy of the present estimation is too low for a firm conclusion. Moreover, the HFW has the following attractive properties:

- Payload capacity growth can be provided by the traditional expedient of lengthening the fuselage with cylindrical body plugs without the necessity of major wing redesign.
- The emergency evacuation problem is more easily solved, especially in the case of a single-deck configuration. In this respect, the evacuation of the twin-deck cabin is easier than for the much larger fuselage of a conventional aircraft.
- The cabin floor has no inclination in level flight.

5.5.5   Span Loader

Radical flying wing designs known as span loaders were advocated by NASA [65] and investigated by Boeing and Lockheed during the period 1975–1985. By distributing cargo inside the primary structure of a huge prismatic wing along its full span, the concept aimed

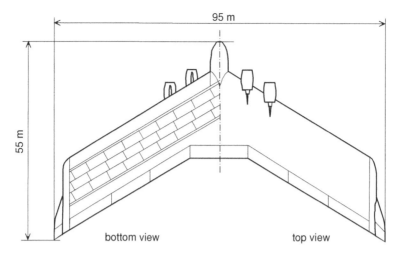

Figure 5.16 Sketch of a large cargo aircraft with span loader configuration

at the development of very large dedicated freighters for civil and military applications. All the cargo of a span loaded freighter can be distributed inside the wing so that a close match is obtained between the aerodynamic and the mass loadings. In such a non-tapered structure, the front and rear spars run parallel and the cargo hold can be loaded and unloaded through doors in the wing tips. The span loading concept is primarily aimed at counteracting the negative effects of the square cube law [62] by taking advantage of its positive effects. The net bending moments and shear forces on the wing structure due to manoeuvre and gust loads are thereby significantly reduced.

In spite of its modest Mach number, a span loader has a swept wing with about half the aerodynamic loading of a conventional freighter. Since it operates at low lift coefficients in all flight regimes, the span loader can take off with a relatively simple flap system – this circumvents the typical problem of trimming a flying wing.[9] Moreover, a span loader has a low geometric aspect ratio and the low aerodynamic loading and Mach number allow a high thickness ratio. Thickness ratios between 18 and 20% for transonic application based on Griffith and Goldschmied sections are proposed in [69, 70]. This unorthodox shape allows a large useful volume in the wing for a given planform area. The lift is concentrated more outboard than usual and the circulation distribution along the span is far from elliptical and induced drag is not ideal. However, due to its large tip chord, a span loader profits from winglets more than a tapered wing. Large winglets increase the aerodynamic aspect ratio considerably and provide directional stability and control.

Dedicated civil freighters will probably be sized to carry $8 \times 8 \times 20$ ft containers. This implies that a transonic span loader will have some 3.50 m thickness which requires a chord length of at least 20 m. The available internal volume is used efficiently with three or four rows of containers between the main spars. Figure 5.16 depicts a span loaded flying wing design featuring a capacity for 48 containers corresponding to a maximum payload of 250 metric tons.

---

[9]Hybrid span loaders have been proposed with a wing as well as a fuselage containing cargo using a foreplane for longitudinal control.

For a typical TOGW of 550 metric tons, the plane requires four engines with 350 kN take-off thrust each. It has a favourable ratio of wetted area to volume, whereas the geometric aspect ratio amounts to just 4.5 (Table 5.2). The aerodynamic efficiency is estimated by assuming conservatively that two 80 m^2 near-vertical winglets increase the aerodynamic aspect ratio by 20% to $A = 5.4$. The estimated maximum aerodynamic efficiency of 24.0 is similar to the values for the FW wing airliners described previously. At 40 000 ft altitude the lift coefficient is 0.26 and the actual L/D is less than those of the other FW designs. For a thickness ratio of 0.16 the maximum speed is limited to Mach 0.75. The favourable effect of span loading on empty weight is the dominant feature of this configuration and it is anticipated that the prismatic structure will be relatively easy to design and manufacture. A major challenge for designers of the span loader is to develop a satisfactory landing gear configuration.

5.6 Flying Wing Design

5.6.1 Hang-Ups or Showstopper?

This chapter has demonstrated that a jetliner with the cabin (partly) integrated in the wing leads to a significant increase of the maximum aerodynamic efficiency. However, the analysis has low fidelity and many other aspects have to be considered as well. For example, compared to the conventional TAW layout, an FW has more surface area that can be treated to obtain a laminar boundary layer. Flying wings therefore have the potential of achieving large surfaces with a laminar flow, leading to an even greater gain in aerodynamic efficiency compared to an all-turbulent TAW plane. Moreover, a wing with considerably more volume than usual has more structural stiffness which reduces the effects of aero-elasticity. On the other hand, formidable hurdles have to be fixed before an integrated wing body vehicle can be certified and offered to the airlines – some critical aspects are mentioned in Figure 5.17. The increasing size of airliners leads to major concerns such as taxiway and runway width limits, gate limits, passenger handling, community noise and aircraft separation to avoid wake vortices. Most of the operational problems are magnified for airliners with an integrated configuration since their accessibility for loading and unloading will be quite different from conventional airplanes. Provisions for a safe evacuation of 500 up to 1 000 passenger in less than 90 seconds in case

- Number, location and structure of emergency exits.
- Comfort in passenger cabin.
- Appreciation by passengers of a windowless cabin.
- Cabin floor inclination in high-speed and low-speed flight.
- Embarking and disembarking of passengers.
- Arrangement and accessability of cargo holds.
- Landing gear wheel base and track.
- Turning the plane on taxiways.
- Community noise and wake vortices.
- Possibility of the family concept: stretching and shrinking.
- Acceptance of the airplane layout by airlines.

Figure 5.17   Critical development aspects of a flying wing design

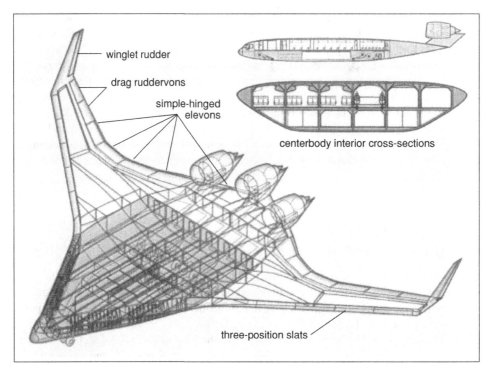

winglet rudder

drag ruddervons

simple-hinged
elevons

centerbody interior cross-sections

three-position slats

Figure 5.18   Structural concept and flight controls of the Boeing BWB-450 baseline. Adapted from [49]

of evacuation from a cabin with many aisles in crosswise arrangement and unusually long distances to exits may even appear to be an insurmountable problem. Stability and control problems inherent to flying wings have to be suppressed by advanced and complex electronic technologies. Application of the unorthodox flying wing concept can be effective only for the largest category of airliners, for which the design process is anyhow very complex.

5.6.2   Structural Design and Weight

A major challenge for designers of a FW structure is the integration of the pressure cabin. Boeing's BWB-450 project (Figure 5.18) features two main structural components: a central lifting body and two outboard wings. The forward part of the central body containing the cabin, flight deck and cargo hold features thick skin-stringer panels carrying the cabin pressure as well as the bending loads. Longitudinal internal ribs interconnect the upper and lower surfaces by carrying the pressure load in bending and tension. Fatigue becomes the design condition and it is anticipated that composite material such as CFRP and/or GLARE will be required for the majority of the pressure vessel. Since there is currently little experience of using highly non-circular pressurized structures that have been subjected to many thousands of flight cycles, the detailed design of the lifting central body requires a major technological development programme.

Table 5.3 Comparison of weight components and fractions for a 480-passenger BWB and a conventional configuration with composite structure [49]

Component	(Lifting) body	Outboard wings	Empennage or winglets	OEW	MTOW
weight ratio BWB/conventional	1.21	0.73	0.18	0.91	0.89
weight fraction, BWB	0.128	0.075	0.003	0.472	1
weight fraction, conventional	0.094	0.091	0.015	0.462	1

With the box-like pressurized passenger cabin, the central lifting body structure will be heavier than the conventional circular-section fuselage structure. Because of this necessary beefing-up, the crucial weight advantage due to bending load alleviation of a flying wing will be negated to some extent. The outboard BWB wings are smaller but structurally similar to that of a conventional transport and the choice of their structural material is a standard design procedure. According to [58], fabrication in aluminium will lead to a typical 20% weight penalty compared to composite wings. Table 5.3 presents a comparison of major weight component and fractions for a BWB project compared with a conventional design with the same payload/range performance. Although the numbers are sensitive to parameters such as wing loading and span, it is noteworthy that the total airframe structure weight and the empty weight fractions of both designs are nearly equal, with a slight advantage for the conventional design. Consequently, the primary effect on the gross weight reduction stems from the improved aerodynamic efficiency resulting in significantly reduced fuel expenditure and installed engine thrust.

5.6.3 The Flying Wing: Will It Fly?

The systematic case study in Section 5.3 has shown that the pure flying wing achieves its superior range performance potential only with a thick airfoil section, at a lower cruise speed and/or at a higher altitude compared to a conventional configuration. The aerodynamic efficiency of the presently dominant configuration cruising at Mach 0.85 is difficult to improve, in particular if its wing aspect ratio can be increased by up to 25%. This observation may not be valid if the useful load is distributed in a more sophisticated way than assumed in this parametric survey. In particular, an improved planform shape reduces the central wing's surface area for a given useful enclosed volume. This feature is applied in the BWB concept which improves the aerodynamic efficiency by at least 20% relative to the TAW. It has also been stated that the possibility exists to design a BWB for near-sonic speeds up to Mach 0.95 [52]. The HFW is a less radical concept with slightly lower aerodynamic performance but with several advantageous properties. The span loader concept is worth considering for a dedicated very large freighter. However, a comparison of merely aerodynamic performance is incomplete in the framework of a realistic design study.

When various aircraft concepts are compared, a difficulty arises in relation to their passenger capacity. This is related to the numerous seating accommodations, passenger-cargo and payload-range combinations that have to be available for large airliners. For example, the Boeing 747-400 has typical tri-class cabin for 416 seats and a twin-class cabin for 524 seats – its maximum structural payload varies from 67 to 123 metric tons for the passenger and freighter versions, depending on design range. A comparison between radically different general the arrangements can only be valid when the airplanes are designed for the same accommodation and payload-range performance, information which is of the utmost importance to airlines. Another complication is that radically different cabin layouts with the same seating capacity may be quite different as regards passenger comfort in terms of available cabin volume and floor area. For example, the outboard seats in a flying wing are likely to have a lower ceiling and less comfort than in a present-day wide-body airliner. And if a family of designs with different capacities is to be developed, the flying wing configuration with its integrated pressure cabin may be a less attractive concept.

Bibliography

[1] Maloney, E.T., *Northrop Flying Wings*, World War II Publications, Planes of Fame Publishers, Inc., Buena Park, CA, 1975.

[2] Woolridge, E.T., *Winged Wonders: The Story of the Flying Wing*, Smithsonion Institution Press, Washington, DC, 1993.

[3] Nickel, K., and M. Wohlfahrt, *Tailless Aircraft in Theory and Practice*, AIAA Education Series, American Institute of Aeronautics and Astronautics, Inc., Washington, DC, USA, 1994.

[4] Sweetman, Bill, *Inside the Stealth Bomber*, MBI Publishing, Osceola, WI, 1999.

[5] Russell, L., *Only the Wing: Reimar Horten's Epic Quest to Stabilize and Control the All-Wing Aircraft*, Smithsonian Institution Press, Washington, DC, 2011.

Flying Wing Controversy

[6] Ashkenas, I.L., "Range Performance of Turbojet Airplanes", *Journal of the Aeronautical Sciences*, Vol. 15, No. 2, pp. 97–101, February 1948.

[7] Foa, J.V., and I.L. Ashkenas, Discussion of [6], in Readers Forum, *Journal of the Aeronautical Sciences*, Vol. 16, No. 4, pp. 253–254, April 1949.

[8] Gepfert, K., "Northrop Claims AF Scuttled Flying Wing", *Los Angeles Times*, December 1980.

[9] Foa, J.V., "The Flying Wing Reconsidered", The George Washington University Technical Report, GWU-TR-83-FW-2, January 1983. Also: *Canadian Aeronautics and Space Journal*, Vol. 30, No. 1, pp. 62–65, March 1984.

[10] Schreiber, T.S., "Subsonic Airplane Configurations for Maximum Range or Endurance", AIAA Paper No. 83-2536, October 1983.

[11] Sears, W.R., "Flying Wing Could Stealthily Reappear", *Aerospace America*, pp. 16–19, July 1987.

[12] Foa, J.V., Letter to the Editor of *Defence Science*, pp. DS6–DS9, March 1990.

[13] Torenbeek, E., "Aerodynamic Performance of Wing-Body Configurations and the Flying Wing", SAE Technical Paper No. 911019, April 1991.

[14] Torvik, P.J., "On the Maximum Range of Flying Wings", AIAA Paper No. 92-4223, August 1992.

Flying Wing Design

[15] Northrop, J.K., "The Development of All-Wing Aircraft", *Journal of the Royal Aeronautical Society*, pp. 481–510, 1947.

[16] Lee, G.H., "The Possibilities of Cost Reduction with All-Wing Aircraft", *Journal of the Royal Aeronautical Society*, pp. 744–749, November 1965.

[17] Caddell, W.E., "On the Use of Aircraft Density in Preliminary Design", SAWE Paper No. 813, May 1969.

[18] Davies, S.D., "The History of the AVRO Vulcan", 14th Chadwick Memorial Lecture, 12 March 1970, *The Aeronautical Journal*, Vol. 74, May 1970.

[19] Andersen, F., "Northrop, An Aeronautical History", Northrop Corporation, 1976.

[20] Sears, W.R., "Flying Wing Airplanes: The XB-35/YB-49 Program", AIAA Paper No. 80-3036, March 1980.

[21] Begin, L., "The Northrop Flying Wing Prototypes", AIAA Paper No. 83-1047, March 1983.

[22] Volkhausen, R., "Historie und Zukunft von Flugzeugen mit mittragender Nutlastzelle", *DGLR Jahrbuch* 1983 II, Paper 83-127, October 1983.

[23] Ashkenas, I.L., and D.H. Clyde, "Tailless Aircraft Performance Improvement with Relaxed Static Stability", NASAz CR-181806, March 1989.

[24] Grellman, H.W., "B-2 Aerodynamic Design", AIAA Paper No. 90-1802, February 1990.

[25] Roskam, J., S.M. Malaek, and W. Anemaat, "AAA (Advanced Aircraft Analysis): A User-Friendly Approach to Preliminary Aircraft Design", ICAS Paper No. 90-2.10.2, Stockholm, September 1990.

[26] Roskam, J., "The B-2 Bomber: Why?; How?; Is it Affordable?", paper presented to the Netherlands Association of Aeronautical Engineers (NVvL) on July 5, 1991.

[27] Fremaux, C.M., D.M. Vario, and R.D. Whipple, "Effect of Geometry and Mass Distribution on Tumbling Characteristics of Flying Wings", *Journal of Aircraft*, Vol. 32, No. 2, p. 404, March–April 1995.

[28] Denisov, V.E., A.L. Bolsunovsky, N.P. Busoverya, B.I. Gurevich, and L.M. Shkadov, "Conceptual Design of Passenger Airplanes of Very Large Passenger Capacity in Flying Wing Layout", ICAS Paper No. 96-4.6.1. September 1996.

[29] Göksel, B., "Conceptual Study for an Advanced Transport Aircraft with Extreme Large Capacity", *Euroavia News*, pp. 7–11, December 1996.

[30] Denning, R.M., J.E. Allen, and F.W. Armstrong, "Future Large Aircraft Design: The Delta With Suction", *The Aeronautical Journal*, Vol. 101, pp. 187–198, May 1997.

[31] Wood, R.M., and S.X.S. Bauer, "Flying Wings/Flying Fuselages", AIAA Paper No. 2001-0311, 2001.

[32] Bolsunovsky, A.L., N.P. Buzoverya, B.I. Gurevich, V.E. Denisov, A.I. Dunaevsky, L.M. Shkadov, O,V. Sonin, A.Yu. Udzhukhu, and Yu,P. Zhurihin. "Flying Wing· Problems and Decisions", *Aircraft Design*, Vol. 4, No. 4, pp. 193–219, December 2001.

[33] Denning, R.M., J.E. Allen, and F.W. Armstrong, "The Broad Delta Airliner", itshape *The Aeronautical Journal*, pp. 547–558, September 2003.

[34] Dmitriev, V.G., L.M. Shkadov, V.E. Denisov, B.I. Gurevitch, S.V. Lyapunov, and O.V. Sonin, "The Flying Wing Concept: Chances and Risks", AIAA Paper No. 2003-2887, September 2003.

[35] Martínez-Val, R., and E. Peréz, "Medium Sized Flying Wings", *Innovative Configurations and Advanced Concepts for Future Civil Aircraft*, von Kármán Institute for Fluid Dynamics, Lecture Series 2005-06, June, 2005.

[36] Bolsunovsky, A.L., N.P. Buzoverya, B.I. Gurevich, V.E. Denisov, and O.V. Sonin, "Flying-Wing: Problems and Decisions", *Innovative Configurations and Advanced Concepts for Future Civil Aircraft*, von Kármán Institute for Fluid Dynamics, Lecture Series 2005-06, June, 2005.

Blended Wing Body

[37] Callaghan, J.T., and R.H. Liebeck, "Some Thoughts on the Design of Subsonic Transport Aircraft for the 21st Century", SAE Paper No. 901987, October 1990.

[38] Liebeck, R.H., M.A. Page, B.K. Rawdon, P.W. Scott, and R.A. Wright, "Concepts for Advanced Subsonic Transports", NASA CR 4624, September 1994.

[39] Liebeck, R.H., M.A. Page, and B.K. Rawdon, "Evolution of the Revolutionary Blended Wing Body Subsonic Transport", Transportation Beyond 2000: Technologies Needed for Engineering Design, NASA Conference Publication 10184, pp. 431–460, February 1996.

[40] M.A. Potsdam, M.A. Page, and R.H. Liebeck, "Blended wing body Analysis and Design", AIAA Paper No. 97-2317, June 1997.

[41] Liebeck, R.H., "Configuration Control Document CCD-3: Blended Wing Body", Final Report Under Contract NAS1-20275, NASA Langley Research Center, October 1997.

[42] Liebeck, R.H., M.A. Page, and B.K. Rawdon, "Blended wing body Subsonic Commercial Transport", AIAA Paper No. 98-0438, January 1998.

[43] Denisov, V.E., A.L. Bolunovsky, N.P. Buzoverya, and B.I. Gurevich, "Recent Investigations of the Very Large Passenger Blended Wing Body Aircraft", ICAS Paper No. 98-4.10.2, September 1998.

[44] Wakayama, S., and I.H. Kroo, "The Challenge and Promise of Blended Wing Body Optimization", AIAA Paper No. 98-4736, September 1998.

[45] Wakayama, S., "Multidisciplinary Optimization of the Blended Wing Body", AIAA Paper No. 98-4938, September 1998.

[46] Wakayama, S., "Blended Wing Body Optimization Problem Setup", AIAA Paper No. 2000-4740, September 2000.

[47] Smith, H., "College of Aeronautics BWB Development Programme", ICAS Paper No. 1.1.4, September 2000.

[48] Österheld, C., W. Heinze, and P. Horst, "Preliminary Design of a Blended Wing Body Configuration Using the Design Tool PrADO", CEAS Conference on Aircraft Design and Optimization, 25–26 June 2001.

[49] Liebeck, R.H., "Design of the Blended Wing Body Subsonic Transport", 2002 Wright Brothers Lecture, AIAA Paper No. 2002-0002, January 2002.

[50] Mukopadhyay, V., J. Sobieszczanski-Sobieski, I. Kosaka, G Quinn, and C. Charpentier, "Analysis and Optimization of Non-cylindrical Fuselage for Blended Wing Body (BWB) Vehicle", AIAA Paper No. 2002-5664, September 2002.

[51] Mialon, B, T. Fol, and C. Bonnand, "Aerodynamic Optimization of Subsonic Flying Wing Configurations", AIAA Paper No. 2002-2931, June 2002.

[52] Gilmore, R., S. Wakayama, and D. Roman, "Optimization of High-Subsonic Blended Wing Body Configurations", AIAA Paper No. 2002-5666, September 2002.

[53] Roman, D., R. Gilmore and S. Wakayama, "Aerodynamics of High-Subsonic Blended Wing Body Configurations", AIAA Paper No. 2003-554, January 2003.

[54] Ko, Y.-Y. A., Leifsson, L.T., J.A. Schetz, W.H. Mason, B. Grossman, and R.T. Halftka, "MDO of a Blended Wing Body Transport Aircraft with Distributed Propulsion", AIAA Paper No. 2003-6732, November 2003.

[55] Wittmann, R., "Passenger Acceptance of BWB Configurations", ICAS Paper No. 2.6.0, September 2004.

[56] Qin, N., A. Vavalle, A. le Moigne, M. Laban, K. Hackett, and P. Weinerfelt, "Aerodynamic Considerations of Blended Wing Body Aircraft", *Progress in Aerospace Sciences*, Vol. 40, No. 6, 2004.

[57] Coppinger, R., "BWB Needs Larger Control Surfaces", *Flight International*, p. 42, 13–19 July 2004.

[58] Liebeck, R.H., "Design of the Blended Wing Body Subsonic Transport", *Journal of Aircraft*, Vol. 41, pp. 10–25, January–February 2004.

[59] Qin, N., A. Vavalle, and A Le Moigne, "Spanwise Lift Distribution for Blended Wing Body Aircraft", *Journal of Aircraft*, Vol. 42, No. 2, pp. 356–365, March–April 2005.

[60] Chambers, J.R., "The Blended Wing Body: Changing the Paradigm", Innovation in Flight: Research of the NASA Langley Research Center on Revolutionary Advanced Concepts for Aeronautics, NASA SP 2005-4539, August 2005.

[61] Nickol, C.L., and L.A. McCullers, "Hybrid Wing Body Configuration System Study", AIAA Paper No. 2009-931, January 2009.

Span-Distributed Loading Cargo Aircraft

[62] Cleveland, F.A., "Size Effects in Conventional Aircraft Design", *Journal of Aircraft*, Vol. 7, No. 6, pp. 483–512, November–December 1970.

[63] Whitehead, A.H., "Preliminary Analysis of the Span-Distributed-Load Concept for Cargo Aircraft Design", NASA Technical Memorandum X-3319, December 1975.

[64] Whitehead, A.H., "The Promise of Air Cargo: Systems Aspects and Vehicle Design", NASA TM X 71981, July 1976.

[65] Whitlow, D.H., and P.C. Whitener, "Technical and Economic Assessment of Span-Distributed Loading Cargo Aircraft Concepts", NASA Contractor Report No. 144963, 1976. Also NASA CR-144962 (NASA), CR-145034 (Lockheed), and CR-145229 (Boeing).

[66] Jernell, L.S., "Preliminary Study of a Large Span-Distributed-Load Flying-Wing Cargo Airplane Concept", NASA Technical Paper TP-1158, 1978.

[67] Whitener, P.C., "Distributed Load Aircraft Concepts", *Journal of Aircraft*, Vol. 16, No. 2, pp. 72–77, February 1979.

[68] Toll, T.A., "Parametric Study of Variation in Cargo-Airplane Performance Related to Progression from Current to Spanloader Designs", NASA Technical Paper No. 1625, April 1980.

[69] Goldschmied, F.R., "Thick-Wing Spanloader All-Freighter: A Design Concept for Tomorrow's Air Cargo", AIAA Paper No. 90-3198, September 1990.

[70] Chaplin, H.R., "Application of Very Thick BLC Airfoils to a Flying Wing Type Transport Aircraft", SAE TP 901992, October 1990.

[71] Morris, S.J., and W.C. Sawyer, "Advanced Cargo Aircraft May Offer a Potential Renaissance in Freight Transportation", International Symposium "The Future of Cargo Transport Aircraft", Strasbourg, France, March 25–27, 1993.

6

Clean Sheet Design

Prize that which is best in the universe; and this is that which useth everything and ordereth everything.

—Marcus Aurelius (AD 121–180), *Meditations*, v. 21

6.1 Dominant and Radical Configurations

Designing the general arrangement of a clean-sheet passenger airplane is the most far-reaching of all conceptual design activities. Although a great variety of existing or feasible configurations may serve as examples, the superficial appearances of recently developed and projected civil planes are of such conformity that even experts find it difficult to distinguish between them. Should we therefore conclude that the aeronautical design world has arrived at the end of its creativity? And if this is not the case, why have alternatives options been rejected?

The history of aviation is not a continuous success story. Of the many attempts to move away from the configuration that was classical in its time only a few were successful, most have ended in failure. The reasons for the failures were not necessary poor technology background or a lack of prospective market understanding. Perhaps the most dramatic example was the Concorde supersonic (Mach 2) airliner which was a brilliant technical achievement of a good concept with an initially good order book. However, a mismatch with the world airliner demand, unforseen fuel price explosions and a long development time killed the project [6]. There are many more not so dramatic stories to tell and the advanced designer's time is well spent when getting acquainted with them. Some proposed mutations in the evolution of airplane concepts are reviewed in this chapter. In addition to radical departures from the tube and wing (TAW) configuration discussed in Chapter 5, attention is paid to several alternative concepts based on unusual combinations of fuselages and lifting systems.

6.1.1 Established Configurations

Conceptual design mostly entails a comparison of an established general arrangement with an unorthodox layout. The relative merits of different proposals are determined by completing the essential sizing process in accordance with the same technology standards and methodical

Advanced Aircraft Design: Conceptual Design, Analysis and Optimization of Subsonic Civil Airplanes, First Edition. Egbert Torenbeek.
© 2013 by Egbert Torenbeek. Published 2013 by John Wiley & Sons, Ltd.

ground rules. The following fundamental questions concerning the topography of major airplane components have to be answered in the first place.

(a) How should the useful load be distributed over the available internal space?
(b) Where to locate lifting surfaces – main wing, horizontal and vertical stabilizer(s) and controls?
(c) How many (propulsion) engines will be used and where will they be installed?
(d) How is the aircraft balanced during all likely loading and flight conditions?
(e) Where should the landing gears and on-board systems be installed?

The majority of modern passenger aircraft have been designed with discrete major components arranged according to the following principles of the TAW concept.

- Airline passengers, flight deck and cabin crew are allocated inside a pressure cabin. The walls of this vessel are formed by the fuselage shell, front and rear pressure bulkheads, wheel bays and a cover for the wing carry through structure. With the exception of a few light business jets, the basic shell is cylindrical with a (near-) circular cross-section. Non-pressurized nose and tail sections provide for the required streamline shape. Passenger cabin and freight holds together occupy between 60 and 80% of the gross fuselage volume. Most of the remaining volume is occupied by the flight deck, bays for electronics and landing gear retraction, wing centre section, auxiliary power unit (APU), and other systems. Long-range business jets may feature a fuel tank in the rear fuselage, outside the pressure cabin. As a rule, unused space is a small fraction of the total fuselage volume.
- The high aspect ratio wing is the main lifting surface located above or below the pressure cabin. The primary structure between the front and rear spars – also known as the wing box – occupies up to 55% of the gross wing volume. A large proportion of the wing box outboard of the wing-to-fuselage attachment contains fuel tanks. Long-range aircraft may also feature fuel tanks inside the central carry through structure. The secondary structure to the fore and aft of the wing box consists of fixed leading and trailing edge structures to which (movable) high-lift devices and flight control surfaces are attached.
- The turbofans of passenger aircraft are installed in nacelles which are attached by means of pylons to the wing box or to the rear fuselage structure. Triple engine aircraft feature a central engine inside or above the rear fuselage. Propeller aircraft have turboprop engines with tractor propellers mostly attached to the fore of the wing box.
- With a total area between 30 and 45% of the wing area, the empennage can be quite large. The vertical tail plane is nearly always attached to the crown of the rear fuselage. The horizontal plane is attached to the rear fuselage or to the vertical fin. Although the empennage has a relatively small volume, some airliners feature fuel tanks inside it for centre of gravity (CG) control.

An important property of the presently dominating configuration is that the areas exposed to the flow of wing and fuselage are of similar magnitude, whereas the useful wing volume is not more than 20% of the fuselage volume.

Since the basic characteristics of the established general arrangements are known in principle, its selection offers the advanced design engineer a safe approach yet still allows some leeway to find a competitive design. However, new technologies and conflicting or extreme design requirements tend to ask for concepts different from the classical ones. For example, new generations of turbofans have ever increasing bypass ratios and fan diameters, whereas the open rotor engine may eventually appear as the ultimate solution for fuel efficient propulsion based on the Brayton cycle. However, with increasing engine sizes, the fuel burn reduction is reversed by installation penalties and adverse conditions for the classical low-wing mounted engine pods. Open rotor engines mounted to a high wing or rear fuselage may offer good alternatives with noise shielding properties; see Chapter 3.

6.1.2 New Paradigms

During the 1980s there was considerable interest in unusual concepts for application to business aircraft and small airliners such as the canard configuration with pusher propellers, the joined wing and biplane systems. But many initially promising projects were eventually cancelled. It is likely that some crucial development risks could have been avoided by making an advance inventory of factors that might appear crucial for certification – typical design failures are documented in [9]. Although a few innovative designs were actually developed and taken into production, they did not trigger a revolution in business aviation. Toward the end of the 1980s, many experts became concerned that traditional airliner configurations might no longer offer the best solution to a significant improvement in energy efficiency and operating costs. This stimulated a worldwide generation of research and technology programmes aimed at radical departures from the established layouts, creating possibilities for developing more economical and environmentally friendly vehicles. In particular, design studies on flying wing (FW) configurations with large seating capacity were carried out by the Russian institute TsAGI. At the instigation of NASA, McDonnell Douglas started work on the revolutionary blended wing body (BWB) concept (Figure 5.3) continued by Boeing after the merger of the two companies. These innovative concepts promised considerable (or even spectacular) performance gains and were convincing enough so that funds for in-depth research and applied design studies could be raised.

Investigation of the BWB configuration has become an on-going technology programme carried out in cooperation between airframe manufacturers, research institutes and university teams. The majority of the investigators came to the conclusion that the BWB concept has considerable potential for large long-range airliners. Several pre-competitive investigations of new civil aircraft concepts have been subsidized by the European Union. Alternative concepts have been studied which deserve consideration for specific applications, such as the transonic boxplane, the joined wing and the C-wing. However, up to the present day none of these concepts has become the starting point of a for-sale aircraft development programme. After all, in an industrial environment novel concepts are only justified when they target a real need in an economically viable manner. Therefore, from the beginning of the design cycle, the design problem must be approached in a cooperative effort between designers, manufacturers and operators. Once a design concept and a requirement match, a final decision can be taken to embrace the novel concept.

6.2 Morphology of Shapes

A scientist discovers that which exists. An engineer creates that which never was

—Theodore von Kármán

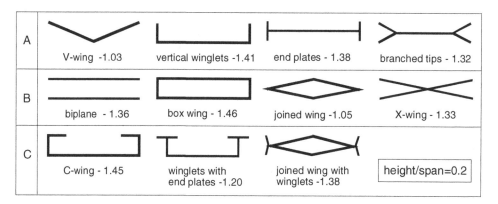

Figure 6.1 Front view of lifting systems and their theoretical span efficiency factors for optimal lift distribution, based on [72, 76, 89]

6.2.1 Classification

Numerous attempts have been made over the past decennia to challenge the classical wing and tail and the TAW design paradigms. In view of the huge variety of existing and imaginable airplane concepts, an effort to make a systematic and complete inventory is doomed to fail. Instead, we will discuss matrices of schematic layouts in front and top views. All concepts are intended to ensure: (1) a favourable allocation of the useful load inside the airplane hull; (2) trimmed and (preferably inherently) stable flight; and (3) a healthy and light structural concept. It is worth noting that, from the aerodynamic point of view, a schematic front view as in Figure 6.1 is most relevant to vortex drag since it depends primarily on the horizontal (span) and the vertical extent of the lifting system. Plan views as depicted in Figure 6.2 are closely associated with the plane's exposed area and skin friction drag. If the front view includes the fuselage, the ratio of frontal to plan view area forms an indication of form drag, mostly a relatively small component.

6.2.2 Lifting Systems

In designing the airplane's general arrangement, the configuration of its lifting system is the most essential feature. This system must ensure that sufficient lift is generated in such a way that the airplane can be trimmed longitudinally in all flight conditions. The traditional solution is a monoplane combining a planar wing and a (horizontal) tailplane. However, many alternative concepts have been used in the past – in particular, the biplane was popular until the 1930s. Although the cantilever monoplane of metal construction became the standard for high-speed flight, continuous attempts have been made to develop more efficient lifting systems based on non-planar wings, biplanes and even triplanes. Multiplanes are mostly non-coplanar, that is, the lifting surfaces are vertically displaced relative to each other. Obviously, it is important how this vertical gap affects parasite and induced drag components and structural weight. Many investigations of the benefits of non-(co)planar lifting systems were made during the second half of the twentieth century and they are still continuing. Biplanes distribute the lift

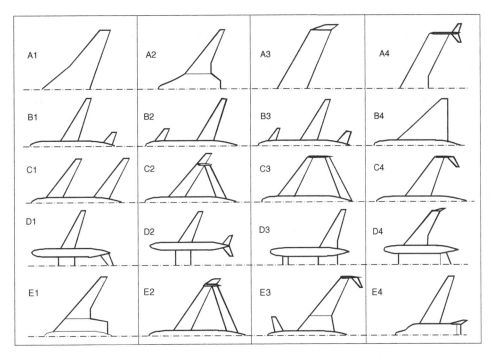

Figure 6.2   General arrangement spectrum for civil jet airplanes

vertically as well as horizontally and thereby reduce vortex drag, whereas certain concepts may also lead to a significant reduction of wing structure weight.

The classification in Figure 6.1 compares the span efficiency factor for several lifting systems of equal span having a vertical lifting extent of 20% span. Most concepts shown have a higher interference drag compared to an elliptically loaded planar monoplane.

- Category A – Non-planar monoplanes.
  In spite of its large dihedral, the V-wing reduces vortex drag by a few percent – vertical winglets and lifting end plates bring about higher efficiencies. Many intermediate shapes between large dihedral and winglets can be obtained. For instance, a wing with branched tips can achieve a high span efficiency due the large vertical distance between the tips [76]. Other related concepts (not shown) are outward canted winglets, wings cambered along the span and gull wings.
- Category B – Biplanes and boxplanes.
  Biplanes and boxplanes are classical aerodynamic concepts for reducing vortex drag (Section 4.3). A highly staggered biplane with wings of approximately equal span is known as a tandem wing (not shown). The rectangle boxplane is a closed system of lifting surfaces which forms the ultimate goal of achieving minimum induced drag for given wing span. Its application for high subsonic speed is known as the transonic boxplane. A joined wing has a modest span efficiency factor. It can be considered a boxplane without the vertical elements at the tips. An essential feature of the joined wing is its self-bracing structural

concept enabled by the diamond shape in front view as well as in plan view. The X-wing is an exotic concept that should not be confused with an X-shaped system of wings in plan view.

• Category C – Hybrid plane systems.
A C-wing has large vertical fins at its tips carrying inward-pointing secondary winglets on top. This yields a span efficiency between a wing with vertical winglets and a boxplane. The wing with end-plated vertical winglets is effectively a monoplane with tip-mounted T-tail empennages. Its span efficiency is lower than that of the C-wing but this is achieved with a main wing of shorter span. The third concept indicates that the efficiency of a joined wing can be considerably improved by symmetric inclined winglets.

When some of the radical concepts shown are adopted for a new design concept, it is probably found that the design is penalized by a significant parasite drag increment and considerable structural and aero-elastic complications.

6.2.3   Plan View Classification

Figure 6.2 classifies airplane general arrangements on their lifting system planform and allocation of the useful load.

• Category A – Flying wings; see Chapter 5.
A1 – All-wing aircraft (AWA) or pure flying wing: carries all the useful load in the wing and lacks a fuselage as well as a tail- or a foreplane. Ideally, the engines are buried inside the wing.
A2 – Blended wing body (BWB): considered as a flying wing since its body functions as a lifting surface.
A3 – Span loader: large freighter with tip-loaded cargo all along the span of a constant-chord wing with (near-)vertical winglets.
A4 – Span loader with end-plated vertical fin at each tip.
• Category B – Planar monoplane, single body.
B1 – Wing and tail (WAT) configuration: the general arrangement of the majority of civil airplanes. Alternative empennage: V-tail.
B2 – Canard configuration: or tail-first aircraft: uses a foreplane for trimming, stability and control. Existing canard aircraft have wing-mounted pusher propellers.
B3 – Three-surface aircraft: (or triplane) has the usual WAT layout augmented by a foreplane.
B4 – Tailless aircraft: lacks a horizontal stabilizer but does have a vertical tail.
• Category C – Non(co)planar lifting system, single body.
C1 – Tandem wing: highly staggered biplane with two wings of equal span generating approximately the same amount of lift.
C2 – Joined wing: closed lifting system with a diamond-like shape in planform as well as front view. The aft-swept front wing with upward V-shape and the forward-swept rear wing with downward V-shape are jointed at or near the tips.
C3 – Boxplane: closed biplane lifting system with wings of equal span and tips connected by vertical fins. A transonic boxplane has an aft-swept front wing and a forward-swept rear wing.

C4 – C-wing: features vertical fin with inboard pointing winglet at each wing tip. These winglets act as a horizontal stabilizer and pitch control device.
- Category D – Planar monoplane, twin- or multi-body. Examples are twin-fuselage aircraft with
  D1 – single horizontal tail,
  D2 – duplicated empennages,
  D3 – foreplane between the fuselages,
  D4 – cabin integrated in two fuselages and the wing centre section.
- Category E – Hybrid configurations combining features of the foregoing concepts.
  E1 – Hybrid flying wing (HFW): integrated cabin inside the body and the wing.
  E2 – Joined wing with symmetric inclined winglets.
  E3 – C-wing combining the characteristics of the HFW, the C-wing and the three-surface layout.
  E4 – Tailless aircraft in lifting fuselage layout featuring elevators at its trailing edge [115].

Figure 6.2 does not show the rhomboidal wing which consists of dual coplanar lifting surfaces of equal span but opposite sweep and horizontal plates connecting their tips – in top view this layout looks like a joined wing. Also omitted is a wing with boom-mounted outboard horizontal stabilizers replacing the empennage of the traditional WAT layout, intended primarily to improve the aerodynamic performance and accessibility for loading and unloading of a small freighter or utility aircraft [24].

6.2.4   Strut-Braced Wings

For many years, strut-braced wings (SBWs) were not unusual on low-speed short-range transport aircraft such as the Short Skyvan. In the early 1950s the French company Hurel Dubois developed several aircraft with very high aspect ratio strut-braced wings, whereas W. Pfenninger at Northrop advocated the idea of using a truss-braced wing for very long-range transonic flight. Around 1980, studies of very large military transports with strutted wings were reported by NASA, Lockheed and Boeing. Reference [48] describes more recent NASA-sponsored work done at Virginia Tech which was followed by an industry/university study reported in [50]. Advantages claimed for the SBW embrace several synergistic design features. The strut provides bending load alleviation to the wing and the structural efficiency of the truss compared to a cantilever structure allows an increased span and/or decreased thickness and sweep without an increase in wing weight. Consequently, parasite drag and/or induced are reduced whereas wing sweep may be reduced in order to promote natural laminar flow. The reduced wing thickness may alternatively be used to increase flight Mach number. The reduced gross weight in combination with increased aerodynamic efficiency permits the installation of smaller engines. Studies carried out at Virginia Tech [51] point to appreciable weight reductions. For a wide body airliner of the Boeing 777 category, up to 9% MTOW reduction, fuel burn savings between 11 and 16% and more than 20% reduced installed thrust relative to a cantilever wing design are quoted. Optimum SBW aspect ratios varied between 9.7 and 12.2 and aero-elastic problems are not foreseen for a composite wing.

The introduction of an SBW design does not necessarily require a radical design paradigm nor a costly R&D programme and Figure 6.3 illustrates that the overall impression is not

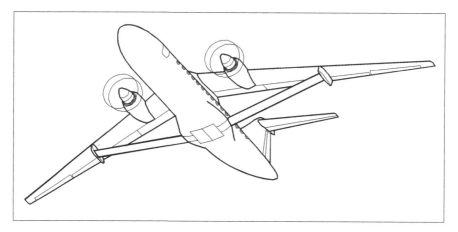

Figure 6.3   Concept of a regional aircraft with strut-braced wing and CRPF engines

particularly unaesthetic. However, the following key issues are inherent to the presence of struts.

- The airplane is bound to be a high wing configuration, which may be acceptable for a freighter or a commuter aircraft but objectionable for a transonic airliner.
- The presence of the strut may dictate an unfavoured engine location. For example, turbofans may have to be installed outboard of the strut-to-wing connection, over the wing or at the rear fuselage sides.
- Strut bracing increases the number of primary structural parts carrying loads comparable to the wing leading to more possible failure modes. A mechanism must be found to relieve the compression load on the strut under negative loads. This leads to additional development risks and increased manufacturing costs.
- Struts add to the number of trailing edges over which the boundary layer passes and may thus contribute to increased airframe noise generation. However, in the case of wing-mounted open rotor engines the noise level will probably not be dominated by airframe noise.

6.2.5   Propulsion and Concept Integration

A realistic attempt to investigate an innovative concept includes consideration of airframe-engine integration – this introduces another degree of freedom not covered by Figure 6.2. Finding a suitable place to install the engines can have a decisive effect on the feasibility of a novel configuration. The following text is devoted to major aspects that need to be considered when an initially favoured general arrangement is developed into a practical design. Instead of covering a comprehensive variety of complete airplane layouts we will discuss a selection of basically sound concepts deserving attention for application in the foreseeable future. This selection does not cover most of the aggressive synergistic airframe-propulsion interactions and integration concepts proposed in [31].

In a typical advanced design environment, the characteristics of an unusual configuration must be analyzed by methods based on first principles rather than empirical and statistics-based methods. Continuing advances in computational capabilities in the major aeronautical disciplines combined with MDO technology enable advanced designers to go into considerable detail before embarking on a detail design development. Since the desirability of a novel concept depends primarily on the airplane size and critical requirements such as cruise Mach number, range or take-off field length, definite conclusions will not be drawn and recommendations about specific concepts will not be made in the following text.

6.3 Wing and Tail Configurations

The traditional layout of a plane with aft-mounted tail has established itself as a sound approach to ensure good flying and operational qualities. Its design process is usually straightforward, converging in a well-balanced baseline having the flexibility to generate derivative designs with a high degree of commonality such as (nearly) identical tailplanes. However, installing an aft tail entails penalties of roughly 10% parasite drag and 3% empty weight. Since the airplane's general arrangement has a great impact on the required tail size, design characteristics that are most influential on the minimum required horizontal tail area have to be carefully selected.

Horizontal tail sizing is an essential aspect of the aircraft balancing process involving the arrangement of major components – wing, body, empennage, landing gears, power plant and other system components – relative to each other so that the centre of gravity (CG) is at all times within limits. The loading diagram of a given aircraft specifies CG limits versus the GW for each useful load in various flight phases – en route, landing/take-off, gear up/down. Most relevant are aerodynamic limits derived from stability and control requirements and ground load limits derived from the strength of landing gear components depending on the location of the undercarriage legs. A typical gear load limit is the maximum ground force acting on the nose gear while taxying, dependent on the GW and the CG location.

6.3.1 Aerodynamic Limits

Although the minimum required tail size is affected by many factors which are outside the conceptual designer's control, the following derivation enables an initial assessment of major effects associated with the airplane's general arrangement. The required tail area is derived from a schematic model – see Figure 6.4 – of a 'wing' (index w) in combination with the horizontal tail (index h). The 'wing' represents the *airplane less horizontal tail*, including the fuselage, engine nacelles and other bodies exposed to the flow. The X-axis coincides with the wing's mean aerodynamic chord (MAC) with length $\bar{c}$ and has its origin at the leading edge (LEMAC). The Y-axis is not depicted since no consideration is given of the effects of the vertical placement of components relative to each other. The lift L_w on the tail-off aircraft is transferred to the aerodynamic centre (AC) by introducing the pitching moment M_{ac}. The tail lift is denoted L_h, its pitching moment is neglected. The GW acts in the CG at a distance x_{cg} behind the LEMAC. Figure 6.4 shows the aerodynamic forces and the pitching moment – definitions are in positive direction – for the trimmed condition as well as lift increments due to an angle of attack disturbance $d\alpha$. Although lift contributions act normal to the local flow, their pitching moment contributions are computed as if they act normal to the X-axis.

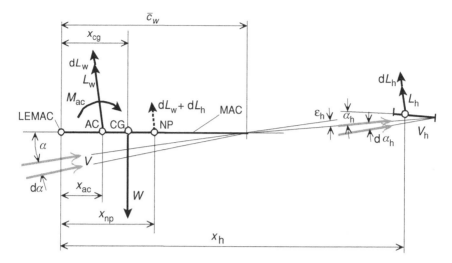

Figure 6.4 Schematic model for deriving stable trimmed flight conditions

In order to illustrate the process of balancing the airplane in combination with horizontal tail design, we consider two basic limits associated with longitudinal flying qualities.

1. Static equilibrium in steady level flight is determined from the pitching moment about the AC,

$$M = M_{ac} + W(x_{cg} - x_{ac}) - L_h(x_h - x_{ac}) \quad \text{for} \quad W = (L_w + L_h) \tag{6.1}$$

Setting $M = 0$ leads to the trim capability required expressed in terms of the horizontal tail volume coefficient

$$C_{L_h} \frac{S_h \, l_h}{S_w \, \bar{c}_w} \frac{q_h}{q} = C_{M_{ac}} + C_L \frac{x_{cg} - x_{ac}}{\bar{c}_w} \tag{6.2}$$

where q_h/q accounts for the dynamic pressure loss at the tail immersed in the airplane's wake. The tail moment arm $l_h = x_h - x_{ac}$ is assumed to be equal to the distance measured along the X-axis between the mean quarter-chord points of the wing and the tail. The aircraft with deflected high-lift devices has a large nose-down (negative) $C_{M_{ac}}$ requiring the tail to generate a download ($L_h < 0$) whereas the AC is backward from the en-route configuration. This is usually the dominant condition for the forward CG limit.

2. If the airplane experiences a disturbance of the trimmed flight caused by an increment or decrement of the angle of attack $d\alpha$, the wing and tail lift increase or decrease by dL_w and dL_h, respectively. The resulting lift change acts in the neutral point (NP); its distance behind the AC derived from the moment equation amounts to $l_h \, dL_h/dL_{w+h}$. For positive

stability the CG must be in front of the NP. Sufficient inherent stability can be imposed by prescribing a (positive, absolute) stability margin Δx_{sm} by locating the CG location at

$$x_{cg} = x_{np} - \Delta x_{sm} = x_{np} + \bar{c}_w \, dC_M / dC_L \qquad (6.3)$$

Substitution in the differential of Equation (6.1) results in the required stabilizing tail contribution versus the CG location

$$C_{L\alpha_h}(1 - d\epsilon_h/d\alpha)\frac{S_h \, l_h}{S \, \bar{c}_w}\frac{q_h}{q} = C_{L\alpha_{w+h}}\frac{x_{cg} - x_{ac} + \Delta x_{sm}}{\bar{c}_w} \qquad (6.4)$$

where $C_{L\alpha}$ denotes the lift curve slope and $d\epsilon_h/d\alpha$ is the downwash gradient at the tail. The tail sizing process of large transonic airliners is complicated by the effects of compressibility and aero-elasticity. High-speed flight is, in general, the dominant condition for the aft CG limit.

6.3.2   The Balanced Design

A typical tail sizing procedure of an aft-tail passenger plane with powered controls is depicted in Figure 6.5. This example is simplified since in reality more than two CG limits have to be considered.

- A crucial forward CG limit is determined by the ability to stall the aircraft with high-lift devices deployed for landing. In this condition the CG is ahead of the tail-off lift vector;

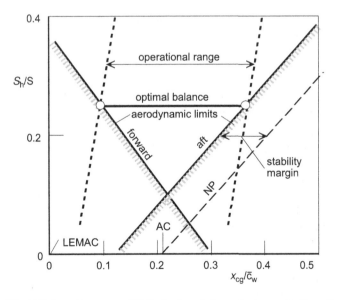

Figure 6.5   Tailplane size versus CG limits for a balanced wing and tail configuration

hence, moment equilibrium requires a download on the tail. Sufficient elevator power must be available for manoeuvring the airplane back to normal flight. Another essential control requirement is the take-off rotation for attaining the minimum unstick speed [121]. The tail download required is then sensitive to the main landing gear location.

- The aft CG limit of an inherently stable airplane is mostly determined by the minimum acceptable longitudinal stability margin in the en-route configuration.
- By increasing the horizontal tail area, the forward and aft CG aerodynamic limits move forward and backward, respectively, thereby increasing the available CG range. The aerodynamic limits have to be matched in an optimum fashion to the CG travel required for varying operational loading conditions.

Figure 6.5 depicts a diagram known as a scissor plot or X-plot. It is obtained by computing the aerodynamic and the operational CG limits as functions of the tail area. For an unbalanced aircraft the intersections of both forward and aft CG limits yield different tail areas so that either the forward or the aft limit dominates. The design is balanced by moving the wing forward or backward along the fuselage until the operational CG travel fits neatly between the aerodynamic limits.[1] The minimum tail size is assumed to represent the best overall design – more refined optimum conditions are considered in [54].

The minimum tail size is found analytically by subtracting the CG limits according to Equations (6.2) and (6.4) and setting the difference equal to the operational range Δx_{cg}. If between low and high speed flight q_h/q is unaltered whereas the AC location changes by a distance Δx_{ac}, the required tail volume coefficient is determined by

$$\frac{S_h\,l_h}{S_w\,\bar{c}_w} = \frac{-C_{M_{ac}}/C_{L_{max}} + (\Delta x_{cg} + \Delta x_{ac} + \Delta x_{sm})/\bar{c}_w}{\left\{(1 - d\epsilon_h/d\alpha)\,C_{L\alpha_h}/C_{L\alpha_{w+h}} - C_{L_h}/C_{L_{max}}\right\}q_h/q} \tag{6.5}$$

where $C_{M_{ac}}$ and $C_{L_{max}}$ refer to the stall with landing flap setting.

6.3.3   Evaluation

Civil jet aircraft design concepts with inherent stability are compared by applying Equation (6.5) with the following typical input:

- Dynamic pressure ratio: $q_h/q = 0.85$ for low tail, 0.95 for T-tail.
- Downwash gradient: $d\epsilon_h/d\alpha = 0.40$ for low tail, 0.30 for T-tail.
- Airplane lift gradient: $C_{L\alpha_{w+h}} = 1.10\,C_{L\alpha_w}$.
- Minimum stability margin: $\Delta x_{sm} = 0.02\,l_h$.
- Variation of the AC location: $\Delta x_{ac} = 0.05$.
- Nose-down pitching moment, flaps/slats down: $C_{M_{ac}} = -0.3\,\Delta C_{L_{\alpha=0}}$.
- Maximum tail download: $C_{L_h} = -0.80$.

---

[1] The balancing procedure is described in detail in [1]. The author is grateful to R. Slingerland for pointing to a mistake in Figure 9-19 of that publication where addition of the horizontal tail weight moves the airframe CG forward instead of backward.

The area ratio S_h/S of existing jet airliners varies between between 0.15 and 0.35; the variation in the tail volume coefficient is smaller but still significant. The statistically determined expression Equation (A.7) demonstrates that the required tail size depends not only on the wing size but is equally dependent on fuselage geometry. Equation (6.5) also explains that much of the variation depends on $l_h/\overline{c}$. The tail arm is increased considerably if the engines are located to the fore of the wing leading edge and by attaching the stabilizer to the tip of a sweptback fin – the T-tail. Furthermore, the operational CG variation is small for aircraft with the cabin centroid close to the empty airplane's CG, which is the case for wing-mounted engines. Aft-fuselage mounted engines lead to a larger CG travel and may suffer from high trim drag due to the tail download, especially in case of a full passenger cabin. Another dominant parameter is the vertical position of the stabilizer. A high-set tail experiences little dynamic pressure loss and the local downwash gradient is less than for a low tail. Since the large CG travel of an aircraft with engines attached to the rear fuselage is partly compensated by the adoption of a T-tail, its tail area ratio is about average, unless its size must be increased to cope with the deep stall problem. High-α pitch control of several business jets with T-tail has been improved by mounting two ventral strakes in an inverted V arrangement below the rear fuselage.

Summarizing, a general arrangement with wing-mounted engines in combination with a high-set horizontal tail seems hard to beat when minimum tail size is the figure of merit. For example, the tail area ratio of the Boeing 747-100 (with low wing and tail) was 1.7 times the value for the Lockheed C-5A (with high wing and tail) and its volume coefficient was 40% larger.[2] It is therefore remarkable that a T-tail has not been adopted on any airliner with engines attached to a low-set wing.

6.3.4   Relaxed Inherent Stability

The traditional airplane concept can be criticized for its property that in most flight conditions the tailplane experiences a download causing trim drag and reduced maximum lift. These penalties are reduced by shifting the CG further backward than usual, for instance, by means of a trim tank in the empennage. The associated reduction of the stability margin can be accepted when the plane is equipped with a stability augmentation system (SAS). This reduces the tail download in high-speed flight, whence it follows that the main wing carries all the lift with low induced drag due to its large span. If such a balancing process is based on the application of relaxed stability, the wing is shifted forward relative to the fuselage and the tail moment arm can be increased. Equation (6.5) demonstrates that this results in a reduced tail size. Consequently, relaxed stability leads to a worthwhile reduction in weight and drag, albeit at the cost of increased system complexity.

6.4   Aircraft Featuring a Foreplane

During the 1980s, several designers of business aircraft as well as NASA and academic researchers investigated alternative solutions to avoid some of the disadvantages associated with the aft-tail layout. Many of these were inspired by the innovative general aviation designs

---

[2]The C-5A has a longitudinal stability augmentation system.

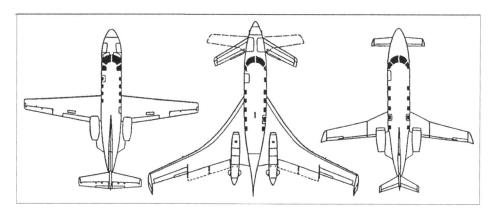

Figure 6.6 Light business aircraft designs with similar passenger cabin and different arrangement of the lifting surface system: wing and tail (left), canard configuration (middle), and three-surface layout (right)

produced during the 1970s by the American designer Burt Rutan [3]. This resulted in the canard and the three-surface configuration, both featuring a foreplane (Figure 6.6).

6.4.1   Canard Configuration

In view of their successful application to fighter aircraft it is not surprising that the canard configuration has its proponents in the civil design world as well. The basic argument quoted mostly goes as follows: The canard layout operates with positive foreplane lift, offering a remedy for the 'disease' of a down-loaded tailplane. However, it should be noted that fighters with a canard layout are short coupled, with the foreplane just in front of and above the wing in order to exploit the favourable aerodynamic interference between the two lifting surfaces at large angles of attack. On the other hand, a passenger plane in canard layout generally has its foreplane attached to the fuselage nose in combination with an aft-located main wing such that interference between the two lifting surfaces is minimized. Actually, the main drive behind the canard in pusher general arrangement is improving cabin comfort. This concept, known as a long-coupled canard, has benefits that are largely ascribed to its layout which allows (a) the wing to be located below the rear end or aft of the passenger cabin with (b) the engines to be placed in a pusher arrangement above or behind the wing's trailing edge. As a side effect, the foreplane trims the aircraft with positive lift, in particular when the wing flaps are deployed; hence, for a given total lift, the main wing can be smaller.

The scissor plot for balancing a canard configuration is depicted in Figure 6.7. With increasing foreplane area S_c, both CG limits move forwards proportional to S_c but the forward limit is more sensitive to foreplane size than the aft limit. The scissor plot demonstrates that the design tends to have a rather large foreplane and a small usable CG shift ahead of the LEMAC. Consequently, designers of a canard aircraft have to cope with several contradictory conditions.

- Different from a tailplane, the foreplane does not operate in the wing downwash which makes it an effective de-stabilizer. This problem can be countered by adopting a small, low

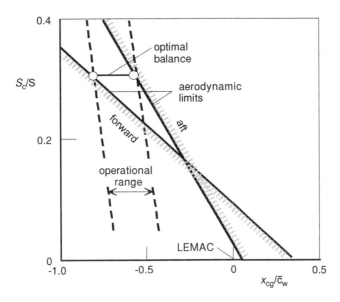

Figure 6.7   Foreplane size versus CG limits for a balanced canard configuration

aspect ratio, sweptback foreplane with a small horizontal stagger between the two lifting surfaces.

- The inevitable induced drag increment due to foreplane lift (Section 4.3) can be reduced by a selecting a vertical gap and large stagger between the two lifting surfaces and by giving the foreplane a large span.

- The foreplane's trailing vortex sheet affects the flow around the main wing in such a way that its lift distribution is far from elliptical. Interference effects are reduced in high-speed flight when a negative vertical gap is selected so that the foreplane's trailing vortex sheet passes below the wing. Since the interference between the two lifting surfaces is complex, its prediction should be validated by wind tunnel experiments.

- Trailing edge (main) wing flaps of a canard configuration tend to be far aft of the CG and their extension causes a large nose-down pitching moment. Consequently, the foreplane must generate a large positive lift for trimming. To obtain the desirable nose-down tendency in the stall, the foreplane must dump some of its lift at a lower angle of attack than the main wing without preventing the wing from achieving its maximum lift potential. Recovery from the stall requires the elevator to stay effective with separated foreplane flow.

- A large CG shift due to burning fuel will occur if all the fuel tanks are inside the primary wing box.

- The vertical fin moment arm tends to be small and the long fuselage is strongly destabilizing directionally.

Different approaches can be taken to resolve the conflicts between these desiderata and requirements with the usual result of an increase in drag, weight and/or complexity. In order to avoid the need for a very large foreplane, the canard concept requires a sizing procedure in close harmony with wing size, leaving no margin for last-minute tailoring if a problem with flying qualities occurs. The CG shift due to burning fuel can be reduced at the cost of extra parasite drag and empty weight by installing a tank inside a forward wing extension at the root. An alternative would be to accept the complication of a trim tank in the fuselage nose or in the foreplane.

This survey leads to the conclusion that the pure canard configuration will not lead to a viable concept for airliner application. However, the Beechcraft Starship (Figure 6.6) has demonstrated that a business airplane with canard layout can be designed, certified and sold to customers. The design was accomplished after flight testing an 85% scale model developed by Rutan which made its maiden flight in 1983. The Starship featured a mechanism for giving the foreplane variable sweep, an ingenious but costly solution to the stability-versus-controllability conflict. The high aspect ratio foreplane had sweepback for adequate stability in high-speed flight and forward sweep for trimming low-speed flight. Large winglets at the tips of the swept wing replaced the conventional fuselage-mounted vertical tail plane and highly swept inboard wing leading edges contained fuel tanks. The Starship's performance and flying qualities did not appear to be competitive, and production came to an end when 45 planes were delivered.

6.4.2 Three-Surface Aircraft

Disregarding the Beechcraft 99 with auxiliary stabilizers mounted to the rear fuselage, the only example of a certified three-surface aircraft blessed with a good market response is the twin-engine business turboprop Piaggio 180 Avanti; see Figure 6.8 (a). The P 180 II is capable of carrying seven to nine passengers and one or two pilots with cabin comfort and speed comparable to that of light business jets. Powered by two P&W PT6A turboprops of 635 kW (850 HP), the Avanti attains a maximum cruise speed of 730 km/h and 12 500 m maximum operating altitude, much faster and higher than existing propeller aircraft such as the Beech King Air 350. Maximum range with IFR reserves is 2 700 km. The Avanti's specific range compares favourably with competing turbofan business jets. Design criteria for a three-surface aircraft and its pros and cons compared with conventional and canard configurations are discussed in [67]. The P 180's aerodynamic design is characterized by a high/mid wing

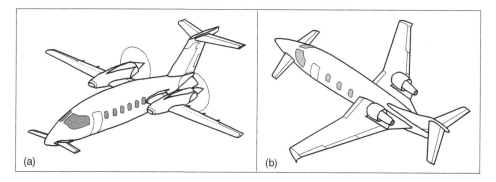

Figure 6.8 Examples of business aircraft with three-surface layout. (a) Piaggio 180 Avanti. (b) Hypothetical small business jet

position aft of the cabin leading to minimum frontal area, wetted area and interference drag. Extensive regions of natural laminar flow (NLF) are obtained over its high aspect ratio straight wing with laminar flow sections and the streamlined fuselage with very smooth skin surfaces. The pusher configuration places all lifting surfaces outside the propeller slipstream leading to low parasite drag as well as low vibration levels and cabin noise.

During the conceptual design phase it was decided that the P 180 would negotiate the wing position behind the cabin by installing a foreplane but still retaining a conventional aft tail for pitch control. Since the foreplane contributes positively to lift, the main wing area can be reduced for the same total lift. However, the CG is located in front of the LEMAC and, compared with the conventional layout, variations of the payload increases the CG shift. On the other hand, the tail sizing process allows more CG variation for a given tail area. Depending on cruise and loading conditions, the configuration may add some lifting surface parasite drag with little effect on induced drag. Compared with the pure canard layout, the balancing problem is reduced since there is adequate space inside the fuselage to install a fuel tank above the wing centre section which obviates the need for fuel tanks in leading edge extensions at the wing root. The flap control system is one of the P 180's most unconventional characteristics. The remote wing position produces a large pitch-down moment when its flaps are extended. The airplane is balanced by synchronized deployment of the main wing flaps and the foreplane flaps so that excessive tail download is avoided. Therefore, the foreplane flaps are used exclusively for balancing whereas the tail has elevators for pitch control. The two outboard main wing flaps are mechanically interconnected by torsional shafts, the two inboard flaps are operated in a similar fashion but they are mechanically independent of the outboard flaps. Similarly, the two foreplane flaps are mechanically independent. Synchronization of flap surface extensions is provided by an electronic control unit.

Figure 6.8 (b) illustrates that the three-surface layout may also be considered for the design of a small business jet. Forward sweep has been chosen in order to bring the wing carry-through structure aft of the cabin and to promote NLF. The wing has distinct dihedral for positive lateral-directional stability and its composite structure is tailored to avoid aero-elastic divergence. In order to avoid the separated flow entering the engine intakes, the wing features a small leading edge flap. Similar to the P 180 the foreplane flaps serve to trim out the nose-down effect of deflected main wing flaps, the conventional aft tail is used for pitch control.

6.5   Non-Planar Lifting Systems

6.5.1   Transonic Boxplane

In his analysis of multiplane lifting systems – see Section 4.3 – Prandtl gave no consideration to wing sweep. However, application of the Munk stagger theorem implies that sweeping the wing does not affect vortex drag on the condition that the lift distribution remains constant when staggering the lifting surfaces. This principle is applied in the transonic boxplane consisting of a forward-mounted sweptback lower wing and a rear-mounted forward swept upper wing, connected by swept vertical planes at their tips. The transonic boxplane concept was invented by L. Miranda of the Lockheed Corporation, who called it the transonic biplane [68]. This configuration has significantly less vortex drag than a monoplane designed with the same span and it allows for a constant section fuselage shape while closely matching an ideal area distribution for cruising at near-sonic Mach numbers. A high-performance boxplane employs

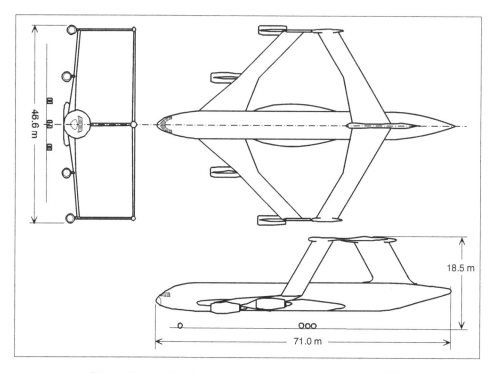

Figure 6.9   Lockheed transonic biplane design for Mach 0.95 [70]

airfoils having substantially different camber from those of competitive monoplanes in order to avoid premature flow separation, unnecessary high profile drag and low maximum lift [71]. A feasibility study of a transonic biplane depicted in Figure 6.9 and wind tunnel tests verified the predicted low induced drag at high speed. However, flutter analysis revealed both symmetric and antisymmetric instabilities to occur well below the required flutter speed that could not adequately be suppressed. No significant reductions in wing and tail weight were realized for the boxplane relative to a monoplane with the same mission capability. These findings led to the abandonment of boxplane research by Lockheed.

Recent engineering applications of the boxplane to high-performance aircraft are under development at Italian universities – they call it the Prandtl Plane concept. A summary of their achievements by A. Frediani [80] claims that the Prandtl Plane offers a suitable configuration for small airplanes, narrow and wide body airliners and freighter aircraft. The study configuration depicted in Figure 6.10 is a large passenger transport with a capacity similar to the Airbus A 380 – however, it features an extra wide single-deck passenger cabin with 14 seats abreast and three aisles. The rear wing is positioned over the fuselage and connected to it by two fins. The front and aft wings contain roughly equal amounts of fuel. The structures of the lifting system are over-constrained to the fuselage and static bending deflection is lower than that of a cantilever wing with the same total lift and span. Although Figure 6.10 does not illustrate a power plant installation, the authors suggest several options claiming that the Prandtl Plane will be a low-noise aircraft. Due to the large moment of inertia about the lateral axis, pitch

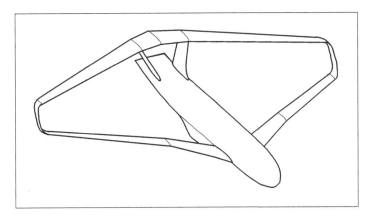

Figure 6.10   Prandtl Plane concept of a large airliner. Adapted from [80]

control must be very efficient. This is obtained by a set of elevons on both wings moving in opposite direction. The wings are also equipped with full-span high-lift devices and the plane is expected to have a very stable stall.

The boxplane has some inherent properties making its feasibility as a competitive transport aircraft questionable. A boxplane is aerodynamically and structurally complex – and probably costly – to develop into an efficient lifting system free from unacceptable aero-elastic properties. The low induced drag (for given span) must be paid for by increased parasite drag and reduced maximum lift due to low chord Reynolds numbers and the presence of non-lifting vertical tip planes. Due to the integrated character, application of the family concept is impossible without the major redesign of all major components.

6.5.2   C-Wing

Compared to the boxplane, a C-wing attains almost the same theoretical span efficiency factor with the same span and vertical extent but without the radical modification of the entire lifting system. The optimal lift distribution of a C-wing (Figure 6.11) shows that all upward lift is is generated by the main wing. The vertical fins carry a substantial inward air load load, the horizontal winglets have a small download. Comparing it with Figure 4.8 (a) shows that this distribution differs considerably from that of a boxplane. Although the C-wing has been investigated primarily for application to a very large passenger transport, it may be considered for other airplane concepts as well.

Integration of a C-wing into the overall aircraft system can have some favourable synergistic effects ameliorating the worst of the very large transport problems [79]. In particular, application of a C-wing may improve the aerodynamic performance and flying qualities of a highly span-constrained very large airliner. The C-wing concept for which J. McMasters and I. Kroo were granted a design patent in 1995 featured a 600-passenger single-deck span loader – Figure 6.12 depicts a modified version. The pressure cabin is housed partly in a conventional wide body fuselage and partly in the inboard section of the relatively thick inboard wing, classifying the aircraft as a hybrid flying wing. The moderately tapered main wing has a low aspect

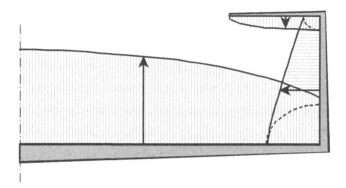

Figure 6.11   Distribution of lift over the elements of a C-wing

ratio. The tip-mounted non-planar lifting surface systems are configured as control surfaces replacing the empennage of a conventional airplane. Each vertical fin has roughly the size of a Boeing 747 vertical stabilizer. The horizontal surfaces are augmented by a foreplane rendering this configuration effectively into a three-surface aircraft. The following basic features make the C-wing configuration a possible candidate for a very large transport:

- The wing span problem associated with the 80 m box airfield limit is solved without incurring a significant induced drag penalty.
- The single-deck cabin reduces the emergency evacuation problem of a very large airliner.

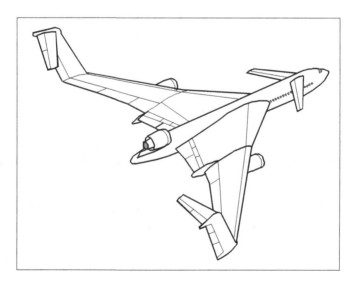

Figure 6.12   C-wing transport configuration for 600 passengers. Reproduced from [79]. Copyright 1998 Elsevier

- Airplane growth can be accommodated by the traditional expedient of lengthening the fuselage with cylindrical body plugs without the necessity of a major wing redesign.
- The trailing vortex field is highly non-planar, a potentially advantageous asset in reducing wake vortex hazards. Much work remains to be done to verify this.

The developers came to the conclusion that, with sufficient further effort, the C-wing configuration proposed can probably be made to work. In [79] they compare the characteristics of the configuration depicted in Figure 6.12 with an optimized conventional WAT airplane. This makes it clear that weight and performance figures of their C-wing giant are inferior to those of the conventional competitor – a conclusion that may hold only for subsonic transports carrying no more than about 600 passengers in a three-class arrangement. Further explanation is not given but it is likely that some of the problems are similar to those of the boxplane. In particular, designers of a C-wing structure are faced with a challenge since its concept entails formidable structural dynamics problems. Other sources of concern are probably the weight of the pressure cabin section in the wing and the drag of the 18% thick inner wing.

6.6   Joined Wing Aircraft

The joined wing configuration has been patented and investigated theoretically and experimentally by J. Wolkovitch [82, 83]. A joined wing aircraft (JWA) can be defined as an airplane configuration incorporating two dual wings arranged to form diamond shapes in both planform and front view – they are also known as bi-diamond or diamond-ring wings. A single vertical tail or twin tails may be used to support the centre section of the aft wing. Subject to aerodynamic and structural optimization, the joints may be located at or inboard of the forward (main) wing tips. The wings can be joined by means of streamline bodies but it is also possible to join the wings at their tips by winglets. Advantages claimed for the joined wing include lightness, stiffness, low vortex drag, low wave drag due to a favourable transonic area distribution, direct lift and side-force control capability, and high trimmed maximum lift. A JWA would probably be quieter in climb-out and landing approach than conventional aircraft. Most of these claims have been supported by independent studies, design analysis and wind tunnel tests described in a comprehensive overview [89], the source of most information in the present overview.

   Integration of the joined wing concept into a civil aircraft results in an unorthodox general arrangement. The example of a hypothetical executive jet in Figure 6.13 shows

- a conventional fuselage with aft-mounted turbofans and a single vertical tail;
- a high aspect ratio sweptback front wing with trailing edge flaps;
- a forward swept aft wing with trailing edge flaps, with its root attached at the top of the vertical tail(s);
- full-span trailing edge flaps on both wings;
- the rear wing's tips jointed just aft of the front wing's trailing edge inboard of its tips;
- a fairing for the main landing gears which are retracted inside the fuselage structure.

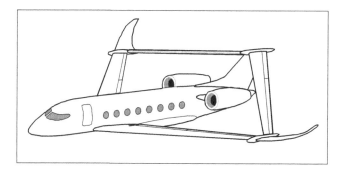

Figure 6.13 Concept for a high speed executive jet with joined wings

6.6.1   Structural Principles and Weight

The joined wing differs from a conventional WAT with cantilever wing in its external appearance as well as in its internal structure. Figure 6.14 shows how the lift loads on each wing are resolved into components acting normal and parallel to the truss structure formed by the joined wings. The in-plane components are well resisted by the truss, the out-of-plane components bend the wing structures about a bending axis that is tilted at an angle θ to the longitudinal axis. The primary structure of each wing involves chordwise tapering of the equivalent panel thickness – an essential property of the concept that yields a significant weight advantage. The minimum amount of structural material is needed if bending is resisted by means of a beam with maximum second moment of inertia about the bending axis. This implies that the material must be concentrated near the upper front spar cap and the lower rear spar cap, as shown schematically in the lower portion of Figure 6.14. Moreover, it is profitable to use the largest possible fraction of the airfoil chord with limits set by the space needed for de-icing,

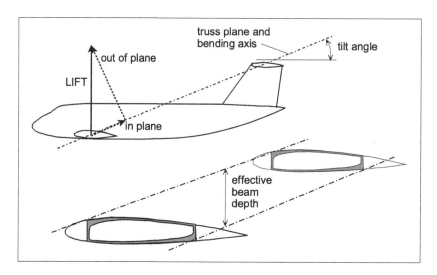

Figure 6.14 Structural principles of the joined wing

high-lift devices and control surfaces. For instance, the joined wing box will typically occupy 5 to 75% of the chord length versus 15 to 65% of the cantilever-wing chord.

The effective beam depth of a joined wing is determined by the chord of its airfoils – their thickness is secondary. Hence, thin airfoils may be employed with less weight and drag than for a cantilevered wing. The torsional stiffness of a joined wing is high since torsion of one wing is resisted by flexure of the other. This yields higher aileron effectiveness and higher flutter speeds than obtainable with a cantilevered wing of comparable weight. Since under positive load factors the rear wing is under compression, overall column buckling must be considered. Although rear wing buckling was not found to be a constraint in [88], it was found to be a critical design issue for a JWA in the DC-9 category in [95]. Joined wings are not invariably lighter than aerodynamically equivalent WAT combinations. Weight is saved only if the wing planform is properly chosen and the internal structure is optimized. By studying a wide range of geometric parameters, [88] reported some joined-wing configurations realizing weight savings between 20 and 35% of the primary structure weight relative to a WAT configuration. If there is no constraint on wing span of a joined wing, it has the best inter-wing joint location at approximately 70% of the semi-span.

6.6.2   Aerodynamic Aspects

Similar to biplane wings of equal span, both tip-jointed wings carry approximately the same lift in order to achieve minimum vortex drag whereas the sensitivity to a non-optimum lift distribution is low. The span efficiency factor depends primarily on the gap; that is, the vertical displacement of the dual wings at their roots. Joined wings together form a closed lifting system similar to the boxplane. Nevertheless, since vorticity at the tips is released in a small area, its span efficiency factor is only slightly higher compared to the monoplane. Figure 6.15 indicates that, for realistic values of the gap/span ratio, the theoretical span efficiency factor e_v is between 1.05 and 1.10. Hence, the minimum vortex drag of tip-jointed wings is 5 to 10% less than the minimum vortex drag of an isolated monoplane with the same span. This

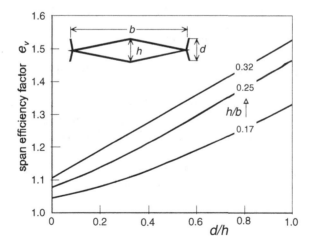

Figure 6.15   Effect of winglets on the span efficiency factor of a joined wing. Adapted from [89]

drag saving falls to a level of only 2–3% for wings with inter-wing joints inboard of the tips. Although the joined wing gains little vortex drag from its non-planarity, it reduces trim drag in cruising flight to less than 1% by exploiting the closed lifting system [81] for its balance. Figure 6.15 also demonstrates that, with symmetric inclined winglets at the tips, the span efficiency factor approaches that of a boxplane with the same ratio of vertical extent to span.

In view of the large number of geometric properties defining a JWA geometry, it is not feasible to make generally valid statements on parasite drag. The following observations form a liberal interpretation of [89].

- Adverse interferences between the front and aft wings are avoided by using non-overlapping airfoils near the inter-wing joints. The junction of the aft wing with the vertical tail should be filleted properly.
- Due to their smoother area progression, a JWA has less zero-lift wave drag than a conventional airplane with comparable leading edge sweep. The application of self-bracing wings permits wings with relatively thin airfoils, giving a further drag saving.
- The lift on each airfoil causes the other airfoil to be immersed in a curved flow field. Since this causes induced flow curvature, airfoils for joined wings should be designed using methods similar to those used for multi-element airfoils. The front wing should have more camber than the rear wing and should be washed out towards the tips. The aft wing should have wash-in.

Wind tunnel tests of a joined wing configuration without high-lift flaps indicate that the trimmed maximum lift coefficient of the joined wing is 4 to 7 percent higher than for a conventional WAT with the same total exposed area. Since the rear wing is immersed in the downwash of the front wing, the front wing of a statically stable JWA is the first airfoil that stalls. Although this property provides good stall recovery, it means that the rear wing may be oversized in providing the required trimmed lift for the low speed configuration. Moreover, the high-lift devices are less effective since they occupy a smaller fraction of the chord than for a cantilever wing. The following measures will help to increase maximum lift:

- Increase the lift curve slope contribution of the rear wing by increasing its chord and giving it less sweep than the front wing.
- Optimize camber, wash-out and setting angles of both airfoils to achieve a positive pitching moment at zero lift.
- Minimize the static margin by moving the CG aft, for instance, by using a trim tank in the empennage.
- Install leading-edge flaps, slats or strakes near the root of the inboard front wing.
- Employ maximum trailing-edge flap deflections on the inboard front wing and the outboard aft wing as illustrated in Figure 6.16. This makes the zero-lift pitching moment more nose-up.

If these measures appear to be insufficient, a low-aspect ratio foreplane may be employed to provide appreciable nose-up pitching with little forward AC shift.

6.6.3   Stability and Control

At low angles of attack, the aerodynamic centre (AC) of a JWA configuration is in front of the joined wing centre of area. The AC moves gradually backwards at moderate and high angles

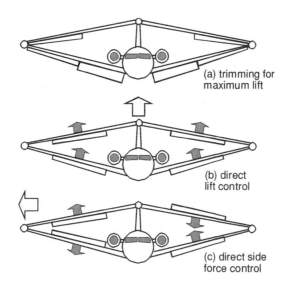

Figure 6.16   Joined wing aircraft lift and control capabilities

of attack below the stall, leading to a pitch-down tendency which can be reduced, if necessary, by proper aerodynamic design. Elevators may be mounted on both the front and rear wings and on the foreplane (if present). If they are deflected by equal angles the resulting lift acts close to the CG which enables direct lift control (Figure 6.16). A combination of direct lift control and front elevator deflection facilitates take-off rotation. Figure 6.16 shows that the feasibility of direct side force control is inherent to the joined wing configuration due to the large dihedral of the front wing and anhedral of the rear wing which is beneficial for controlling the aircraft in cross-wind landings. In conclusion, a well-designed JWA design is likely to have stability and control characteristics as good as or better than a conventional aircraft.

6.6.4   Design Integration

Although it is imaginable to mate a joined wing to the body and vertical tail of an existing airplane, this is not likely to produce the best result. However, the conceptual design of a completely new JWA airframe involves some novel considerations with regard to fuel tankage, landing gear and wing and body geometry, as well as the load and balance aspect. For instance, a tip-jointed wing would normally carry fuel in both the front and rear wings. Although the volume of two wings of equal area and geometry would be less than that of a cantilever monoplane, the larger box chord fractions may compensate for this. And the following aspects are relevant to fuselage design:

- It appears desirable for minimum weight that the fuselage has a low fineness ratio. This is compatible with the aerodynamic requirements of a JWA primarily because it does not require a long fuselage to provide a sufficiently large tail moment arm.

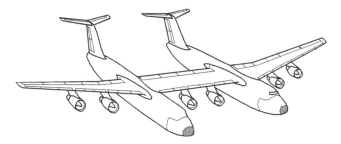

Figure 6.17   Lockheed design of a very large cargo aircraft with twin-fuselage configuration [103]

- Both wings of a JWA generate upward lift when the aircraft experiences a high upward load. Hence, the fuselage is supported near both ends. By contrast, a conventional wing supports the fuselage midway between nose and tail with the tail applying a download. Parts of the JWA fuselage designed to resist the air loads may therefore be lighter than usual. On the other hand, the body must be reinforced where the main landing gears are installed.
- The airliner family concept requires a body that can be stretched or shrunk to suit different payload capacities. It would be a challenge to incorporate this flexibility in a JWA without major design modifications.

The JWA has several inherent advantages over a conventional configuration classifying it as a serious candidate for designing small passenger planes as well as airliners of widely different capacities. However, a JWA represents a highly integrated concept with more complex lifting and flight control systems than usual and is probably more costly to manufacture and maintain.

6.7   Twin-Fuselage Aircraft

A twin-fuselage aircraft (TBA) employs the favourable structural effect of span distributed loading (SDL) to reduce wing bending by carrying its payload in two fuselages located at equal distances outboard from the plane of symmetry. SDL leads to a significantly lower wing structural weight for a given span, resulting in a larger optimum aspect ratio compared to a traditional layout. Figure 6.17 depicts a very large twin-body cargo aircraft conceived by the Lockheed Georgia Company. Although a tri-body aircraft with one central and two outboard fuselages is imaginable, it is not likely to offer sufficient advantages to make it a candidate for civil airplane design.[3]

Comparison between a conventional airplane and a TBA with the same payload/range capabilities explains that significant empty weight reductions are inherent to the twin-body layout. Most significant is the effect of the unusual SDL illustrated in Figure 6.18. Disregarding the lift losses due to the fuselage(s), the two configurations have a nearly identical distribution of lift along the span and substantially different mass distributions. Both wings experience the maximum bending moment where the wing is connected to the fuselage(s). However, maximum bending moments on a typical twin-fuselage airplane are at least 50% lower than usual

---

[3]American designer Burt Rutan used tri-body configurations for his Voyager aircraft that circled non-stop around the world in 1989 and, more recently, for WhiteKnightTwo, launcher of the sub-orbital Virgin Galactic SpaceShipTwo.

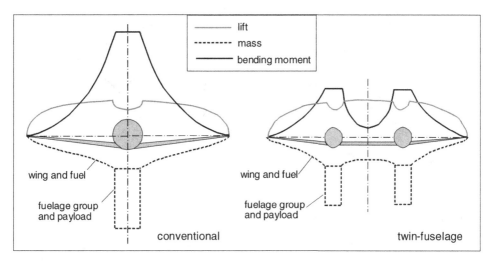

Figure 6.18 Distribution along the span of lift, mass and bending moment for a twin-fuselage compared to a conventional aircraft

whereas the bending load between the fuselages decreases rapidly towards the centreline.[4] On the other hand, torsion constitutes the dominating load on a large portion of the central wing. Overall, the central wing has a significantly lower weight per unit span than the traditional inboard wing and carry-through structure. Since the area under the bending moment distribution forms an indication for the material required to withstand bending, a considerable weight reduction is obtained for a given span. In view of the sensitivity of fuel weight to induced drag – hence, wing span – the TBA has a larger optimum span than the traditional airplane.

A criterion for sizing a large portion of a cabin shell structure is that it must withstand a specified number of pressurization cycles. An overpressure Δp in a cylindrical cabin with diameter d causes the skin to experience a tensile force equal to $P = \frac{1}{2} l d \Delta p$ (Figure 6.19). The required skin thickness for a given hoop stress σ is $t = \frac{1}{2} d \Delta p / \sigma$. Hence, skin thickness is proportional to cabin diameter and the skin weight amounts to

$$W_{\text{skin}} = \frac{\pi}{2} l d^2 \rho g \frac{\Delta p}{\sigma} = 2 Q \rho g \frac{\Delta p}{\sigma} \tag{6.6}$$

with ρ and Q denoting the material density and cabin volume, respectively. Consequently, the skin weight of a pressure cabin is proportional to its volume. This result can be utilized in different ways.

1. A traditional airliner is compared with a TBA having the same total pressure cabin volume and similar geometry. Equation (6.6) indicates that they have the same skin weight.
2. The TBA cabins are designed to have the same total floor area as the traditional airplane cabin. For similar geometries, each cabin has a length $l/\sqrt{2}$, a diameter $d/\sqrt{2}$ and a volume

---

[4]The bending moment due to lift is zero in the plane of symmetry when the bodies are suspended in the centre of pressure of each half-wing.

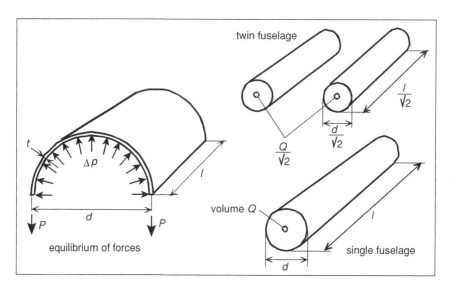

Figure 6.19 Weight advantage for the overpressure-sized skins of two fuselages compared to a single pressure cabin with the same floor area

$Q/(2\sqrt{2})$ (Figure 6.19). Using Equation (6.6) it is found that the two skins together have a weight equal to that of the single-body cabin divided by $\sqrt{2}$, a 29% gain. A similar result is found for the front and rear pressure bulkheads.

The volume and weight savings of the second option are ascribed to the unused space of a wide body airliner with single-deck cabin – the twin-body layout eliminates this loss. A negative consequence is that the total cargo hold volume of the two narrow bodies is reduced relative to the single-body layout. If this were unacceptable, the twin-body cross-sections would have to be modified to match the required freight hold capacity. The TBA skin weight saving is then less than 29%. The reverse argument applies to the volumetric utilization of a twin-deck cabin which is better than that of two single-deck cabins with the same total floor area.

It is concluded from the previous discussion that two narrow bodies have a smaller total wetted area and less parasite drag than a wide body with the same total number of cabin seats. Consequently, if the number of seats is in the range of 200 to 400, it appears that the use of two fuselages will yield a drag saving. This observation is confirmed by statistical information on fuselage gross wetted area per cabin seat [112]:

- single deck, single aisle: 2.6 to 2.9 m^2;
- single deck, two aisles: 3.0 to 3.5 m^2;
- two decks, four aisles: 2.4 to 2.6 m^2.

This confirms that the wetted area of two narrow bodies is 15 to 20% less than that of a single wide body with an equal number of seats. This difference is augmented by the absence of a flight deck in one of the fuselages of a twin-body airplane. However, this beneficial effect is

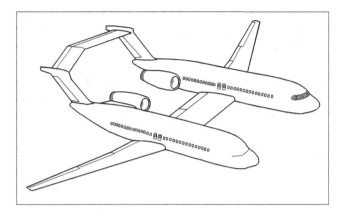

Figure 6.20 Medium range airliner design for 250-300 passengers with twin-fuselage configuration [114]

defeated by the scale effect on skin friction drag causing the parasite drag coefficient of a wide body to be less than that of standard body.

6.7.1 Design Integration

Figure 6.20 depicts a clean-sheet airliner design featuring two narrow bodies with a single-aisle cabin [114]. Table 6.1 compares this design with a traditional wide body airliner with the same passenger capacity designed according to the same ground rules. This shows that savings in the design weights, fuel consumed and installed engine thrust are significant. Gross weight reductions between 10 and 16% and improvements of about 20% in fuel efficiency are suggested in the available literature for medium-sized airliners and even more for large aircraft. Cost reductions accrue not only from lower empty weight, installed thrust and fuel usage. Several publications have indicated the feasibility of pairing fuselages from existing transport designs to provide inexpensive bodies [113], [105]. Reduced maintenance costs are associated with parts commonality in the two fuselages and the empennage(s). Especially

Table 6.1 Characteristics of the design on Figure 6.20 compared with a conventional design with equal payload/range capability

Mass or force	Units	Conventional	Twin fuselage	Δ %
Maximum take-off weight	kg	155 000	134 000	−13.5
Maximum landing weight	kg	128 000	113 000	−11.7
Maximum zero fuel weight	kg	120 000	106 000	−11.7
Operating empty weight	kg	84 000	70 000	−16.7
Maximum structural payload	kg	36 000	36 000	0
Mission fuel for 8 000 km range	kg	40 715	34 245	−15.9
Engine thrust (SL ISA)	kN	2×222.5	2×178.0	−20.0

for short- to medium-range applications, the TBA concept can be an attractive alternative for doubling the passenger capacity of narrow body airliners with significantly less than twice the operating cost. In general, a TBA has the following promising characteristics:

- Different from radical unconventional concepts such as the BWB or the C-wing, the TBA configuration is based on conventional technology and offers a low-risk alternative with distinct advantages. The advantages lie in the general arrangement itself which combines conventional technology with favourable synergistic effects.
- A TBA needs significantly less installed thrust than a (large) traditional design. This may favour application of three engines instead of four with freedom for power plant installation at the central wing enabling application of centreline thrust. The arrangement depicted in Figure 6.20 also has the potential of engine noise shielding.
- The non-swept and relatively thin central wing can be designed as a simple prismatic structure, an ideal shape for NLF or HLFC and for adapting the wing span to growth versions of the baseline.
- The configuration is ideally suited to the family concept, especially if the wing between fuselages is prismatic.

The twin-fuselage concept can be considered for passenger aircraft of any size. However, for very large aircraft a single twin-deck fuselage may be superior to two single-deck wide body. Since the optimum span of a TBA is larger than the conventional 80 m box span constraint, its performance is degraded more than for the traditional configuration. A configuration with cabins in each outboard fuselage and in the wing-between-fuselages might be considered as an option for a very large passenger transport aircraft – see configuration D4 in Figure 6.2. Due attention must be paid to aero-elasticity, approach and landing flight characteristics and emergency evacuation. Results of several studies ([113], [104]) lead to the conclusion that aero-elastic behaviour and flight characteristics of a TBA are conventional when modern control technology is applied.

There exist several technical unknowns concerning how to configure a clean sheet TBA design. Efficient ground handling is possible provided two-door loading and unloading facilities are available. The emergency evacuation procedure is unconventional and in certain areas between the fuselages exits may not be possible. However, safety aspects are not considered as a potential hindrance. In spite of many positive findings, manufacturers have not yet embraced the TBA concept for their clean sheet or derivative designs. One of their concerns is the expectation that lateral displacement of the cabins from the centreline generates an unfavourable opinion about cabin comfort.

6.8   Hydrogen-Fuelled Commercial Transports

In the mid-1950s, NASA successfully flight-tested a B-57 equipped with a liquid hydrogen (LH2) fuel system. The oil crisis of the 1970s brought about a concern that within several decades there would be a serious shortage of crude oil resources. This prompted NASA-supported studies by Lockheed and Douglas of aircraft configured to use LH2 as a gas turbine engine fuel. The Russians were the first to use hydrogen on a civil transport with the purposely developed Tupolev Tu-155, which first flew in 1988 and made successful flights with both

Figure 6.21 DASA Cryoplane wide body airliner with LH2 propulsion. Reproduced with permission from Daimler-Benz Aerospace Airbus GmbH

LH2 and liquid natural gas. During the 1990s, DASA in Germany cooperated with Tupolev in cryogenic fuel system technology leading to the Cryoplane project (Figure 6.21). The EU-sponsored Cryoplane research aimed at identifying the feasibility of future application to commercial aviation [118].

6.8.1 Properties of LH2

The principal benefit accruing from the use of liquid hydrogen is that it is essentially non-polluting. Compared to jet-fuelled engines delivering the same thrust, hydrogen combustion products consist of 2.6 times the amount of H_2O in contrails contributing less to the greenhouse effect[5] whereas up to 80% NOx can be saved by lean combustion [118]. Hydrogen offers a very high calorific value ($H = 120$ kJ/g), that is, 280% of current jet engine fuel. Especially for long-range aircraft, this results in substantial fuel burn and gross weight reductions. The main disadvantage of LH2 is its low density and its extremely low temperature. In the liquid state, liquid hydrogen has a temperature of about 20 K and a specific mass of only 70.8 kg/m^3 C compared to 800 kg/m^3 for kerosine.[6] Consequently, for a given amount of internal energy,

---

[5]H_2O is a greenhouse gas with a lifetime in the upper atmosphere of only half a year, whereas CO_2 emitted by a kerosene-burning engine stays aloft for more than 100 years.
[6]When hydrogen vaporizes and the gas is heated to 15° C, its density increases to 83.8 kg/m^3.

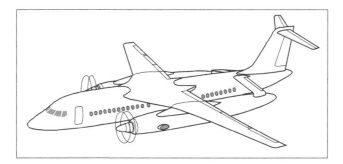

Figure 6.22 Hypothetical regional airliner with LH2 propulsion

the storage of LH2 requires a pressurized tank with four times the volume of a normal jet fuel tank – far too large to be accommodated in a traditional airplane wing.

The result of these characteristics is that a hydrogen-powered aircraft will look quite different from a conventional fuel-powered aircraft. The main differences are in the use of an unusually large fuselage or in the presence of large wing-mounted streamline bodies to provide the extra tankage (Figure 6.22); other solutions are discussed in [121]. In view of the cryogenic fuel properties, LH2 tanks have to be extensively isolated and pressurized. LH2-fuelled aircraft are therefore characterized by a relatively low aerodynamic efficiency and increased empty weight. In spite of it, the large reduction in the aircraft gross weight brings the fuel energy requirement down by approximately 5 to 20% and the fuel weight by 65 to 70% [116]. Moreover, the reduced take-off gross weight permits smaller engines to be installed.

6.8.2 Fuel System

Although hydrogen fuel systems have been developed for space mission applications, the requirements for aircraft are so different in regard to mission duration, system life, operating cycles, and safety aspects that entirely different design problems have to be solved. Hydrogen can be used as a fuel for a normal aero gas turbine with minor modifications to the power plant and its systems. The turbo-machinery would be unchanged, the combustor and the fuel control system must be redesigned and a heat exchanger to vaporize the fuel must be added. Hydrogen must be stored in liquid form at a temperature of 20.2 to 22.3 K and a pressure of 1.2 to 2.0 bar. It is then heated to around 150 K and twice the engine combustor pressure in order to drive the fuel injectors. The heat exchanger has little effect on the engine layout, weight and drag since it can be wrapped around the gas generator exhaust pipe – utilizing this otherwise wasted heat improves the cycle efficiency compared to conventional fuel. However, the two-phase flow in the fuel system exhibits compressibility. The internal pressure varies continuously depending on the balance between the tank insulation characteristics and combustion inlet conditions. Consequently, a hydrogen fuel system may exhibit unexpected dynamics and critical failure modes.

The weight of the large tank(s) is a major issue which requires special attention. In order to make the insulation light, effective, reliable and damage-tolerant, the tank can be produced from foam or super-insulation type multi-layer materials. In the stored condition, there is a continuous phase change from liquid to gas and vice versa, and some hydrogen will inevitably be lost due to boil-off when the aircraft is parked between flights. Integration into the aircraft of the pressurized fuel tank(s) is the greatest challenge for hydrogen application. The best tank layout depends on the aircraft category and general arrangement. The fuselage seems suitable as it is already a pressure vessel, but the presence of passengers next to a hydrogen tank might raise public concerns. For wide body transports, the fuel may be stored both fore and aft of the passenger compartment. In order to minimize the CG travel due to fuel burn, the fuselage length/diameter ratio and the tank length can be reduced and/or the engines located at the rear end of the fuselage. A three-surface configuration could be a good layout for this type of aircraft. Special attention should be paid to the risk that a disk burst might lead to an explosion of the LH2 tank.

6.8.3   Handling Safety, Economics and Logistics

Considering general safety aspects of handling and using hydrogen for aviation, studies have shown that the overall safety level is at least as high as when using kerosene, on the provision that safety procedures are strictly observed [121]. The overall airport layout and procedures do not have to be changed and the aircraft can be serviced at the same positions as kerosene-fuelled aircraft. Although the larger fuel volumes may double refuelling time, this does not necessarily increase turnaround time and airport operations will not require significant adaptations to the aircraft itself. However, unless a local LH2 production to fulfil the demand for cryogenic fuel is erected at each airport, a huge and entirely new logistic system would be required for the transportation, storage and handling of fuel throughout the world. This will prevent the use of this radical change in fuelling technology for a long time to come.

The results of the Cryoplane project confirmed that LH2, from the technical side, could be an alternative future fuel; however, the economics of LH2-fuelled aircraft is a different story. Although the cryogenic fuel system penalizes the hydrogen aircraft, the technological challenges in the application of hydrogen to aviation are more related to cheap, environmentally acceptable and large-scale production of hydrogen than to the adaptation of the aircraft concept. The electrolysis process plus liquefaction to produce LH2 from water requires four to five times the amount of energy produced by burning LH2. Using electricity produced from fossil fuel or gas for the production of hydrogen is several times more expensive than that of jet-A fuel and does not appear the way to conserve energy resources. It seems likely that the missing infrastructure and the risk and investment of the hydrogen system will stall its implementation until cheap, unlimited energy becomes available from fission, fusion, wind or solar resources.

6.9   Promising Concepts

Although the TAW has been under attack from innovative concepts ever since its emergence in the 1950s, it will continue to be the favoured general arrangement of most clean sheet designs. The cylindrical body shell forms an ideal pressure vessel enabling a passenger cabin with flexible seating arrangements, a high volumetric efficiency and a straightforward realization of the family concept. Satisfactory provisions for embarking and disembarking, ground handling and emergency evacuation are achievable and there exists a wealth of design experience for safety, reliability, durability and maintenance. The aerodynamics of the high-aspect ratio wing are thoroughly optimized for long-range transonic cruise with maximum efficiencies up to

$L/D = 20$, as well as excellent low speed performance. Balancing the dominant configuration is generally satisfactory with inherent stability and low trim drag in most loading conditions.

The choice between improvements obtained from more integration versus discrete major components is a major issue. Chapters 5 and 6 demonstrate how the presently dominant configurations may be improved by unusual or innovative concepts leading to the following observations:

- The BWB and the HFW are strong contenders for very large aircraft with seating capacities in excess of 600 seats. The BWB is a highly integrated flying wing concept offering an aerodynamic efficiency up to $L/D = 25$. Its design process is complicated and in spite of the considerable effort already expended in research, many aspects remain to be investigated when developing a for-sale aircraft. The span constraint of the 80m box is a serious handicap for these very large aircraft, which can be countered by adopting a non-planar wing configuration or by using hinged wing tips. Both solutions leads to considerable weight penalties.
- The twin-fuselage configuration deserves to be considered for short- to medium-range aircraft with seating capacities between 200 and 400 seats. The span loading feature enables a (much) larger wing span resulting in considerably less vortex drag with no increase of wing structure weight. The twin-fuselage layout offers a good platform for application of laminar flow and very high bypass ratio turbofans or open rotor engines and does not require a costly technology development.
- Installing open rotor engines on regional aircraft may lead to higher cruise speeds up to Mach 0.75. Strut bracing allows a high-wing airliner to increase its span with little or no weight increase.
- An innovative business jet configuration can be considered if it improves cabin comfort without weight and drag penalties. A three-surface layout with a mid wing positioned behind the cabin may be favoured for a light business jet. Application of a joined wing to a long-range executive jet improves its aerodynamic efficiency in cruising flight and reduces the empty weight.

Any major excursion from the dominant configuration requires an all-out overall aircraft system optimization, taking into account all the driving factors affecting economy, environmental issues, operational and safety aspects. Consequently, the advanced designer should be familiar with multidisciplinary optimization (MDO).

Bibliography

[1] Torenbeek, E., *Synthesis of Subsonic Airplane Design*, Chapters 2 and 9, Springer Verlag, Heidelberg, 1981.
[2] Stinton, D., *The Design of the Aeroplane*, Granada Publishing Ltd., London, 1983.
[3] Downie, D. and J. Downie, *The Complete Guide to Rutan Aircraft*, Third Edition, Airlife Publishing Ltd., Shrewsbury, 1987.
[4] Sterk, F.J., and E. Torenbeek, 'Unconventional Aircraft Concepts', in *Proceedings of a Symposium of the Netherlands Society of Aeronautical Technology*, Delft University Press, 1987.
[5] Stinton, D., *Flying Qualities and Flight Testing of the Aeroplane*, Blackwell Science Ltd., Oxford, 1996.
[6] Fielding, J.P., *Introduction to Aircraft Design*, Cambridge University Presss, Cambridge, 1999.
[7] Torenbeek, E., and H. Deconinck (eds), *Innovative Configurations and Advanced Concepts for Future Civil Aircraft*, von Kármán Institute for Fluid Dynamics, Lecture Series 2005–06, June 2005.

[8] Whitford, R., *Evolution of the Airliner*, The Crowood Press, Ramsbury, 2007.
[9] Roskam, J., *Lessons Learned in Aircraft Design*, DAR Corporation, Lawrence, Kansas, 2007.

Advanced Concepts and Technology

[10] Black, R.E., and J.A. Stern, "Advanced Subsonic Transports: A Challenge for the 1990's", *Journal of Aircraft*, Vol. 13, No. 5, pp. 321–326, May 1976.
[11] Clay, C.W., and A. Sigalla, "Shape of the Future Long-Haul Airplane", *Journal of Aircraft*, Vol. 13, No. 8, pp. 551–558, August 1976.
[12] Lange, R.H., and E.S. Bradley, "Parametric Study of Advanced Long Range Military/Commercial Cargo Transports", AIAA Paper No. 77–1221, 1977.
[13] Conner, D.W., and J.C. Vaughan, "Multi-Role Cargo Aircraft Options and Configuration", NASA TM 80177, October 1979.
[14] Klug, H.G., "Neue Konfigurationen und ihr Beitrag zur Reduzierung der Betriebskosten von Verkehrsflugzeugen", *Jahrbuch der DGLR*, 1981.
[15] Arata, W.H. Jr., "Very Large Vehicles: Technology Looking for a Need", *Astronautics and Aeronautics*, April 1982.
[16] Lange, R.H., and J.W. Moore, "Application of Composite Materials and New Design Concepts for Future Transport Aircraft", ICAS Paper No. 82–2.7.3, September 1982.
[17] Rosenthal, L., "Nouvelle Architecture des Avions", *L'Aéronautique et l'Astronautique*, 1986–1 – No. 116, 1986.
[18] Lange, R.H., "A Review of Advanced Turboprop Transport Aircraft", *Progress in Aerospace Sciences*, Vol. 23, pp. 151–166, 1986.
[19] Lange, R.H., Review of Unconventional Design Concepts", *Journal of Aircraft*, Vol. 25, No. 5, pp. 385–392, May 1988.
[20] Roskam, J., "What Drives Unique Configurations", SAE Technical Paper 881353, October 1988.
[21] Kentfield, J.A.C., "Aircraft Configurations with Outboard Horizontal Stabilizers", Department of Mechanical Engineering, University of Calgary, January 1990.
[22] Roskam, J., and C. Gomer, "Design Developments for General Aviation Aircraft", SAE Paper No. 911022, April 1991.
[23] Weisshaar, T., and C. Allen, "Design of a Turbofan Powered Regional Transport Aircraft", AIAA Paper No. 91–3080, September 1991.
[24] Kentfield, J.A.C., "Aircraft Configurations with Outboard Horizontal Stabilizers", *Journal of Aircraft*, Vol. 28, No. 10, pp. 670–672, October 1991.
[25] Oelkers, W., "High Capacity Aircraft", ICAS Paper No. 92–1.10.3, September 1992.
[26] Bushnell, D.M. (ed.), "Potential Impacts of Advanced Aerodynamic Technology on Air Transportation System Productivity", NASA Technical Memorandum 109154, September 1994.
[27] Morris, S.J., "Advanced Aerodynamic Configurations and Their Integration into the Airport Environment", Session II of [26], September 1994.
[28] Kentfield, "Case for Aircraft with Outboard Horizontal Stabilizers", *Journal of Aircraft*, Vol. 32, No. 2, pp. 398–403, March-April 1995.
[29] Hefner, J.N., W.L. Sellers III, J.L. Thomas, R.W. Wlezien and R.N. Antcliff, "Challenges and Opportunities in Fluid Mechanics Research", ICAS Paper No. 96–2.1.1, September 1996.
[30] Peel, C.J., "Advances in Materials for Aerospace", *The Aeronautical Journal*, Vol. 100, No. 1000, pp. 487–503, December 1996.
[31] Yaros, S.F., M.G. Sexstone, L.D. Huebner, J.E. Lamar, R.E. McKinley, A.O. Torres, C.L. Burley, R.C. Scott, and W.J. Small, "Synergistic Airframe-Propulsion Interactions and Integrations," White Paper prepared by the 1996–1997 Langley Aeronautics Technical Committee, March 1998.
[32] Sandoz, P.L., "The Next Careful Steps in Commercial Aircraft Structures", *Structures Technology; Historical Perspective and Evolution*, A.M. Noor (ed.), pp. 91–106, American Institute of Aeronautics and Astronautics, 1998.
[33] Bushnell, D.M., "Frontiers of the "Responsibly Imaginable" in (Civilian) Aeronautics", The 1998 Dryden Lecture, American Institute of Aeronautics and Astronautics, 1998.
[34] Greitzer, E.M. et al., "N+3 Aircraft Concept Designs and Trade Studies", Final Report, Volume 1 and 2., NASA/CR – 2010–216794/VOL1 and VOL2, December 2010.

Flying Qualities and Airplane Design

[35] Trubshaw, E.B., "Low Speed Handling with Special Reference to the Super Stall", *Journal of the Royal Aeronautical Society*, Vol. 70, pp. 695–704, July 1966.

[36] Sliwa, S.M., "Impact of Longitudinal Flying Qualities Upon the Design of a Transport with Active Controls", AIAA Paper No. 80–1570, August, 1980.

[37] Sliwa, S.M., "Economic Evaluation of Flying-Qualities Design Criteria for a Transport Aircraft with Relaxed Static Stability", NASA Technical Paper No. 1760, December 1980.

[38] Hood, R.V., S.M. Dollyhigh, and J.R. Newsom, "Impact of Flight Systems Integration on Future Aircraft Design", AIAA Paper No. 84–2459, 1984.

[39] Huber, B., "Center of Gravity Control on Airbus Aircraft – Fuel, Range and Loading Benefits", SAWE Paper No. 1843, May 1988.

[40] Roskam, J., "Evolution of Airplane Stability and Control: a Designer's Viewpoint", *Journal of Guidance, Control, and Dynamics*, Vol. 14, pp. 481–491, May-June 1991.

[41] Slingerland, R., "Prediction of Tail Downwash, Ground Effect and Minimum Unstick Speed of Jet Transport Aircraft", doctoral thesis, Delft University of Technology, June 2005.

Airplanes with Strut-braced Wings

[42] Kulfan, R.M., and J.D. Vachal, "Wing Planform Geometry Effects on Large Subsonic Military Transport Airplanes", AFFDL-TR-78–16, February 1978.

[43] Carrier, G., O. Atinault, S. Dequand, et al., "Investigation of a Strut-Braced Wing Configuration for Future Commercial Transport", ICAS Paper No. 1.10.2, September 2012.

[44] Park, P.H., "The Effect on Block Fuel Consumption of a Strutted vs. Cantilever Wing for a Short-Haul Transport Including Strut Aeroelastic Considerations", AIAA Paper No. 78–1454, August 1978.

[45] Jobe, C.E., and R.M. Kulfan, "Wing Planforms for Large Military Transports", *Journal of Aircraft*, Vol. 16, pp. 425–432, July 1979.

[46] Turriziani, R.V., W.A. Lovell, G.L. Martin, J.E. Price, E.E. Swanson, and G.F. Washburn, "Preliminary Design Characteristics of a Subsonic Business Jet Concept Employing an Aspect Ratio 25 Strut-Braced Wing", NASA CR-159361, October 1980.

[47] Smith, P.M., J. DeYoung, W.A. Lovell, J.E. Price, and G.F. Washburn, "A Study of High-Altitude Manned Research Aircraft Employing Strut-Braced Wings of High Aspect Ratio", NASA CR-159262, February 1981.

[48] Grasmeyer, J.M., et al., "Multidisciplinary Design Optimization of a Strut-Braced Wing Aircraft with Tip-Mounted Engines", Virginia Polytechnic Institute and State University, MAD Center Report 98–01-01, Virginia Tech., January 1998.

[49] Grasmeyer, J.M., et al., "Multidisciplinary Design Optimization of a Strut-Braced Wing Aircraft", MSc thesis, Virginia Polytechnic Institute and State University, April 1998.

[50] Martin, K.C., and B.A. Kopec, "A Structural and Aerodynamic Investigation of a Strut-Braced Wing Transport Aircraft Concept", Lockheed Martin Aeronautical Systems, Report No. LG98ER0431, November 1998.

[51] Gundlach, J.F., et al., "Conceptual Design Studies of a Strut-Braced Wing Transonic Transport", *Journal of Aircraft*, Vol. 37, No. 6, pp. 976–983, November–December 2000 (Also: AIAA Paper 2000–0420).

Two-Surface and Three-Surface Airplanes

[52] Anderson, S.B., "Handling Qualities of Canards, Tandem Wings, and other Unconventional Configurations", SAE Paper No. 830763, April 1983.

[53] Butler, G.F., "Effect of Downwash on the Induced Drag of Canard-Wing Combinations", *Journal of Aircraft*, Vol. 19, pp. 410–411, May 10982.

[54] Kroo, I.H., "Tail Sizing for Fuel-Efficient Transports", AIAA Paper No. 83–2476, October 1983.

[55] Butler, G.F., "An Analytical Study of the Induced Drag of Canard-Wing-Tail Aircraft Configurations with Various Levels of Static Stability", *Aeronautical Journal*, Vol. 87, pp. 293–300, October 1983.

[56] McGeer, T., and I.M. Kroo, "A Fundamental Comparison of Canard and Conventional Configurations", *Journal of Aircraft*, Vol. 20, pp. 983–992, November 1983.

[57] Kendall, E.R., "The Minimum Induced Drag, Longitudinal Trim and Static Longitudinal Stability of Two-Surface and Three-Surface Airplanes", AIAA Paper No. 84–2164, 1984.

[58] Kendall, E.R., "The Aerodynamics of Three-Surface Airplanes", AIAA Paper No. 84–2508, 1984.

[59] Yip, L.P., "Wind-Tunnel Investigation of a Full-Scale Canard Configuration General Aviation Airplane", NASA TP No. 2383, March 1985.

[60] Arbuckle, P.D., and S.M. Sliwa, "Parametric Study of a Canard-Configured Transport Using Conceptual Design Optimization", NASA TP No. 2400, March 1985.

[61] Rokhsaz, K., and B.P. Selberg, "Analytical Study of Three-Surface Lifting Systems", SAE Paper No. 850866, April 1985.

[62] Selberg, B.P., and K. Rokhsaz, K., "Aerodynamic Trade-Off Study of Conventional, Canard, and Tri-surface Aircraft Systems", *Journal of Aircraft*, Vol. 23, No. 10, October 1986.

[63] Muchmore Jr., C.B., "The Effects of Canard-Wing Flow-Field Interactions on Longitudinal Stability, ERffective Dihedral and Potential Deep-Stall Trim", AIAA Paper 88–2514 CP, 1988.

[64] Goodrich, K.W., S.M. Sliwa, and F.J. Lallman, "A Closed-Form Trim Solution Yielding Minimum Trim Drag for Airplanes with Multiple Longitudinal-Control Effectors," NASA Technical Paper 2907, 1989.

[65] Pompeis, R de, P. Cinquetti, and S. Martini, "Development and Certification Flight Test on the Piaggio P.180 Avanti Aircraft: A General Overview", AIAA Paper No. 911003, April 1991.

[66] Middel, J., *Development of a Computer Assisted Toolbox for Aerodynamic Design of Aircraft at Subcritical Conditions with Application to Three-Surface and Canard Aircraft*, Doctoral Thesis, Delft University Press, ISBN 90–6275-768–5 / CIP, April 1992.

[67] Sacca, G., and C. Lanari, "The Three Lifting Surface Concept and Lessons Learned from the Piaggio P180", *Innovative Configurations and Advanced Concepts for Future Civil Aircraft*, von Kármán Institute for Fluid Dynamics, Lecture Series 2005–06, June 2005.

Aircraft with Non-planar Lifting Surfaces

[68] Miranda, L.R., "Boxplane Configuration: Conceptual Analysis and Initial Experimental Verification", Lockheed California Company Report, LR 25180, March 1972.

[69] Miranda, L.R., and G.L. Dougherty, "Transonic Wind Tunnel Testing of a Low Induced Drag Lifting System", Vols. I and II, Lockheed Aircraft Corp., Burbank, CA, February 1974.

[70] Lange, R.H., J.F. Cahill, et al., "Feasibility Study of the Transonic Biplane Concept for Transport Aircraft Application", NASA Contractor Report CR-132462, June 1974.

[71] Addoms, R.B., and F.W. Spaid, "Aerodynamic Design of High Performance Biplane Wings", *Journal of Airaft*, Vol. 12, pp. 629–630, August 1975.

[72] De Young, J., "Induced Drag Ideal-Efficiency Factor for Arbitrary Lateral-Vertical Wing Forms", NASA Contractor Report 3357, December 1980,

[73] Rhodes, M.D., and B.P. Selberg, "Benefits of Dual Wings over Single Wings for High-Performance Business Airplanes", *Journal of Aircraft*, Vol. 21, February 1984.

[74] Kroo, I.M., "A General Approach to Multiple Lifting Surface Design and Analysis", AIAA Paper No. 84–2507, 1984.

[75] Gall, P.D., and H.C. Smith, "Aerodynamic Characteristics of Biplanes with Winglets", *Journal of Aircraft*, Vol. 24, pp. 518–522, August 1987.

[76] Kroo, I.H., J. McMasters, and S.C. Smith, "Highly Non-Planar Lifting Systems", *Transportation Beyond 2000: Technologies Needed for Engineering Design*, NASA TP 10184, pp. 331–370, 1996.

[77] McMasters, J.H., D.J. Paisley, R.J. Hubert, I. Kroo, K.K. Bofah, J.P. Sullivan, and M. Drela, "Advanced Configurations for Very Large Subsonic Transport Airplanes", NASA CR-198351, October 1996.

[78] Grasmeyer, J., "A Discrete Vortex Method for Calculating the Minimum Induced Drag and Optimal Load Distribution for Aircraft Configurations with Noncoplanar Surfaces", Report VPI-AOE-242, January 1997.

[79] McMasters, J.H., and I.M. Kroo, "Advanced Configurations for Very Large Transport Airplanes", *Aircraft Design*, Vol. 1, No. 4, pp. 217–242, 1998; also AIAA Paper No. 98–0439.

[80] Frediani, A., "The Prandtl Wing", in *Innovative Configurations and Advanced Concepts for Future Civil Aircraft*, von Kármán Institute for Fluid Dynamics, Lecture Series 2005–06, June 2005.

[81] Kroo, I.H., "Nonplanar Wing Concepts for Increased Aircraft Efficiency", *Innovative Configurations and Advanced Concepts for Future Civil Aircraft*, von Kármán Institute for Fluid Dynamics, Lecture Series 2005–06, June 2005.

Joined Wing Aircraft

[82] Wolkovitch, J., Joined Wing Aircraft, U.S. Patents 3 942 747 (1976) and
[83] Wolkovitch, J., "Principles of the Joined Wing", Engel Engineering Report No. 80–1, December 1980.
[84] Samuels, M.F., "Structural Weight Comparison of a Joined Wing and a Conventional Wing", *Journal of Aircraft*, Vol 19, pp. 485–491, June 1982.
[85] Kuhlman, J.M., and T.J. Ku, "Numerical Optimization Techniques for Bound Circulation Distribution for Minimum Induced Drag of Non-Planar Wings: Computer Program Documentation", NASA CR-3458, August 1982.
[86] Hajela, P., "Weight Evaluation of the Joined Wing Configuration", Final Report, Department of Engineering Sciences, University of Florida, Gainesville, FL, 1983.
[87] Wolkovitch, J., and W.H. Bettes, "Low-Speed Wind Tunnel on Joined Wing and Monoplane Configurations", Vol. I: Analysis of Results, Vol. II: Test Data, ACA Report 82–2, 1984.
[88] Miura, H., A. Shyu, and J. Wolkovitch, "Parametric Weight Evaluation of Joined Wings by Structural Optimization", AIAA Paper No. 85–0642, April 1985.
[89] Wolkovitch, J., "The Joined Wing: An Overview", *Journal of Aircraft*, Vol. 23, No. 3, pp. 161–178, March 1986. (also AIAA Paper No. 85–0274, 1985).
[90] Selberg, B.P., and D.L. Cronin, "Aerodynamic-Structural Optimization of Positive/Negative Stagger Joined Wing Configurations", AIAA Paper No. 86–2686, October 1986.
[91] Wolkovitch, J., "A Second Look at the Joined Wing", in *Proceedings of the NVvL Symposium "Unconventional Aircraft Concepts"*, Delft University Press, 1987.
[92] Gallman, J.W., I.M. Kroo, and S.C. Smith, "Design Synthesis and Optimization of Joined-Wing Transports", AIAA Paper No. 90–3197, September 1990.
[93] Hashimoto, M., M. Ishikawa, N. Hirose, and T. Ohnuki, "A Computational and Experimental Analysis of Joined-Wing Aerodynamics", ICAS Paper No. 90–2.6.1, September 1990.
[94] Gallman, J.W., Structural and Aerodynamic Optimization of Joined Wing Aircraft", Doctoral Thesis, Stanford University, 1992.
[95] Gallman, J.W., S.C. Smith, and I.M. Kroo, "Optimization of Joined-Wing Aircraft", *Journal of Aircraft*, Vol. 30, No. 6, pp. 897–905, November–December 1993.
[96] Gallman, J.W., and I.M. Kroo, "Structural Optimization for Joined Wing Synthesis", *Journal of Aircraft*, Vol. 33, No. 1, pp. 214–223, January 1996.
[97] Bagwill, T., and B.P. Selberg, "Aerodynamic Investigation of Joined Wing Configurations for Transport Aircraft", AIAA Paper No. 96–2373, June 1997.
[98] Nangia, R.K., and M.E. Palmer, and C.P. Tillman, "Unconventional High Aspect Ratio Joined-Wing Aircraft with Aft- & Forward-Swept Wing-Tips", AIAA Paper No. 2003–0605, January 2003.

Multi-Body and Lifting-Body Aircraft

[99] Lange, R.H., "Trends in Very Large Aircraft Design and Technology", AIAA Paper No. 80–0902, 1982.
[100] Houbolt, J.C., "Why Twin-Fuselage Aircraft?", *AIAA Astronautics and Aeronautics*, pp. 26–35, April 1982.
[101] Arata Jr., W.H., "Very Large Vehicles: Technology Looking for a Need", *AIAA Astronautics and Aeronautics*, pp. 42–43, April 1982.
[102] Moore, J.W., and D.V. Maddalon, "Design Analysis and Benefit Evaluation of Multi-Body Aircraft", AIAA Paper No. 82–0810, May 1982.
[103] Moore, J.W., E.P. Craven, B.T. Farmer, J.F. Honrath, R.E. Stephens, and R.T. Meyer, "Multibody Aircraft Study Volume I and II., NASA Contractor Report 165829, July 1982.
[104] Grantham, W.D., P.L. Deal, G.L. Keyser Jr., and P.M. Smith, "Simulator Study of Flight Characteristics of a Large Twin-Fuselage Cargo Transport Airplane During Approach and Landing", NASA TP-2183, 1983.
[105] Nangia, R.K., "Multi-Fuselage Fuel-Efficient Aircraft Concepts (Including Conversions of Existing Types)", RKN/AERO/REP 85–09, Issue 2, September 1985.
[106] Udin, S.V., and W.J. Anderson, "Wing Mass Formula for Twin Fuselage Aircraft", *Journal of Aircraft*, Vol. 29, No. 5, pp. 907–914, September-October 1992.
[107] Lauter, C.E. et al., "Design Study: A Global Range Large Subsonic Military Transport", ICAS Paper 94–1.3.3, September 1994.

[108] Hahl, R., and J. Katz, "Lifting Fuselage/Wing Aircraft Having an Elliptical Forebody", U.S. Patent Application 08–642997, May 13, 1996.

[109] Spearman, M.L., "A High-Capacity Airplane Design Concept Having an Inboard-Wing Bounded by Twin Tip-Mounted Fuselages", AIAA Paper No. 97–2276, June 1997.

[110] Katz, J., S. Byrne, and R. Hahl, "Stall Resistance Features of Lifting-Body Airplane Configurations", *Journal of Aircraft*, Vol. 36, No. 2, pp. 471–474, 1999.

[111] Roskam, J., "Design and Economic Challenges of 10–22 Passenger, Jet-Powered Regional Transports", *Aircraft Design*, Vol. 3, No 1, pp. 33–48, 2000.

[112] Torenbeek, E., "The Case for Twin-Fuselage Configurations and the H-Cabin", National Aerospace Laboratory NLR-TR-2000–095, February 2000.

[113] Schoemaker, F.C., "Aeroelastic Characteristics of Twin-fuselage Configurations", National Aerospace Laboratory NLR-TR-2001–046, July 2001.

[114] Torenbeek, E., and E. Jesse, "Exploratory Design Study of a Transonic Airliner with Twin Fuselage Configuration", National Aerospace Laboratory NLR-TR-2001–309, July 2001.

[115] Kalyguina, G.V., et al., "The Dolphin: A New 100-Seat Aircraft in Lifting-Fuselage Layout", ICAS 2002 Congress, pp. 152.1–152.8, September 2002.

Liquid Hydrogen-Fuelled Aircraft

[116] Brewer, G.D., et al., "Study of Fuel Systems for LH2-Fueled Subsonic Transport Aircraft", NASA CR 145369, Lockheed California Co., July 1978.

[117] Klug, H.G., "Cryoplane: Quantitative Comparison of Contributions to Anthropogenic Greenhouse Effect of Liquid Hydrogen Aircraft Versus Conventional Kerosene Aircraft", EGS XXI General Assembly, Den Haag. May 1996.

[118] Westenberger, A., "Liquid Hydrogen Fuelled Aircraft: System Analysis, Cryoplane", Final Technical Report, GRD1–1999-10014, submitted to the European Commission, 2003.

[119] Svensson, F., A. Hasselrot, and J. Moldanova, "Reduced Environmental Impact by Lowered Cruise Altitude for Liquid Hydrogen-Fuelled Aircraft", *Aerospace Science and Technology*, Vol. 8, pp. 307–320, 2004.

[120] Svensson, F., "Potential of Reducing the Environmental Impact of Civil Subsonic Aviation by Using Liquid Hydrogen", Ph.D. thesis, Cranfield University School of Engineering, April 2005.

[121] Slingerland, R., "Aircraft with LH2 Propulsion", *Innovative Configurations and Advanced Concepts for Future Civil Aircraft*, Von Kármán Institute for Fluid Dynamics, Lecture Series 2005–06, June 6–10, 2005.

7

Aircraft Design Optimization

Instead of cutting the engineer out of the picture, optimization puts him squarely in the middle. It is another design tool; one that is particularly applicable when the interaction of more than one variable is involved.

—W.Z. Stepniewski and C.F. Kalmbach [17] (1970)

7.1 The Perfect Design: An Illusion?

In the early days of aircraft development, conceptual design was based on personal experience with previous projects and company tradition. Vehicle weight, flight performances and operating costs were estimated with the help of slide rules or calculators by means of rules of thumb and design handbooks. Geometry was defined on paper using drawing boards. Optimization was limited to relatively superficial investigations to study the effects of varying a few major design parameters. Since this approach allows an experienced designer to generate a new design in a short time, handbooks methods are still used in ad hoc studies. However, it has always been the engineer's dream to have all aspects of analysis done in a relatively short time period so that many different configurations can be examined and the best suitable product can be delivered on time. Although this may still be a dream, actual design turn-around time has become shorter due to the use of mathematical optimization techniques which have been introduced into the design process. How and when to use these techniques is arguably the key factor for advanced engineering operations [66]. One condition for being successful is that a proven computer augmented design synthesis program in combination with a suite of optimization algorithms is available.

Since the 1970s, there has been a tremendous expansion of optimization strategies and algorithms supporting advanced designers. The introduction of automated optimization has enabled designers to go into much greater in depth and fidelity of analysis than before. Synthesis programs effectively connecting the inputs and outputs of the functional group disciplines by means of an automatic control logic have been developed at aircraft manufacturers, research establishments and academia. Sophisticated computer assisted design (CAD) systems for defining three-dimensional body geometries and computer graphics tools for rapidly preparing parametric surveys are available at a modest cost. System engineering methods have brought

Advanced Aircraft Design: Conceptual Design, Analysis and Optimization of Subsonic Civil Airplanes, First Edition. Egbert Torenbeek.
© 2013 by Egbert Torenbeek. Published 2013 by John Wiley & Sons, Ltd.

about a paradigm shift in project development towards integrated product development (IPD) and – at least for traditional designs – this approach is highly refined and widely accepted. For an unusual aircraft concept, however, existing synthesis programs will have to be thoroughly modified as the risk of the results being unreliable is high since methods cannot be calibrated with statistical data. Moreover, advances in the field of practical optimization do not depend exclusively on the availability of fast computers or efficient optimization algorithms but on the overall company-wide development of computational frameworks geared toward flexibility, automation, and exploitation of high-fidelity analysis systems [93].

In spite of these developments, little progress has been made in creating computational tools to aid in the concept finding and early conceptual design phases. A great challenge to engineers carrying out this sensitive activity is not necessarily to conceive a realistic clean sheet airplane design but to convince project management that the best feasible concept complying with market needs is being proposed. Although it is often argued that there exists no substitute for the expert's insight based on former projects, widespread effort is devoted to improving formalized optimization methods. A basic problem is that the choice of a single figure of merit (FOM) defining the design quality is inadequate and even multi-objective optimization is not always the panacea. At the same time, the availability of sophisticated numerical analysis methods and optimization algorithms running on fast and cheap computers is tempting the inexperienced novice to try exercises generating irrelevant or even misleading results. In order to avoid this situation, advanced design managers should have an understanding of modern optimization tools for multidisciplinary tasks. The aim of this chapter is to offer an elementary introduction for non-specialists to approaches that have proven effective in conceptual design.

7.2   Elements of Optimization

The techniques for design applications of numerical optimization/search are very challenging. Although an explanation of optimization algorithms is beyond the scope of this book, an overview of elements and terminology important to advanced design engineers is presented. Textbooks [5, 6, 11] can be consulted for more detailed information. Optimization techniques for aircraft design synthesis use the same engineering process described for parametric surveys (Section 1.5). The basic difference between parametric and optimization studies lies in the introduction of an iterative process control system – this is the optimization driver in Figure 1.8 – which interprets numerical results and then iteratively assesses variables to seek the global optimum of an objective function. In the aircraft design context, the objective function is used to decide which configuration is considered the best combination of design variables. Typical examples are figures of merit, such as (specific) range, maximum take-off gross weight, energy efficiency and operating costs.

7.2.1   Design Parameters

The results of an optimization effort are largely determined by the problem structure which is set up to carry out the process. Consequently, it is essential to specify unambiguously how design parameters, defining the values of physical properties and quantities, are to be categorized during the computational process. In particular, the distinction between constants, independent and dependent variables has to be made carefully. The following terminology is used in the context of aircraft design:

- Pre-assigned parameters define properties or quantities which are fixed at the outset to remain constant during the optimization. Typically, pre-assigned parameters are derived from design requirements, accepted principles and assumptions regarding the technological state of the art, or previous design experience. In the case of a clean sheet design, for instance, it is likely that the design mission payload and range, properties of structural materials and the gas turbine engine cycle are listed as pre-assigned parameters. And in a design effort where the airframe is matched to a specified engine, engine characteristics such as the variation of thrust and fuel consumption with altitude, engine dimensions and weight fall into this category. However, in this particular case, design mission payload and range form the varying outcome of the optimization.
- Independent variables are parameters with a range of values in the modification procedure seeking the fittest aircraft for its task. In the context of design, they are are mostly called selection variables. A further subdivision distinguishes between (1) integer optimization when selection variables are integers – examples are the number of cabin seats and the number of propulsion engines; and (2) continuous optimization of selection variables that can be defined by any real number in a specific interval. Typical continuous variables are engine take-off power or thrust, wing area and aspect ratio. The number of independent variables reflects the design freedom and is called the dimensionality of the design space, their upper and lower values limit the boundary domain.
- Dependent variables – also known as behaviour variables – are parameters generated by the design (optimization) process. Forming the outcome of design analysis, their values are controlled by the selection variables. Typical dependent design variables are geometric parameters derived from geometric selection variables, weight and inertia moments of airframe components, aerodynamic parameters such as lift and drag coefficients and stability derivatives, and numbers characterizing the impact on the environment of aircraft operation.

In principle, every combination of selection variables leads to a design analysis resulting in a set of behaviour variables which together represent a unique aircraft design. Addition of a new selection variable will increase the design space by one dimension, that is, another set of designs is generated and included in the optimization process. In spite of the availability of powerful computers, the large amount of relationships to be enumerated and the iterative character of the analysis process can easily cause an explosion of computations. The use of analytically derived optima, advanced optimization techniques and efficient programming systems will help to avoid this situation. Engineering judgement is a prerequisite in making practical assumptions to direct the process.

7.2.2 Optimal Control and Discrete-Variable Optimization

Two fundamentally different types of optimization problems [17] can be distinguished with respect to aircraft design synthesis in general and optimization in particular.

1. Optimal control problems are encountered in aeronautics when searching for the most beneficial process mathematically described by a set of ordinary differential equations. The

goal of the problem is usually a requirement to find the optimum variation within a certain interval of several control parameters, for instance, angle of attack and/or engine control setting. The objective is a functional expressing a pay-off quantity – typically, an integral such as distance travelled or fuel consumed during a given time interval. Typically, optimal control problems are found in flight mechanics, for example, optimal trajectories for specified range [28, 49] and optimal time-manoeuvres. Another example is finding an airfoil section shape which minimizes the drag for a given lift. As opposed to design, aircraft flight mechanics and aerodynamics are mono-disciplines based on a well-established set of equations to be solved without relying on simplifying assumptions. Trajectory optimization is complicated by the flight dynamics aspect and in principle requires application of techniques such as the calculus of variations or dynamic programming.

2. Discrete-variable optimization is defined as a process where the objective to be optimized is expressed as a function of continuously variable parameters. This is done by means of a closed form analytical expression or by numerical elaboration of a set of linear or non-linear algebraic equations. The optimum is defined by discrete values of all variables, known as the optimizer. Most optimizations involved in the design of an aircraft or its components belong to the discrete-variable category. An example is finding a set of characteristics such as engine take-off thrust, wing area, span, sweep angle, root section thickness/chord ratio and taper ratio resulting in the lowest take-off gross weight.

Introduction of flight path optimization in conceptual design synthesis has been investigated by several authors [22, 62, 42]. Such a program can be incorporated in the mission analysis (Figure 1.5) which introduces an additional inner loop into the synthesis program. However, this approach complicates the computational system considerably and the rewards are mostly marginal. In subsonic civil airplane design, the dynamic segments of a flight path can often be ignored or simplified by (quasi-)analytical approximations. For example, in the conceptual design stage, Bréguet's range equation for cruising flight in combination with analytical approximations for non-cruising segments is generally accurate enough to compute the mission range (Chapter 12). The kinematic aspects can then be parameterized which reduces the problem to discrete optimization. The following discussion applies to discrete-variable optimization.

7.2.3   Basic Terminology

The topic of non-linear constrained optimization determined by enumeration of constituent equations defining a technical system is treated in many books, e.g [5, 6]. We furnish here a brief summary of the terminology and the basic mathematical formulation of an optimization problem.

- The objective function is a scalar function of the design variables that is to be minimized or maximized during the optimization. Alternative terms are cost function, figure of merit (FOM) and effectiveness criterion.
- Constraints are functions of the design variables representing limitations imposed upon the design. Based on sound engineering judgement, constraints are derived from top-level

requirements, airworthiness regulations, operational rules and other practical or physical restrictions. A distinction is made between equality, inequality and side constraints.

- An equality constraint may be imposed on a behaviour variable such as the design mission range. Such a constraint reduces the number of variables by one and thereby eliminates one dimension from the design space. Many sizing conditions are translated into equations acting as equality constraints, although they are not always recognized as such since many of them are nothing but generic aeronautical relationships. For example, the condition that in straight and level flight $T = D$ can be interpreted as: 'In a specified flight condition and cruise rating, engines are sized to deliver the thrust required to balance drag.'
- An inequality constraint is a practical or physical condition that distinguishes feasible from unfeasible designs. An inequality constraint reduces the size of the design space so that certain combinations of independent variables need not be considered. In aeronautics, the set of inequality constraints nearly always contains some complicated and implicit functions of the independent variables. Typical examples are upper limits on take-off field length required and approach speed and a lower limit on the initial cruise altitude. A typical example of a geometric inequality constraint is the condition that the wing must have enough volume to contain all the fuel required for a specified long range mission. Depending mainly on wing planform shape and mean thickness ratio, this constraint leads to a lower limit for the wing area. Feasible designs are defined as combinations of the selection variables which comply with all the inequality constraints.
- A side constraint on a selection variable can be interpreted as follows: For selection variable x_1, consider only the range of values between specified x_L and x_U.

An active constraint is an equality constraint or an exactly satisfied inequality constraint. A feasible design is a vector of values for the selection variables that satisfies all the constraints. The complete collection of feasible points is called the feasible region.

7.2.4 Single-Objective Optimization

Many optimization problems in conceptual aircraft design are solved by minimizing a cost function – such as a component of weight or operating costs – or by maximizing a merit function – such as payload or range. This type of problem is called a single-objective optimization since a single cost or merit function is considered as representative for the overall quality of the design. The problem formulation is as follows:

Independent variables are represented by a set of N real numbers known as the control vector

$$\vec{X} = (x_1, x_2, \ldots, x_n, \ldots, x_N)^T, \quad \text{with} \quad x_U \geq x_n \geq x_L \tag{7.1}$$

They are bounded by upper and/or lower side constraints x_U and x_L, respectively, defining a closed boundary domain. The optimization task is to find the control vector which minimizes the objective function $F(\vec{X})$ which is assumed to be continuous inside the boundary domain. The control vector is subject to a set of inequality constraint functions,

$$U \geq g_i(\vec{X}) \geq L \quad i = 1, 2, \ldots, n, \ldots, I \tag{7.2}$$

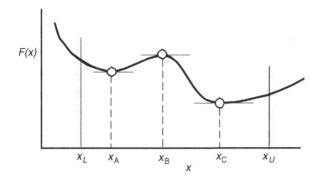

Figure 7.1   Side constraints and optimizers in linear optimization

where U and L denote the upper and lower bounds of the control vector, respectively. An equality constraint function g_n can be specified by setting $U_n = L_n$. In order to avoid the problem becoming over-constrained, the number of equality constraints must comply with $I < N$. The optimization thus formulated serves to find the optimum of a convex function $F(\vec{X})$. Maximization of a concave function $F(\vec{X})$ is executed by minimizing $-F(\vec{X})$ or $1/F(\vec{X})$ for $F(\vec{X}) > 0$.

In the application to a design problem, $x_1, x_2, \ldots, x_n, \ldots, x_N$ are selection variables which can be interpreted as components of $\vec{X}$ defining the N-dimensional Euclidian design space. Their variation results in a set of $F(\vec{X})$ values which may be represented and visualized by an $(N + 1)$ - dimensional response surface. Chapters 8 through 10 treat examples of optimization for problems with two variables in the form of three-dimensional response surfaces. The following text explains how the geometry of a response surface can be interpreted.

The objective function $F(x)$ and the constraint functions $g(x)$ are related to the design variables through a set of non-linear equations included in the modules of the design sizing process. These equations are solved numerically using the dataflow depicted in Figure 1.8. The simple example of a two-dimensional response surface is the result of a linear model as illustrated in Figure 7.1. A local optimizer is a feasible design with an objective function value at least as good as any other nearby feasible point. A local minimizer of $F(x)$ is x_A. The first derivative $\partial F(x)/\partial x$ is negative to the left and positive to the right of x_A and zero in x_A; hence, the objective function is convex. The same applies to x_C. Since the second derivative of $F(x)$ is positive in A and C, points x_A and x_C are local minimizers. Point x_B is a local maximizer since the second derivative is negative in B where the objective function is concave. A global optimizer is a local optimizer that yields at least as good a value for the objective as any other local optimizer in the boundary domain. Point x_C in Figure 7.1 is the global optimizer of $F(x)$.

7.2.5   Unconstrained Optimizer

Application of Newton's theory of extremes is the simplest possible approach to optimization if there are no inequality constraints on the variables and the objective function is differentiable

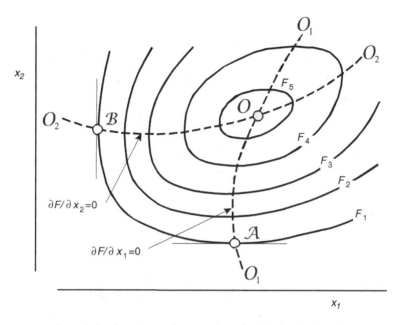

Figure 7.2 Optimizers of a two-dimensional objective function

by analysis. If $F(\vec{X})$ is continuous and differentiable, its partial derivatives can be set equal to zero and the resulting set of equations is solved for the optimizer. The character of the solution is determined by the sign (positive, negative, or zero) of the second partial derivatives as defined by the Hessian matrix. A local minimizer is obtained if in the immediate neighbourhood of the optimizer the objective function everywhere has a higher value.

Figure 7.2 depicts the objective function $F(\vec{X})$ of a two-dimensional optimization with independent variables x_1 and x_2 varied along orthogonal axes in the plane of the drawing. $F(\vec{X})$ forms a three dimensional response surface represented by indexed curves connecting points with equal values of F. Their projections in the drawing plane visualizes the response surface and is known as a contour plot – alternative names are iso-merit plot or thumbprint plot. The example shows the typical hilly character often found in conceptual design optimization. For a given x_2, point $\mathcal{A}$ defines the partial optimum $\partial F / \partial x_1 = 0$ where the contour has a horizontal tangent. The collection of points $\mathcal{A}$ for variable x_1 is denoted by curve O_1. Similarly, point $\mathcal{B}$ defines $\partial F / \partial x_2 = 0$ in the vertical tangent to an F-contour, with curve O_2 forming the collection of these partial optima. The minimizer of $F(\vec{X})$ is point $\mathcal{O}$ where O_1 and O_2 intersect. The minimum objective function in a three-dimensional design space is defined by two zero first derivatives and two positive second derivatives.

In the literature on optimization, the situation is sometimes mentioned of an objective function with a saddle point. In the saddle point shown in Figure 7.3, both partial derivatives are zero, whereas the second derivative is positive for x_1 and negative for x_2. Although this situation is mathematically well defined, its interpretation in the framework of design optimization is problematic: a saddle point defines a local maximum and a minimum at the same time of the same objective function. A unique example of a saddle surface is the response

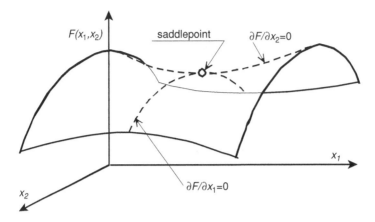

Figure 7.3 Objective function with a saddle point

surface of the aerodynamic efficiency L/D in Figure 10.11. The addition of more independent design variables increases the dimensionality of the problem to become a hyperspace which is difficult – if not impossible – to visualize. The global optimizer with respect to all selection variables is the intersection in the same point of the partial optimizers $\mathcal{O}_1, \mathcal{O}_2, \ldots, \mathcal{O}_N$.

When partial optimizers do not intersect, they are incompatible with each other. In that situation there exists no global minimizer and the solution requires introduction of one or more constraints. An example of incompatible partial derivatives is experienced when computing the most fuel-economical cruise speed and altitude of a jet airplane for which the aerodynamic model is oversimplified by ignoring the compressibility drag (Chapter 12). Another complication in design optimization occurs when the objective function is obtained via the iterative solution of a large set of equations, making analytical differentiation unfeasible. Practical solutions are the use of calculus-based methods using local (numerically obtained) derivatives or direct search methods based on approximation of response surfaces (Section 7.3).

7.2.6 Constrained Optimizer

In principle, imposing an equality constraint can be treated by adding an additional equation to the system to be solved which effectively reduces the number of independent variables by one. For instance, if J equality constraints are incorporated into the system, the dimensionality of the problem is reduced to $(N\text{-}J)$. The reduced problem is then formulated as follows:

Minimize the objective function $F(\vec{X}')$ where the modified control vector of $N - J$ selection variables is defined as

$$\vec{X}' = (x_1, x_2, \ldots, x_{N-J})^T \tag{7.3}$$

subject to inequality constraints

$$U \geq g_i(\vec{X}') \geq L \quad i = 1, 2, \ldots, (I - J) \tag{7.4}$$

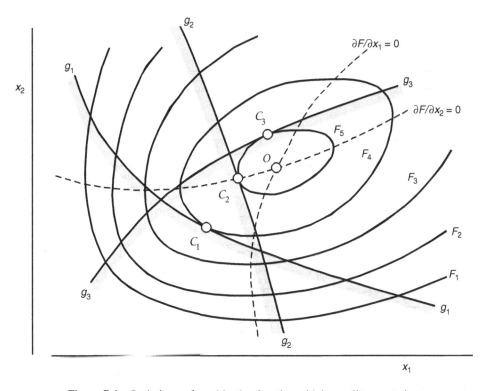

Figure 7.4   Optimizers of an objective function with inequality constraints

Let us assume that the unconstrained problem of Figure 7.2 is augmented by adding several inequality constraints. Each constraint subdivides the design space into a feasible region and an unfeasible region, as indicated by the shaded curves in Figure 7.4. The feasible region is to the right of constraint g_2 and above constraint g_3. The new situation leads to the following observations:

- Global optimizer $\mathcal{O}$ is unfeasible since it is located in the unfeasible region. This invalidates the method of intersecting partial optima to find the (feasible) global optimum.
- Constraint g_1 is inactive in the complete design domain and its optimizer $\mathcal{C}_1$ is unfeasible (and irrelevant).
- Optimizer $\mathcal{C}_2$ on constraint g_2 is unfeasible since it is not located in the feasible region.
- Optimizer $\mathcal{C}_3$ on the active constraint g_3 is on the boundary of the feasible region. Consequently, $\mathcal{C}_3$ represents the constrained optimizer of the objective function.

The present example features a global optimizer determined by one active constraint. In aircraft design optimization it is often found that the best feasible optimizer is found on the intersection of two inequality constraints. On the other hand, a feasible design cannot always be identified. More explicitly, in the case of very ambitious top-level requirements it may be impossible to conceive a design that complies with all constraints. If it appears that constraints

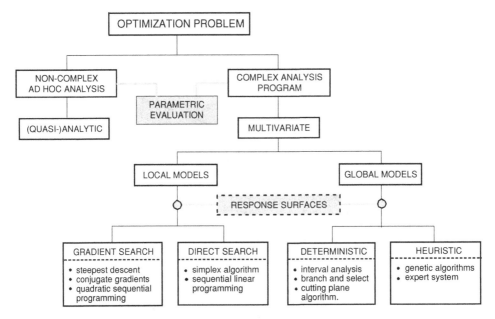

Figure 7.5 Classification of methods for single-objective optimization in aircraft design

have an unfavourable influence on the overall design quality, the advanced design team should have the authority to decide that, instead of aiming at an overambitious optimum, aiming at a more practical or economical design by relaxing constraints is preferable. As a result, the constrained design becomes a closer approximation of the unconstrained global optimum.

7.3 Analytical or Numerical Optimization?

7.3.1 Analytical Approach

The choice of an optimization strategy is related to the type of problem. The classification in Figure 7.5 makes a distinction between non-complex problems dealing with a small design space and complex non-analytic problems with many independent variables. This classification corresponds with different stages of advanced design practice where computational complexity increases from statistics-based superficial studies via more detailed parametric and trade-off studies to complex high-fidelity multidimensional analysis and multivariate optimization (MVO).

In order to build up confidence in computer-generated answers, it can be useful to use a model which approximates parts of the solution by means of an analytical method. For instance, ad-hoc problems may be solved by using Newton's theory of extrema which requires the objective function and constraints to be analytical functions. The global optimizer is the design which satisfies all partial optima. Equality constraints may either be substituted into the set of system equations or eliminated by adding pseudo-variables known as Lagrange multipliers to the original variables. A minimizer for the constrained problem is then found by minimization of the original objective function increased by the Lagrangian function.

Table 7.1   Assessment of parametric surveys

Advantages
Traditional method rooted in industrial practice
Relatively simple programming
Designer keeps control over decisions
A priori choice of objective function is not required
Sensitivity of the design to off-optimum conditions remains visible

Disadvantages
Few selection variables can be handled – four to five at most
Insufficient to guarantee that (near-)optimum is identified
Many designs are computed, few are actually used
The analyst is not encouraged to extend the search
Changes in design (requirements) make most results obsolete

An interesting strategy is geometric programming (GP) which applies to a special category of problems based on monomial and posynomial functions. A monomial has the form

$$f(x) = c\, x_1^{a_1} x_2^{a_2} \ldots x_n^{a_n} \tag{7.5}$$

where $c > 0$ and $a_i \in \mathbf{R}$. A sum of one or more monomials is called a posynomial function. An early publication on GP is found in [1], a recent overview is published in [60]. An application of GP to aircraft design is imaginable in the form of minimization of a posynomial consisting of weight components derived from regression analysis of statistical data. However, the use of statistical equations is not always recommended since they may give a false impression of design sensitivity.

Parametric surveys (Table 7.1) are derived from a non-complex and well-organized set of equations enabling the analyst to compute and visualize response surfaces with constraints included. An example is depicted in Figure 1.9. This approach allows a flexible way of explicit optimization and decision-making. Compared to automated design which gives little insight, parametric surveys are useful to explain the sensitivity of a design to non-optimal design variables. They are also useful for investigating the effects of varying design constraints and for technological or trade-off studies. However, such investigations are feasible only for a problem with not more than five selection variables.

7.3.2   Multivariate Optimization

Multivariate optimization (MVO) is a powerful tool if the AD team uses a validated computer-assisted design sizing and analysis system. For comprehensive optimization tasks it is necessary to use a mathematical technique which can handle a large range of selection variables – between ten and thirty, typically. The module containing a suite of available MVO algorithms should be loosely coupled to the design sizing program. The scheme in Figure 7.5 makes a distinction between local and global models.

- Local models proceed in a stepwise manner. They start at an initial guessed point, compute a new control vector pointing in a direction which improves the design. A design under

Table 7.2   Assessment of multivariate aircraft design optimization

Advantages
Rigorous approach, may potentially lead to improved design quality
Minimum occurrence of biased decisions
The only viable approach in the case of a non-conventional configuration
Changes in the baseline design/requirements are readily accommodated

Disadvantages
Programming and debugging are complicated and time-consuming
Convergence problems may occur
The program may be stuck in a local instead of the global optimum
Several algorithms may have to be tested before a sensible result is found
There is no insight in the design sensitivity to primary design variables
Inexperienced designers may produce unrealistic results

---

analysis during this process is called the active point. The search is ended when the objective function becomes stationary in a local optimizer.

• Global models are aimed at identification of local as well as global optimizers. They make direct evaluations of the objective function of (many) carefully selected designs distributed within the closed boundary domain. The process may proceed with optimization of a response surface approximation or by using a heuristics-based algorithm.

Automatic optimization greatly reduces the number of configurations to be analyzed and avoids the enormous waste of generating configurations that have to be rejected afterwards. Although an abundance of strategies for MVO have been investigated and published, not many of them have been tested in a realistic aircraft design environment. Introduction of MVO is often associated with so many computational pitfalls and failures that finding a mathematically exact and reliable solution appears elusive. Although the use of automated optimization has found applications in many technical design disciplines, including aerospace design, MVO methods continue to have disadvantages. They are intrinsically difficult to program and debug, the optimization algorithms are not always effective, and convergence problems may mean that no solution is found. Dependent on the computational expense per configuration, the analyst may run into the problem of size which is associated with the large-scale and complexity. It is concluded from the overview of the pros and cons of MVO in Table 7.2 that a decision to apply MVO to an optimization problem must be thoroughly prepared.

Certain complex non-analytic problems are well behaved; that is, the objective function varies smoothly with each independent variable. In this case, algorithms known as indirect or gradient search are likely to locate the optimum most rapidly. Gradient methods use calculus-based information to find local derivatives of the objective function known as sensitivity information. This is used to form a local model of the problem in a region near the active point. Local models are low-degree polynomials designed to approximate the true objective and constraint functions. They are used to derive the control vector pointing in the direction of the nearest local minimizer. Every step to a new active point is normal to the local contour curve of the objective function.

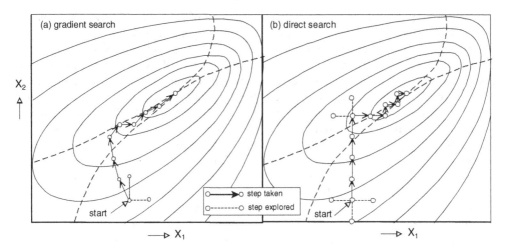

Figure 7.6 Comparison between steepest descent and direct search methods for unconstrained optimization

7.3.3   Unconstrained Optimization

Unconstrained optimization using gradients is illustrated in Figure 7.6 (a) showing a stepwise search starting at an initial guessed set of independent variables. The explored steps are used to find the optimizer by means of a local approximation model. Moving along the steepest slope of the local response surface approximation improves the local objective and is found by one-dimensional probing in the orthogonal directions of the variables. If quadratic polynomials are used to approximate the response surface, its local curvature can be used to derive the Hessian matrix and the optimal step size. This leads to a more efficient search at the cost of increased computational effort. The process proceeds along steps with continuously varying directions and step size, until no improvement is obtained. This method is known as the steepest descent algorithm. To a large extent, the gradient search path is determined by the location of the partial optima, depicted by the dotted lines.

Dependent on the location of the starting point, the gradient search can be a slowly converging process. The conjugate gradient algorithm uses information from the previous iteration steps which improves the convergence considerably. In principle, gradient search identifies an exact solution of the local optimizer. Gradient methods are often applicable in conceptual design because many problems are well behaved due to the simple and continuous mathematical models used. In more complex programs, however, indirect search methods may get stuck in a local optimum. It is advantageous to start the gradient search with a design point that is expected to be feasible. Since a gradient method identifies a local optimizer, it is advisable to repeat the process by starting the search at different points. A more serious problem occurs when the response surface exhibits kinks or discontinuities caused by a transition between different design criteria applying in different regions of the design space. Another difficulty is computational noise that occurs when complex multidisciplinary computational systems are involved. Many engineering optimization problems are of these awkward types, and predicting their behaviour is difficult. In these situations some form of direct search may perform better since it uses only (objective) function information to find a global optimum.

Direct search methods are useful when local design sensitivity information is lacking since they do not make use of objective function derivatives. Instead, direct search methods merely generate a number of designs and use a direct search algorithm to identify a local optimum. A typical direct method is sequential linear programming (SQL) which approximates the objective function and the constraints by a first-order Taylor expansion. The resulting linear constrained problem can be solved with standard methods such as the widely used simplex algorithm [25]. Figure 7.6(b) shows an example of direct local search with steps taken in orthogonal directions of the independent variables. Starting at the initial guessed point, initial steps are a line-search in the direction of one axis until the objective is no longer improving. The line-search then proceeds along the other axis. At the point where the object function becomes stationary, the first variable is no longer on its partial optimum and the next step is resumed in the original direction. In the present example, the final steps are zigzagging between the two partial optima, indicating slow convergence. Although direct search methods do not provide an exact solution, they are reliable and robust in terms of convergence. However, the direct method approach needs expensive design analysis for every step of the optimization algorithm.

7.3.4   Constrained Optimization

An efficient method of constrained optimization using gradient search is sequential quadratic programming (SQP). A calculus-based algorithm is used to find a local quadratic approximation of the objective function subject to a linear approximation of the constraints (Figure 7.7). SQP uses the gradient method to search for the steepest gradient at the (feasible) active point S and then performs a line search in that direction. As soon as the active point reaches the unfeasible region of a constraint, the search direction is changed and continued along the local constraint function until the design reaches the constrained optimizer C. This process generally leads rapidly to the optimizer, provided the starting point is suitably chosen. An implementation of the SQP code can be found in [38]. Like all gradient methods, SQP is

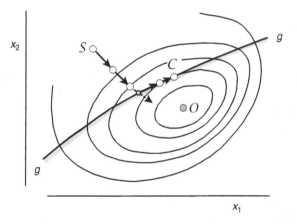

Figure 7.7   Geometric interpretation of the SQP method for two independent variables

only useful in problems which do not produce discontinuities of the response surface and its derivatives such as ridges and sharp bends. The method is also sensitive to a computational noise error on the derivatives inherent to the solution of complex numerical analysis such as CFD and FEM methods. Applications in aircraft design are reported in [41, 50].

Several algorithms try to cast the constrained problem as an unconstrained one and then use one of the algorithms from that category. They make use of a penalty function which adds a penalty on the objective in the region where constraint violation occurs. A pseudo-objective function created from the original objective by adding the imposed penalty is subsequently optimized. This method is known as a sequential unconstrained minimization technique (SUMT). An alternative approach uses the Kreisselmeier-Steinhauser (KS) method to replace the constraint and objective function boundaries in the design space with a single surface [43]. The KS method is based on a continually differentiable function,

$$\text{KS}(X) = \frac{1}{\rho} \ln \Sigma_{k=1}^{K} \exp\{f_k(X)\} \tag{7.6}$$

where $f_k(X)$ is a set of K objective and constraint functions. The parameter ρ controls the distance of the KS function surface from the maximum value of this set of functions evaluated at X. This envelope function formulation effectively converts a constrained problem to an unconstrained single-objective formulation which can also be useful for multilevel optimization to be discussed later.

7.3.5 Response Surface Approximation

Direct methods are used for searching for the global optimizer of the complete design domain. They are not based on a stepwise improvement of active points but evaluate a (usually large) number of design points obtained by carrying out exact analysis. However, since computation of the complete response surface increases exponentially with the size of the problem, it may be considered (a) to simplify the design analysis model or (b) to simplify the search process. The first option may lead to an inaccurate response surface – which is probably unacceptable – whereas the second option may or may not lead to an inaccurate optimizer, an uncertainty that may be acceptable. Therefore, methods have been developed which potentially generate an approximation of the 'real optimum'. Compared to integration of the detailed and expensive high fidelity analysis into the optimization algorithm, approximation models require a modest number of direct and exact evaluations which is computationally inexpensive. By performing the simulation *a priori*, the simulation model is effectively separated from the surface fitting and optimization efforts. This enables the analyst to run these processes sequentially and makes the complete procedure versatile. An early example of an application to turbofan engine selection is found in [26].

Approximation models for response surfaces replace the expensive simulation codes in the analysis process and thereby reduce computational costs. They approximate the true response surface by a simple function of the independent variables. The approximation starts by computing the objective and constraint functions at a number of sample points, carefully selected by a suitable method of experimental design;[1] thereafter, they are all approximated using different response surface fits. Approximation models are typically based on low-degree

---

[1] In this context, 'design' means the method for selecting sample points rather than the original engineering problem.

polynomials with coefficients for the terms derived from regression on the true function values. The surface fit is tested by means of statistical evaluation and after its acceptance the optimization itself proceeds with gradient search on the response surface fit using one of the algorithms from that category. The global response surfaces identify local as well as global optimizers and provide designers with insight into the behaviour of complex phenomena. For instance, a screening process can be used to eliminate variables from the design problem which appear to have little effect on the objective.

7.3.6   Global Models

Global optimization is the field of mathematics that studies extremal locations of non-convex functions which have many different local optima. Due to the high complexity of global optimization, its execution is computationally very costly and the method has not yet found widespread application to advanced airplane design. Therefore, the following overview is superficial and concentrates on the two basic classes of algorithms.

1. Deterministic methods take advantage of the analytical properties of the problem to generate points that converge to a global optimizer. Examples are branch and select algorithms, cutting plane algorithm, interval analysis and dynamic programming.
2. Heuristic methods are basically rules of thumb that hopefully will provide a good answer. They have been found to be more flexible and efficient than deterministic methods.

Since the quality of the solution obtained with a heuristic method cannot be guaranteed, application is not advised for problems where good conventional methodologies exist. This is especially true if the knowledge can be expressed in terms of mathematical models. Nevertheless, successful applications of heuristic optimization techniques to aircraft design have been published for genetic algorithms and expert systems.

Genetic algorithms perform design optimization by using Darwin's survival of the fittest model of manipulating long bit strings as the analogy for the genes of replicating individuals. Analogous to a natural population, the starting point is a collection of designs, each with a different set of selection variables mapped into the bit strings. Progress towards a 'fitter population' is made by iteratively 'mating' promising bit strings to form new bit strings and by randomly 'mutating' bit strings. Genetic algorithms are relatively easy to implement and have the advantage that they do not get stuck in local optima of convoluted response surfaces. However, they do not guarantee a global optimum and may require too many function evaluations to yield a satisfactory approximate solution. See [9] for a detailed description of genetic algorithms and [74] for an application to a supersonic airliner design.

An expert system is a method from the field of artificial intelligence (AI) explicitly based on the problem-solving ability of human experts. Experts systems are based on an amorphous object-oriented data base known as a rule base or knowledge base[2] and on a single module called an inference engine. The inference engine has control over the execution of rules containing declarative statements about objects relevant to the application. The relevance of expert systems to optimization lies in the potential to manipulate formalized knowledge about

---

[2]Expert systems for application in the design field are also known as knowledge-based engineering (KBE).

objects. Objects can be physical (aircraft) components, data structures, values of parameters or design procedures. See [94] for an application to aircraft design.

7.4   Large Optimization Problems

> There is the risk that you gradually expand all your variables and parameters and then find that most of them are irrelevant anyhow and that the most important one is somewhere else. It will be a long job to establish the viability of total optimization.
>
> —R. Cockburn [23]

Since the early 1980s, a new field of research has been explored to overcome the challenge of integrated analysis and optimization of artifacts represented by complex engineering systems, with multidisciplinary optimization (MDO) of an aircraft design forming a prominent example. The outcome is inherently compatible with the way engineers cooperate in a design organization with distributed computing capabilities. Prominent contributions to MDO technology have been delivered by J. Sobieszczanski-Sobieski and his co-workers at NASA Langley. During the past decennia, MDO has been developed on the basis of a considerable theoretical framework and a multitude of practical applications into a powerful tool in the hands of designer teams. MDO has sparked intense interest from industry as well as the research field.[3] MDO has thoroughly changed the way in which design is being organized and it is justified to refer to it as a new engineering discipline [72].

Computer simulations of steadily increasing power are becoming available for modelling a wide variety of physical phenomena. In the advanced design (AD) environment, these simulations enable the engineer/analyst to learn more about the processes and physics involved than they have been able to learn from physical experiments. However, in view of their complexity, sizing and optimization programs have traditionally been developed and maintained by specialists in isolation from what is done in other technical fields. As a result, the optimization process consisted of sequential disciplinary activities leading to sub-optimization, computational inefficiency and project delays. In today's state of practice, the methodology of multidisciplinary analysis and optimization coherently exploits the synergism of mutually interacting computational domains to improve the design of complex engineering systems.

7.4.1   Concept Sizing and Evaluation

Conceptual design involves the exploration of alternate concepts to satisfy vehicle design requirements. Concept sizing can be described as a program for determining the geometry, weight, performances and other physical characteristics of a design complying with a specified mission, including constraints on its feasibility. Basic constituents of a concept sizing program are: (1) a data processing system; (2) contributing disciplinary analysis programs; (3) a user interface for configuration control and evaluation; and (4) a suite of optimization algorithms. Figure 7.8 illustrates how a broad range of disciplines are interacting in a conceptual sizing program. The outcome of the process is defined in terms of flight performances, flying qualities and economical figures of merit.

---

[3]The steadily growing body of literature on MDO is too large to be completely cited in this chapter's bibliography. Progress is reported every two years at AIAA/ISSMO conferences.

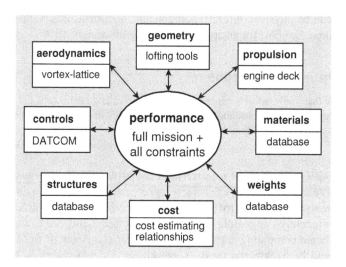

Figure 7.8   Conceptual design disciplines and tools. Copyright NATO STO - CSO

When selecting or developing a concept sizing tool, the emphasis is on freedom in representing the general arrangement and technology. The program should enable designers to freely vary selection parameters, constraints and objectives. Disciplinary analysis programs are based on low or modest fidelity computations of proven quality derived from handbooks, in-house knowledge and statistics. The supporting computer assisted modelling system should be able to generate parametric geometries enabling several competing designs to be defined by three-dimensional models. Evaluation of design quality is enabled by parametric surveys, trade-off and 'what-if' studies.

The process of airplane concept sizing is in the hands of a relatively small AD team and generally has the character of a single-discipline optimization. However, a concept sizing program is mostly drawn up and maintained in co-operation between the AD team and functional groups contributing knowledge from their fields of specialization. Therefore, concept sizing has a distinct multidisciplinary character and optimization at this stage is considered a class of MDO problems, namely, single-level optimization. Different from large MDO systems, the essence of computational support is the provision of a flexible, robust and user-friendly system rather than computer-intensive simulations. A major task is making an evaluation of the figure of merit on which selection of the best design can be based. This may ask for multi-objective optimization.

7.4.2   Multidisciplinary Optimization

After a design concept is selected, the design and analysis process evolves toward the preliminary design phase during which actual components and subsystems are specified. Specialists from various functional groups of the company become involved in the design and analysis of different subsystems. They use increasing detail in their specialism resulting in a more limited interaction with other disciplines. Nevertheless, interactions between different disciplines and

Table 7.3 Information science and technology associated with MDO based on the discussion by Sobieszczanski-Sobieski [72]

Information science and technology	Design-oriented multi-disciplinary analysis	MDO
product data models	mathematical modelling	discipline optimization
data and software standards	trading cost vs. accuracy	decomposition
data management, storage, and visualization	smart re-analysis	design space search
	sensitivity analysis	approximations
software engineering practices		optimization procedures
human interface		

between disciplines and the total vehicle system remain essential. Multidisciplinary optimization (MDO) can be described as a methodology for the optimum design of systems where the interaction between several functional groups must be considered and where a designer in one functional group is free to significantly affect the performance in other disciplines [73] and at the system level. In the design of complex engineering systems such as aircraft, multidisciplinary interactions are prominent in the organization of integrated product development (IPD) in general and MDO in particular. Subjecting such a system to all-out optimization would lead to an unmanageable data flow and/or a very expensive computational process.

Table 7.3 is a taxonomy of the MDO discipline. A prerequisite for MDO is the availability of parametric product data models (PDMs) forming the basis for the geometry and discretization models that are consistent across the disciplines [76]. In order to integrate MDO programs into an aircraft design effort, an analysis and optimization process has to be realized that accounts for the effects of mutual interactions of several engineering disciplines. An essential feature required for MDO is that a good system control software exists for communication and transfer of data between the functional groups involved. In principle, the MDO methodology can be used at any design stage although its complexity does not justify application in the early conceptual phase. A great deal of effort is presently being devoted to making MDO methods utilitarian for preliminary aircraft design by including manufacturing considerations and constraints, and estimations of cost implications in the analysis.

7.4.3 System Decomposition

The MDO methodology is based on a formal method for analysis and optimization of large engineering systems by system decomposition introduced in the 1980s [61, 65]. A large optimization task is decomposed into a set of smaller (less dependent) subtasks that can be solved concurrently while preserving the interactions between them (Figure 7.9). Dependent on the problem structure, the subtasks are arranged in hierarchical levels with the assembled system at the top level. Sub-domain processes are carried out receiving inputs from other sub-domains and optimizing on a local level. The system level process controls the exchange of object and constraint function information between sub-domains. Decomposition methods are aimed at identifying the best sequence of disciplinary sub-processes resulting in maximum

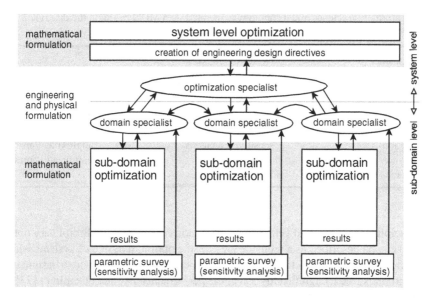

Figure 7.9 Schematic overview of a large industrial bi-level integrated system optimization process. In reality, the number of sub-domains is much higher than three. Adapted from [80]

computational efficiency through minimizing the number of iteration steps. The principle of decomposing enables the analysis and (gradient-based) optimization to be performed concurrently and independently, accounting for the effects of changing a parameter in one subsystem on parameters of the other subsystem. It turns out that decomposing the optimization problem has the following advantages:

• enhanced insight into the communications between subsystems;
• enhanced insight into the design space and how the optimum is achieved;
• functional groups handle optimizers and sensitivity information of subsystems;
• flexibility in the choice of dependent and independent variables; and
• flexibility in choosing the methods of analysis for each subsystem.

System decomposition is in line with procedures in the industry where collaborative engineering is a fact of life and it can be facilitated by using tools from system engineering. The design process of a technical system can be represented by a compact matrix form that visualizes and optimizes the flow between subsystems. A well-known representation is the design structure matrix (DSM) or N^2 diagram, with N denoting the number of subsystems. As an example, Figure 7.10 depicts a small DSM with subsystems (or design tasks) arranged on the diagonal. The order of execution of the subsystems is reflected by their position on the diagonal, starting top left and ending bottom-right. Feed-forward relations are characterized by dots above the diagonal, dotted feedback entries are located below the diagonal. The position of the subsystems along the diagonal has a significant impact on the effectiveness of the computational process. Partitioning algorithms and/or common-sense judgement are used to rearrange the positions of the subsystems in order to remove as many feedback relations as

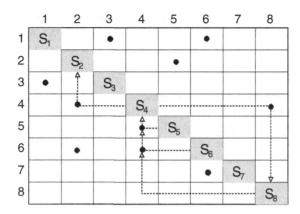

Figure 7.10   Example of a simple DSM [92]. Subsystem 4 provides input to subsystems 2 and 8, and it receives input from subsystems 5, 6 and 8

possible, thereby minimizing the number of computational iterations. For instance, when the positions on the diagonal of tasks 6 and 7 are reversed, the feedback is transformed into a more preferable feed-forward. However, real-life design problems can be too complicated to capture the computational process in a nicely arranged DSM.

7.4.4   Multilevel Optimization

It appears beneficial to separate the design variables and constraints local to a sub-domain from those that govern the entire system. Multilevel optimization is a methodology to exploit disciplinary optimization of a decomposed system. In a collaborative setting, multilevel optimization is becoming part of the MDO toolbox aimed at dividing the total optimization into sub-domain optimizations and coordinating them at the system level. This method coordinates the information flow from various sub-domains, all having their specific definitions of the objective function and of the sensitivity of one disciplinary optimum to varying the input from other disciplines.

The number of analysis variables and design variables is adding up with each additional discipline. Although a single-objective function may be used for each sub-domain optimization, the entire system optimization requires multiple objectives with an attendant increase in computational expenses. This problem may be further magnified when MDO has to be used at the system level. Consequently, it is not surprising that in the industrial environment much attention is paid to bi-level optimization, featuring a system level that coordinates a set of subsystems which are not decomposed. Examples of bi-level optimization strategies are bi-level integrated system synthesis (BLISS) and collaborative optimization [83, 89].

In spite of significant progress made so far, the organizational challenges to realize MDO optimization in the distributed and highly diversified environment such as the aircraft industry remain formidable. An industrial setting of bi-level optimization as depicted in Figure 7.9 taking advantage of interdisciplinary interactions requires specific briefing processes for coordination between specialists in the joint disciplines. The basis for any optimization task is a well thought-out problem definition. Such a strategy is needed because the computational multidisciplinary capability is usually difficult to develop and expensive. A clear statement of the design variables and their allowable ranges, constraints and the objective function(s) is therefore necessary. The appropriate strategy depends on such factors as:

• the mix of continuous and discrete variables;
• the strength of the interdisciplinary interactions;
• separability of the constraints with respect to the design variables;
• susceptibility of the analysis tools to algorithmic noise;
• computational requirements of the analysis; and
• compatibility of the MDO strategy with the existing organizational structure and culture.

7.4.5   Multi-Objective Optimization

In order to reduce design time, conceptual sizing programs (Section 1.5) are equipped with automated communication between disciplines. Each discipline may select its own set of design objectives and constraints. In such a setting it is unavoidable that conflicts between the interests of different disciplines must be settled. Obviously, this problem has to be solved at the system level where a suitable weighting method must be conceived. The challenge is to develop a strategy for making a trade-off between different sub-domain optimization objectives and simultaneously complying with all constraints applying to these sub-domains. This inherently complicated problem, known as multi-criteria or multi-objective optimization (MOO), is even more difficult to solve when components of the system objective have different units of measure. Since it is easier to compare objective values than to compare design vectors, MOO deals with the objective space rather than the design space. Designers must be aware that aircraft designs optimized for different objectives may have quite different geometric, technical and performance properties.

Multi-objective optimization in aircraft design considers all conflicting design objectives and constraints simultaneously to meet mission requirements and other design drivers, such as those mentioned in Table 2.1. An approach to this topic is discussed in [45] which includes an application to the conceptual design of a typical wide body transport aircraft. One of the solutions studied is based on the Kreisselmeier-Steinhauser (KS) function defined by Equation (7.6). Multidisciplinary methods for preliminary design require MOO at the system level to trade off the multiple objectives used in different participating sub-domains [63]. Arguably, adaptation of the KS method to application in MDO appears to be a viable approach [64].

Early applications of MOO to pre-conceptual technology assessment are discussed in [32] and [33]. The second of these publications compares figures of merit such as direct operating costs (DOC) – which emphasizes speed – and acquisition costs which emphasizes empty weight. These figures are affected in a different way by selection variables such as cruise Mach number, wing planform parameters and engine thrust. Figure 7.11 depicts a number of designs optimized for different figures of merit in a two-dimensional objective space with mission fuel and life cycle costs (LCC) divided by productivity (PROD) along the axis. What is referred to as a 'merit trade boundary' is comparable to the widely known concept of the Pareto front. In the present example, two Pareto-optimum designs are at the extreme ends of the trade boundary. The minimum fuel design has a large wing of very high aspect ratio, cruises at a low Mach number and has a high LCC compared to its productivity. The design with the lowest LCC/PROD ratio achieves this at the cost of a 25% penalty in mission fuel and a high thrust/weight ratio.

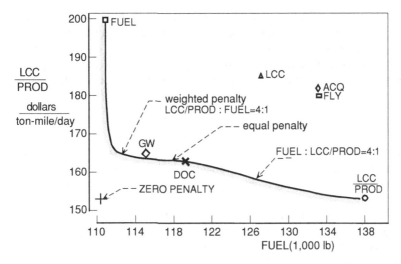

Figure 7.11 Merit trade boundary with different designs optimized for seven objective functions. Adapted from [33]

Designs optimized for minimum fly-away costs (FLY), acquisition costs (ACQ) and LCC are underpowered, cruise at a low altitude and are far away from the trade boundary of best designs. Designs with the lowest GW and direct operating costs (DOC) are very similar to each other and have the lowest overall penalty relative to the hypothetical zero penalty design. This example illustrates clearly that multi-objective optimization requires the designer to have a good insight into the various trades. Especially when deciding design technologies, the choice of a representative (weighted) objective function is of paramount importance.

7.5   Practical Optimization in Conceptual Design

> The most elegant automated optimization procedure is of very limited use to the designer who is not familiar with setting an objective function and making allowances for uncertainties.
>
> —U. Haupt, AIAA Paper No. 77-1244

7.5.1   Arguments of the Sceptic

In the industrial environment it is often argued that, in automation of a complex optimization problem, the danger exists that the designer will lose contact with the basic realities underlying the computerized procedures. This may result in failures of judgement and the acceptance of unsuspected errors, leading in extreme instances to management by – rather than of – the computer. However, these kinds of objections made are a direct consequence of the greatest advantages of rigorous optimization: the ability to deal with a large design space and identify an optimum design outside the experience of an advanced design group.

Automated optimization programs generate the best design for a single objective function whereas, in general, the designer likes to know more about the design than just the mathematical optimum. For example, if the optimum is highly constrained by some field performance requirement, it makes sense to look at progressively relaxing this constraint, improve the technology level or change some design parameter that has been treated as a pre-assigned parameter. One desirable feature is therefore that the computer program allows visual inspection of the response surface in the region near the optimum design.

7.5.2   Problem Structure

The design sizing process discussed in Section 1.4 does not provide enough detail to draw up a universally valid sequence of steps according to which synthesis problems can be attacked. The engine-to-airframe matching problem may serve as a typical example.

- If the top-level requirements have to be fulfilled with pre-selected engines of an existing (or projected) type, the process will proceed differently from the case where study engines can be varied in power to match the airframe in an optimum fashion.
- In the real world of civil aircraft design, the situation may be a mixture of both cases: a clean-sheet aircraft or a derivative of an existing type is developed alongside existing airliners or business jets which have similar engines installed.
- Aircraft are rarely designed in complete isolation and their engines may be future derivatives of types optimized for existing fleets.
- Airliners may also be offered to the customer with a choice of alternative (competing) engine types.
- A study of turbofan engine cycle parameters – for instance, bypass ratio and overall pressure ratio – can be carried out in order to generate information for an airplane design concept study.

The concept sizing program should be sufficiently flexible to cope with these and other problem structures.

7.5.3   Selecting Selection Variables

The number of selection variables increases greatly as the designer goes into more detail. Since designers cannot afford to dissipate their resources on optimization of secondary parameters, a judicious classification of selection variables is important. To take an example: why optimize wing rib pitch if the wing aspect ratio is still being varied? It appears useful to make a distinction between the selection variables affecting the aircraft's configuration definition and the control parameters affecting the operation of a given aircraft. For example, it is tempting to treat the lift coefficient in cruising flight as a selection variable since experienced designers usually have a good feeling for its (optimum) value. This leads to the following considerations.

- Vertical equilibrium in steady level flight requires $W = L = 1/2\rho V^2 S C_L$. For a given gross weight, cruise altitude and flight speed, the lift coefficient and wing area are interrelated and cannot be used simultaneously as independent variables.
- Since the take-off gross weight is different for flight profiles deviating from the design mission, there exists no unique cruise lift coefficient.

- The gross weight varies with time during the flight.
- If 'the lift coefficient' is treated as a selection variable, it is not obvious how wing drag and structure weight are affected by its variation.

Consequently, treating a control parameter as a selection variable may introduce unforeseen effects if one is not fully aware of the problem structure. A similar argument applies to the (initial) cruise altitude. The present example shows that it is recommendable to consider (gross) wing area and engine thrust available as the primary selection variables.

Selection of the cruise Mach number deserves special attention because it can be seen as a selection variable as well as a control parameter. Treating the cruise Mach number as an independent design variable has a distinct effect on the design sensitivity information, the (optimum) airframe geometry, the best type of propulsion and the installed engine power. Economic performances such as DOC and productivity are also sensitive to cruise Mach number. On the other hand, derivation of the best cruise Mach number of a fully specified aircraft is not a trivial task, as demonstrated in Chapter 12. Considering the cruise Mach number as a selection variable as well as a control variable adds an additional iteration loop to the synthesis process increasing computational expenses considerably. This explains why the desirable cruise Mach number is usually selected in the pre-conceptual study phase – henceforth, it is considered a pre-assigned parameter. However, flight performance analysis treats the Mach number separately as a control variable which may take different values dependent on the mission.

Certain design parameters are not usually selected by means of formal optimization for the following reasons.

- The design sensitivity to their variation is difficult (or even impossible) to compute.
- The impact of their variation on technical/economical quality of the aircraft is negligible.
- There is no generally accepted criterion for design quality; hence, a single figure of merit is of no value.

The selection of this category of design parameters is normally based on previous experience and engineering judgement rather than rigorous computations. A typical example is the wing taper ratio. Design experience shows that, from the aerodynamic/structural point of view, the taper ratio usually optimizes at unrealistically low values. Instead of making a considerable effort to establish the design sensitivities, it makes sense to assign a lower constraint to taper which is likely to avoid premature outer wing stalling or shock stall. Although such an approach may slightly penalize empty weight and drag, it is unrealistic to expect that, for example, acceptable stalling behaviour can be incorporated in an analytical objective function.

During the first optimization cycle of a civil airliner or business aircraft, the following parameters are the most appropriate candidates to become selection variables.

- Total engine take-off power P_{to} or thrust T_{to} normalized to sea level static (SLS) conditions in the ICAO standard atmosphere (ISA). Alternatively, thrust-to-weight ratio T_{TO}/W_{TO} or thrust loading W_{TO}/T_{TO} is often used for jet aircraft. The power-to-weight ratio P_{TO}/W_{TO} or the power loading W_{TO}/P_{TO} is mostly used for propeller aircraft.
- Wing area S_w or take-off wing loading W_{TO}/S_w.
- Wing span b_w or aspect ratio $A_w = b_w^2/S_w$.
- Wing sweep angle Λ_w.
- Wing thickness ratio t/c at the root and at the tip, or a mean value of these.

Even if only these five or six parameters are varied with ten values per variable, the design space will consist of 10^5 or 10^6 different designs. It goes without saying that designing and evaluating so many designs – even at the conceptual level – would require an immense effort. It is therefore obvious that alternative approaches are applied such as those introduced in the following chapters.

7.5.4   Design Sensitivity

Optimization is useful only when all the relationships between the design variables and the objective function affected by these variables are sensitive to the variations in the design during the sizing process. Although this condition forms a significant complication to the optimization process, much can be done to reduce computational expenses. Analysis methods built into sizing analysis routines are often of low to medium fidelity but they must be sufficiently detailed to provide meaningful trade-offs and sensitivities. Although weight prediction methods based on statistics are commonly used in conceptual design, they may be inappropriate for optimization since the statistical variation of component weights is based on regression analysis which is not necessarily functionally related to the selection variables. The use of statistics is therefore acceptable only in computations which do not affect the results of the optimization. For instance, statistics can be used to predict the weight of components such as cabin furniture, operational items and several on-board system components which are not affected by variation of the wing shape.

Design sensitivity can be defined as the percentage change in a performance measure due to a percentage change in some parameter. An important output of any synthesis program is sensitivity information permitting an assessment of critical areas. Subsystem inputs to configuration optimization must be of sufficient technical depth to accurately reflect the effects of subsystem changes on geometry and behaviour variables. To take one (admittedly rather extreme) example: airliner wing structural weight is statistically close to 12% of the MTOW, a simple rule of thumb which can be useful for an initial estimation. Such a rule of thumb does, however, not reflect the effects on wing weight of geometry variations such as wing size, low or high aspect ratio, and thick or thin sections. Wing design optimization demands a much finer tolerance than statistical equations can guarantee in spite of their proven absolute accuracy. Consequently, a weight prediction method which is to be used for optimum wing design must be essentially design sensitive. This does not exclude the use of statistics since empirical and statistical data are indispensable to calibrate the absolute accuracy to reality.

7.5.5   The Objective Function

The all-important selection of the objective function can be problematic, as illustrated by the following example. Suppose that the designer attempts to reduce the fuel consumption by increasing the wing span. From the calculations he will conclude that a low wing structural and empty weight are obtained for a small span. Minimum fuel and empty weight are both desirable properties, but since they are conflicting objectives, they cannot be realized in the same design. The traditional figure of merit in conceptual design has been the MTOW, which solves the problem to some extent. Moreover, the MTOW is an essential property which has to be computed accurately for many purposes other than optimization.

During the optimization it will also be observed that the installed engine thrust required and propulsion system weight follow similar trends as the fuel weight. Minimizing the MTOW is, therefore, rational if mission fuel and empty weight are assumed to be of equal importance. However, the economical importance of engine and fuel weights are different and minimizing their sum forms an oversimplification. Therefore, it is more rational to weigh them on the basis of their contribution to operational costs. This requires a lot more depth of analysis, information and assumptions, and introduces the problem that fuel prices sharply fluctuate in time. And if two dissimilar criteria such as operational economy and atmospheric pollution are competing, it appears desirable to consider both properties as objective functions. Again, this leads to multi-objective optimization.

Bibliography

[1]   Duffin, R., E. Peterson, and C. Zener, *Geometric Programming: Theory and Application*, Wiley and Sons, New York, 1967.
[2]   Wilde, D.J., and C.S. Beightler, *Foundations of Optimization*, Chapter 4, Polynomial Inequalities: Geometric Programming, Prentice-Hall, Inc., Englewood Cliffs, NJ, 1967.
[3]   Fox, R.L., *Optimization Methods for Engineering Design*, Addison-Wesley Publishing Company, Inc., Menlo Park, CA, 1971.
[4]   Burley, D.M., *Studies in Optimization*, Intertext Books, ISBN 0 7002 0248 X, 1975.
[5]   Gill, P.E., W. Murry, and M.H. Wright, *Practical Optimization*, Academic Press, New York, 1981.
[6]   Vanderplaats, G.N., *Numerical Optimization Techniques for Engineering Design: with Applications*, McGraw-Hill Book Company, New York, 1984.
[7]   Box, E.P., and N.R. Draper, *Empirical Model-Building and Response Surfaces*, John Wiley & Sons, New York, 1987.
[8]   Farin, G., *Curves and Surfaces for Computer Aided Geometric Design: A Practical Guide*, Academic Press Professional, Inc., San Diego, CA, USA, 1988.
[9]   Goldberg, D.E., *Genetic Algorithms in Search, Optimization and Machine Learning*, Addison-Wesley, Inc., Reading, MA, 1989.
[10] Arora, J.S., *Introduction to Optimum Design*, Second Edition, Elsevier, San Diego, CA, 2004.
[11] Horst, R., and P.M. Pardalos, *Handbook of Global Optimization*, Kluwer Academic Publishers, Dordrecht, 1995.
[12] Vanderplaats Research & Development, Inc., *DOT User's Manual*, Version 4.20, Colorado Springs, CO, 1995
[13] Nocedal, J., and S. Wright, *Numerical Optimization*, First Edition, Series in Operational Research, Springer Verlag, New York, 1999.
[14] Hönlinger, H. (ed.), *CEAS Conference on Multidisciplinary Aircraft Design and Optimization*, Maternushaus, Köln, 25–26 June, 2001.

Optimization Techniques

[15] Krzywoblocki, M.Z., von, and W.Z. Stepniewski, "Application of Optimization Techniques to the Design and Operation of V/STOL Aircraft", in Proceedings of the International Congress of Subsonic Aeronautics", the New York Academy of Sciences, NY, April 1967.
[16] Hague, D.S., and C.R. Glatt, "An Introduction to Multivariate Search Techniques for Parameter Optimization", NASA CR-73200, April 1968.
[17] Stepniewski, W.Z., and C.F. Kalmbach, "Multivariate Search in Aircraft Design Optimisation", *Journal of the Royal Aeronautical Society*, Vol. 74, pp. 419–432, May 1970.
[18] Piggott, B.A.M., and B.E. Taylor, "Application of Numerical Optimization Techniques to the Prelimnary Design of a Transport Aircraft", Royal Aircraft Establishment Technical TR-71074, April 1971.

[19] Kirkpatrick, D.L.I., "Review of Two Methods of Optimising Aircraft Design", Royal Aircraft Establishment TM Aero-1423, June 1972; also AGARD LS-56, Paper No. 12, April 1972.

[20] Kirkpatrick, D.L.I., and D.H. Peckham, "Multivariate Analysis Applied to Aircraft Optimization – Some Effects of Research Advances on the Design of Future Subsonic Aircraft", RAE Technical Memorandum Aero 1448, September 1972.

[21] Beale, E.M.L., "Optimization Methods Based on Linear Programming", Paper No. 2, Symposium on Optimization in Aircraft Design, The Royal Aeronautical Society, November 15, 1972.

[22] Hague, D.S., C.R. Glatt, and R.T. Jones, "Integration of Aerospace Vehicle Performance and Design Optimization", AIAA Paper No. 72-948, September 1972.

[23] Bishop, A.W., "Optimisation in Aircraft Design: the Whole Aircraft" Paper presented at the Symposium on Optimisation in Aircraft Design, Royal Aeronautical Society, November, 1972.

[24] Kirkpatrick, D.L.I., and M.J. Larcombe, "Initial Design Optimisation of Civil and Military Aircraft", AGARD CP-147, Vol. 1, Paper No. 19, October 1973.

[25] Olsson, D.M., "A Sequential Simplex Program for Solving Minimization Problems", *Journal of Quality Technology*, Vol. 6, No. 1, pp. 53–57, January 1974.

[26] Healy, M.J., J.S. Kowalik, and J.W. Ramsey, "Airplane Engine Selection by Optimization on Surface Fit Approximations", *Journal of Aircraft*, Vol. 12, No. 7, pp. 593–599, July 1975.

[27] Vanderplaats, G.N., "Automated Optimization Techniques for Aircraft Synthesis", AIAA Paper No. 76-909, September 1976.

[28] Barman, J.F., and H. Erzberger, "Constrained Optimum Trajectories for Specified Range", *Journal of Aircraft*, Vol. 30, No. 10, pp. 748–754, October 1976.

[29] Edwards, J.B., "The Use of Computer Based Optimisation Methods in Aircraft Studies", AGARD CP-280, Paper No. 4, September 1979.

[30] Torenbeek, E. "Some Fundamental Aspects of Transport Aircraft Conceptual Design Optimization", AGARD-CP-280, Paper No. 5, 1979.

[31] Dixit, C.S., and T.S. Patel, "Multivariate Optimum Design of a Subsonic Jet Passenger Airplane", *Journal of Aircraft*, Vol. 17, No. 6, pp. 429–433, June 1980.

[32] Sliwa, S.M., "Use of Constrained Optimization in the Conceptual Design of a Medium-Range Subsonic Transport", NASA Technical Paper 1762, December 1980. Also AIAA Paper No. 80-1095, 1980.

[33] Jensen, S.C., I.H. Rettie, and E.A. Barber, "Role of Figures of Merit in Design Optimization and Technology Assessment", *Journal of Aircraft*, Vol. 18, No. 2, pp. 76–81, January 1981.

[34] Ashley, H., "On Making Things the Best: Aeronautical Uses of Optimization", AIAA 44th Wright Brothers Lecture in Aeronautics, AIAA Paper No. 81-1738, 1981; also *Journal of Aircraft*, Vol. 19, No. 1, January 1982.

[35] Collingbourne, J., "Multivariate Optimisation Applied to the Initial Design of Transport Aircraft", RAE TR 84044, 1984.

[36] Haberland, C., J. Thorbeck, and W. Fenske, "A Computer Augmented Procedure for Commercial Aircraft Preliminary Design and Optimization", ICAS Paper 4.8.1, 1984.

[37] Chacksfield, J.E., "Multivariate Optimization as Applied to Aircraft Project Design", ICAS Paper 4.8.2, 1984.

[38] Gill, P.E., W. Murry, M.A. Saunders, and M.H. Wright, "User's Guide for NPSOL (Version 4.0): A Fortran Package for Nonlinear Programming", Technical Report SOL 86-2, Department of Operations Research, Stanford University, January 1986.

[39] Alsina, J., J.P. Fielding, and A.J. Morris, "Progress Towards an Aircraft Design Expert System", *Computer Applications in Aircraft Design and Operation*, Computational Mechanics Publications, Springer Verlag, Berlin, 1987.

[40] Britton, C.L., and W.W. Jenkinson, "A Computer-Aided Design System for Airplane Configuration", *Computer Applications in Aircraft Design and Operation*, Computational Mechanics Publications, Springer Verlag, Berlin, 1987.

[41] Bil, C., *Development and Application of a Computer-Based System for Conceptual Aircraft Design*, doctoral dissertation, Delft University of Technology, Delft University Press, 1988.

[42] Simos, D., and L.R. Jenkinson, "Optimization of the Conceptual Design and Mission Profiles of Short-Haul Aircraft", *Journal of Aircraft*, Vol. 25, No. 7, pp. 618–624, July 1988.

[43] Wrenn, G.A., "An Indirect Method for Numerical Optimization Using the Kreisselmeier-Steinhauser Function", NASA CR-4220, March 1989.

[44] Takai, M., "A New Architecture and Expert System for Aircraft Design Synthesis", Stanford PhD thesis, June 1990.

[45] Dovi, A.R., and G.A. Wrenn, "Aircraft Design for Mission Performance using Non-Linear Multi-objective Optimization Methods", NASA CR-4328, 1990. Also AIAA Paper No. 89-2078.

[46] Jenkinson, L.R., "Regional Fanjet Aircraft Optimization Studies", *Journal of Aircraft*, Vol. 30, No. 2, March–April 1993.

[47] Velden, A. van der, "Tools for Applied Engineering Optimization", AGARD Report No. 803, AGARD-FDP-VKI Special Course, April 1994.

[48] Gallman, J.W., R.W. Kaul, R.M. Chandrasekharan, and M.L. Hinson, "Optimization of an Advanced Business Jet", AIAA Paper No. 94-4303-CP, September 1994.

[49] Lee, H., "Optimal Aircraft Trajectories for Specified Range", Report ARC-11282, NASA, 1994.

[50] Gage, P.J. "New Approaches to Optimization in Aerospace Design", NASA Contractor Report 196695, March 1995.

[51] Wakayama, S., and I.M. Kroo, "Subsonic Wing Planform Design Using Multidisciplinary Optimization", *Journal of Aircraft*, Vol. 32, No. 4, pp. 746–753, July–August 1995.

[52] Cramer, E.J., "Numerical Methods for Aircraft Performance", AIAA Paper No. 95-3879, September 1995.

[53] Anderson, M.B., and G.A. Gebert, "Using Pareto Genetic Algorithms for Preliminary Subsonic Wing Design," AIAA Paper No. 96-4023, September 1996.

[54] Obayashi, S., Y. Yamaguchi, and T. Nakamura, "Multi-objective Genetic Algorithm for Multidisciplinary Design of Transonic Wing Planform", *Journal of Aircraft*, Vol. 34, No. 5, pp. 690–693, May 1997.

[55] Roth, G.L., and W.A. Crossley, "Commercial Transport Aircraft Conceptual Design Using a Genetic Algorithm Based Approach", AIAA Paper No. 98-4934, 1998.

[56] Pant, R.K., and J. Fielding, "Aircraft Configuration and Flight Profile Optimization Using Simulated Annealing," *Aircraft Design*, Vol. 2, No. 4, December 1999.

[57] Jameson, A., "Re-Engineering the Design Process Through Computation", *Journal of Aircraft*, Vol. 36, No. 1, pp. 36–50, January–February 1999.

[58] Willcox, K, and S. Wakayama, "Simultaneous Optimization of a Multiple-Aircraft Family", AIAA Paper No. 2002-1423, 2002.

[59] Buonanno, M.A., and D.N. Mavris, "Aerospace Vehicle Concept Selection Using Parallel, Variable Fidelity Genetic Algorithms", paper presented at AIAA/ISSMO Multidisciplinary Analysis and Design Conference, August/September 2004.

[60] Boyd, S., S-J Kim, L. Vandenberghe, and A. Hassibi, "A Tutorial on Geometric Programming", *Optimization and Engineering*, (2007)-8, pp. 67–127, 2007.

Multidisciplinary Design and Optimization

[61] Sobiezczanski-Sobieski, J., "A Linear Decomposition Method for Large Optimization Problems: Blueprint for Development", NASA TM 83248, 1982.

[62] McCullers, L.A., "FLOPS: Flight Optimization System", Recent Experience in Multidisciplinary Analysis and Optimization, Part 1, NASA CP-2327, pp. 395–412, April 1984.

[63] Sobiezczanski-Sobieski, J., J-F.M. Barthlemy, and G.L. Giles, "Aerospace Engineering Design by Systematic Decomposition and Multilevel Optimization", ICAS Paper 84-4.7.3, September 1984.

[64] Sobiezczanski-Sobieski, J., A.R. Dovi, and G.A. Wrenn, "A New Algorithm for General Multiobjective Optimization", NASA TM-100539, March 1988.

[65] Sobiezczanski-Sobieski, J., "Sensitivity Analysis and Multidisciplinary Optimization for Aircraft Design: Recent Advances and Results", ICAS Paper 88-1.7.3, September 1988.

[66] Abdi, F., H. Ide, M. Levine, and L. Ausrel, "The Art of Spacecraft Design: a Multidisciplinary Challenge", NASA Langley Research Center, Recent Advances in Multidisciplinary Analysis and Optimization, NASA CP-3031, Part 3 pp. 1137–1155, 1989.

[67] Kirkpatrick, D.L.I, and J.S. Smith, "Multidisciplinary Optimisation in Aircraft Design", ICAS Paper No. 90-2.3.1, September 1990.

[68] Consoli, R.D., and J. Sobiezczanski-Sobieski, "Application of Advanced Multidisciplinary Analysis and Optimization Methods to Vehicle Design Synthesis", ICAS Paper No. 90-2.3.4, September 1990.

[69] Kroo, I., S., S. Altus, R. Braun, P. Gage, and I. Sobieski, "Multidisciplinary Optimization Methods for Aircraft Preliminary Design", AiAA Paper No. 94-4325, September 1994.

[70] Martens, P., "Airplane Sizing Using Implicit Mission Analysis", AIAA Paper No. 94-4406, September 1994.
[71] Malone, B., and W.H. Mason, "Multidisciplinary Optimization in Aircraft Design Using Analytic Technology Models", *Journal of Aircraft*, Vol. 32, No. 2, pp. 431–438, March–April 1995.
[72] Sobieszczanski-Sobieski, J., "Multidisciplinary Design Optimization: An Emerging New Engineering Discipline", *Advances in Structural Optimization* (483–496), Kluwer Academic Publishers, 1995.
[73] Sobieszczanski-Sobieski, J., and R.T. Haftka, "Multidisciplinary Aerospace Design Optimization: Survey of Recent Developments", AIAA Paper No. 96-0711, January 1996.
[74] Bos, A.H.W., *Multidisciplinary Design Optimization of a Second-Generation Supersonic Transport Aircraft Using a Hybrid Genetic / Gradient-Guided Algorithm*, PhD thesis, Delft University of Technology, Faculty of Aerospace Engineering, Delft, the Netherlands, ISBN 90-5623-041-7, June 1996.
[75] Kroo, I.H., "Multidisciplinary Optimization Application in Preliminary Design: Status and Direction", AIAA Paper No. 1997-1408, April 1997.
[76] Korte, J.J., R.P. Weston, and T.A. Zang: "Multidisciplinary Optimization Methods for Preliminary Design", AGARD CP-600, Vol. 3, April 14–17, 1997.
[77] Rogers, J.L., "Tools and Techniques for Decomposing and Managing Complex Design Projects", *Journal of Aircraft*, Vol. 36, No. 1, January–February 1999.
[78] Sobieszczanski-Sobieski, J., "Multidisciplinary Design Optimization Methods: Their Synergy with Computer Technology in the Design Process", *The Aeronautical Journal*, Vol. 103, No. 1026, pp. 373–383, August 1999.
[79] Morris, A., "Distributed MDO: The Way of the Future", *CEAS Conference on Multidisciplinary Aircraft Design and Optimization*, Maternushaus, Köln, pp. 3–18 of [14], June 2001.
[80] Meussen, M., "An Industrial Approach to Multidisciplinary Optimization", *CEAS Conference on Multidisciplinary Aircraft Design and Optimization*, Maternushaus, Köln, pp. 57–62, June 2001.
[81] Kriszler, T., M. Dugas, and N. Kresse, "Multidisciplinary Wing Analysis and Optimization in the Predesign Phase", *CEAS Conference on Multidisciplinary Aircraft Design and Optimization*, Maternushaus, Köln, pp. 155–164, June 2001.
[82] Raymer, D.P., "Enhancing Aircraft Conceptual Design Using Multidisciplinary Optimization", doctoral thesis, KTH-Royal Institute of Technology, Department of Aeronautics, Stockholm, Sweden, ISBN 91-7283-259-2, May 2002.
[83] Sobieszczanski-Sobieski, J., T.D. Altus, M. Phillips, and R.R. Sandusky Jr., "Bi-Level Integrated System Synthesis (BLISS) for Concurrent and Distributed Processing", AQIAA Paper No. 2002-5409, September 2002.
[84] Belie, R., "Nontechnical Barriers to Multidisciplinary Analysis and Optimisation in the Aerospace Industry", AIAA Paper No. 2002-5439, September 2009.
[85] Laban, M., P. Arendsen, W.F.J.A. Rouwhorst, and W.J. Vankan, "A Computational Design Engine for Multidisciplinary Optimisation with Application to a Blended Wing Body Configuration", AIAA Paper No. 2002-5446, September 2002.
[86] Morris, A., P. Arendsen, G. La Rocca, M. Laban, R. Voss and H. Honlinger, "MOB: A European Project on Multidisciplinary Design Optimisation", in Proceedings of the 24th ICAS Congress, Yokohama, Japan, September 2004.
[87] Perez, R.E., and H.H.T. Liu, "Evaluation of Multidisciplinary Optimization Approaches for Aircraft Conceptual Design", AIAA Paper No. 2004-4537, August/September 2004.
[88] Kim, H., S. Ragon, G. Soremekun, B. Malone, and J. Sobieszczanski-Sobieski, "Flexible Approximation Model Approach for Bi-Level Integrated System Synthesis", AIAA Paper No. 2004-4545, August/September 2004.
[89] Kroo, I.H., "Distributed Multidisciplinary Design and Collaborative Optimization", *VKI Lecture Series on Optimization Methods & Tools for Multicriteria/Multidisciplinary Design*, Von Kármán Institute for Fluid Dynamics, Rhode Saint Genèse, Belgium, November 2004.
[90] Perez, R.E., H.H.T. Liu, and K. Behdinan, "Relaxed Static Stability Aircraft Design via Longitudinal Control-Configured Multidisciplinary Design Optimization Methodology", *Canadian Aerospace and Space Journal*, Vol. 52, No. 1, pp. 1–14, March 2006.
[91] Alonso, J.J., "Requirements for MA&O in the NASA Fundamental Aeronautics Program", 12th AIAA/ISSMO Multidisciplinary Anlysis and Optimization Conference, Reston, VA, USA, 2008.
[92] Gerwen, D. van, "Multidisciplinary Design Optimization", in Advanced Design Methodologies, TU Delft, Department of Aerospace Engineering, Lecture Notes AEA-232, January 2009.

[93] La Rocca, G., and M.J.L. van Tooren, "Knowledge-Based Engineering Approach to Support Aircraft Multidisciplinary Design and Optimization", *Journal of Aircraft*, Vol. 46, No. 6, pp. 1875–1885, November–December 2009.

[94] La Rocca, G., and M.J.L. van Tooren, "Knowledge-Based Engineering to Support Aircraft Multidisciplinary Design and Optimization", *Proc. IMechE* Vol. 224, Part G: *J. Aerospace Engineering*, February 2010.

Geometry Representation

[95] Samareh, J.A., "Survey of Shape Parameterization Techniques for High-Fidelity Multidisciplinary Shape Optimization", *AIAA Journal*, Vol. 39, No. 5, pp. 877–884, May 2002.

[96] Kulfan, B.M., "Universal Parametric Geometry Representation Method", *Journal of Aircraft*, Vol. 45, No. 1, pp. 142–158, January–February 2008.

[97] Ciampa, P.D., T. Zill, and B. Nagel, "CST Parametrization for Unconventional Aircraft Design Optimization", ICAS Paper No. 2010-1.4.3, September 2010.

8

Theory of Optimum Weight

Without doubt, weight and weight distribution, or balance, are of more importance in airplane design than in any other branch of engineering.

—T.P. Wright, quoted in [4], 1999

8.1 Weight Engineering: Core of Aircraft Design

Nowhere in the industry is the role of weight engineers in the design process as essential as in aircraft design. Weight engineering activities are initiated during the crystal ball phase of concept finding and they continue to form an essential part of conceptual, preliminary, and detail design. During the manufacturing, testing and certification phases of an aircraft programme, weight information is still being generated and used by all the program participants. Weight engineers are also active in supporting airlines during the operational deployment of delivered aircraft and they are involved in the development of modifications and derivatives of the basic aircraft. The aircraft industry has traditionally derived relationships for airplane weight estimation in the design stage using accumulated statistical data for their validation and regression methods for correlating weight components to the most influential design parameters. Semi-empirical methods calibrated by statistics are still in use – they are normally augmented by analytical relationships between component function, geometry, loads and sizing criteria. A great deal of effort has been expended in developing analytical weight prediction methods for major structural elements, in particular, the wing. Dependent on the design phase, various levels of prediction accuracy vary between about 10% standard deviation in the early conceptual phase down to a few percent in the preliminary phase.

Weight engineering activities in a new project gear up as soon as configuration details become available, in particular during the aircraft balancing process. They become most intensive when the general arrangement of the airplane has been defined and subsystems are in the detail design phase. Information obtained from the complete spectrum of disciplinary groups involved in the development process are centralized and controlled by the weight management group. A weight problem may appear during the detailed design phase often consisting of excess subsystem weights compared to predictions. Such a situation must be addressed by means of a comprehensive weight reduction program in which virtually the

Advanced Aircraft Design: Conceptual Design, Analysis and Optimization of Subsonic Civil Airplanes, First Edition. Egbert Torenbeek.
© 2013 by Egbert Torenbeek. Published 2013 by John Wiley & Sons, Ltd.

complete organization participates. Although weight engineers usually do not create new aircraft designs, they deliver essential contributions to new aircraft programmes as well as to the development of derivative versions.

Weight prediction is important when making a sound decision on the feasibility of the project before the costly detail design phase begins and must therefore be accurate. The selection of the design characteristics such as the wing geometry and engine power or thrust has a major effect on empty weight, fuel consumption, operating costs and other properties contributing to the airplane's economic effectiveness. This chapter gives an overview of the early weight prediction for a complete civil airplane and demonstrates how a typical initial sizing of the wing and the engines is carried out. For this purpose, design-sensitive empty weight components at the system level and simplified relationships for power plant thrust and the total fuel load are derived. The present approach is unique in that it aims at developing a closed form solution which enables explicit optimization in the form of response surfaces for significant weights such as the operating empty weight (OEW), useful load (UL) and maximum take-off weight (MTOW). Effects of varying basic selection variables on the weight distribution and analytic conditions for unconstrained optima are also derived. Finally, conditions are derived for achieving maximum energy efficiency of the airplane in cruising flight. The effects of design constraints imposed by low speed performance and other requirements are treated in Chapter 9.

8.1.1   Prediction Methods

> The writer is convinced after many years of work generating, using, and improving generalized weight estimating methods that more of the aircraft components are size dependent than load dependent.
>
> —W.E. Caddell [9]

Weight predictions made during conceptual design must be reasonably accurate at the major component level but should also provide reliable trends for trade-off and optimization studies. The desired accuracy is ensured by calibration with statistical data. However, the statement cited above implies that the conceptual designer should use geometry of major components and design weights as soon as they are sufficiently settled. As soon as a (provisional) three-view drawing becomes available of a configuration which is to be evaluated on its merits, it is recommended to apply more detailed methods which are sensitive to specific vehicle geometry and design requirements. Functional sensitivity is obtained by using first principles for component weight sizing related to major component size, design requirements and conditions such as aerodynamic loads and cabin overpressure. Provisional design weights obtained from a Class I method are used as input. For major components such as the wing box and fuselage shell structure, a choice has to be made between a quasi-analytical method and a multiple station analysis.

- A quasi-analytical approach is based on functional considerations, a schematic structural model and a single load case of the component. The derived expression is often linearized by means of regression analysis and calibrated with the help of empirical data of certified aircraft. For each weight component, the prediction consists of a single term with known accuracy. The method is validated for past or contemporary technology which may have the disadvantage that its design sensitivity raises doubts. Examples are found in [16] and [27].

- A multiple station method defines a suitable number of stations along the component length. Design criteria and applied loads are determined at each station and the amount of material required per unit length to resist the dominating load at each station is computed. Integration of the local material along the length yields the component's weight. A multiple station method is preferably used for large structures such as the primary structures of fuselages and wings. It has the potential of being accurate, design sensitive and versatile to cope with unconventional concepts and (local) application of advanced structural materials. Compared to the quasi-analytical approach, a station analysis method is more elaborate to develop, requires detailed knowledge of the geometry and an extensive data collection for calibration. Examples are found in [13, 28].

The quasi-analytical approach is mostly applied in Class II weight predictions. A multiple station method requires much more input and is therefore considered a typical Class III weight prediction, applicable in the preliminary design phase.

8.1.2   Use of Statistics

Initial weight estimates are required for starting up airplane sizing and feasibility studies. However, in the early design stage it is impossible to calculate the weight of every piece of the structure and every airborne system. It takes months of design effort before detailed drawings become available and enough detailed information is available to make accurate weight predictions. Moreover, there is no point in spending the effort on detailed design of structures and systems for an airplane that may never be built. Nevertheless, the empty weight of an airplane must be estimated with sufficient accuracy before a decision can be taken about the project's feasibility. Careful use of statistics forms a panacea for this dilemma.

When developing a weight prediction methodology the engineer may be confronted with many pitfalls and sources for inaccuracies.

- Methods presented in design handbooks may contain terms originating from different (industrial) sources, leading to an inaccurate result when they are combined.
- Calibration of a quasi-analytical prediction method requires a weight breakdown of past and existing aircraft – these are not readily available for reasons of proprietorship. Moreover, weight breakdowns do not always indicate the production phase to which they apply.
- Although standardized weight breakdown forms exist, many manufacturers use their own version of these standards. For instance, manufacturers may consider an aerodynamic wing-to-fuselage fairing as a wing component, others count it as a body component. This results in incomparable weight breakdowns of airframes produced by different manufacturers, although these inaccuracies cancel out when all terms are added to yield the empty weight.
- Weight prediction is also problematic for a design that deviates significantly from existing aircraft if a new technology or an unusual configuration is inaugurated.

Since the use of statistics in conceptual design is inescapable, component weight predictions mostly have a standard deviation between 5 and 10%. Fortunately, the standard deviation of a full-configuration OEW prediction can be considerably less than the average deviation of the constituents [31]. It is generally observed that methods which are calibrated for a specific airplane category are more reliable than general methods drawn up for a large but diffuse data collection.

Simple statistical wing weight prediction formulas exist with a standard deviation up to 10%. If this is not satisfactory, the method can be improved as follows:

1. Increase the number of influential parameters – in particular, wing size, design weights and the critical load factor.
2. Increase the number of terms decomposing the structure into components designed for different functions – for example, wing box upper and lower panels, shear webs, ribs, secondary structure and non-optimum weight.

Weight engineers have experienced that the first strategy is not always successful, thereby confirming the statement made as early as 1950: the accuracy of a weight prediction does not necessarily improve when more influencing parameters are involved [5]. The present author is not aware of a published rational explanation for this remarkable observation. However, even if the accuracy achieved for a purely statistical wing weight prediction is satisfactory, the sensitivity to variation of the primary design parameters is essential for optimization. The choice of selection variables and the objective function must be logical otherwise designers have no guidance in their effort to improve the aircraft's characteristics. The predictive accuracy can be optimized by using a multiple station method for primary structural components such as the wing box or the pressure cabin, in combination with a statistics-based method for secondary structures that are mostly sized by many different functional requirements.

8.2   Design Sensitivity

8.2.1   Problem Structure

The example of sensitivity analysis treated in Chapter 5 explains the influence of useful volume distribution on aerodynamic performance of an airframe with constant total useful volume. The present chapter deals with the more comprehensive problem of finding an optimum weight distribution for a jet transport with a conventional tube and wing (TAW) configuration. Different from the previous case, cabin arrangement and external fuselage geometry are assumed to be settled. Wing loading and initial cruise altitude for the design mission are treated as selection variables. Geometry and useful volume of the fuselage are fixed whereas wing area, wing volume and total airframe useful volume are varying.

An investigation of cruising flight and low speed performance capabilities dependent on wing size and engine thrust forms an essential element of the design synthesis. This activity has the objective of defining a concept that yields the most economical aircraft within the constraints imposed by the top level requirements. In accordance with Section 7.2, the designer may structure the problem as follows:

- A choice is made of the design condition defining a scenario for which operational characteristics and economics have to be optimized. In optimization jargon, the design condition is specified by pre-assigned parameters. A typical example is a representative combination of payload, range and cruise speed or Mach number.

- An objective function is selected defining 'the best aircraft'. Many designers consider the MTOW as the prime figure of merit (FOM) for a commercial aircraft since it affects most operational and economical properties. Moreover, the MTOW has to be predicted for sizing the airframe and recomputed for each design iteration. Another objective of the present analytical study is to compare designs optimized for different objective functions. For example, the energy efficiency is considered a useful alternative FOM.
- The sensitivity of major design characteristics to the selection variables is established. In particular, the variation of aerodynamic drag, installed engine thrust, major component weights and flight conditions are established. Although wing size and installed engine thrust have an influence on development, manufacturing and operating costs, these sensitivities are difficult, if not impossible, to quantify in the early design stage.
- A set of design constraints is derived from top level requirements (TLRs), certification rules and technological restrictions. The best design is mostly derived from constrained optimization.

Solving a constrained design optimization problem, even a relatively basic one such as finding the best cruise altitude, forms a major effort. Even though the suitability of using a single FOM is disputable, it may be decided that a comprehensive solution does not have a high priority and that a simplified problem statement is acceptable for getting a quick insight into the major design problems and possible solutions.

8.2.2   Selection Variables

The approach of this chapter illustrates an unconstrained optimization based on a quasi-analytical weight prediction – introduction of design constraints is postponed to Chapters 9 and 10. The derivation of major weight components and the MTOW uses input that is normally available during early design work. The analysis is illustrated by the example of a hypothetical medium range jetliner design depicted on Figure 8.1.

In the stage of conceptual wing design a distinction can be made between (1) the (gross) wing planform area S_w; (2) the wing planform shape in terms of aspect ratio, taper ratio and sweep angle; and (3) variation along the span of airfoil sections and their incidence. The present study concentrates on optimization of the TOGW and the major weight components required to comply with the design mission – wing shape optimization forms the subject of Chapter 10. The following parameters are selection variables:

1. The (maximum) wing loading W_{MTO}/S_w. It is worth noting that – although the GW of a given aircraft decreases during the flight – the maximum TOGW is a primary dependent design variable.
2. The initial cruise altitude defined by the (relative) ambient pressure δ (Appendix B). This altitude is not a design characteristic of the airplane but it determines the thrust required to attain the specified cruise Mach number and, for a specified cruise rating, the take-off thrust to be installed.

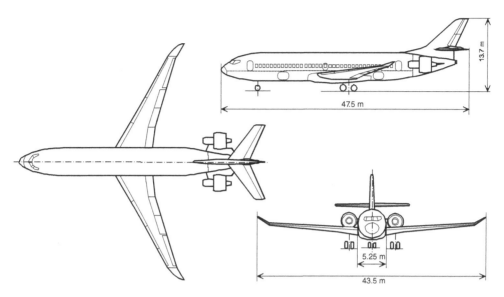

Figure 8.1 Baseline design of a medium-range airliner with a twin-aisle cabin for 180 to 210 seats. Harmonic range: 7 300 km @ Mach 0.8

The use of these variables allows analysis and optimization resulting in closed-form solutions. However, if a design synthesis program is used for the optimization, it is more appropriate to consider the wing area and the engine take-off thrust as the selection variables. Maximum TOGW and cruise altitude are then dependent variables.

8.3   Jet Transport Empty Weight

The concise weight prediction methodology exposed in this section serves as an illustration of the approach used for early sizing and optimization studies. More accurate methods are found in [1, 2, 16] and in SAWE papers. Although the proposed equations have been calibrated with a (diffuse) database, a high overall prediction accuracy is not claimed. Significant deviations may occur when an unusual aircraft configuration, advanced design technology or structural material is used. The analyst should be alert to verify statistics-based coefficients with in-house information.

8.3.1   Weight Breakdown

In the present context, the design payload/range and cruise Mach number are pre-assigned parameters, the MTOW and the fuel load are dependent variables affected by the aerodynamic efficiency and the overall engine efficiency in cruising flight. The manufacturer's empty weight (MEW) is broken down into group weights of airframe structure, power plant, airframe systems, and furnishings and equipment. Addition of the operator's items (OIs) prepares the aircraft for a condition ready for flight. Figure 8.2 shows how group weights can be further divided into sub-groups.

1. AIRFRAME STRUCTURE	2. POWERPLANT
10. Wing 11. Fuselage 12. Horizontal Tailplane 13. Vertical Tailplane 14. Landing Gear 15. Engine Pylons	20. Equipped Engines (Incl. Nacelles) 21. Bleed Air System 22. Engine Control System 23. Fuel System
3. AIRFRAME SYSTEMS	**4. FURNISHING AND EQUIPMENT**
30. Auxiliary Power Unit 31. Hydraulic and Pneumatic Systems 32. Environmental Control System 33. De-Icing and Anti-Icing 34. Fire Protection 35. Flight Controls 36. Flight Deck Instruments 37. Navigation and Communication 38. Electrical and Electronic Systems	40. Thermal and Acoustic Insulation 41. Crew Seats 42. Flight Deck and Cabin Furnishing 43. Fixed Emergency Oxygen 44. Lighting 45. Water Installation
	5. OPERATOR'S ITEMS
	50. Standard Items (Incl. Passenger Seats) 51. Operational Items

Figure 8.2   Jet transport empty weight breakdown scheme

One should be aware that advancements in technology may not be visible in a component weight reduction when at the same time design requirements have become more stringent. An example is passenger seat weight which is reduced by application of advanced materials but increased by safety provisions and increased comfort. Conservative as well as optimistic assumptions should be avoided because they may lead to cumulative prediction errors. The present approach illustrates how an elementary Class II weight prediction of a jet transport can be used to develop a sensitivity analysis by linearizing empirical (statistical) equations containing terms with non-integer exponents. The item numbers refer to Figure 8.2.

8.3.2   Wing Structure (Item 10)

Trade-off studies and optimization of wing geometry require a structure weight prediction reflecting the influence of design variables such as MTOW, wing area, aspect ratio, thickness ratio and angle of sweep. Chapter 11 describes a Class II wing weight prediction method developed to be used in the preliminary design stage. Since its application requires knowledge of design data that are not yet available in the early conceptual stage, the use of a simplified version which correctly reflects the design sensitivities is preferred. The semi-empirical prediction method presented in Figure 18.18 of [2] satisfies this requirement. It is approximated by a two-term equation,

$$W_{\mathrm{w}} = 0.0013\, n_{\mathrm{ult}}(W_{\mathrm{MZF}} W_{\mathrm{MTO}})^{0.5}\, \frac{\eta_{\mathrm{cp}} b_{\mathrm{w}}}{b_{\mathrm{ref}}}\, \frac{A_{\mathrm{w}}}{(t/c)_{\mathrm{w}}(\cos \Lambda_{\mathrm{w}})^2} + \Omega_{\mathrm{S}} S_{\mathrm{w}} \tag{8.1}$$

The factors b_{ref} and Ω_{S} are dependent on structural topography and material(s), geometry and actuation of high-lift devices, scale effects and many other factors. They can be derived from information such as [2] and calibrated with in-house data. Values used for the examples in this chapter are $b_{ref} = 100$ m (328 ft) and $\Omega_S = 210$ N m^{-2} (4.4 lbf/ft^2). These numbers apply to a predominantly metal wing.

The first term of Equation (8.1) represents the primary structure weight. The ultimate load factor equals 1.5 times the limit load factor. For MTOW exceeding 50 000 lbf (222.5 kN), $n_{lim} = 2.50$. If the manoeuvre load exceeds the maximum gust load, the structure is designed for $n_{ult} = 3.75$ – see Section 11.2. The primary structural components are sized to withstand wing bending, shear and torsion. Bending is resisted by the upper and lower covers with weight proportional to b_w; hence, this weight component as a fraction of the MTOW increases with wing size. This consequence of the square cube law is defeated by the proper choice of structural materials and, to a lesser extent, by distributing fuel and engine pods along the span. The inertia relief due to fuel is represented by using the gross weight $(W_{MZF} W_{MTO})^{0.5}$ in Equation (8.1) instead of the MTOW. The dimensionless lateral coordinate of the centre of pressure can be approximated as a function of the wing taper ratio: $\eta_{cp} = 0.36 (1+\lambda_w)^{0.5}$. The factor $\overline{(t/c)}_w$ is the average thickness ratio weighted by chord. The term $\cos^2 \Lambda_w$ allows for the effect of sweep on the structural span.

The second term of Equation (8.1) is the weight of secondary structures: trailing and leading edge structures, high-lift devices and flight control surfaces, including their supports and controls. This weight depends primarily on the planform area of these structures and on the complexity of the deflection systems and support structures. For a given design, the specific weight Ω_S is hardly affected by wing area variation although its magnitude depends statistically on the airplane MTOW. The present analysis considers wing area variation for constant aspect ratio, thickness ratio, taper ratio and sweepback angle, ultimate load factor and the ratio W_{MZF}/W_{MTO}. The following wing weight fraction follows from Equation (8.1):

$$\mu_w \stackrel{\text{def}}{=} \frac{W_w}{W_{MTO}} = \left(\frac{\Omega_W S_w}{W_{MTO}} \right)^{0.5} + \frac{\Omega_S S_w}{W_{MTO}} \tag{8.2}$$

with $\Omega_W = 30$ Nm^{-2} applying to the baseline design on Figure 8.1.

8.3.3 Fuselage Structure (Item 11)

Fuselage geometry is considered to be independent of wing geometry. Its structure weight estimation is based on first principles resulting in a quasi-analytical expression applying to predominantly metal fuselages,

$$W_{fus} = C_{shell} d_{fus}^2 \{ l_{fus} + l_{ref} \} + \Omega_{fl} n_{ult}^{0.5} d_{fus} l_{fus} \tag{8.3}$$

where $l_{ref} = 1.50$ m (5 ft). The equivalent diameter (Figure A.1) is defined as $d_{fus} = 0.5 (w_{fus} + h_{fus})$. Calibration factors for bodies with a single-deck cabin are $C_{shell} = 60$ N m^{-3} (0.38 lbf ft^{-3}) and $\Omega_{fl} = 160$ N m^{-2} (3.3 lbf ft^{-2}). The terms of Equation (8.3) are interpreted as follows:

1. The major part of the first term is shell weight which is proportional to pressure cabin volume. The theoretical ideal Al-alloy skin weight of a hypothetical cylindrical shell equals 25 $l d^2$ for 0.5 bar overpressure and a tensile stress of 84 MPa for a fatigue-resistant

skin. If the weight of stiffeners and longerons is assumed to be equal to 50% of the skin weight and frames weight equal to 20% of the stiffened skin weight, the result is an ideal gross shell weight of 45 ld^2. However, nose and tail sections are not cylindrical, additional skin thickness is required in areas where the overpressure is not the dominant design case and the gross shell weight must be corrected for cutouts and openings. This explains the statistically determined ratio 60/45.

2. The second term is the weight of pressure bulkheads which is independent of the cabin length.
3. The third term summarizes the weight of cabin floors and floor supports, doors, windows, cargo hold structures, wheel bays, the fairing around the wing carry-through structure and attachment structures for the wing, tail and nose landing gear. Attachment structure of aft fuselage mounted engines is not taken into account. Most of these components are closely related to cabin floor area and, to some extent, to the ultimate load factor.

The first and the last terms are of the same order of magnitude, the second one amounts to no more than 5% of the shell weight.

Equation (8.3) illustrates that a large part of the fuselage weight per unit floor area increases proportional to the body diameter. This is another consequence of the square cube law explaining that airplanes get heavier when they get heavier. A twin deck or an alternative cabin cross-section is almost unavoidable for cabins with more than ten seats abreast. The square cube law is defeated by application of shell material with a higher ratio of allowable hoop stress to specific weight. This can be very effective in areas where the tensile stress due to the combination of stochastic 1.5-g down bending and cabin pressurization-depressurization cycles results in a fatigue-critical case. A fuselage weight reduction between 5 and 10% is achievable for wide body airliners by replacing Al-alloy skin with fibre metal laminate. It is shown in [40] that the fatigue-critical case mainly occurs in rear panels of the fuselage crown. Although application of Glare 3 reduces their weight by 20 to 30% the weight gain as a fraction of the fuselage weight is modest. A thinner shell with significantly less weight can be realized with an all-composite structure.

8.3.4 Empennage Structure (Items 12 and 13)

Al-alloy tail surfaces have a typical specific weight of 250 N per square metre of planform area. For tail surface areas estimated with Equations (A.7) and (A.8) the empennage weight can be approximated as

$$W_{emp} = W_{12} + W_{13} = 250 \left(0.2\, S_w^{1.5} + l_{fus}\, d_{fus}^2 \right) (l_h l_v)^{-0.5} \tag{8.4}$$

An alternative approach considers the horizontal tail weight as a fraction of the wing weight and the vertical tail as a fraction of the fuselage weight,

$$W_{emp} = r_h W_w + r_v W_{fus} \tag{8.5}$$

Typical values are $r_h = 0.10$ and $r_v = 0.07$, subject to calibration with data on similar aircraft. All-composite tail surfaces are up to 20% lighter than metal ones.

8.3.5   Landing Gear Structure (Item 14)

Primary undercarriage design criteria are the heat generated during an aborted take-off and the landing touch-down load. Consequently, the landing gear weight depends on the MTOW as well as on the MLW typically as follows:

$$W_{\mathrm{lg}} = W_{14} = 0.025\,W_{\mathrm{MTO}} + 0.016\,W_{\mathrm{ML}} \qquad (8.6)$$

with constants based on statistics. The following empirical equation for the landing gear weight fraction of low-wing aircraft can be used if the MLW is not known:

$$\mu_{\mathrm{lg}} \stackrel{\mathrm{def}}{=} \frac{W_{\mathrm{lg}}}{W_{\mathrm{MTO}}} = 0.039 \left(1 + \frac{l_{\mathrm{fus}}}{l_{\mathrm{ref}}} \right) \qquad (8.7)$$

with $l_{\mathrm{ref}} = 1\,100$ m (3 600 ft). The undercarriage of a high-wing aircraft tends to be heavier compared to a low wing if the external retraction bay weight is included.

8.3.6   Power Plant and Engine Pylons (Items 2 and 15)

The bare weight of a turbofan engine is approximately proportional to the engine airflow in the take-off condition. However, since several engine component weights are not affected by thrust variation, there is a favourable scale effect. For example, the weight of one large engine is roughly 15% less than the weight of two engines together delivering the same thrust. A similar effect applies to the pylon structure between the engine pod and the airframe which has a weight of 14% of the power plant weight. Statistics indicate that the total weight of a fully equipped and installed turbofan engine is proportional to $T_{\mathrm{TO}}^{0.8}$. This scale effect on specific engine weight can be taken in to account by linearizing the total power plant weight. It is then found that, for thrust per engine between 5 and 50 metric tonnes,

$$W_{\mathrm{pp}} = W_2 + W_{15} = 0.25\,T_{\mathrm{TO}} + N_{\mathrm{eng}}(W_{\mathrm{pp}})_{\mathrm{fix}} \qquad (8.8)$$

where N_{eng} is the number of installed propulsion engines. The scale effect is quantified by the term $(W_{\mathrm{pp}})_{\mathrm{fix}}$ which amounts to 8 000 N (1 800 lbf) per engine.

8.3.7   Systems, Furnishings and Operational Items (Items 3, 4 and 5)

The following weight components are approximations of statistical equations in [31], containing the fuselage length and diameter in metres.

Airframe systems:

$$W_{\mathrm{sys}} = W_3 = 270\,l_{\mathrm{fus}}\,d_{\mathrm{fus}} + 150\,l_{\mathrm{fus}} \qquad (8.9)$$

Table 8.1 Empty weight breakdown for the baseline design in Figure 8.1 derived for an 185-seat cabin. $W_{MTO} = 1\,100$ kN. $T_{TO} = 275$ kN

Component	Method	Weight (kN)	Symbol
wing structure	Equation (8.2)	121	W_w
fuselage structure	Equation (8.3)	142	W_{fus}
empennage structure	$0.10W_w + 0.07\,W_{fus}$	22	W_{emp}
landing gear	$0.038\,W_{MTO}$	42	W_{lg}
power plant	Equation (8.8)	85	W_{pp}
systems	Equation (8.9)	66	W_{sys}
furnishings, equipment	Equation (8.10)	48	W_{fur}
operator's items	400 N per seat	74	W_{ops}
operating empty weight	$340 + 0.1955\,W_{MTO} + 231S_w$	600	W_{OE}

Furnishings and equipment:

$$W_{fur} = W_4 = 12\,l_{fus}\,d_{fus}(3\,d_{fus} + 0.5N_{deck} + 1) + 3\,500 \tag{8.10}$$

where N_{deck} denotes the number of cabin decks.

Operator's items weight per seat is typically as follows:
   2-class standard body, short/medium range: 350 N.
   2-class wide body, medium range: 500 N.
   3-class wide body, long range: 700 N.

8.3.8   Operating Empty Weight: Example

Decomposition of empty weight into fixed and variable components must be adapted to the problem structure. In the present context the sensitivity to wing area, installed engine thrust and MTOW is made explicit in all weight components. Results of a simplified empty weight prediction are summarized in Table 8.1. This approach may have to be refined and extended if the optimization problem requires it. The resulting OEW for the present case is as follows:

$$W_{OE} = \Sigma\,W_{fix} + \big\{(1 + r_h)\mu_w + \mu_{lg}\big\}\,W_{MTO} + \mu_T\,T_{TO} \tag{8.11}$$

The term $\Sigma\,W_{fix}$ summarizes the weight components related to the body group: fuselage and vertical tail structure, airframe systems, furnishings and equipment, and OIs. The second term summarizes the weight components that are proportional to the MTOW. The last term is the power plant weight. The take-off thrust is decomposed into terms related to the cruise thrust required to balance wing drag and fuselage drag – see Section 8.5.

8.4   Design Sensitivity of Airframe Drag

Wing area and installed engine thrust are basic selection variables during conceptual sizing. Their variation affects the aerodynamic efficiency in cruising flight, fuel weight, engine weight and TOGW. It is emphasized that wing area variation has a significant effect on the drag polar, an essential element of design optimization. Using the background information of Chapter 4, this section describes the dependence on wing area and flight altitude of the drag polar and thrust required in high-speed flight. The variation of installed engine thrust is based on the rubberizing concept and the engine nacelles are scaled to match the required thrust. Accordingly, the total drag is split up into airframe drag $\overline{D}$ and nacelle drag which is considered a loss of installed thrust and overall efficiency (Section 8.5).

8.4.1   Drag Decomposition

Several methods of decomposing airplane drag are discussed in Chapter 4. For steady level flight in the en route configuration we use the two-term parabolic approximation of the drag polar,

$$C_D = C_{D_0} + C_{D_L} = C_{D_0} + K_L C_L^2 \tag{8.12}$$

The drag due to lift of the complete aircraft and the airframe can be assumed to be equal on the provision that the engine nacelles do not contribute to the lift and its distribution along the span. Since variation of the wing aspect ratio is not considered in this chapter, the factor K_L is constant; see Equation (4.22). The zero-lift drag area is obtained by adding the parasite drag areas of all N airplane components exposed to the flow,

$$C_{D_0} S_w = \Sigma_1^N (C_{D_p} S)_i \tag{8.13}$$

The zero-lift drag area is broken down into a variable drag area depending on the wing area S_w and a fixed drag area.

- The wing drag area is proportional to S_w if the profile drag coefficient is constant.
- The drag area of the fuselage is constant since its geometry is assumed to be independent of the wing geometry.
- The planform areas of horizontal and vertical tailplanes are affected by fuselage and wing geometry (Appendix A). The empennage drag area therefore consists of a constant term dependent on fuselage geometry and a variable term dependent on the wing area. This subdivision is simplified by considering the vertical tail (index v) drag as a fixed term related to fuselage geometry and the horizontal tail (index h) drag as a variable term related to wing geometry.
- The drag area of the engine nacelles is proportional to the total surface area of all engines. Equation (A.10) shows this to consist of a term proportional to the total engine take-off thrust and a surface area per engine. The first term is translated into a loss of installed thrust, the latter is counted as a (fixed) airframe drag component.

The airframe surface area is less than the total gross surface areas of all components. Although fairings are used to minimize interference drag they contribute additional wetted area. Quantifying the net effect would require considerable effort. The airframe parasite drag is therefore

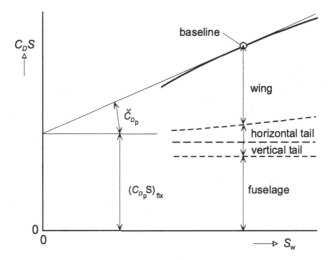

Figure 8.3   Variation of airframe parasite drag area with wing area

considered as the summation of drag components based on component gross surface areas. This leads to a decomposition of the airframe drag area into a varying term

$$(C_{D_p}S)_{\text{var}} = (C_{D_p}S)_{\text{w}} + (C_{D_p}S)_{\text{h}} \tag{8.14}$$

and an invariable term

$$C_{D_p}S = (C_{D_p}S)_{\text{fus}} + (C_{D_p}S)_{\text{v}} + (C_{D_p}S)_{\text{eng}} \tag{8.15}$$

The (small) term S_{eng} – originating from a linearized expression of the nacelle wetted area – is the second term of Equation (A.10). At first sight, the varying drag areas are proportional to the wing and horizontal tail areas. However, area variation affects chord Reynolds number, and hence C_{D_p}, and the drag areas the appear to be proportional to $S_{\text{w}}^{0.92}$. Figure 8.3 illustrates that the drag can be linearized for a limited wing area deviation from the baseline design. This is effected by an 8% reduction of the variable wing and tail drag and an increment of the fixed drag area equal to 8% of the baseline (index base) wing and tail drag area. The total airframe drag area is thus

$$C_{\overline{D}_0}S_{\text{w}} = \check{C}_{D_p}S_{\text{w}} + (C_{D_p}S)_{\text{fix}} \tag{8.16}$$

where

$$\check{C}_{D_p} = 0.92\{(C_{D_p})_{\text{w}} + (C_{D_p})_{\text{h}}\,S_{\text{h}}/S_{\text{w}}\} \tag{8.17}$$

and

$$(C_{D_p}S)_{\text{fix}} = (C_{D_p}S)_{\text{fus}} + (C_{D_p}S)_{\text{v}} + (C_{D_p}S)_{\text{eng}} + 0.08\,\{(C_{D_p}S)_{\text{w+h}}\}_{\text{base}} \tag{8.18}$$

Fuselage and empennage are mostly designed so that their drag divergence occurs above the cruise Mach number. Compressibility drag is therefore counted as an increment of wing profile and drag due to lift.

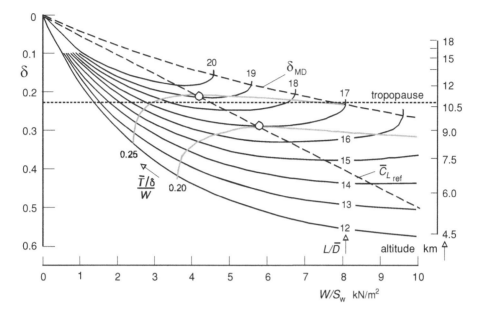

Figure 8.4   Airframe aerodynamic efficiency at Mach 0.80 affected by wing loading and cruise altitude for the design depicted in Figure 8.1. Aerodynamic data: $\check{C}_{D_p} = 0.008$, $(C_D S)_{\text{fix}} = 1.9 \text{ m}^2$ and $K_L = 0.040$

8.4.2   Aerodynamic Efficiency

The variation of airframe drag in straight and level flight with given gross weight is derived from Equations (8.12) and (8.16),

$$\frac{L}{D} = \left\{ \frac{\check{C}_{D_p}}{C_L} + \frac{\delta\, q_{\text{sl}}\, (C_{D_p} S)_{\text{fix}}}{W} + K_L\, C_L \right\}^{-1} \qquad (8.19)$$

where $q_{\text{sl}} = 0.5\, \gamma\, p_{\text{sl}} M^2$ denotes the hypothetical dynamic pressure when flying with the cruise Mach number at sea level and $\delta = p/p_{\text{sl}}$ is the relative ambient pressure. Figure 8.4 depicts the sensitivity of $L/\overline{D}$ to variation of the selection variables[1] for a hypothetical jetliner design flying at given Mach number and gross weight. The following partial optima are relevant:

1. Constant altitude: the maximum $L/\overline{D}$ occurs for minimum wing-related drag defining a partial optimum lift coefficient and wing loading,

$$\overline{C}_{L_{\text{ref}}} = \sqrt{\check{C}_{D_p} K_L^{-1}} \quad \text{and} \quad (W/S_{\text{w}})_{\text{ref}} = q_{\text{sl}}\, \delta\, \overline{C}_{L_{\text{ref}}} \qquad (8.20)$$

---

[1]The unrealistic range of wing loadings and altitudes in Figure 8.4 serves for illustration – $L/\overline{D}$ contours are only accurate for modest deviations from the baseline design.

For this condition, the wing-related profile drag equals the drag due to lift and the maximum aerodynamic efficiency amounts to

$$\frac{L}{D} = \left\{ 2\sqrt{\check{C}_{D_p} K_L} + \frac{q\,(C_{D_p} S)_{\text{fix}}}{W} \right\}^{-1} \tag{8.21}$$

Figure 8.4 indicates that the effect of selecting a wing loading up to 20% above or below the optimizer has little effect on the airframe aerodynamic efficiency.

2. Constant wing loading: the zero-lift drag coefficient is constant and the conditions for minimum airframe drag are

$$\overline{C}_{L_{\text{MD}}} = \sqrt{C_{\overline{D}_0} K_L^{-1}} \quad \text{and} \quad \frac{L}{D} = \left(2\sqrt{C_{\overline{D}_0} K_L} \right)^{-1} \tag{8.22}$$

corresponding to the altitude for minimum drag determined by

$$\delta_{\text{MD}} = \frac{W}{q_{\text{sl}}\,\overline{C}_{L_{\text{MD}}} S_{\text{w}}} \tag{8.23}$$

These two optimizers are incompatible; hence, there exists no unconstrained maximum aerodynamic airframe efficiency. Figure 8.4 shows that the achievable $L/\overline{D}$ is improved by increasing the flight altitude on the provision that the wing loading is reduced concurrently. For example, the design with optimum wing loading at 11 km altitude has a 12% higher $L/\overline{D}$ than the optimum at 9.5 km. However, the higher flying aircraft needs 25% more installed thrust and a 45% bigger wing. The overall system-optimal design has a higher wing loading than the reference value according to Equation (8.20) and flies below the altitude for minimum drag. In other words, the best C_L is located between $\overline{C}_{L_{\text{ref}}}$ and $\overline{C}_{L_{\text{MD}}}$. Even though aerodynamic efficiency contributes significantly to fuel economy, it is not a suitable FOM for overall system optimization. These remarks are elaborated in the next sections.

8.5 Thrust, Power Plant and Fuel Weight

Most system design optimization studies aim at reducing airframe structural weight since it amounts to approximately 30% of the MTOW. However, the combination of the power plant and total fuel weighs between 25 and 50% of the MTOW. It is therefore instructive to consider this as a propulsion weight penalty that should be minimized. Since both weight components depend primarily on the MTOW, this section considers the optimization for fixed MTOW.

8.5.1 Installed Thrust and Power Plant Weight

Except for relatively small (regional) aircraft, it is unusual that off-the-shelf engines will be chosen to be installed in transport aircraft. Instead, engine selection is based on the outcome of engine-to-airframe matching studies and negotiations between the airframer and engine manufacturer(s). In the present study context this implies that engines are subject to variation of performance capabilities. In addition to installed power or thrust, certain engine cycle parameters may be considered as selection variables. Typical examples are the bypass ratio of a turbofan and the propeller diameter of a turboprop engine (Chapter 3). The following analysis exclusively investigates the variation of installed turbofan thrust by means of the rubberizing concept which implies that airflow and thrust of a baseline engine are scaled up and down by varying the inlet diameter while keeping operating pressures and temperatures constant. Engine rubberizing requires a set of scaling laws relating basic engine characteristics – diameter, cowl length, weight – to take-off thrust. Although engine manufacturers generally use non-linear scaling laws, linearized relations are considered acceptable for the present study.

The installed take-off thrust is selected so that the condition $\overline{T} = \overline{D}$ is complied with in steady level flight at the initial cruise condition (index cr) after reaching the top of climb (TOC). The minimum installed thrust is required when the engines operate at their maximum cruise rating or a specified percentage of it.[2] Hence, the required sea level static (SLS) thrust is

$$T_{\text{TO}} = \frac{T_{\text{TO}}}{\overline{T}_{\text{cr}}} \left(W \frac{C_{\overline{D}}}{C_L} \right)_{\text{cr}} \tag{8.24}$$

The thrust lapse with altitude is derived from the corrected lapse rate defined by Equation (3.10), modified to include nacelle (index nac) drag,

$$\overline{\tau} = \frac{\overline{T}/\delta}{T_{\text{TO}}} = \frac{T - D_{\text{nac}}}{\delta\, T_{\text{TO}}} = \tau - q_{\text{sl}} \frac{(C_D S)_{\text{nac}}}{T_{\text{TO}}} \tag{8.25}$$

Since the temperature in the stratosphere is constant, turbofan thrust at a given rating and Mach number is proportional to the ambient pressure. Consequently, τ as well as $\overline{\tau}$ are independent of the altitude[3] and, dependent on the engine cycle, their value in cruising flight is not very different from one. Substitution of Equation (8.25) into (8.24) yields

$$T_{\text{TO}} = \left(\frac{W}{\overline{\tau}\, \delta} \frac{C_{\overline{D}}}{C_L} \right)_{\text{cr}} \tag{8.26}$$

Assuming 2% TOGW fuel burn-off during take-off and climb yields the power plant (index pp) weight

$$W_{\text{pp}} = \mu_T \left(\frac{W}{\overline{\tau}\, \delta} \frac{C_{\overline{D}}}{C_L} \right)_{\text{cr}} + N_{\text{eng}} W_{\text{ref}} \tag{8.27}$$

with typical values $\mu_T = 0.26$ and $W_{\text{ref}} = 8$ kN. Equations (8.26) and (8.8) show that the installed engine thrust and power plant weight are sensitive to increasing the cruise altitude.

---

[2] If engines are sized to comply with a dominating take-off requirement, the cruise rating may be less than the maximum allowed value (Chapter 9).

[3] The corrected thrust lapse rate is not exactly constant in the troposphere, an irrelevant aspect for the present application.

8.5.2   Mission Fuel

Similar to τ and η_{ov}, the installed overall efficiency of a turbofan depends on the Mach number only,

$$\bar{\eta}_o \overset{\text{def}}{=} \frac{(T - D_{nac})V}{\dot{W}_f H/g} = \eta_o \left(1 - \frac{D_{nac}}{T}\right) = \eta_o \frac{\bar{\tau}}{\tau} \tag{8.28}$$

The loss of installed efficiency due to nacelle drag is significant for high bypass turbofans at transonic Mach numbers whereas the horizontal equilibrium requires the installed thrust to be equal to the airframe drag. Hence, the two expressions for the range parameter are equal: $\eta_o L/D = \bar{\eta}_o L/\overline{D}$. The symbology used in this chapter is a refinement necessary because engines are scaled up and down in size. However, all derived equations are equally valid if the traditional terminology – aircraft drag and uninstalled engine thrust – is used instead of airframe drag and installed thrust.

The simplified mission fuel weight estimation derived in Section 2.3 is used here in a slightly modified form by introducing the equivalent range,

$$R_{eq} \overset{\text{def}}{=} (R_{mis} + R_{lost})(1 - 0.5\, W_{misf}/W_{MTO}) \tag{8.29}$$

This concept is used since many detailed computations need not be repeated in the framework of optimization. The second bracketed term of Equation (8.29) allows for the continuously reducing thrust and fuel flow during cruising due to the reducing gross weight. Near the optimum condition, this term is practically identical to the value for the baseline design. The lost range R_{lost} representing the reduced specific range during take-off, climb and descent, amounts to a few hundred kilometres. This is a relatively minor term for long-range flight; but for short-range flight it may be as much as 10% of the cruise range. The terms R_{lost} and R_{eq} are treated as constants. Mission fuel weight is written as a fraction of the take-off gross weight,

$$\frac{W_{misf}}{W_{MTO}} = \frac{R_{eq}}{\bar{\eta}_o H/g} \frac{C_{\overline{D}}}{C_L} \tag{8.30}$$

Computation of the MTOW uses the harmonic range to be substituted for the mission range in combination with the maximum payload.

8.5.3   Propulsion Weight Penalty

We define the combination of power plant weight and mission fuel weight as the propulsion weight penalty. It is found by combining Equations (8.12), (8.27) and (8.30),

$$\frac{W_{prop}}{W_{MTO}} = F_{prop} \left\{ \frac{\check{C}_{D_p}}{C_L} + \frac{\delta\, q_{sl}\,(C_D S)_{fix}}{W_{MTO}} + K_L C_L \right\} \tag{8.31}$$

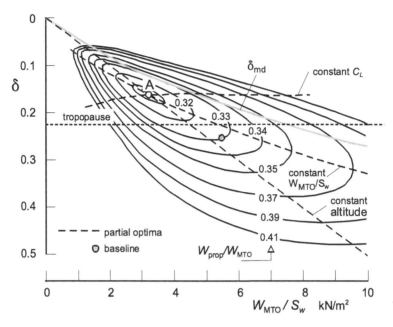

Figure 8.5 Propulsion weight fraction affected by wing loading and altitude. Data: $\bar{\eta}_o = 0.33$, $\mu_T = 0.25, \bar{\tau} = 0.8$

where the propulsion function is defined as

$$F_{\text{prop}} \stackrel{\text{def}}{=} \frac{R_{\text{eq}}}{\bar{\eta}_o H/g} + \frac{\mu_T}{\bar{\tau}\,\delta} \tag{8.32}$$

The effects of varying wing loading and cruise altitude are manifest in Figure 8.5 depicting propulsion weight penalty fraction contours. Also shown are the following partial optima of the selection variables:

1. For constant altitude, F_{prop} is constant and the optimum wing loading is identical to the reference wing loading defined by Equation (8.20).
2. For constant wing loading, the optimum altitude is below the minimum drag altitude defined by Equation (8.22). Setting the partial derivative of Equation (8.31) equal to zero yields a simple but transcendental equation

$$\delta = \delta_{\text{MD}} \left(1 + 2\frac{W_{\text{pp}}}{W_{\text{misf}}}\right)^{0.5} = \delta_{\text{MD}} \left(1 + 2\frac{\mu_T}{\bar{\tau}\,\delta}\frac{\bar{\eta}_o H/g}{R_{\text{eq}}}\right)^{0.5} \tag{8.33}$$

from which the cruise altitude is solved by iteration. The optimum altitude in Figure 8.5 appears to be 1 500 m (5 000 ft) below the minimum drag altitude.

As opposed to the case of aerodynamic efficiency and mission fuel, the partial optima of the propulsion weight penalty are compatible in the intersection of the two partial optima – point A in Figure 8.5. This global optimizer is obtained analytically by introduction of the following constant terms:

- The mission fuel fraction required to overcome the minimum wing-related drag,

$$\mu_1 \overset{\text{def}}{=} 2 \frac{R_{\text{eq}}}{\bar{\eta}_\text{o} H/g} \sqrt{\check{C}_{D_\text{p}} K_\text{L}} \tag{8.34}$$

- The power plant weight fraction required to overcome the invariable parasite drag,

$$\mu_2 \overset{\text{def}}{=} \frac{\mu_T}{\bar{\tau}} \frac{q_{\text{sl}}(C_D S)_{\text{fix}}}{W_{\text{MTO}}} \tag{8.35}$$

Closed-form analytical expressions for the optimizers are found by substitution of μ_1 and μ_2 into Equation (8.31) and setting the derivative with respect to the altitude equal to zero:

$$\text{wing area:} \quad S_\text{w} = \frac{(C_D S)_{\text{fix}}}{2\,\check{C}_{D_\text{p}}} \sqrt{\frac{\mu_1}{\mu_2}} \tag{8.36}$$

$$\text{altitude:} \quad \delta = \frac{\mu_T}{\bar{\tau}} \frac{\bar{\eta}_\text{o} H/g}{R_{\text{eq}}} \sqrt{\frac{\mu_1}{\mu_2}} \tag{8.37}$$

$$\text{thrust:} \quad \frac{T_{\text{TO}}}{W_{\text{MTO}}} = \frac{\mu_2 + \sqrt{\mu_1 \mu_2}}{\mu_T} \tag{8.38}$$

The corresponding mission fuel and power plant weight fractions are

$$\frac{W_{\text{misf}}}{W_{\text{MTO}}} = \mu_1 + \sqrt{\mu_1 \mu_2} \quad \text{and} \quad \frac{W_{\text{pp}}}{W_{\text{MTO}}} = \mu_2 + \sqrt{\mu_1 \mu_2} \tag{8.39}$$

The optimum ratio of power plant to mission fuel weight is

$$\frac{W_{\text{pp}}}{W_{\text{misf}}} = \sqrt{\frac{\mu_2}{\mu_1}} \tag{8.40}$$

and the minimum propulsion weight penalty is defined by the following remarkable expression:

$$\frac{W_{\text{prop}}}{W_{\text{MTO}}} = \left(\sqrt{\mu_1} + \sqrt{\mu_2}\right)^2 \tag{8.41}$$

In spite of their simplicity, these results must be carefully interpreted. For example, Figure 8.5 indicates that point A corresponds to a wing loading of only 3 200 N/m^2 and an unusually high initial cruise altitude of 13 000 m. This is reflected in an unpractical large and heavy wing, driving up empty and take-off weights. It is therefore appropriate to extend the optimization by including wing weight. Due consideration must also be given to constraints on wing and thrust loading, the subject of Chapter 9.

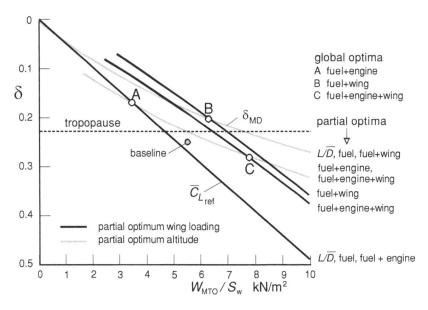

Figure 8.6 Optimum wing loading and cruise altitude treating several weight fractions as the objective function

8.5.4   Wing and Propulsion Weight Fraction

Addition of wing weight according to Equation (8.2) (excluding wing attachment weight) to the propulsion weight penalty yields

$$\frac{W_{\text{w}} + W_{\text{prop}}}{W_{\text{MTO}}} = \left\{ \frac{\Omega_{\text{W}}}{W_{\text{MTO}}/S_{\text{w}}} \right\}^{0.5} + \frac{\Omega_{\text{S}}}{W_{\text{MTO}}/S_{\text{w}}} + \frac{W_{\text{prop}}}{W_{\text{MTO}}} \tag{8.42}$$

where the last term is defined by Equation (8.31). The contours of this weight fraction have a shape similar to Figure 8.5 but they are shifted to significantly higher wing loadings. Partial optima are summarized in Figure 8.6, leading to the following observations:

- For constant altitude, the optimum wing loading is found by setting the partial derivative of Equation (8.42) equal to zero resulting in

$$\frac{W_{\text{MTO}}}{S_{\text{w}}} = \left(\frac{W_{\text{MTO}}}{S_{\text{w}}} \right)_{\text{ref}} \left(1 + \frac{0.5\sqrt{\Omega_{\text{W}}\,W_{\text{MTO}}/S_{\text{w}}} + \Omega_{\text{S}}}{\delta\,q_{\text{sl}}\,\check{C}_{D_{\text{p}}}\,F_{\text{prop}}} \right)^{0.5} \tag{8.43}$$

where $(W_{\text{MTO}}/S_{\text{w}})_{\text{ref}}$ is defined by Equation (8.20). This equation must be solved by successive iterations. An initial approximation is found by inserting 1.4 times the reference wing loading into the square root term.

- Optimum altitude: since wing structure weight is independent of the cruise altitude, the partial optimum is equal to the altitude for minimum propulsion weight according to Equation (8.33).

8.5.5   Optimum Weight Fractions Compared

Figure 8.6 shows that the three unconstrained optimizers A, B and C define quite different airplane designs. Associated values of altitude and wing loading are obtained from Equations (8.20), (8.33) and (8.43). Design A (minimum fuel plus power plant weight fraction) is unrealistic since it ignores the effect of the varying wing size on empty weight. Design B for minimum fuel plus wing weight fraction – not derived explicitly in this text – is a more practical solution. It has a lower optimum altitude and a considerably higher wing loading than design A. Design C for minimum fuel plus power plant plus wing weight fraction is derived from the most complete figure of merit and is representative for the design with the lowest MTOW. However, design C has a high wing loading and flies at a low altitude.

The selection of one of these unconstrained optimizers does not necessarily lead to a good design since wing and thrust loading constraints imposed by low-speed performance requirements are ignored. This need not be a problem since the combination of propulsion weight penalty and wing weight fractions is not sensitive to non-optimum conditions. A more serious objection is that the economic impact of wing structure weight variation is less significant than engine thrust and mission fuel weight variation. In principle, this effect can be incorporated by multiplying the three terms involved by weight factors derived from their economic impact.

8.6   Take-Off Weight, Thrust and Fuel Efficiency

The previous optimizations were derived for a given MTOW. In the real world of conceptual design the payload and range are usually specified and the MTOW forms the outcome of adding all the empty weight components and the useful load. The next step is therefore to investigate the various optimizers for the case of a specified payload and range capability. Since many empty weight components depend on the MTOW, the objective of the present quasi-analytical approach is to derive a closed form equation for the MTOW for a specified payload and range capability.

8.6.1   Maximum Take-Off Weight

The equations derived in the previous sections are used to add all empty weight components, payload and fuel. Solving for the MTOW yields

$$W_{\text{MTO}} = \frac{W_{\text{pay}} + \Sigma\, W_{\text{fix}} + F_{\text{prop}}\, \delta\, q_{\text{sl}}\, (C_{D_{\text{p}}} S)_{\text{fix}}}{1 - \left\{ \mu_{\text{resf}} + \mu_{\text{lg}} + (1 + r_{\text{h}})\, \mu_{\text{w}} + F_{\text{prop}}\, (\check{C}_{D_{\text{p}}}/C_L + K_L\, C_L) \right\}} \qquad (8.44)$$

The propulsion function F_{prop} is defined by Equation (8.32) and the lift coefficient is proportional to the wing loading: $C_L = (W_{\mathrm{MTO}}/S_{\mathrm{w}})/q$. In accordance with Table 8.1, the following weight component groups are identified in Equation (8.44):

- The numerator summarizes fixed weight components that are (practically) independent of the MTOW – they include the design payload, the cabin and its contents, systems, operator's items, the fuselage and vertical tail structures, and a small term dependent on the number of engines. Since fuselage and vertical tail geometry are considered to be independent of wing geometry, their weights are fixed. The term proportional to F_{prop} represents the mission fuel and power plant weight attributed to the thrust required to balance the fuselage and vertical tail drag along the mission distance. It is easy to show that this engine weight component is independent of the altitude. However, since the associated fuel weight component is proportional to the ambient pressure, it decreases with increasing altitude.
- The denominator summarizes all weight components treated as fractions of the MTOW. Statistics indicate that the reserve fuel load can be considered a fraction of the MTOW: $\mu_{\mathrm{resf}} = 0.045$. The landing gear weight fraction according to Equation (8.7) is $\mu_{\mathrm{lg}} = 0.040$, typically. The term $r_{\mathrm{h}} \approx 0.10$ is a correction factor accounting for the horizontal tailplane weight and Equation (8.2) is substituted for the wing weight fraction μ_{w}. The term proportional to F_{prop} represents the propulsion weight penalty attributed to the thrust required to balance the wing and horizontal tail drag along the mission distance. The fuel weight fraction is independent of the altitude, the engine weight fraction is inversely proportional to the ambient pressure and increases (significantly) with increasing altitude. Both terms are independent of fuselage geometry.

Equation (8.44) forms a suitable tool for explicit optimization of the wing loading and the initial cruise altitude – it can readily be adapted to cope with variation of the wing planform geometry and thickness ratio. Figure 8.7 shows the MTOW to have an unconstrained optimizer for a wing loading of 7 kN/m^2 at 10 000 m altitude initial cruise altitude (design D). This optimum is insensitive to the wing loading, especially if the altitude is optimized for each

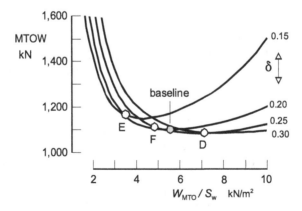

Figure 8.7 Effect of wing loading and initial cruise altitude on MTOW. Baseline: Figure 8.1

wing loading. Another useful feature of Equation (8.44) is that it enables a quick computation of important characteristics such as the required engine thrust and mission fuel load by means of closed-form solutions.

It is often stated that the MTOW constitutes the most complete figure of merit (FOM) for assessing the quality of a new airliner design. This point of departure leads to focussing on a single criterion whereas other factors also contributing to the design's viability are not considered explicitly. Examples are the required installed engine thrust and fuel energy efficiency.

8.6.2 Installed Thrust and Fuel Energy Efficiency

The installed thrust of engines sized for cruising is derived from the initial cruise condition,

$$T_{TO} = \frac{\overline{D}}{\overline{\tau}\,\delta} = \overline{\tau}^{-1}\left\{ q_{sl}\,(C_{D_p}S)_{fix} + \delta^{-1}(\check{C}_{D_p}/C_L + K_L C_L)\,W_{MTO}\right\} \qquad (8.45)$$

Since the fixed parasite drag lapse equals the thrust lapse with altitude, the thrust required to balance the parasite drag is independent of the cruise altitude. However, the wing drag contribution in Equation (8.45) causes the thrust required to balance the wing drag to be sensitive to cruise altitude. Figure 8.8 shows that the wing loading for minimum thrust is 500 N m^{-2} higher than the reference wing loading defined by Equation (8.20). Increasing the altitude along this partial optimum results in a progressively increasing thrust required with altitude. The present thrust minimization must be augmented by constraints derived from low-speed performances, the subject of Chapter 9.

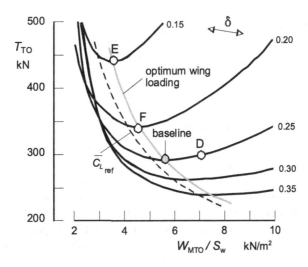

Figure 8.8 Effect of wing loading and cruise altitude on installed thrust. Baseline: Figure 8.1

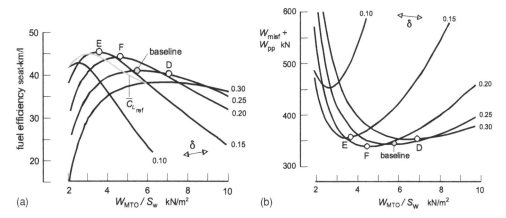

Figure 8.9 Effect of wing loading and cruise altitude on propulsion characteristics. (a) Fuel energy efficiency (b) Propulsion weight penalty

The energy efficiency (Section 2.5) is a useful figure for comparing the fuel economy of different commercial transport aircraft. It can be expressed as an instantaneous performance but in actual operation it varies during the flight due to the varying GW and the reduced fuel efficiency during non-cruising flight segments. The average fuel energy efficiency for a complete mission is derived from Equations (2.26) and (8.30),

$$E_{\text{EN}} = \mathcal{E} \, \bar{\eta}_{\text{o}} \frac{L}{D} \frac{W_{\text{pay}}}{W_{\text{MTO}}} \frac{R_{\text{mis}}}{R_{\text{eq}}} \tag{8.46}$$

with $\mathcal{E} = \rho_{\text{f}} H N_{\text{s}} / W_{\text{pay}} = 37$ seat-km/litre (75.5 seat-nm/gallon). Figure 8.9 (a) demonstrates that fuel efficiency is sensitive to wing loading and cruise altitude.

The propulsion weight penalty is defined as the combined weight of the mission fuel and the power plant (Section 8.5). Different from expressing it as a weight fraction, Figure 8.9 (b) depicts its absolute value.

8.6.3 Unconstrained Optima Compared

Figure 8.10 depicts three unconstrained optimizers: design D for minimum MTOW, design E for minimum mission fuel and and design F for a minimum propulsion weight penalty. Design D has a slightly lower MTOW and a much higher wing loading than the other optimizers. Design E achieves 45-seat-km production per litre at 13 500 m altitude for a wing loading of 3.6 kN/m^2, 12% higher than the fuel efficiency of design D. However, this high fuel efficiency is achieved at the cost of 47% more installed thrust. The minimum propulsion weight design F is an unconstrained optimum for a wing loading of 4.6 kN/m^2 at an initial cruise altitude slightly above the tropopause. This design has 13% more installed thrust than design D and a few percentage lower fuel efficiency than design E. Design F appears to represent a good compromise, provided wing loading and thrust loading comply with low-altitude performance

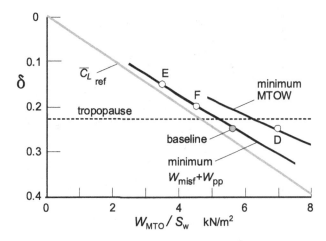

Figure 8.10   Optimum wing loading and cruise altitude for several figures of merit

constraints. These numbers demonstrate how important it is to consider several figures of merit and their sensitivity when optimizing a commercial aircraft.

8.6.4   Range for Given MTOW

The usual approach to optimization studies is to consider the maximum payload and the harmonic range (or another design payload/range combination) as pre-assigned parameters. The MTOW is then obtained via an iterative process. If the MTOW is the objective function in an optimization study subject to many selection variables, the use of a design synthesis program is unavoidable. However, the numerous iterations result in a computer-intensive process. A useful alternative is to consider the MTOW obtained for the baseline design as a pre-assigned quantity and consider the range as the objective function. This approach enables a computationally cheaper process but the question is intriguing whether the design with minimum MTOW for a specified range is identical to the design with a maximum range for constant (baseline) MTOW. The author is not aware of any published proof that this simpler problem structure leads to (nearly) the same optimum design.

The equivalent range for a given MTOW follows from Equation (8.44),

$$\frac{R_{eq}g}{\bar{\eta}_0 H} = \frac{1 - \left\{ \mu_{lg} + \mu_{resf} + (1 + r_h)\,\mu_w \right\} W_{MTO} - W_{pay} - \Sigma W_{fix}}{\delta\, q_{sl}(C_D S)_{fix} + (\check{C}_{D_p}/C_L + K_L C_L) W_{MTO}} - \frac{\mu_T}{\tau\,\delta} \tag{8.47}$$

Contours of constant range depicted in Figure 8.11 indicate that the maximum range design for a given MTOW has a 10% higher wing loading than the minimum MTOW design for a given range and flies at a lower altitude. It is also clear that for designs flying near the tropopause the achievable range is only a few percentage lower on the provision that the wing loading and cruise altitude are adapted. Although a theoretical effort to prove that Equations (8.44) and (8.47) have the same optimizers is likely to fail, the simplified approach may be acceptable in conceptual design practice.

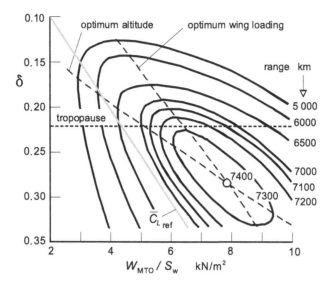

Figure 8.11   Achievable range for constant MTOW. Baseline: Figure 8.1

8.6.5   Extended Range Version

A useful application of Equation (8.44) is a first-order weight analysis of a derived version of a baseline design. Figure 8.12 shows the weight distribution of an extended range (ER) version of the design depicted in Figure 8.1. The harmonic range is increased by 43% from 7 000 to 10 000 km. One possibility is to assume that both versions have the same wing area. This causes the MTOW to increase by 19% from 5.6 kNm^{-2} to 6.7 kNm^{-2}; the mission fuel increases by 49%. Although the increased fuel weight fraction drives the optimum wing loading down, the penalty of the increased wing loading is negligible due to the flatness of the MTOW optimum. Another option is to increase the wing area so that both designs have a wing loading of 5.5 kNm^{-2} and the same initial cruise altitude of 10 500 m. This approach does not affect the mission fuel increment of 49% but enables the plane to use the same fields for taking off and landing. In both cases the fuel efficiency decreases from 40 to 36 seat-km per litre.

8.7   Summary and Reflection

8.7.1   Which Figure of Merit?

A general observation from this chapter is that weight fractions and absolute weights of the considered components have different optimum conditions. Also, the absolute weights are more sensitive to non-optimum conditions than the weight fractions. It is therefore advisable to relate weight components to invariable quantities such as the maximum payload or number of seats. It is also an open question which weight component should be considered a suitable figure of merit (FOM) for the aircraft's sizing process.

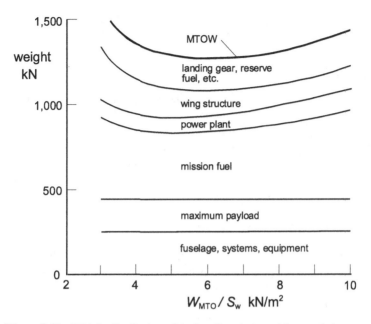

Figure 8.12   Weight distribution of the baseline design with extended range

The accumulated costs of fuel consumed during the service life of a medium-range commercial aircraft is generally of the same magnitude as the aircraft's first cost; the design mission fuel weight is about 50% of the OEW. On the other hand, the engines' first cost amounts to about 20% of the aircraft's first cost versus a weight of 12% of the OEW, typically. Saving engine weight through reduced installed thrust appears to be as valuable as saving the same fuel weight. On the other hand, it is unlikely that saving the same wing structure weight has a comparable direct effect on the aircraft operating costs, while the indirect effect on fuel consumed and thrust required (via wing loading and MTOW) is far more significant. This leads to the proposition that – to optimize wing loading and initial cruise altitude – the (mission-based) propulsion weight component is a more useful FOM than MTOW.

From the economical and environmental points of view, the various weight components are not equally important. How should their weight be weighed? For example, mission fuel costs are very substantial and will dominate any design decision, while reserve fuel is seldom consumed and contributes to costs indirectly through its contribution to gross weight. Fuel burn is responsible for atmospheric pollution since CO_2 production is proportional to it. On the other hand, although (gas turbine) engines are very costly, it may turn out that the availability of a suitable engine type may be decisive, rather than the installed thrust.[4] It is also significant that the number of engines to be installed has not yet entered the analysis up to now, but this situation will change in Chapter 9 when design constraints are introduced. And finally, although wing weight considerations dominate the location of the combined optimum, the

---

[4]Installed engine costs are often assumed to be proportional to take-off thrust. This assumption is based on statistical evidence, which may not be valid in actual design practice. In particular, a highly fuel-efficient engine is more costly than an engine with older technology with the same thrust.

costs of designing and building a wing structure are much lower than the costs related to engines and accumulated fuel expenditure. This makes the MTOW a less suitable FOM.

8.7.2 Conclusion

The sensitivity to variations in wing loading and initial cruise altitude of the aerodynamic efficiency, the fuel load, weight and thrust of the installed power plant, wing weight, and the aircraft gross weight at take-off has been examined. A rational breakdown of drag and weight into contributions from the wing, the engine installation, and (mainly payload-related) invariable items has been proposed in order to derive analytical expressions for the fuel load, power plant and wing structural weight as fractions of the MTOW. Although simplifications were required to obtain closed-form equations leading to unconstrained optimum values for the wing loading and the initial cruise altitude, it is concluded that realistic results were found within the validity range of these assumptions.

- The aerodynamic efficiency L/D in cruising flight has partial optima with respect to the wing loading and the altitude which are mutually incompatible to define a combined unconstrained optimum for both variables. Although a high L/D is a very important property of an airliner, it is not suitable as a figure of merit for the overall system. The same applies to the mission fuel weight fraction since it is inversely proportional to L/D whereas the overall engine efficiency is independent and practically constant in the stratosphere.
- Installed engine thrust and weight increase significantly when the initial cruise altitude is increased for the same engine rating. The required cruise thrust at a given altitude has a minimum value which is insensitive to the wing loading for fairly large excursions above or below the optimum.
- Optimum values for cruise altitude and wing loading have been derived resulting in minimum combined weight fractions of fuel and engine, referred to as the propulsion contribution. Partial and global unconstrained optima have been expressed as closed-form analytical expressions. Since wing weight is excluded from this optimization, the result is a design with an unrealistically low wing loading.
- Adding the wing and the propulsion weight fractions leads to optimum conditions that differ significantly from the previous optima. Although the combined weight can be computed readily, the unconstrained optimum has not been expressed in explicit form.
- The OEW is obtained from addition of invariable – mostly payload-related – weight components and weight components that are considered proportional to MTOW. The resulting expression for the MTOW appears to provide a non-iterative solution which is useful for minimizing the MTOW as well as the constituting weight components, in particular, the propulsion contribution to weight.
- By-products of the MTOW equation were derived and applied to optimize the example design, yielding realistic results. For example, it appeared that the conditions for maximum design range for given MTOW and the minimum MTOW for given range were almost – but not exactly – identical. Another property of the quasi-analytical approach is a quick computation of the weight distribution of an extended-range version of the baseline design. And finally, the fuel efficiency in terms of productivity delivered per fuel burn rate and its sensitivity to range excursions from the baseline design are readily obtained. It was found that fuel efficiency degrades considerably with ranges in excess of about 6 000 km.

8.7.3 Accuracy

Several assumptions made in the interest of deriving simplified optimum conditions may have an effect on the accuracy of the results.

- Constraints on the wing area and engine take-off thrust due to low-speed performance requirements, available fuel tank volume and other practical considerations have not been considered.
- Supercritical compressibility effects on drag have been taken into account formally, but high wing loadings and/or altitudes lead to high cruise lift coefficients. These are feasible only in combination with increased wing sweep and/or decreased thickness. Both modifications will increase wing structural weight, and the best wing loading may be below the present prediction.
- Overall engine efficiency is assumed to be independent of the stratospheric altitude. The effects of the Reynolds number variation can be incorporated readily into the analysis by means of numerical solutions. It is likely that more accurate solutions will shift the optima to cruise altitudes slightly closer to the tropopause, where the highest overall engine efficiency is obtained.
- Increasing the cruise altitude will require a larger cabin pressure differential and this will increase the fuselage structural weight, as well as several system component weights. This is another argument to decrease the design cruise altitude to a sub-optimum value.

The first two points are most essential and will be addressed in Chapters 9 and 10, the other points are worth being studied in any comprehensive optimization study but are beyond the scope of the present book.

Bibliography

[1] Torenbeek, E., *Synthesis of Subsonic Airplane Design*, Chapter 9 and Appendix D, Springer Verlag, Heidelberg, 1981.
[2] Shevell, R.S., *Fundamentals of Flight*, Chapter 18, Second Edition, Prentice-Hall, Englewood Cliffs, NJ, 1989.
[3] Roskam, J., *Airplane Design, Part V: Component Weight Estimation*, Roskam Aviation and Engineering Corporation, Ottawa, Kansas, 1989.
[4] Schatzberg, E.M., *Wings of Wood, Wings of Metal: Culture and Technical Choice in American Airplane Materials, 1914–1945*, Princeton University Press, Princeton, NJ, 1999.

Design Sensitivity of Weight

[5] Carayette, J.F., "Aircraft Wing Weight Estimation", *Aircraft Engineering*, January pp. 8–11 and April p. 119, 1950.
[6] Burt, M.E., "Structural Weight Estimation for Novel Configurations", *Journal of the Royal Aeronautical Society*, pp. 15–30, January 1962.
[7] Küchemann, H., and Weber, J., "An Analysis of Some Performances of Various Types of Aircraft Designed to Fly over Different Ranges at Different Speeds", *Progress in Aeronautical Sciences*, Vol. 9, Pergamon Press, 1968, pp. 329–456.
[8] Coker, B.B., "Problems in Airframe Development Associated with Weight and Balance Control in Heavy Logistics Transport Vehicles such as the C-5A Transport", Short Course in modern theory and practice of weight optimization and control for advanced aeronautical systems, University of Tennessee, November 1968.
[9] Caddell, W.E., "On the Use of Aircraft Density in Preliminary Design", SAWE Paper No. 813, May 1969.

[10] Gerend, R.P., and J.P. Roundhill, "Correlation of Gas Turbine Engine Weights and Dimensions", AIAA Paper No. 70-669, June 1970.

[11] Howe, D., "The Prediction of Empty Weight Ratio and Cruise Performance of Very Large Subsonic Jet Transport Aircraft", Cranfield Report Aero 3, 1971.

[12] Marsh, D.P., "Post-Design Analysis for Structural Weight Estimation", SAWE Paper No. 936, 1972. (Douglas Paper 6021).

[13] Schneider, W., "Project Weight Prediction Based on Advanced Statistical Methods", AGARD CP-147 Vol. 1, Paper No. 17, 1973.

[14] Patterson, R.W., "Weight Estimates for QUIET/STOL Aircraft", SAWE Paper No. 1001, 1974.

[15] Waters, M.H., and E.T. Schairer, "Analysis of Turbofan Propulsion System Weight and Dimensions", NASA TM X-73199, January 1977.

[16] Beltramo, M.N., D.L. Trapp, B.W. Kimoto, and D.P. Marsh, "Parametric Study of Transport Aircraft Systems Cost and Weight", NASA CR-151970, April 1977.

[17] Storey, R.E., "Dynamic Loads and Airplane Structural Weight", SAWE Paper No. 1153, 1977.

[18] Onat, E., and G.W. Klees, "A Method to Estimate Weight and Dimensions of Large and Small Gas Turbine Engines", NASA CR-159481, 1979.

[19] Anonymus, "Requirements for Short/Medium Range Aircraft for the 1980s", Association of European Airlines, document G.2110/R2, Third Edition, July 1981.

[20] Scott, P.W., "Developing Highly Accurate Empirical Weight Estimating Relationships: Obstacles and Tactics", SAWE Paper No. 2091, May 1992.

[21] Martin, J.F., "Étude Preliminaire d'un Avion de Transport à Reaction", *l'Aeronautique et l'Astronautique*, 1992-6, No. 157, pp. 62–70, 1992.

[22] Scott, P.W., "Conceptual Estimation of Moments of Inertia", SAWE Paper No. 2171, May 1993.

[23] Schneegans, A., "Weight Analysis of Commercial Aircraft Dependent on Cruising Altitude", *DGLR Jahrbuch 1993*, Paper No. 93-03-096, 1993.

[24] Read, A., "Stretching an Aircraft: A Case Study", SAWE Paper No. 2165, May 1993.

[25] Giles, G.L. 'Equivalent Plate Modelling for Conceptual Design', AIAA Paper No. 95–3945, September 1995.

[26] Liebeck, R.H., D,A, Andrastek, J. Chau, R. Girvin, R. Lyon, B.K. Rawdon, P.W. Scott, and R.A. Wright, "Advanced Subsonic Airplane Design and Economic Studies", NASA CR-195443, April 1995.

[27] Ardema, M.D., M.C. Chambers, A.P. Patron, A.S. Hahn, H. Miura, and M.D. Moore: "Analytical Fuselage and Wing Weight Estimation of Transport Aircraft", NASA TM-110392, 1996 (also AIAA Paper No. 96-5583, October 1996).

[28] Schmidt, A., M. Läpple, and R. Kelm, "Advanced Fuselage Weight Estimation for the New Generation of Transport Aircraft", SAWE Paper No. 2406, May 1997.

[29] Saeedipour, H.R., and T.N. Stevenson, "The Effects of Small Changes to the Design Specification of a Jet Civil Transport Aircraft", *Aircraft Design*, Vol. 1, No. 1, pp. 25–41, 1998.

[30] Shustrov, Y.M., "Starting Mass: A Complex Criterion of Quality for Aircraft On-Board Systems", *Aircraft Design*, Vol. 1, No. 4, pp. 193–203, 1998.

[31] Dorbath, F., "Civil Aircraft Mass Analysis (MTOM> 40t): Statistical Mass Estimation", LTH Mass Analysis, UL-442.0(T), Ausgabe A, DLR-LY/Airbus, 2011.

Structures Technologies and Materials

[32] Lovell, D.T., and M.A. Disotell, "Structural Material Trends in Commercial Aircraft", AIAA Paper No. 781552, August 1978.

[33] Wissel, W.D., "Advanced Material Application on the European Wide Body Transport Aircraft Airbus", SAWE Paper No. 1484, 1982.

[34] Lange, R.H., and J.W. Moore, "Systems Study of Application of Composite Materials for Future Transport Aircraft", AIAA Paper No. 82-0812–1982.

[35] Burns, J.W., "Advanced Composites Sizing Guide for Preliminary Weight Estimates", *SAWE Journal*, pp. 44–61, Fall 1992.

[36] Giles, G.L., "Equivalent Plate Modeling for Conceptual Design of Aircraft Wing Structures", AIAA Paper No. 95-3945, September 1995.

[37] Peel, C.J., "Advances in Materials for Aerospace", *The Aeronautical Journal*, Vol. 100, No. 1000, pp. 487–503, December 1996.

[38] Anonymous, "Metallic Materials and Processes", http://www.nap.edu/openbook/0309053900/html/26.html, copyright 1996, The National Academy of Sciences, 2000.

[39] Rudolph, P.K.C., "High-Lift Systems on Commercial Subsonic Airliners", NASA Contractor Report 4746, September 1996.

[40] Slingerland R., F. van Reijn van Alkemade, and B. Vermeulen, "A Preliminary Prediction Method for the Effect of New Fuselage Materials on Transport Aircraft Weight," AIAA Paper 2007-2289, April 2007.

[41] Sandoz, P.L., "The Next Careful Steps in Commercial Aircraft Structures", *Structures Technology: Historical Perspective and Evolution*, A.M. Noor, (ed.), AIAA, pp. 91–106, 1998.

[42] Curran, R., A. Rothwell, and S. Castagne, "Numerical Method for Cost-Weight Optimization of Stringer-Skin Panels", *Journal of Aircraft*, Vol. 43, No. 1, pp. 264–274, January–February 2006.

[43] Ciampa, P.D., B. Nagel, and M.J.L. van Tooren, "Global Local Structural Optimization of Transport Aircraft Wings", AIAA Paper No. 2010-3098, April 2010.

9

Matching Engines and Airframe

The Wright brothers, among their other notable achievements, discovered a design truism still valid today: if an aircraft engine is sized to optimum airspeed, pilots will decide that the aircraft is underpowered.

—B.H. Carson, in AIAA Paper 80–1847 (1980)

9.1 Requirements and Constraints

Design sensitivity to variations in wing loading, initial cruise altitude and installed engine thrust was the subject of the previous chapter. The emphasis was on unconstrained optimization of several basic objective functions: aerodynamic efficiency, mission fuel, fuel plus power plant weight, and maximum take-off weight (MTOW). It was found that optimization of these objective functions results in widely divergent optimum values of the wing area, engine thrust and cruise altitude. In view of the importance of engine selection, the present chapter focusses on design constraints reducing the designer's freedom of choice. The most relevant constraints are top level and safety requirements such as available take-off and landing distances and climb or cruise performance with an inoperative engine. Performance constraints are mostly translated into minimum required values of the installed engine thrust and wing area. Engine thrust is irrelevant for constraints imposing a lower limit on wing size, such as the condition that the maximum fuel load must be contained inside the primary wing box and the cruise lift coefficient must be limited to avoid high speed buffet. These constraints are treated in Chapter 10.

Modern mature gas turbine engines are extremely reliable – their in-flight shutdown (IFSD) rates have reduced from about one per 10 000 engine hours for the first generation of turbofans to one per 100 000 hours for current engines. Nevertheless, a fleet with an average yearly utilization of 4 000 hours per airliner will occasionally experience a significant in-flight power or thrust reduction or even a complete engine failure. Regardless of the number of propulsion engines installed, a safe continuation of the flight must be possible in such a situation albeit at a reduced performance level. And when the pilot decides to abort the take-off when the plane is still on the runway, the emergency stop must be a safe event as well. Engine failure is likely to occur during take-off and initial climb when the engines operate at their maximum rating, but a safe continuation of the flight en route can be equally critical during a flight over water in view

Advanced Aircraft Design: Conceptual Design, Analysis and Optimization of Subsonic Civil Airplanes, First Edition. Egbert Torenbeek.
© 2013 by Egbert Torenbeek. Published 2013 by John Wiley & Sons, Ltd.

of its long duration. Airworthiness requirements associated with engine failure are dependent on the operational conditions – field lengths and elevations, ambient temperature and pressure, etc. – for which civil airplanes must comply with one engine inoperative (OEI) requirements. Obviously, the consequences of engine failure on performance are sensitive to the number of engines to be installed and their position relative to the airframe. Both subjects are considered outside the context of this book – the number of installed engines is assumed to be pre-assigned.

Top level requirements (TLRs) specify flight performances which dominate the engine to airframe matching process – the most relevant parameters involved are installed thrust and wing size. Constraints on these selection variables are derived from the following performance constraints:

- The cruise speed or cruise Mach number – occasionally in combination with an initial cruise altitude capability (ICAC) – following the take-off with MTOW, ascending to the top of climb (TOC) and acceleration to cruise speed.
- Climb gradients achievable after engine failure for several flight segments and aircraft configurations.
- Engine-out altitude capability (EOAC): the altitude achievable in level flight with OEI. This requirement is relevant for airplanes having the capability to cross extensive uninhabited terrains and oceans.
- The take-off field length (TOFL) required for taking off with engine failure at the critical speed.
- The approach speed and/or the landing field length (LFL) required.

Low-speed flight, field performances and cruise speed capability may pertain to standard and/or non-standard ambient conditions defined by the ICAO standard atmosphere (ISA) defined in Appendix B.

9.2   Cruise-Sized Engines

Having reached the TOC, the airplane accelerates to the initial cruise speed and the ICAC is attained when a specified rate of climb (ROC) is achievable with maximum climb rating. The present analysis simplifies this requirement by assuming that, after throttling back from maximum climb to cruise rating, the engines have sufficient thrust to fly horizontally at the prescribed speed and altitude. A 1 m/s (200 ft/min) ROC is equivalent to a gradient of 0.4%, which requires 8% of the cruise thrust, typically. Hence, if the maximum climb and cruise ratings produce a thrust ratio of 1.08, the two criteria yield the same thrust requirement. The thrust lapse rate for cruise conditions is used to translate this requirement into an installed take-off thrust at sea level ISA.

9.2.1   Installed Take-Off Thrust

The matching relationship between engine thrust and airplane drag makes use of the rubber-izing concept explained in Section 8.5. The essential characteristic is the corrected lapse rate defined by Equation (3.10)

$$\tau \stackrel{\text{def}}{=} \frac{T/p}{T_{\text{TO}}/p_{\text{sl}}} = \frac{T/\delta}{T_{\text{TO}}} \tag{9.1}$$

which is derived from basic engine performance information. For instance, the corrected cruise thrust lapse is derived from

$$\tau^* = \frac{T^*/p^*}{T_{\mathrm{TO}}/p_{\mathrm{sl}}} \tag{9.2}$$

denoting properties at the tropopause by an asterisk. Since the thrust at high (stratospheric) altitudes is (theoretically) proportional to the ambient pressure, the corrected lapse rate depends on the engine cycle and cruise Mach number only. Hence, it is fair to assume a constant $\tau = \tau^*$ for altitudes not too far above or below the tropopause. The relationship between (all-engines) take-off thrust and initial cruise altitude is obtained from Equation (8.45), assuming that the initial cruise weight is equal to the TOW,

$$T_{\mathrm{TO}} = \frac{T/\delta}{\tau} = \frac{\overline{D}/\delta}{\overline{\tau}} = \frac{1}{\overline{\tau}} \left[q_{\mathrm{sl}} \left\{ (C_{D_{\mathrm{p}}} S)_{\mathrm{fix}} + \check{C}_{D_{\mathrm{p}}} S_{\mathrm{w}} \right\} + K_{\mathrm{L}} \frac{(W_{\mathrm{MTO}}/\delta)^2}{q_{\mathrm{sl}} S_{\mathrm{w}}} \right] \tag{9.3}$$

with $q_{\mathrm{sl}} = 0.5\gamma p_{\mathrm{sl}} M^2$ and δ denotes the relative ambient pressure at the initial cruise altitude. The installed thrust lapse rate $\overline{\tau}$, referring to engine thrust minus nacelle drag, and other symbols are defined in Sections 8.4 and 8.5. Equation (9.3) determines how the required thrust/weight ratio varies with the cruise altitude and wing loading at take-off. The presence of the MTOW is a minor complication which is solved by substitution of Equation (8.44).

9.2.2 The Thumbprint

The design selection chart Figure 9.1 – some designers call it a 'performance thumbprint' – is useful for explicit optimization; an application to high-lift system technology is described in [21]. This diagram depicts the ratio of installed thrust to MTOW versus the take-off wing loading. An alternative diagram with installed thrust or power versus wing area has a considerable advantage if the MTOW is fixed and the design range is considered a figure of merit (FOM) as discussed in Section 8.6. Figure 9.1 depicts the case of unconstrained optimization, the complete selection chart to be discussed later visualizes several performance, geometric and aerodynamic constraints in combination with objective functions such as MTOW or mission fuel weight. The following remarks apply to the present example:

- The constant altitude curves are more or less horizontal for usual combinations of wing loading and altitude. Hence, if small excursions are considered from a baseline wing size, the required cruise thrust required needs little or no revision.
- Although wing structure weight has a major effect on the location of the unconstrained optimizer (point D), the MTOW is more sensitive to altitude than to wing loading. Point D combines a rather high wing loading with a low initial cruise altitude and low-speed performance constraints are likely to render design D unfeasible.
- Fixed (fuselage-related) drag decreases with increasing cruise altitude, minimum wing drag is insensitive to altitude. These effects are counteracted by increasing installed engine thrust and weight with increasing cruise altitude. Curve I defines the partial optimum thrust for given wing loading.

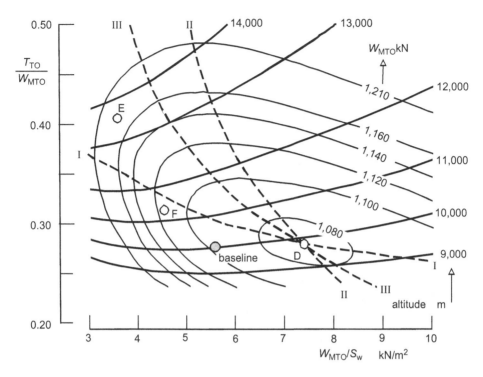

Figure 9.1 Design selection chart with MTOW contours for the design in Figure 8.1. Partial optima: I – constant wing loading, II – constant thrust/weight, III – constant altitude

- The optimum wing loading for a given thrust is defined by Curve II. This can be seen as the minimum MTOW for the case of a given engine type.
- The optimum wing loading and thrust loading for a given altitude are defined by Curve III. It is found that the optimum wing loading decreases with altitude such that the lift coefficient is $C_L = 0.63$. The thrust to be installed increases considerably with the altitude; see also Figure 8.8.

Design points E and F in Figure 9.1 are unconstrained optimizers for minimum mission fuel and minimum propulsion weight penalty, respectively. Due to its initial cruise altitude of nearly 13 000 m, design E needs extremely large engines in combination with a large wing, leading to a 6% higher gross weight than aircraft D. Design F with the lowest weight penalty of two costly items and a mere 2% MTOW penalty might be an attractive compromise. Whether this design satisfies constraints other than the cruise speed requirement remains to be investigated.

The computation of thrust and wing size effects on MTOW contours in the design selection chart are derived for the MTOW. However, the performances treated in the following sections apply to aircraft gross weights equal to or less than the maximum GW for each case considered. The notations are therefore general: W_{TO} for take-off and W_L for landing.

9.3   Low Speed Requirements

Critical climb performances of civil aircraft can be categorized as (a) minimum climb gradients available during critical phases of the flight, in particular after failure of an engine, and (b) obstacle clearance requirements for the net flight path, obtained by subtracting gradient deviations from the gross flight path. Although both performances are subject to safety requirements for all operations, only certification requirements affect the engine-to-airframe matching process. The operational variables involved are the aircraft gross weight (GW), the pressure altitude and the ambient temperature. The permissible TOGW is the highest GW $\leq$ MTOW for which all climb requirements are satisfied. In case of an elevated airfield with higher than standard ambient temperature, the permissible TOGW may be significantly less than the MTOW, resulting in an undesirable reduction of payload or range. Similar arguments for landing are usually less restrictive. Consequently, engine sizing is often based on 'hot and high' conditions representative of the most elevated airfields to be served.

9.3.1   Stalling Speed

The safety speeds during and after the take-off and before landing are related to the stalling speed V_S with deflected high-lift devices. The 1-g stalling speed $(V_S)_{1-g}$ is derived from level flight tests during which the airplane decelerates at a rate of 1 kt/s until g-break occurs. The stall develops into a descent followed by recovery of the non-stalled condition.[1] Reference speeds are based on the minimum speed occurring during the stall or on $0.94\ (V_S)_{1-g}$. Assuming the latter to be the highest, the (equivalent) stalling speed in flight with TOGW is

$$V_S = 0.94 \left(\frac{2W_{TO}}{\rho_{TO}\, S_w C_{L\,\mathrm{max}}} \right)^{1/2} \tag{9.4}$$

The factor 0.94 is equivalent to accounting for the g-break by using a normal load factor of 0.88 during the take-off manoeuvre.[2] Alternatively, the maximum lift coefficient obtained from aerodynamic data can be increased by 13%. The free air safety speed V_2 is the speed to which the airplane accelerates while climbing away after lift-off with one engine inoperative. For jet aircraft this speed must comply with $V_2 \geq 1.20\ V_S$. The safety speed with one engine inoperative at $V_2 = 1.20\ V_S$ amounts to

$$V_2 = \left(\frac{2W_{TO}}{\rho_{TO}\, S_w\, C_{L_2}} \right)^{1/2} \tag{9.5}$$

with $C_{L_2} = 0.786\ C_{L\,\mathrm{max}}$. Instead of the (aerodynamic) maximum lift coefficient, the lift coefficient at V_2 can be treated as a basic parameter. The conceptual stage offers the possibility of optimizing C_{L_2} and V_2, a dominating condition for sizing the engines. The flap deflection corresponding to the optimum C_{L_2} is one of the design cases of aerodynamic and mechanical flap design to be settled during the downstream design stages.

---

[1]Many airliners have a flight envelope protection system preventing the plane from entering the stalled condition. Nevertheless, the stalling speed is subject to certification.
[2]This effect is ignored when using $C_{L_2} = C_{L\,\mathrm{max}}/1.2^2$, leading to a significant error in the stalling speed and the take-off distance.

Table 9.1 Summary of FAR/JAR 25 climb requirements for jet transports. N: number of installed engines. OEI: one engine inoperative; AEO: all engines operating

Phase of flight		Airplane configuration				Climb gradient%		
		Flaps	uc	Eng. rating	Flight speed	$N = 2$	$N = 3$	$N = 4$
lift-off	1st seg.	TO	↓		TO	$V_{\text{LOF}} \to V_2$	0	0.3	0.5
take-off	2nd seg.	TO	↑	OEI	TO	$V_2 \geq 1.20 V_S$	2.4	2.7	3.0
path	final seg.	↑	↑		MC	$V \geq 1.25 V_S$	1.2	1.5	1.7
approach climb		APP	↑		TO	$V \leq 1.5 V_S$	2.1	2.4	2.7
landing climb		LAND	↓	AEO	TO	$V \leq 1.3 V_S$	3.2	3.2	3.2

9.3.2 Take-Off Climb

Most pertinent climb requirements for civil transport aircraft reproduced in Table 9.1 pertain to the continued take-off (CTO) after failure of one engine. The flight path after lifting off (LOF) is subdivided into four segments.

- The first segment commences at the moment of lifting off with OEI. Undercarriage retraction begins at least three seconds after lift-off, the condition of full retraction denotes the end of the first segment. Although flight performance during this benefits from ground effect, the required climb gradient pertains to flying out of ground effect.
- During the second segment, engine rating and flap position are the same as for the first segment, but the climb gradient has improved due to the undercarriage retraction. This segment lasts until flap retraction begins at a height of at least 400 ft (120 m).
- The third segment (not indicated) is the flight phase during which the aircraft is accelerated to the flap retraction speed, flaps are retracted and the engine rating is reduced to maximum continuous. The speed is finally increased until the flaps-up safety speed is reached.
- The final segment refers to the aircraft in the en route configuration, with the operating engines at maximum continuous rating.

The minimum take-off (or free air) safety speed is $V_2 = 1.20 V_S$ for most multi-engine aircraft, except for four-engine propeller aircraft and a certain class of jet aircraft; see FAR/JAR 25.107 (b). Since the best speed to achieve a high climb gradient V_X is higher than $1.20 V_S$ the aircraft improves its climb performance for $V_2 > 1.20 \ V_S$. This overspeed situation is achieved by delaying the rotation. In the design stage, overspeed is not an option since the TOFL requirement is usually a dominant requirement.

9.3.3 Approach and Landing Climb

The climb requirements in Table 9.1 pertain to the approach climb after engine failure and the climb-out after an aborted landing.

1. A missed approach is one that is discontinued at or above the decision height; the subsequent flight segment is referred to as the approach climb. The decision to go around is made after

engine failure with flaps in the approach setting. At this point undercarriage is retracted and the remaining live engines are set to take-off power or thrust.

2. If after landing touch-down the ground roll is discontinued, the thrust levers off all engines are advanced to the maximum take-off power position without altering the flap setting. Climbing out after a baulked landing is called the landing climb. It is not likely that this case will affect the engine thrust unless the flap angle is very large. In that case the landing flap angle may have to be reduced which obviously increases the stalling speed and the landing distance.

9.3.4 Second Segment Climb Gradient

The following derivation concentrates on the often critical second segment climb. The climb angle in steady flight at speed V_2 with OEI follows from the equations of motion for a small climb angle,

$$\gamma_2 = \frac{(1 - N_{\text{eng}}^{-1})\, T_{V_2}}{W_{\text{TO}}} - \left(\frac{C_D}{C_L}\right)_{V_2} \tag{9.6}$$

with T_{V_2} denoting the total thrust of N_{eng} installed engines. Equation (4.36) is substituted to obtain the thrust-to-weight ratio for a specified climb gradient,

$$\frac{T_{V_2}}{W_{\text{TO}}} = \frac{N_{\text{eng}}}{N_{\text{eng}} - 1}\left(\gamma_2 + C_0 + \frac{C_{L_2}}{\pi A_{\text{w}} E}\right) \quad \text{for} \quad 1.2 \le C_{L_2} \le 1.8 \tag{9.7}$$

Apart from the number of engines, the lift coefficient at V_2 and the wing aspect ratio A_{w} are the most influential parameters affecting the required thrust. The take-off thrust required is derived from the engine's thrust lapse with flight speed between the static condition and the climb-out condition at V_2. For standard ambient conditions, the engine thrust lapse rate depends mainly on the specific thrust (or bypass ratio) and the safety speed. A significant increase of the required thrust must be applied if the take-off is specified at a 'hot and high' airfield. This requires information on the flat rating characteristics of the engine; that is, how take-off thrust varies with ambient temperature and elevation of the field.

Equation (9.7) makes it clear that the thrust required is reduced when the climb-out lift coefficient is low. However, this leads to a high lift-off speed and a long ground run. The next section explains how the optimum C_{L_2} is derived for a combined take-off and climb away performance analysis.

9.4 Schematic Take-Off Analysis

The required take-off performance capability of a commercial transport aircraft is essential for sizing its wing and engines – however, it is the most complicated of all flight performance requirements. This becomes clear when reviewing definitions and procedures in the relevant airworthiness codes, for example, FAR/JAR 25.103 through 25.121. A detailed explanation of how these rules are used in the operation of civil aircraft is found in [4].

9.4.1   Definitions of Take-Off Field Length

The required take-off field length (TOFL) for an airliner in operation is prescribed by the flight manual (FM) which has been derived from certification flight tests on a dry concrete runway – operational take-off characteristics on wet runways are assessed separately. A take-off will be safe when the runway is long enough for the normal situation that all engines are operating continuously and if that engine failure occurs at the most critical speed during the take-off. An airplane may take off if the available runway is long enough to accommodate the longest of the following distances complying with a nominal piloting technique.

1. The distance required for taking off with all engines operating (AEO). Acceleration on the runway from standstill to rotation at speed V_R is followed by the lift-off at speed V_{LOF} and climb to 35 ft (10.7 m) obstacle height. In order to account for technical and operational variations from the nominal situation, the runway length required is equal to the regular take-off distance (RTOD) multiplied by 1.15.
2. The distance required for acceleration on the runway from standstill to the speed at which one engine fails, followed by a continued take-off (CTO) with OEI. This distance is called the continued take-off distance (CTOD).
3. The distance required for acceleration on the runway from standstill to the engine failure speed V_{EF} followed by deceleration to standstill using maximum braking effort. During this rejected take-off (RTO) the aircraft stays on the ground. The runway length required is called the accelerate-stop distance (ASD).

For ensuring a safe take-off from a given airfield, the longest of the three distances mentioned above is determined (Figure 9.2). Since the second and third depend on V_{EF}, the critical situation occurs when the CTOD and the ASD are equal. The decision speed V_1 is scheduled so that the pilot can act appropriately when an engine fails. If during the ground run an engine fails at a speed below V_1, the take-off is aborted, otherwise it is continued. The requirement $V_1 \leq V_R$ implies that the take-off must not be rejected once the rotation for taking off has been initiated. The balanced field length (BFL) refers to the situation where the CTOD and the ASD are equal.

Take-off analysis in the concept design stage can be complicated by a lack of detailed aerodynamic data and emergency braking characteristics. Hence, it is often assumed that the

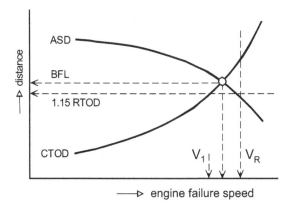

Figure 9.2   Rotation speed, decision speed and balanced field length

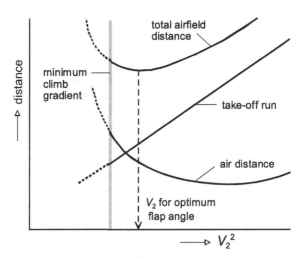

Figure 9.3   Take-off distances affected by the take-off safety speed

BFL is the dominant requirement for deriving the installed take-off thrust. This is likely to be correct for twin-engine airplanes since failure of an engine causes the power or thrust to be reduced by 50%. However, large four engine jetliners have a long take-off run compared to the air distance and the 15% take-off distance reserve may well exceed the effect of 25% thrust loss due to failure of one engine. The following take-off analysis applies to a take-off with engine failure during the take-off run followed by a continued take-off with OEI.

9.4.2   Take-Off Run

The take-off run consists of an acceleration from standstill to V_{LOF} which is about 1.15 V_S. The take-off run length $\mathcal{L}_{RUN}$ is computed from the energy equation,

$$\mathcal{L}_{RUN} = \frac{\text{kinetic energy at lift-off}}{\text{mean accelerating force}} = \frac{(E_k)_{LOF}}{\overline{F}_{acc}} \tag{9.8}$$

When the plane starts rolling, the accelerating force equals the AEO static thrust T_{TO} minus the runway friction. The thrust decreases gradually with speed until engine failure occurs at V_{EF}; thereafter it decays rapidly to the OEI thrust. Since V_2 is only slightly higher than V_{LOF}, the thrust at lift off T_{LOF} is assumed equal to the thrust at the safety speed T_{V_2} which is treated as the dominant parameter determining the acceleration during taking off. Aerodynamic drag increases proportional to V^2, stays relatively small but increases progressively between the rotation speed V_{ROT} and V_{LOF}. The length of the take-off run is written as

$$\mathcal{S}_{RUN} = \frac{W_{TO}^2}{\rho_{TO}\, g\, S_w C_{L_2}\, k_T T_{V_2}} \quad \text{where} \quad k_T \overset{\text{def}}{=} \left(\frac{V_2}{V_{LOF}}\right)^2 \frac{\overline{F}_{acc}}{T_{V_2}} \tag{9.9}$$

The parameter k_T allows for the variation of thrust, friction drag and aerodynamic drag and for the speed increment after lift-off. Its order of magnitude is 0.85, subject to statistical validation.

9.4.3   Airborne Distance

The angle of attack and the lift are building up rapidly during the rotation until the plane lifts off. During the climb segment the plane follows a curved path with the angle of pitch controlled by the pilot until the steady-state climb angle is reached. Due to the engine-out condition there is little acceleration. The minimum speed at the obstacle height V_2 is 1.2 times the power-off stall speed V_S for jet transports and twin-engine turboprops. During airborne flight the remaining failed engine thrust and wind milling drag depend on engine characteristics which are usually unknown during conceptual design. The yawing moment due to thrust asymmetry is negated by a lateral force on the vertical tail which causes induced drag, whereas landing gear retraction and ground effect lead to another variation of the drag. Accurate analysis of a CTO after engine failure must be considered elusive during the conceptual design phase and a quasi-analytical approximation is proposed as an alternative. In particular, it is assumed conservatively that the thrust of the operating engines stays constant and that the failed engine delivers no thrust after lifting off.

The climb segment can be modelled as a manoeuvre with constant rate of pitch [5] or as a circular path, followed by a steady climb. A typical outcome of the analysis is that the air distance equals the distance travelled during a steady climb with gradient γ_2 to a reference height h_{TO} which equals twice the obstacle height. Accordingly, the horizontal distance between lifting off and passing the obstacle derived from Equation (9.6) is written as

$$\mathcal{L}_{AIR} = \frac{h_{TO}}{\gamma_2} = h_{TO} \left\{ \frac{(1 - N_{eng}^{-1}) T_{V_2}}{W_{TO}} - \left(\frac{C_D}{C_L} \right)_{V_2} \right\}^{-1} \tag{9.10}$$

with $h_{TO} = 21$ m.

9.4.4   Take-Off Distance

The field length required for a CTO is found by adding Equations (9.9) and (9.10) and substitution of C_{D_2}/C_{L_2} according to Equation (4.36),

$$\mathcal{L}_{TO} = \frac{W_{TO}^2}{\rho_{TO}\, g\, S_w\, C_{L_2}\, k_T T_{V_2}} + h_{TO} \left\{ \frac{(1 - N_{eng}^{-1}) T_{V_2}}{W_{TO}} - \left(C_0 + \frac{C_{L_2}}{\pi A_w E} \right) \right\}^{-1} \tag{9.11}$$

This expression makes it clear that the minimum distance required for a CTO is obtained by making a compromise between a shorter take-off run and a longer air distance when C_{L_2} increases. The optimum is found from Equation (9.11) by minimizing with respect to C_{L_2}, resulting in

$$\frac{C_{L_2}}{\pi A_w E} = \frac{(1 - N_{eng}^{-1}) T_{V_2} - C_0 W_{TO}}{W_{TO} + (h_{TO}\rho_{TO}g\, \pi A_w E S_w\, k_T T_{V_2})^{0.5}} \tag{9.12}$$

on the provision that $1.20 < C_{L_2} < 1.80$. Consequently, the best value of C_{L_2} is high for a high aspect ratio wing, when many engines are installed or when the thrust-to-weight ratio is high.

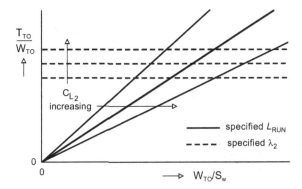

Figure 9.4   Thrust-to-weight ratio required for taking off with specified lift coefficient at V_2

Although the optimum according to Equation (9.12) is not necessarily accurate, the minimum take-off distance is insensitive to a deviation of C_{L_2} from its (real) optimum.

9.4.5   Generalized Thrust and Span Loading Constraint

Top-level requirements specify an upper constraint on the take-off field length from which the advanced designer must derive feasible combinations of installed engine thrust and wing area. This forms a typical example of a reverse engineering problem. Several handbook methods offer a solution based on statistics by considering the TOFL to be determined by the same combination of design variables that determine the take-off run distance. This leads to the conclusion that, in order to minimize the thrust required to cope with a given TOFL constraint, C_{L_2} (and $C_{L\max}$) should be high. Although such a method may be valid when C_{L_2} is selected in the usual range, the inexperienced designer is tempted to try a very high $C_{L\max}$. Indeed, Figure 9.4 shows that the thrust required to lift off within a given runway length is small when C_{L_2} is high but this requires a high thrust to obtain the specified positive climb gradient. Conversely, a low C_{L_2} is favourable to reduce the air distance but leads to a high thrust to reduce the run length. A closed form solution for selecting the correct C_{L_2} is obtained by combining two terms in Equation (9.11),

$$k_{\mathrm{D}} \stackrel{\text{def}}{=} 1 - C_0 \frac{W_{\mathrm{TO}}}{T_{V_2}} \quad \text{with} \quad C_0 \frac{W_{\mathrm{TO}}}{T_{V_2}} \ll 1.0 \tag{9.13}$$

and inserting the factor k_{D} into the following dimensionless parameters:

$$\text{span loading:} \qquad \Omega \stackrel{\text{def}}{=} \frac{\mathcal{L}_{\mathrm{TO}}}{h_{\mathrm{TO}}} \frac{(k_{\mathrm{D}} - N_{\mathrm{eng}}^{-1})}{k_{\mathrm{T}} \, \rho_{\mathrm{TO}} \, g \, h_{\mathrm{TO}}} \frac{W_{\mathrm{TO}}}{b_{\mathrm{w}}^2 \, \pi \, E} \tag{9.14}$$

$$\text{thrust-to-weight ratio:} \quad \Theta \stackrel{\text{def}}{=} \frac{\mathcal{L}_{\mathrm{TO}}}{h_{\mathrm{TO}}} (k_{\mathrm{D}} - N_{\mathrm{eng}}^{-1}) \frac{T_{V_2}}{W_{\mathrm{TO}}} \tag{9.15}$$

$$\text{lift coefficient at } V_2\text{:} \qquad \Lambda \stackrel{\text{def}}{=} \frac{\mathcal{L}_{\mathrm{TO}}}{h_{\mathrm{TO}}} \frac{C_{L_2}}{\pi \, A_{\mathrm{w}} E} \tag{9.16}$$

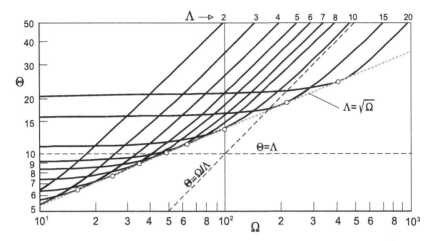

Figure 9.5 Generalized solution for combinations of span loading Ω and thrust-to-weight ratio Θ resulting in a given take-off field length

Rewriting Equation (9.11) in terms of these parameters yields

$$\Theta \Lambda \left\{ 1 - (\Theta - \Lambda)^{-1} \right\} - \Omega = 0 \qquad (9.17)$$

It is worth noting that the span loading W_{TO}/b_w^2 rather than the wing loading appears to be the primary selection variable determining the thrust loading required for a given TOFL. The aspect ratio features only in combination with the climb-out lift coefficient. Equation (9.17) is visualized in Figure 9.5. In order to show that the generalized thrust constraints are curves blending between two asymptotes, the scale of parameter variation is (much) larger than needed in practical applications.

1. The thrust-to-weight ratio for high span loadings tends to become proportional to Ω and its slope approaches Λ^{-1}.
2. The left part of each constraint is dominated by the requirement that the climb-out angle after must be positive, defining the horizontal asymptote $\Theta = \Lambda$. In other words: the derived TOFL constraint also satisfies the first segment climb gradient requirement.
3. The thrust required for a given span loading has a minimum value corresponding to the optimum lift coefficient at the safety speed determined by the enveloping curve

$$\Theta - s\sqrt{\Theta} = \sqrt{\Omega} = \Lambda \qquad (9.18)$$

Figure 9.5 represents optimizers by discrete points where the curves for constant Λ are tangential to the enveloping curve.

It is not surprising that Equation (9.18) is identical to Equation (9.12) defining the climb-out lift coefficient for minimizing the TOFL of an aircraft with given spanand thrust loading.

9.4.6   Minimum Thrust for Given TOFL

Equation (9.18) is a quadratic equation for the minimum thrust required for a specified take-off field length. It can be approximated as $\Theta = 1.15\sqrt{\Omega} + 2$, or

$$\frac{T_{V_2}}{W_{TO}} = 1.15 \left\{ \frac{W_{TO}/b_w^2}{(k_D - N_{eng}^{-1})\mathcal{L}_{TO}\, k_T \rho_{TO}\, g\, \pi\, E} \right\}^{0.5} + \frac{42\,(m)}{(k_D - N_{eng}^{-1})\mathcal{L}_{TO}} \qquad (9.19)$$

with k_D determined by Equation (9.13). This expression shows that the span loading and the modified Oswald factor E are essential variables rather than the wing loading and the maximum lift coefficient, provided the climb-out lift coefficient is optimized. For practical application, Equation (9.19) can be simplified as follows:

$$\frac{T_{V_2}}{W_{TO}} = 0.78 \left(\frac{N_{eng}}{0.89\, N_{eng} - 1} \frac{W_{TO}/b_w^2}{\mathcal{L}_{TO}\, \rho_{TO}\, g} \right)^{0.5} + \frac{42\,(m)}{\mathcal{L}_{TO}} \frac{N_{eng}}{0.89\, N_{eng} - 1} \qquad (9.20)$$

The accuracy of this expression can be improved by calibrating it with statistical information. Similar to Equation (9.6), the engine thrust lapse rate must be used to find the all-engines take-off thrust required.

9.5   Approach and Landing

Top-level requirements for the landing field length (LFL) pertain to transport aircraft with maximum landing weight (MLW), (usually fully) deflected flaps for landing and extended undercarriage. The airplane makes a stabilized final approach with a calibrated air speed of at least 130% of the stalling speed. The available runway for carrying out a landing on a dry runway in airliner operation must be at least 5/3 times the nominal landing distance derived from certification tests. An additional safety factor of 1.15 is applied to landing on a wet runway. The field length factor takes into account that airliner landings in day-to-day operation deviate considerably from certification flight tests. For instance, certification tests are carried out in zero wind conditions by a hard-working pilot using a relatively steep approach, a high vertical speed at touch-down and maximum braking effort. No credit is given for reverse thrust since controllability of the aircraft and the effectiveness of thrust reversing are adversely affected by (asymmetric) engine failure. However, an airliner in normal flight approaches the threshold at 2.5 to 3° descent angle and the flare-out is usually followed by a short float. On the other hand, the plane may land in a headwind and braking is usually augmented by applying reverse thrust.

9.5.1   Landing Distance Analysis

The airborne distance $\mathcal{L}_{AB}$ is the horizontal distance covered during a steady state descent with angle γ from a 50 ft (15 m) screen height h_S followed by a flare-up and touch-down with speed V_{TD}. The energy equation is used to derive

$$\mathcal{L}_{AB} = \frac{W_L}{(D-T)_{av}} \left(\frac{V_{APP}^2 - V_{TD}^2}{2g} + h_S \right) \qquad (9.21)$$

The touch-down speed after a (hypothetical) circular flare with constant incremental load factor Δn between the descent and the horizontal runway is

$$V_{\text{TD}} = V_{\text{APP}}\sqrt{1 - \gamma^2/\Delta n} \qquad (9.22)$$

The ground run length $\mathcal{L}_{\text{GR}}$ is the distance required to come from touch-down to a full stop with an average deceleration a_{av},

$$\mathcal{L}_{\text{GR}} = \frac{V_{\text{TD}}^2}{2 a_{\text{av}}} \qquad (9.23)$$

Addition of Equations (9.21) and (9.23) yields the landing distance

$$\mathcal{L}_{\text{LD}} = \frac{W_{\text{L}}}{(D - T)_{\text{av}}} h_{\text{S}} + \frac{V_{\text{APP}}^2}{2g} \left\{ \frac{1 - \gamma^2/\Delta n}{a_{\text{av}}/g} + \frac{\gamma^2}{\Delta n} \frac{W_{\text{L}}}{(D - T)_{\text{av}}} \right\} \qquad (9.24)$$

A good prediction for jet transports is obtained for $\Delta n = 0.10$, $\gamma = 0.08$ and $W_{\text{L}}/(D - T)_{\text{av}} = 10$ and substitution of h_{S} yields the factored LFL,

$$\mathcal{L}_{\text{LF}} = 250 \,(\text{m}) + \frac{V_{\text{APP}}^2}{2g} \left(\frac{1.56}{a_{\text{av}}/g} + 1.07 \right) \qquad (9.25)$$

Prediction of the average deceleration during the landing run requires knowledge of the effectiveness of lift dumpers and/or air brakes and the operation and performance of wheel brakes. If such information is lacking, the use of statistics is unavoidable. A comparison of jet airliner and business jet landing performance reveals that the deceleration depends primarily on the use of automatic spoilers and on the main landing gear configuration [2]. For jet aircraft with automatic ground spoilers, Equation (9.25) provides a realistic approximation for $a_{\text{av}}/g = 0.40$ for two-wheel trucks, and $a_{\text{av}}/g = 0.32$ for four-wheel trucks.

9.5.2   Approach Speed and Wing Loading

Primary parameters affecting the landing distance are the approach speed and the deceleration during the braked run. Both are affected by the level of technology.

- For given wing geometry and landing weight, the stalling speed is determined by the maximum lift coefficient with flaps in the landing position. Apart from careful aero-mechanical design, $C_{L\max}$ can be increased by applying leading edge high lift devices, multiple trailing edge flap slots and/or large-span flaps.
- Modern commercial aircraft are equipped with anti-skid devices, many also with carbon brakes. For given total landing gear load, the deceleration is increased by applying brakes on both main and nose landing gears. Lift dumpers increase the main gear load and air drag during the ground run. Some regional aircraft use aerodynamic speed brakes both in the air and on the ground.

Using sophisticated technology leads to higher development, production and maintenance costs and complex flap systems increase approach drag and external noise. The need for flaps generating very high maximum is less stringent for long-range aircraft since their low value of MLW/MTOW makes the LFL requirement less critical than the TOFL. However, many regional airliners have effective trailing-edge flap systems, effective brakes and ground spoilers to develop a low approach speed and high deceleration during the landing run.

The initial wing sizing process for approach and landing is likely to start with assumed values for $C_{L\,max}$ and the average deceleration. For a specified LFL the approach speed is then obtained from Equation (9.25) and the stalling speed is 0.77 times the approach speed. Similar to the take-off case, the stalling speed is 0.94 times the 1-g stalling speed; hence, $V_{APP} = 1.22\,(V_S)_{1-g}$. This leads to the following constraint on the landing wing loading:

$$\frac{W_L}{S_w} \leq 0.335\,\rho_{sl}\,V_{APP}^2\,C_{L\,max} \tag{9.26}$$

9.6   Engine Selection and Installation

9.6.1   Identifying the Best Match

Figure 9.6 includes constraints on thrust and wing loading superimposed on the MTOW contours (Figure 9.1). The present design represents a typical situation where design D with the minimum MTOW is unfeasible since all constraints are active. Design P has the lowest MTOW and complies with the cruise altitude constraint but it does not comply the TOFL and LFL constraints. Design Q has the lowest MTOW and complies with all constraints – it forms the MTOW-optimum feasible design. Although the baseline design is unfeasible, it

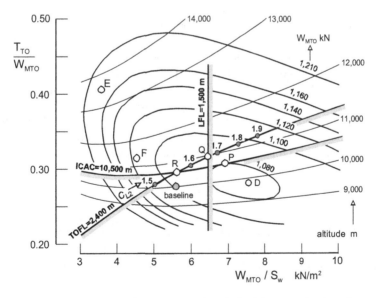

Figure 9.6   Design selection chart including constraints derived from top-level performance requirements. Baseline: Figure 8.1

can be improved for the same wing loading by increasing its installed thrust by 7%. It then becomes practically identical to design R on the intersection of the ICAC and TOFL constraints. Although design R does not represent the constrained MTOW-optimizer, its lower wing loading allows it to have 8% less thrust than design Q for a small take-off weight penalty. Arguably, design R can be considered the best solution for the present performance requirements since its lower wing loading gives it some weight growth potential. Design E with the highest fuel efficiency is not a practical option whereas design F, with the minimum propulsion weight contribution, does not offer gross weight and thrust advantages over design R.

The engine-to-airframe matching chart allocates various constrained and unconstrained optima and clarifies design sensitivities. It can be enriched by overlaying energy efficiency and direct operating costs (DOC) contours. Other important design criteria may also be added such as a buffet boundary and the minimum wing size required to contain the maximum fuel volume (Chapter 10). If the best feasible design is penalized by constraints pushing it too far from unconstrained optima, the designer may introduce (more) advanced design technology or even an unusual general arrangement. And if a particular requirement appears to be very dominant, the associated top-level requirement may have to be reconsidered. When all requirements are met in what is considered to be a well-tempered design, the choice of a suitable engine and further wing design refinement will be the advanced designer's next task. The design selection chart is thus a versatile tool for explicit multi-objective optimization.

9.6.2   Initial Engine Assessment

A straightforward approach to initial engine selection is to compare engine types complying with the thrust or power requirements within the feasible design space. For example, Figure 9.6 indicates that the installed thrust of each engine should amount to approximately 0.15 MTOW. If existing or projected engines with (approximately) this thrust are available, their manufacturers will be contacted for basic information to be used for comparing potential candidates. The propulsion weight penalty introduced in Chapter 8 is a useful figure of merit for comparing the effect of different engine options on the airplane's MTOW.

The propulsion system consists of the (dry) propulsion engines, their nacelles, and the associated systems and accessories. Mission fuel and MTOW are determined by the overall engine efficiency in cruising flight, nacelle drag and power plant weight. These properties are combined in Section 8.5 to derive the propulsion weight penalty; that is, the combination of mission fuel and power plant weight. Irrespective of the airframe aerodynamic efficiency, this large weight component is proportional to the propulsion function

$$F_{\text{prop}} \overset{\text{def}}{=} \frac{R_{\text{eq}}}{\bar{\eta}_o \, H/g} + \frac{\mu_T}{\bar{\tau} \, \delta} \tag{9.27}$$

with $H/g = 4\ 350$ km. The lowest value of F_{prop} can serve as a criterion for comparing different engines. The propulsion function has a more universal significance if we multiply it by H/g and R_{eq}^{-1}. The result is the dimensionless engine figure of merit

$$\text{EFM} = \left(\frac{1}{\text{overall efficiency}} + \frac{H/g}{R_{\text{eq}}} \frac{\text{p.p. weight}}{\text{cruise thrust}} \right) \left(1 - \frac{\text{nacelle drag}}{\text{cruise thrust}} \right)^{-1} \tag{9.28}$$

The usual engine performance parameters are adapted to define the complete installation, nacelles included. The initial cruise condition defines speed, installed thrust and overall efficiency. The equivalent range is used in the denominator of the second term. Although the ratio of power plant weight to cruise thrust depends on altitude, the maximum cruise thrust at the tropopause (11 000 m, 36 000 ft) can be used as a standard. Equation (8.25) proves that the ratio of nacelle drag to thrust is independent of the altitude since, for a given Mach number, both are proportional to ambient pressure.

It is worth noting that the ratio of the second to the first term of Equation (9.28) is identical to the ratio of power plant weight to mission fuel weight. If the manufacturer of a new engine type has taken measures to increase the overall efficiency – for instance, by increasing the BPR or the OPR – this will usually be negated by an engine weight increase. The EFM effectively weighsup these effects and is useful for the engine selection process and for revealing general trends. For example, for engines intended to be applied to short-range aircraft the second term is as important as the first one, which could lead to selecting moderately high values of bypass and/or overall pressure ratios.

9.6.3 Engine Selection

Engines are not only selected for their favourable thrust (or power) delivery and fuel economy.

- If the engine SFC is reduced by increasing the OPR, a penalty has to be paid in the engine price[3] and NO_x emissions.
- Very high BPR engine fans are driven by small cores, making it increasingly difficult to match the fan and turbine rotational speeds. This problem is solved in the geared turbofan (GTF) by using a gearbox to reduce the fan speed.
- For wing-mounted engines with a very large diameter, the airframe design may be penalized because a longer and heavier landing gear is required to provide enough ground clearance. Moreover, the airflow around the wing is influenced by the engine so that adverse interactions may lead to high drag. This problem can be avoided by attaching engines to the rear fuselage.

Selecting a current production engine minimizes the risk associated that engine performance predictions are overoptimistic so that tight performance guarantees can be made to the customer. Application of an engine that is already in service is always more attractive than introducing a new engine with its added logistics and teething problems. Moreover, airlines consider it as an advantage when the engines to be installed are in their inventory.

New engine development is a very costly undertaking and the designer of a commercial aircraft is usually bound to select a suitable engine from only a few candidates. In the case of a small general aviation aircraft, the engine requirements may be fulfilled by off-the-shelf engines. The engine-to-airframe matching will then proceed differently from the case where projected engines can be varied in thrust or power to match the aircraft performance requirements in an optimum fashion – the case considered in this chapter. The situation in the civil aircraft design world may be a mixture of these cases. For example, a new aircraft or a derivative of an existing type is developed alongside aircraft which have similar engines

---

[3] Statistical information in [3] suggests that for 1% TSFC reduction the engine price will increase by almost 3%.

installed. After all, aircraft are rarely designed in complete isolation and their engines may be (derivatives of) existing types. On the other hand, a new airliner generation will probably be offered with alternative engines of (very) similar thrust.

Bibliography

[1] Torenbeek, E., *Synthesis of Subsonic Airplane Design*, Chapter 5 and Appendix K, Springer Verlag, Heidelberg, 1981.
[2] Shevell, R.S., *Fundamentals of Flight*, Chapter 15, Second Edition, Prentice-Hall, Englewood Cliffs, NJ, 1989.
[3] Jenkinson, L.R., P. Simpkin, and D. Rhodes, *Civil Jet Aircraft Design*, Chapter 9: Powerplant and Installation, Arnold, London, 1999.
[4] Swatton, P.J., *Aircraft Performance Theory for Pilots*, Blackwell Science, Oxford, 2000.

Take-Off and Landing Performance

[5] Perry, D.H., "The Airborne Path During Takeoff for Constant Rate-of-Pitch Manoeuvres", Aeronautical Research Counsil, CP No. 1042, 1969.
[6] Perry, D.H. "A Review of Methods for Estimating the Airfield Performance of Conventional Fixed-Wing Aircraft", RAE Technical Memorandum Aero 1264, 1970.
[7] Torenbeek, E, "An Analytical Expression for the Balanced Field Length." AGARD Lecture Series 56, Paper 9, April 1972.
[8] Anonymous, "First Approximation to Take-off Field Length of Multi-engined Transport Aeroplanes", ESDU Item Number 76011, May 1976.
[9] Anonymous, "First Approximation to Landing Field Length for Civil Transport Aeroplanes (50 ft, 15.24 m screen)", ESDU Data Item No. 84040, December 1984.
[10] Chandrasekharan, R.M., "Improved Method of Analyzing Takeoff Performance Data", AIAA Paper No. 88–4509, September 1988.

Matching Engines to the Airframe

[11] Pakendorf, H., "Zur Optimierung der Triebwerksanlage eines Kurzstreckenfleugzeuges, unter Besonderer Berücksichtigung der Höher Bypass Verhältnisse", *WGLR Jahrbuch*, pp. 261–270, 1966.
[12] Whitlow, J.B., and G.A. Kraft, "Optimization of Engines for Commercial Air Transports Designed for Cruise Speeds ranging from Mach 0.9 to 0.98", NASA TM X-67906, 1971.
[13] Dugan, J.F., "Engine Selection for Transport and Combat Aircraft", NASA TM X-68009, 1971. Also in AGARD Lecture Series No. 56, Paper No. 6, April 1972.
[14] McIntire, W.L., and P.E. Beam Jr., "Engine and Airplane: Will It Be a Happy Marriage?", SAWE Paper No. 910, May 1972.
[15] Healy, M.J., J.S. Kowalik, and J.W. Ramsey, "Airplane Engine Selection by Optimization on Surface Fit Approximations", *Journal of Aircraft*, Vol. 12, No. 7, July 1975, pp. 593–599.
[16] Bowles, J.V., M.H. Waters, and T.L. Galloway, "Thrust and Wing Loading Requirements for Short Haul Aircraft Constrained by Engine Noise and Field Length", NASA TN D-8144, 1976.
[17] Keith Jackson Jr., S., "Propulsion System Sizing ... No Longer Exists", Annual Aircraft Design Short Course, University of Dayton, Ohio, August 1977.
[18] Hopps, R.H., and E.C.B. Danforth, "Correlation of Wind-Tunnel and Flight-Test Data for the Lockheed L-1011 Tristar Airplane", AGARD Conference Proceedings No. 242, Paper No. 21, 11–13 October 1977.
[19] D.B. Morden, "Engine/Airplane Performance Matching", The Aerodynamics of Aircraft Gas Turbine Engines, Chapter 9, AFAPL-TR-78-52, The Boeing Commercial Airplane Company, 1978.
[20] Conlon, J.A., and J.V. Bowles, "Powered Lift and Mechanical Flap Concepts for Civil Short-Haul Aircraft", *Journal of Aircraft*, Vol. 15, No. 3, pp. 168–174, March 1978.
[21] Sullivan, R.L., "The Size and Performance Effects of High Lift System Technology on a Modern Twin Engine Jet Transport", AIAA Paper No. 79–1795, August 1979.

[22] Loftin Jr., L.K., "Subsonic Aircraft, Evolution and the Matching of Size to Performance", NASA Reference Publication 1060, August 1980.
[23] Snyder, J.R., "Advanced Technology Tactical Transport", AIAA Paper No. 86–2668, October 1986.
[24] Lavelle, T.M., R.M. Plencner, and J.A. Seidel, "Concurrent Optimization of Airframe and Engine Design Variables", AIAA Paper No. 92–4713, September 1992.
[25] Guha, A. D. Boylan, and P. Gallagher, "Determination of Optimum Specific Thrust for Civil Aero Gas Turbine Engines: A Multidisciplinary Synthesis and Optimization", *Journal of Aerospace Engineering*, Proceedings of the Institution of Mechanical Engineers, Part G: 0(0) pp. 1–26, March 2012.

Engine Integration

[26] Swan, W.C., and A. Sigalla, "The Problem of Installing a Modern High Bypass Engine on a Twin Jet Transport Aircraft", AGARD CP-124, Paper No. 17, April 1973.
[27] Barche, J., "Tragflügelentwurf am Beispiel des Verkehrsflugzeuges VFW 614", *Zeitschrift für Flugwissenschaften*, Vol. 22, pp. 101–115, April 1974.
[28] Krenz, G., and B. Ewald, "Airframe-Engine Interaction for Engine Configurations Mounted above the Wing", AGARD CP-150, Paper 26, 1975.
[29] Smyth, R., "Computer Programs for the Design and Performance of Nacelles for High Bypass-Ratio Engines", AGARD CP-280, Paper No. 25, 1979.
[30] Reubush, D.E., "Effect of Over-the-Wing Nacelles on Wing-Body Aerodynamics", *Journal of Aircraft*, Vol. 16, No. 6, pp. 359–365, June 1979.
[31] Krenz, G., "Engine/Airframe Interference", AGARD-FDP-VKI Special Course, May 1983.
[32] Henderson, W.P., and J.C. Patterson, "Propulsion Installation Characteristics", AIAA Paper No. 83–0087, 1983.
[33] Szodruch, J., and J. Kotschote, "Wind Tunnel Tests of Over-the-Wing Nacelles", *Journal of Aircraft*, Vol. 20, No. 7, pp. 606–611, July 1983.
[34] Awker, R.W., "Application of Propfan Propulsion to General Aviation", AIAA Paper No. 86–2698, October 1886.
[35] Mirat, J.J., R. Perin, and C. Castan, "Engine Installation Design for Subsonic Transport Aircraft", ICAS Paper No. 90–2.7.4, 1990.
[36] Greff, E., K. Becker, M. Karwing, and S. Rill, "Integration of High By-pass Ratio Engines on Modern Transonic Wings for Regional Aircraft", Paper 91–160, *DGLR Jahrbuch* 1991.
[37] Rossow, C.C., J.L. Goddard, H. Hoheisel, et al., "Investigation of Propulsion Integration Interference Effects on a Transport Aircraft Configuration", *Journal of Aircraft*, Vol. 31, No. 5, pp. 1022–1030, 1992.
[38] Bacon, R.J. and J. Roskam, "The 3X Jet: An Advanced Engine Configuration for Two-Engine Aircraft", AIAA Paper No. 95–3963, September 1995.
[39] Rüd, K., and H.J. Lichtfuss, "Trends in Aero-Engine Development, Aspects of Engine Airframe Integration for Transport Aircraft", Proceedings of the DLR Workshop, Braunschweig, Germany, 6–7 March 1996.
[40] Spreen, J.S., R. Slingerland, and E. Jesse, "The 3X Propulsion Concept Applied to a Business Jet Aircraft: Integration of Two Dissimilarly Sized Turbofan Engines", AIAA Paper No. 2007–1047, January 2007.

10

Elements of Aerodynamic Wing Design

The airline is interested in the wing only in so far as it enables it to carry fare paying passengers over the required mission with minimum cost . . . and with adequate safety (as also demanded by the certification authorities).

—J.A. Jupp [27]

10.1 Introduction

Wings produce the lift generated to sustain flight at the expense of major drag and empty weight components – they form the primary focus of attention by aircraft designers. A wing design evolves to the unique configuration necessary to satisfy the needs of a particular application, and hence there are no universal 'best wings'. Even if we restrict ourselves to civil subsonic aircraft, the variety of wing shapes that has been brought forth during the evolution of aircraft is overwhelming. Figure 10.1 depicts a selection of geometric permutations that have been selected from historical documents. It would be easy to increase the size of this collection and if we would produce a list of words characterizing wing concepts and descriptors it would count more than 60 items [11]. However, during conceptual design it is not necessary (or desirable) to consider many details that will characterize the final geometry established during the (downstream) detail design phase. In fact, the main contribution of the wing to the aerodynamic efficiency L/D in cruising flight is determined to a large extent by two key parameters: planform area and span or aspect ratio. Both features contribute to achieve a high L/D and they are also helpful to fulfil take-off and landing field length requirements. Wings for long-range cruise at high altitudes are characterized by long spans and lots of area. Wing sweep does not contribute directly to L/D but is essential to comply with a requirement of obtaining a cruise speed exceeding Mach 0.7. On the other hand, wing span and area are strong drivers of structural weight and the best solution is obtained from a compromise between high aerodynamic efficiency and low structural weight.

Advanced Aircraft Design: Conceptual Design, Analysis and Optimization of Subsonic Civil Airplanes, First Edition. Egbert Torenbeek.
© 2013 by Egbert Torenbeek. Published 2013 by John Wiley & Sons, Ltd.

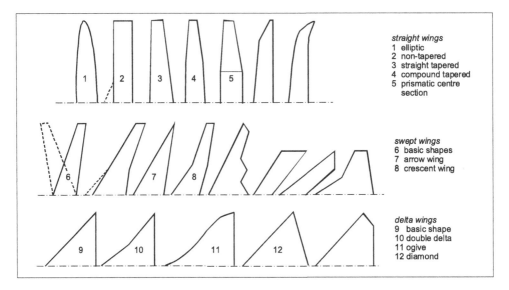

straight wings
1 elliptic
2 non-tapered
3 straight tapered
4 compound tapered
5 prismatic centre
  section

swept wings
6 basic shapes
7 arrow wing
8 crescent wing

delta wings
9 basic shape
10 double delta
11 ogive
12 diamond

Figure 10.1   A collection of wing planform permutations

10.1.1   Problem Structure

The aerodynamic design of a wing can be decomposed into more or less mutually independent activities.

- Selection of the basic planform – area, span, taper ratio and angle(s) of sweep.
- Generating the distribution of airfoil sections and angles of twist along the span.
- Conceiving the basic concepts and geometry of high-lift devices and flight control surfaces.

The first two items are primarily concerned with performance and flying qualities at high speeds, the third item aims at obtaining desirable low-speed characteristics for the missions of which the aircraft must be capable. The main subject of this chapter is selection of the best planform geometry which, in general, precedes the conception of the lateral airfoil section and twist distribution. For wings of subsonic airplanes this is not a major effort – handbooks and other publications on design can be consulted for advice. However, high-speed wing design is complicated by compressibility effects in general and the occurrence of shock waves in particular. This work is a challenge to aerodynamic designers, involving advanced CFD methods for predicting detailed pressure distributions in design and off-design conditions leading to improved wing shapes. High-speed wing design is also complicated by interactions with the flow around the fuselage and wing-mounted engine nacelles. Since this subject in itself deserves a separate textbook such as [4], only a few elements are touched upon here.

The present optimization applies to classical aircraft configurations, also known as tube and wing (TAW) airplanes. Consequently, the useful load inside the wing is exclusively fuel and the accommodation of payload in the fuselage has hardly any influence on wing design optimization. Although no explicit attention is paid to structural design, the topology of the wing structure is assumed to be flexible so that it can be adapted to the external wing geometry. If the wing has a high aspect ratio in combination with sweepback or forward sweep, aero-elastic properties are likely to be important and a structural mass penalty may have to be imposed to avoid excessive wing deformation. Moreover, the lateral stiffness distribution determines the wing shape – and therefore drag – in the loaded condition, which deviates from the jig shape. Since aero-elastic effects are normally covered downstream of the conceptual design phase they are considered to be outside the scope of this text.

10.1.2 Relation to Engine Selection

Sizing and selection of the engines to be installed are the subject of previous chapters. The wing design process can be started as soon as a provisional engine selection has been made. In practical design the situation can be more complicated because engine sizing criteria and aerodynamic wing design are intertwined. For example, the choice of wing span has a significant effect on engine thrust required in the case of constrained performances such as a cruise altitude/speed or a take-off distance constraint. It is therefore useful to classify the design process according to one of the following approaches:

A. Variations in wing geometry are studied for constant flight performance constraints and engine (cruise) rating. The engine thrust to be installed is adapted to wing design variations, often by adopting the rubberizing concept (Section 8.5). The engine selection is made after completion of the wing design.
B. Wing design is executed for a given engine type with specified power or thrust charac- teristics, weight and geometry. Variations in wing geometry then result in varying flight performance.

Option A was the subject of previous chapters in which take-off wing loading and the ratio of installed thrust to MTOW were treated as primary selection variables. An example of a selection diagram illustrated how the initial cruise altitude, MTOW and low-speed performance constraints affected the best choice of the wing loading and the ratio of thrust to weight. The example demonstrated that, for given wing loading, the take-off thrust required is sensitive to the initial cruise altitude, whereas variation of the wing loading for a given cruise altitude has little effect on it. For the present study, the initial operating conditions (altitude and speed) in cruising flight are considered to be pre-assigned. The wing shape is subject to optimization, which requires an extension of the sensitivity analysis used for the case discussed previously. Aircraft figures of merit (FOM) to be optimized are mission fuel and MTOW.

Option B uses the same sensitivity analysis as option A; however, the optimization is basically a weight and performance analysis for designs with different drag and weight char- acteristics. The objective function can be (a) MTOW for a given mission fuel and range; (b) mission fuel required for a given range and MTOW; (c) mission range for a given fuel and MTOW; or (d) a more elaborate economical figure of merit. The preliminary design stage activity uses detailed aerodynamic and engine performance information required to compare competing engine types and/or fine-tuning the wing design.

10.2 Planform Geometry

When unusual wings are presented, you can suspect that they were driven by a set of performance specifications that could not be fulfilled by straightforward means.

—J. Chuprun [11]

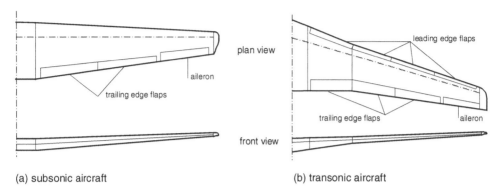

(a) subsonic aircraft   (b) transonic aircraft

Figure 10.2   Schematic basic planforms of two classes of civil airplane wings

The approach to aerodynamic design of a wing is, to a large extent, dependent on the operational requirements – in particular the high speed regime and the available airfield length. Civil aircraft wings can be subdivided according to the following classes of aircraft for which they are designed.

1. Subsonic aircraft wings do not experience critical compressibility effects at speeds up to Mach 0.65. They are designed according to aerodynamical and structural principles that have been applied to civil airplanes since World War II. Their wings do not have significant sweep and are mostly straight-tapered in both plan view and front view; see Figure 10.2 (a). A non-tapered section between the nacelles is occasionally applied. The aspect ratio is typically between 9 and 13 and section shapes are often selected from a standard airfoil catalogue. Thickness ratios vary between 0.15 and 0.20 at the root and between 0.12 and 0.15 at the tip. The wings of subsonic airplanes have a structural box predominantly made of Al-alloys; several recent designs feature a composite outboard wing.
2. Transonic aircraft fly at speeds up to Mach 0.95 and experience supercritical flow around the wing in high speed flight. Although the foundations of their design stem from the period 1945–1955, progress made since then has resulted in considerably improved aerodynamic and structural properties. This has resulted in wing shapes with complex geometries resulting from CFD-based procedures for obtaining favourable pressure distributions; see Figure 10.2 (b). Wings of civil aircraft flying at Mach 0.70 or more are mostly sweptback with a quarter-chord sweepback angle between 15 and 40 degrees.[1] Aspect ratios between 7 and 10 are mostly combined with purpose-designed airfoils. Typical thickness ratios are up to 0.15 at the root and 0.10 to 0.12 at the tip. Many transonic wings feature compound taper, with distinct inboard and outboard sections separated by pronounced kinks in the leading and trailing edges. In the framework of NASA-sponsored Subsonic Ultra Green Aircraft Research (SUGAR), Boeing proposed a design with a novel non-swept very high aspect ratio strut-braced wing (Figure 10.3).

The first case considered is the optimum planform geometry of straight-tapered wings with (aerodynamically) negligible sweep designed for subsonic flow. Thickness and taper ratios

---

[1] A forward swept wing has been applied to the HFB 320 Hansa Jet, an executive aircraft developed in the early 1960s.

Figure 10.3   High span strut-braced folding wing concept described in NASA/CR-2011-216847

are constant and the only selection variables to be optimized are (gross) wing area and span and it is assumed that the span can be varied without hard constraints. The next case is the optimum aerodynamic design of transonic wings. This is complicated by the requirement that the aircraft has to cruise at a specified Mach number without experiencing a large drag penalty due to supercritical compressibility of the flow. For a given airfoil technology, the critical Mach number of a wing is affected by the thickness ratio, sweep angle and lift coefficient. Hence, for a given design cruise Mach number, variation of the wing loading has an effect on drag and weight. This is accounted for in analyzing the design sensitivity.

10.2.1   Wing Area and Design Lift Coefficient

For a conceptual sizing problem, the wing planform area S_w is arguably a primary design parameter.[2] Although the wing loading W_{MTO}/S_w is often treated as an independent variable, it is strictly a dependent variable since the MTOW is generally determined by the requirement that the plane has to carry a given payload over a specified distance. Large weight components such as structure and fuel weight depend on the MTOW which itself results from the addition of many weight contributions. This complicates the sizing and optimization problem, unless closed-form design sensitive equations can be derived for design weights. The previous and present chapters aim at deriving explicit relationships between the selection variable(s), the MTOW and other weight components that are considered as objective functions. A non-iterative solution of the governing sizing equations allows the designer to compute and visualize the design space enabling multi-objective optimization.

The primary selection variable used in the present chapter is the design lift coefficient

$$\hat{C}_L = \frac{W_{MTO}}{\hat{q}\, S_w} \tag{10.1}$$

The present chapter treats the dynamic pressure $\hat{q} = 0.5\,\rho V^2 = 0.5\,\gamma p\, M^2$ as a pre-assigned parameter referring to the initial cruise condition. In view of the fuel burnt during the ascent to cruise altitude, the initial cruise GW is 2 to 3% less than the MTOW which is taken into

---

[2]Airframe manufacturers use different methods to extend leading and trailing edges to the centreline; hence, there exists no international standard for defining the gross wing area.

account by increasing the actual dynamic pressure by the same percentage. The design lift coefficient $\hat{C}_L$ is proportional to the wing loading and is a uniquely defined independent design variable – it forms a demanding condition for the aerodynamic wing design, especially for a transonic airliner.

10.2.2 Span and Aspect Ratio

The wing span b_w is a well-defined geometric property; namely, the distance between the wing tips.[3] Vortex drag is inversely proportional to the span loading W_{MTO}/b_w^2 and a low span loading results in a low take-off distance (Section 9.4). However, a long wing span results in a high structure weight unless unconventional measures are taken, such as strut bracing.

Aerodynamic properties such as the aerodynamic efficiency L/D and the lift gradient $dC_L/d\alpha$ are usually expressed in terms of the aspect ratio

$$A_w \stackrel{\text{def}}{=} \frac{b_w^2}{S_w} \qquad (10.2)$$

Aircraft designers are familiar with the aspect ratio since it is the foremost parameter by which a wing can be matched to performance requirements in an optimum fashion. Different from other selection variables, the aspect ratio can often be varied within a fairly wide range with no hard constraints. An exception is due to an upper limit on wing span specified in ICAO airfield classifications.

Figure 10.4 depicts a matrix of wing and body combinations representing members of a two-dimensional design space of planform areas and aspect ratios. All designs have the same body, wing sweepback angle and taper ratio. The angle of sweep and wing size variation suggest that these wings could be members of the design space for a large long-range jetliner. In a realistic study it is very likely that different airfoil sections and high-lift systems will be matched to the range of cruise lift coefficients and wing loadings. As a starting point, the design space will be studied without paying attention to practical constraints such as fuel tank capacity. Although the wide variation of wing geometry in Figure 10.4 seems rather unrealistic, it provides a good overview of the global design sensitivity and the penalty due to off-optimum choices. Incidentally, the present design case differs fundamentally from the progression in wing and body combinations with the same total volume as depicted in Figure 5.5.

10.3 Design Sensitivity Information

For the purpose of realistic optimization, it is imperative to get a grip on the functional sensitivity of the airplane's drag and weight to relevant selection variables. The methodology discussed in Chapters 8 and 9 forms the basis of the present chapter but the design space is extended with applications to optimize other geometric properties. Different from the previous chapters, the initial cruise altitude remains constant and variation of $\hat{C}_L$ is effected by varying the wing loading.

---

[3]The definition of span may have to be adapted when winglets are attached to the tips.

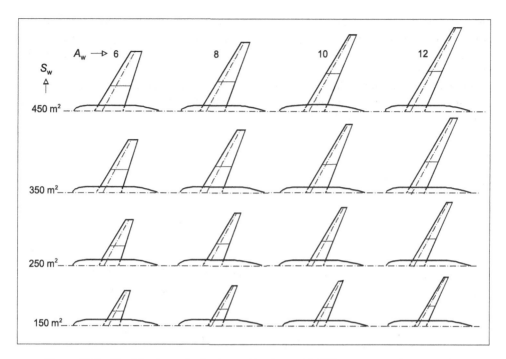

Figure 10.4   Combinations of a fuselage with wings with different areas and aspect ratios

10.3.1   Aerodynamic Efficiency

The airframe drag components essential for the present optimization are profile drag of the wing and horizontal tail, parasite drag of the fuselage and vertical tail and drag due to lift. The variation of drag components with the wing area is discussed in Section 8.4. The profile drag coefficient $\check{C}_{D_p}$ according to Equation (8.17) takes into account that the horizontal tail area is scaled up proportional to wing area. The drag areas of the fuselage and vertical tail are assumed to be fixed; that is, independent of the wing geometry. In view of the variation of installed engine thrust or power, nacelle drag is treated as a reduction of the engine thrust. However, in this chapter, engines are sized for constant cruise altitude and the variation of engine thrust and nacelle size is small. Although the distinction between aircraft drag and airframe drag can still be made, we refrain from this for simplicity and treat the nacelle drag area as fixed.

The inverse of the aerodynamic efficiency in the design condition according to Equation (8.19) is rewritten with the drag due to lift factor K_L dependent on the aspect ratio,

$$\frac{\hat{C}_D}{\hat{C}_L} = \frac{C_{D_0}}{\hat{C}_L} + \frac{\hat{C}_L}{\pi A_w \check{e}} = \frac{\check{C}_{D_p}}{\hat{C}_L} + \hat{q}\,\frac{(C_{D_p}S)_{\text{fix}}}{W_{\text{MTO}}} + \frac{\hat{C}_L}{\pi A_w \check{e}} \qquad (10.3)$$

Section 8.4 explains that $\check{C}_{D_p}$ is slightly sensitive to wing area variation due to the scale effect on chord Reynolds number and skin friction drag. This becomes more complicated when both $\hat{C}_L$ and A_w are design variables – it can be shown that $\check{C}_{D_p} \propto (\hat{C}_L A_w)^{1/12}$. This refinement can be handled by numerical analysis but we ignore it for simplification of the present quasi-analytical approach. The aerodynamic wing design is assumed to be adapted to the variation of $\hat{C}_L$ by selecting an optimum camber and wash-out distribution (Figure 4.3). This is reflected in the modified Oswald factor $\check{e}$ which is higher than the usual Oswald factor; for example, between 0.90 and 0.95.

The drag coefficient due to lift continually decreases when the aspect ratio increases; hence, there exists no aspect ratio resulting in maximum aerodynamic efficiency. However, the design lift coefficient has a partial optimum

$$\hat{C}_{L_{\text{ref}}} \stackrel{\text{def}}{=} \sqrt{\check{C}_{D_{\text{p}}} \pi A_{\text{w}} \check{e}} \tag{10.4}$$

resulting in the minimum obtainable drag for a given GW and flight condition,

$$\left(\frac{\hat{C}_D}{\hat{C}_L} \right)_{\text{ref}} = 2 \sqrt{\frac{\check{C}_{D_{\text{p}}}}{\pi A_{\text{w}} \check{e}}} + \hat{q} \frac{(C_D S)_{\text{fix}}}{W} \tag{10.5}$$

It is important to notice that Equations (10.4) and (10.5) differ from the minimum drag condition of a (fully defined) operational aircraft with a parabolic drag curve,

$$\left(\frac{C_D}{C_L} \right)_{\text{min}} = 2 \sqrt{\frac{C_{D_0}}{\pi A_{\text{w}} e}} \qquad \text{for} \qquad C_{L_{\text{MD}}} = \sqrt{C_{D_0} \pi A_{\text{w}} e} \tag{10.6}$$

The explanation is that $C_{L_{\text{MD}}}$ defines an optimum flight condition for a given wing geometry, whereas $\hat{C}_{L_{\text{ref}}}$ defines a wing loading for maximum $\hat{C}_L / \hat{C}_D$ at a given flight condition. Both criteria do not represent optimum design and operating conditions for the aircraft as a complete system. Even though the two optimizers are similar in their appearance, they apply to different objectives. This fundamental aspect has far-reaching consequences for the theory developed in this chapter.

10.3.2 Propulsion Weight Contribution

Results derived in Section 8.5 are summarized for the present application, as follows:

• Mission fuel is written as a fraction of the MTOW,

$$\frac{W_{\text{misf}}}{W_{\text{MTO}}} = \frac{R_{\text{eq}}}{\eta_0 H/g} \frac{\hat{C}_D}{\hat{C}_L} \tag{10.7}$$

with $H/g = 4\,350$ km (2 349 nm). The equivalent range R_{eq} is the mission range corrected for gross weight reduction during cruising and for lost fuel during take-off, climb, descent, approach and landing.

• The power plant weight as a fraction of the total take-off thrust is derived from the thrust required to balance cruise drag,

$$\frac{W_{\text{pp}}}{W_{\text{MTO}}} = \frac{\mu_T}{\tau \delta} \frac{\hat{C}_D}{\hat{C}_L} \tag{10.8}$$

where μ_T is the power plant weight per unit take-off thrust. The parameter τ denotes the corrected thrust lapse rate; that is, the corrected maximum cruise thrust T/δ divided by the take-off thrust T_{TO} with δ denoting the relative ambient pressure.

- The propulsion weight penalty is the combination of mission fuel and power plant weight,

$$\frac{W_{prop}}{W_{MTO}} = F_{prop}\frac{\hat{C}_D}{\hat{C}_L} = \left(\frac{R_{eq}}{\eta_o H/g} + \frac{\mu_T}{\tau\delta}\right)\left\{\frac{\check{C}_{D_p}}{\hat{C}_L} + \frac{\hat{q}\,(C_D S)_{fix}}{W_{MTO}} + \frac{\hat{C}_L}{\pi A\check{e}}\right\} \quad (10.9)$$

For specified cruise conditions, the propulsion function F_{prop} is a constant which does not depend on the airframe characteristics.

10.3.3 Wing and Tail Structure Weight

A subsonic wing will generally have a near-zero sweep angle, its thickness ratio has an upper limit set by the objective to avoid critical compressibility effects.

In principle, loading cases for all structural components need to be considered which determine the material thicknesses, an approach including a number of steps and iterations. For example, structural components are sized to withstand several load distributions – combinations of fuel and payload, design speeds and altitudes, etc. – in order to find a critical condition. Moreover, the wing structural mass distribution, which is not known *a priori*, affects the inertia relief effect. An example of such a methodology is described in Chapter 11. A more refined multiple station method is required to take into account that for each element the critical loading varies along the span. For components such as the upper and lower box panels a selection must be made of skin/stringer configurations and rib distances. Moreover, all major components of the wing box are subject to sub-optimization and since each wing geometry requires an adaptation of the structural topology, this approach will entail a laborious computational process. Prediction of the horizontal tail weight requires knowledge of the tail geometry and critical tail loads and follows the same structural sizing method as for the wing. The entire weight prediction is time-consuming, computer-intensive and degrades the efficiency of the airplane system optimization with no guarantee of improved accuracy.

For the present analytical optimization, Equation (8.1) is used as a starting point for computing the wing weight sensitivity. The assumption is made that the manoeuvre load is critical for all values of the wing area and the aspect ratio. Horizontal tail weight is closely related to wing area; hence, the tail weight is accounted as a fraction of wing weight,

$$r_h \stackrel{\text{def}}{=} \frac{W_h}{W_w} \quad (10.10)$$

with a typical value $r_h = 0.10$. Substitution into Equation (8.1) yields the wing and horizontal tail weight fraction

$$\mu_{w+h} \stackrel{\text{def}}{=} \frac{W_w + W_h}{W_{MTO}} = \Phi_1\,A_w\sqrt{\frac{A_w}{\hat{C}_L}} + \frac{\Phi_2}{\hat{C}_L} \quad (10.11)$$

The dimensionless weight parameters are defined as

$$\Phi_1 = 0.0013\,(1 + r_h)\,\frac{\sqrt{W_{\mathrm{MZF}}/\hat{q}}}{b_{\mathrm{ref}}}\,\frac{n_{\mathrm{ult}}\,\eta_{\mathrm{cp}}}{(\overline{t/c})_w\,(\cos\Lambda_w)^2} \tag{10.12}$$

with $b_{\mathrm{ref}} = 100$ m (328 ft) and

$$\Phi_2 = (1 + r_h)\frac{\Omega_S}{\hat{q}} \tag{10.13}$$

Typical values used for the baseline design on Figure 8.1 are: $\Phi_1 = 0.0020$ and $\Phi_2 = 0.025$.

10.3.4 Wing Penalty Function and MTOW

Similar to the propulsion weight penalty, we consider a weight penalty associated with the indispensable wing. The wing penalty function (WPF) is composed of three mutually independent weight components:

(1) wing and tail structure weight;
(2) the engine weight required to balance wing and horizontal tail drag; and
(3) the mission fuel weight required to balance wing and horizontal tail drag along the mission range.

The WPF is found from Equation (10.9) by leaving out the drag due to the fixed drag area and adding the wing and horizontal tail structure weight

$$F_{\mathrm{wp}} \stackrel{\mathrm{def}}{=} \Phi_1\,A_w\sqrt{\frac{A_w}{\hat{C}_L}} + \frac{\Phi_2}{\hat{C}_L} + F_{\mathrm{prop}}\left(\frac{\check{C}_{D_p}}{\hat{C}_L} + \frac{\hat{C}_L}{\pi\,A\check{e}}\right) \tag{10.14}$$

The selection variables $\hat{C}_L$ and A_w are explicit in this equation – a convenient situation for deriving analytical optimizers.

The maximum take-off weight (MTOW) is an important figure of merit (FOM) for civil aircraft and is often used as an objective function in design optimization. The MTOW derived in Section 8.6 is modified as follows:

(a) nacelle drag is not considered as a thrust loss and $\overline{\tau}$ is replaced by τ, and
(b) the fixed drag area $(\check{C}_{D_p}S)_{\mathrm{fix}}$ includes the nacelle drag area.

The resulting expression for the MTOW

$$W_{\mathrm{MTO}} = \frac{W_{\mathrm{pay}} + \Sigma\,W_{\mathrm{fix}} + F_{\mathrm{prop}}\,q\,(\check{C}_{D_p}S)_{\mathrm{fix}}}{1 - (\mu_{\mathrm{resf}} + \mu_{\mathrm{lg}} + F_{\mathrm{wp}})} \tag{10.15}$$

shows that the wing penalty function F_{wp} is the only term reflecting the variation of $\hat{C}_L$ and A_w. Hence, for given cruise altitude and speed, minimization of the WPF yields the same optimizers as minimization of the MTOW. In other words: it is appropriate to optimize wing design on the basis of the WPF without knowledge of all weight components in the numerator of Equation (10.15). This implies that *optimum wing area and aspect ratio are independent of all drag and weight components not included in the WPF. Moreover, the accuracy of optimizers depends on the prediction accuracy of the WPF only.*

10.4   Subsonic Aircraft Wing

10.4.1   Problem Structure

A subsonic wing has mostly a near-zero sweep angle and its thickness ratio has an upper limit set by the objective to avoid critical compressibility effects. The sweep angle and thickness ratio are therefore constant pre-assigned parameters. Optimization of the wing is carried out for constant flight condition $\bar{q}$. In principle, the MZFW, centre of pressure location and Ω_S are dependent variables but their deviations from the baseline values are negligible. Consequently, the parameters Φ_1 and Φ_2 determining the wing and horizontal tail weight are treated as constants.

The theory is illustrated for a turboprop-powered freighter design depicted in Figure 10.5. Although the previous analysis is generally applicable, the propulsion function F_{prop} must be adapted to turboprop propulsion. In particular, the overall efficiency η_o in the mission fuel weight term equals the product of engine efficiency based on equivalent power P_{eq} and propeller efficiency η_{pr}; see Equation (3.28). The power plant weight contribution is defined by the dimensionless constant term

$$\frac{\mu_l}{\tau\delta} = \frac{\text{power plant weight}}{\text{available cruise thrust}} = \frac{W_{pp}}{T} = \frac{(W_{eng} + W_{nac} + W_{pr})\,V}{\eta_{pr}P_{eq}} \qquad (10.16)$$

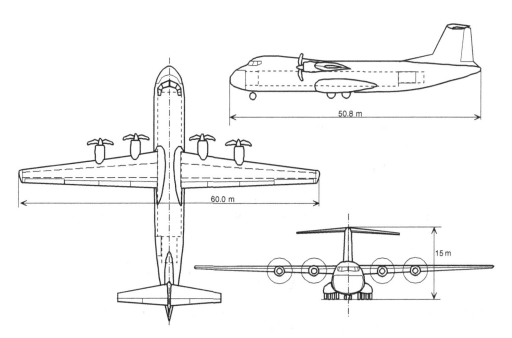

Figure 10.5   Baseline of a propeller-powered cargo aircraft designed to carry 20 tons of payload over 4 000 km. Cruise speed 600 km/h @ 9 150 m altitude. Maximum range 6 000 km

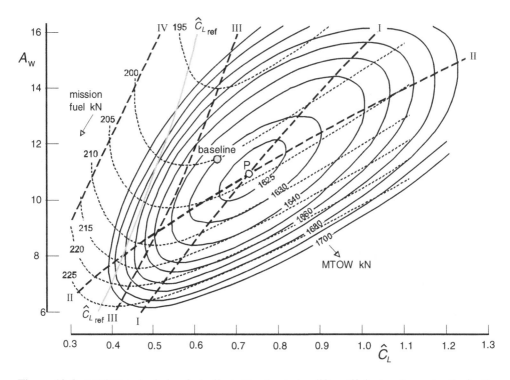

Figure 10.6 MTOW and mission fuel affected by the design lift coefficient and aspect ratio for the design on Figure 10.5

where the index pr denotes the propeller. This section is devoted to deriving conditions for the minimum achievable MTOW and augmented by conditions for minimum mission fuel.

10.4.2 Unconstrained Optima

MTOW contours for the example aircraft are depicted in Figure 10.6. Several optimizers are indicated by curves and points.

1. The optimum design lift coefficient for given aspect ratio (Curve I) is defined by the condition $\partial F_{wp}/\partial \hat{C}_L = 0$, resulting in a transcendental equation which is identical to Equation (8.43),

$$\hat{C}_L = \sqrt{\check{C}_{D_p} \pi A_w \check{e}} \left(1 + \frac{0.5\,\Phi_1\,A_w\,\sqrt{A_w \hat{C}_L} + \Phi_2}{\check{C}_{D_p} F_{\text{prop}}} \right)^{0.5} \quad (10.17)$$

Figure 10.6 clearly shows the difference between the present optimum and the lift coefficient $\hat{C}_{L_{\text{ref}}}$ for minimum wing drag. The difference between the corresponding wing loadings increases when the aspect ratio increases which is caused by the increasing wing weight relative to the fuel weight. Since the response surface is flat near the global optimum, the

MTOW is not very sensitive to off-optimum values. For instance, plus or minus 20% above or below the optimum $\hat{C}_L$ penalizes the MTOW by 1.5%.

2. The optimum aspect ratio for given lift coefficient (Curve II) is obtained from $\partial F_{wp}/\partial A_w = 0$, resulting in

$$A_w = \hat{C}_L^{0.6} \left(\frac{2 F_{prop}}{3 \, \Phi_1 \pi \check{e}} \right)^{0.4} \tag{10.18}$$

Although this aspect ratio increases sensitively for a high $\hat{C}_L$, the MTOW is insensitive to deviations from the partial optimizer.

3. Unconstrained global optimizer. The intersection of curves I and II (design P) is defined by closed-form analytical equations for the design lift coefficient

$$\hat{C}_L = (\check{C}_{D_p} + \Phi_2/F_{prop})^{5/7} (1.5\pi \check{e})^{3/7} (F_{prop}/\Phi_1)^{2/7} \tag{10.19}$$

and the aspect ratio

$$A_w = (\check{C}_{D_p} + \Phi_2/F_{prop})^{3/7} (1.5\pi \check{e})^{-1/7} (F_{prop}/\Phi_1)^{4/7} \tag{10.20}$$

The optimum selection variables are combined into

$$\frac{\hat{C}_L}{\sqrt{A_w}} = \sqrt{1.5\pi \check{e} \, (\check{C}_{D_p} + \Phi_2/F_{prop})} \tag{10.21}$$

whereas the optimum ratio of wing and tail structure weight to mission fuel weight is

$$\frac{W_w + W_h}{W_{misf}} = \frac{2 \check{C}_{D_p} F_{prop} + 4 \Phi_2}{5 \check{C}_{D_p} F_{prop} + 3 \Phi_2} \tag{10.22}$$

The minimum wing penalty function amounts to

$$F_{wp} = 3.5 \left(\frac{F_{prop}}{1.5\pi \check{e}} \right)^{3/7} \left\{ \Phi_1 \, (\check{C}_{D_p} F_{prop} + \Phi_2) \right\}^{2/7} \tag{10.23}$$

The following observations on these results illustrate their usefulness:

- Since the response surface is quite flat near the global optimum, F_w is accurately approximated if $\hat{C}_L$ and A_w are selected close to the optimum. Equation (10.23) is therefore useful for a first prediction of the MTOW if the wing loading and aspect ratio have not yet been selected.
- The sensitivities of the wing penalty function to various aerodynamic and (wing) weight parameters are readily computed, which offers an effective tool for technology assessment. For example, application of laminar flow control (LFC) to the wing reduces parasite drag substantially, and the optimum wing loading and aspect ratio for minimum MTOW will be lower than for a conventional wing. This observation is correct even when secondary effects such as wing weight increments and suction power requirements are taken into consideration.

- The longer the design range, the larger is F_{prop}. For a given cruise altitude, a very long range aircraft wing should have a high design lift coefficient and a high aspect ratio. The global optimum wing and tail structure weight are between 50 and 60% of the mission fuel weight.
- Equations (10.21) and (10.22) indicate optimum ratios which are independent of the primary wing weight parameter Φ_1. For long-haul airplanes, the secondary wing weight parameter Φ_2 has a relatively small influence on these ratios. For $\Phi_2 = 0$ the optimum lift coefficient is 22% higher than $C_{L_{\text{ref}}}$, the lift coefficient for minimum wing drag.

10.4.3   Minimum Propulsion Weight Penalty

In Section 8.7 it was argued that the combined cost of mission fuel and power plant installation constitutes an important DOC penalty. Hence, it makes sense to minimize the absolute value of the propulsion weight penalty

$$W_{\text{prop}} = F_{\text{prop}} \left\{ \left(\frac{\check{C}_{D_{\text{p}}}}{\hat{C}_L} + \frac{\hat{C}_L}{\pi A_{\text{w}} \check{e}} \right) W_{\text{MTO}} + \hat{q}\,(C_D S)_{\text{fix}} \right\} \tag{10.24}$$

where F_{prop} is defined by Equation (10.9). Substitution of Equation (10.15) indicates that the partial optima of W_{prop} are equal to those for

$$(1 - \mu_{\text{resf}} - \mu_{\text{lg}} - \mu_{\text{w+h}})^{-1} \left(\frac{\check{C}_{D_{\text{p}}}}{\hat{C}_L} + \frac{\hat{C}_L}{\pi A_{\text{w}} \check{e}} \right) \tag{10.25}$$

Analytical derivation of the partial optima leads to transcendental equations. The following approximations are proposed for the partial optimum design lift coefficient:

$$\hat{C}_L = \hat{C}_{L_{\text{ref}}}(1 + \mu_{\text{w+h}}) \tag{10.26}$$

and the partial optimum aspect ratio

$$A_{\text{w}} = \hat{C}_L \left(\frac{1 - \mu_{\text{resf}} - \mu_{\text{lg}} - \mu_{\text{w+h}}}{3\,\Phi_1\,\check{C}_{D_{\text{p}}} \pi \check{e}} \right)^{0.4} \tag{10.27}$$

The optima for minimum propulsion weight penalty are identical to those for minimum mission fuel depicted for the example design in the form of contour curves in Figure 10.6. The (fairly flat) optimum $\hat{C}_L$ (curve III) is some 12 to 16% higher than $\hat{C}_{L_{\text{ref}}}$. The optimum aspect ratio for given wing loading (curve IV) is too high for a realistic design. Very high aspect ratios are not advantageous for the example design.

10.4.4   Accuracy

The derivation of unconstrained partial and global optimizers is based on simplified quasi-analytical equations. Since estimation of Φ_1 and Φ_2 is partly based on statistics, errors of 5 to 10% have to be anticipated. The derived equations show that similar errors will be found in the optimizers for wing loading and aspect ratio. Although the wing penalty function may have an error between 1 and 2%, the outcome is still realistic. Wing box weight prediction accuracy can be improved by applying station analysis and/or a finite element method (FEM) whereas secondary structure weight is difficult to predict accurately during the conceptual design stage (Chapter 11). Several published optimization studies simply treat secondary wing weight as a percentage of the primary structure weight. However, if the primary wing weight is 2/3 of the total wing weight, Φ_1 is multiplied by 1.5 and $\Phi_2 = 0$, leading to erroneous results for all optimizers. For instance, the global optimum wing loading and aspect ratio are both reduced by 20 to 35%, wing and tail weight goes down to 40%, mission fuel increases to 60%, although the wing penalty function itself is practically unaffected. An even cruder simplification is to compute wing weight as a constant specific weight multiplied by wing area. In the present theory this means $\Phi_1 = 0$, whereas Φ_2 increases by a factor of three. In this case the optimum aspect ratio becomes infinite, while for a realistic $A_w = 11$, the (partial) optimum wing loading increases by 25%. Although some of these problems may be avoided by imposing inequality constraints, such an artificial remedy degrades the credibility of the results.

10.5   Constrained Optima

The unconstrained global optimum wing area and aspect ratio may be unfeasible due to practical design requirements. In particular, a limit on the take-off distance imposes a constraint on wing area and span. If this becomes an active constraint, the feasible optimum span loading is affected. Constraints and constrained optima for minimum MTOW are depicted in Figure 10.7. Their position relative to each other and relative to the unconstrained optimum finally determines the best feasible design. If performance requirements are not too severe and

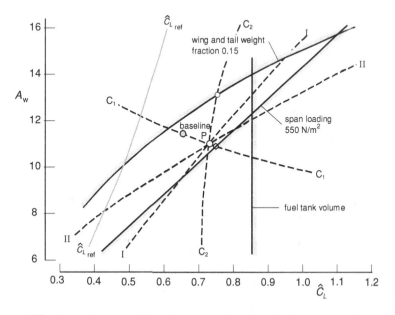

Figure 10.7   Inequality constraints and constrained optima for the design on Figure 10.5

design technology is adequate, combination of all constraints indicates a feasible region of the design space. An optimum wing design can be identified on every constraint. If none of them satisfies all constraints, the global optimum is located on the intersection of two constraints.

10.5.1 Take-Off Field Length

Section 9.4 demonstrates that the engine thrust required for taking off from a given airfield is primarily determined by the span loading $W_{\mathrm{MTO}}/b_{\mathrm{w}}^2$. In the present case the installed take-off thrust-to-weight ratio is considered to be known for the baseline aircraft. Hence, a field length constraint is identical to a constraint on the span loading

$$\frac{\hat{C}_L}{A_{\mathrm{w}}} = \frac{W_{\mathrm{MTO}}}{\hat{q}\, b_{\mathrm{w}}^2} \tag{10.28}$$

is represented by a straight line in the design selection chart (Figure 10.7). The constrained optimum is the point of tangency with the local iso-MTOW contour. It is derived analytically by substitution of Equation (10.28) into Equation (10.14) and setting its differential equal to zero,

$$A_{\mathrm{w}}\hat{C}_L^{1/3} = \left(\frac{\check{C}_{D_{\mathrm{p}}} F_{\mathrm{prop}} + \Phi_2}{\Phi_1}\right)^{2/3} \tag{10.29}$$

This result is depicted as curve C_1 which intersects all partial optimizers in design P.

10.5.2 Tank Volume

For civil aircraft a frequently imposed requirement is that all fuel shall be carried inside the wing. Using a typical relationship between the available and the required tank volume, the available fuel tank volume for given average thickness ratio is

$$V_{\mathrm{tank}} = 0.90\, \eta_{\mathrm{tank}} \overline{(t/c)}_{\mathrm{w}}\, S_{\mathrm{w}}^{1.5} A_{\mathrm{w}}^{-0.5} \tag{10.30}$$

Depending mainly on the wing box planform area as a fraction of the planform area, the volumetric efficiency η_{tank} of a completely 'wet wing' is approximately 0.55. The tank volume is a few percentage larger than the total fuel volume required to fly the maximum range R_{M} (Figure 2.3) which follows from the mission fuel and reserve fuel

$$(W_{\mathrm{f}})_{\mathrm{max}} = \frac{(R_{\mathrm{M}})_{\mathrm{eq}}}{\eta_{\mathrm{o}}\, H/g} \left\{ \left(\frac{C_{D_{\mathrm{p}}}}{\hat{C}_L} + \frac{\hat{C}_L}{\pi\, A_{\mathrm{w}}\, \check{e}}\right) W_{\mathrm{MTO}} + \hat{q}\, (C_D S)_{\mathrm{fix}} \right\} + C_{\mathrm{resf}}\, W_{\mathrm{MTO}} \tag{10.31}$$

with the equivalent range defined by Equation (8.29). If all fuel is to be carried in wing tanks, the upper limit of the take-off wing loading is solved by combining these two equations. Since wing volume and maximum L/D ratio are both inversely proportional to $\sqrt{A_{\mathrm{w}}}$, the wing loading constraint is insensitive to the aspect ratio. The baseline freighter aircraft design used as an example is a medium-range aircraft and the wing tank volume constraint appears to be inactive in most of the design space.

The wing loading constraint varies proportional to $W_{\mathrm{MTO}}^{1/3}$ and the volume constraint can be problematic for relatively small long-range aircraft. If the resulting weight penalty is objectionable, an increased cruise altitude can be considered, leading to a larger optimum wing area and less fuel consumed at the cost of more installed engine thrust. Another solution applied in several business jets is installation of fuel tanks in the rear fuselage.

10.5.3 Wing and Tail Weight Fraction

The unconstrained optimizer according to Equations (10.19) and (10.20) indicates that optimum wing loading and aspect ratio both increase with increasing range. This is partly explained by the fact that the optimum wing and tail weight appears to be a certain fraction of the fuel weight. Arguably, since development and production costs are related to empty weight, a large wing is undesirable. If a limit is imposed on the weight fraction $\mu_{\mathrm{w+h}}$ an upper limit on the aspect ratio is derived from Equation (10.11)

$$A_{\mathrm{w}} \leq \hat{C}_L^{1/3} \left(\frac{\mu_{w+h} - \Phi_2/\hat{C}_L}{\Phi_1} \right)^{2/3} \tag{10.32}$$

which is depicted in Figure 10.7 for $\mu_{w+h} = 0.15$. Application of Equation (10.14) proves that the constrained optimum is identical to a maximum L/D-ratio for constant wing and tail weight fraction. In other words: the following condition applies to minimum MTOW as well as minimum fuel weight:

$$\hat{C}_L = \sqrt{1.5 \, \check{C}_{D_\mathrm{p}} \pi A_{\mathrm{w}} \check{e}} \left(1 - \frac{\Phi_2}{\Phi_1 A_{\mathrm{w}}^{1.5} \hat{C}_L^{0.5}} \right)^{-0.5} \tag{10.33}$$

depicted in Figure 10.7 as curve C_2. Although the aspect ratio is prominent in this equation, the optimum $\hat{C}_L$ appears to be insensitive to it.

10.5.4 Selection of the Design

The combination of Figures 10.6 and 10.7 is called a design selection chart which is similar to Figure 9.6 but not depicted here. For the cargo aircraft designer, this chart leaves little to be desired since there exists a convenient feasible region which contains the unconstrained global optimum, design P. In the present example the baseline design has a slightly higher aspect ratio and a wing loading 10% below design P. The baseline design needs 2% less mission fuel and has better field performances than the global optimum, at the cost of a 0.4% higher MTOW. In the more usual situation, the unconstrained optimum is outside the feasible design space or there exists no feasible design. In the first case, the design point is selected on a constraint or an intersection of two constraints. In the second case it may be necessary to reconsider the engine selection, introduce technology improvements and/or relax the most critical requirement(s) which makes a new design iteration unavoidable.

10.6 Transonic Aircraft Wing

This work also demonstrates very clearly how much more insight into what really matters may be obtained from approximate solutions rather than from purely numerical solutions obtained by a computer, where the main trends may be difficult to detect and well hidden and may easily be missed altogether.

—D. Küchemann [1] page 419 (1978)

10.6.1 Geometry

The aerodynamic design of a transonic wing requires compressibility effects to be taken into account. This requires variation of the wing geometry dependent on the design Mach number in cruising flight. Most high-speed wings feature a positive angle of sweep and a kink in the chord and thickness distributions along the span (Figure 10.2). The wing outboard of the kink has less taper and a smaller thickness ratio compared to the inboard wing. The root thickness ratio is, typically, a factor 1.2 higher. This configuration forms a compromise between conflicting aerodynamic and structural objectives by combining (a) a high thickness ratio at the root, were structural loads are highest, with (b) a reduced, near-constant thickness ratio of the outboard wing where compressibility effects are more demanding. For conceptual design of a transonic wing, the sweep angle Λ_w and the average thickness ratio $\overline{(t/c)}_w$ are additional selection variables.

- Instead of the quarter-chord line mostly used for computing low-speed aerodynamic properties, we use the mid-chord line of the outboard wing to define the sweep angle Λ_w. The rationale is that the mid-chord line is more representative to characterize the supercritical pressure distribution of the highly loaded outboard sections.
- The average thickness ratio $\overline{(t/c)}_w$ of a straight tapered wing as depicted in Figure 10.2 (a) is the mean value of the root and the tip thickness ratios. For a complex wing such as shown in Figure 10.2 (b) it is preferable to use the thickness ratio at the leading edge kink.

Apart from the design lift coefficient and the aspect ratio, the sweep angle and thickness ratio are primary selection variables for a transonic civil aircraft wing. This requires a modification of the wing and horizontal tail weight defined by Equations (10.11) and (10.12) as follows:

$$\mu_{\text{w+h}} = \frac{\Phi_3}{\overline{(t/c)}_w \, (\cos \Lambda_w)^2} \, A_w \sqrt{\frac{A_w}{\hat{C}_L}} + \frac{\Phi_2}{\hat{C}_L} \qquad (10.34)$$

with

$$\Phi_3 = 0.0013 \, (1 + r_h) \, \eta_{\text{cp}} \, n_{\text{ult}} \, \frac{\sqrt{W_{\text{MZF}}/\hat{q}}}{b_{\text{ref}}} \qquad (10.35)$$

The design featuring in Figure 8.1 has $\Phi_3 = 0.0002$, one order of magnitude smaller than Φ_1.

10.6.2   Wing Drag in the Design Condition

The treatise of aircraft drag in Sections 4.5 and 4.4 explains that the drag coefficient depends on Mach number and lift coefficient. There is no alternative to CFD methods for making a reliable prediction of the complete transonic drag characteristics depicted in Figure 4.11. However, the present application concentrates on drag in the design condition for which detailed wing design is to be optimized during the preliminary design stage. This requires primarily the selection of a sweep angle and thickness ratio resulting in a specified compressibility drag in the design condition: cruising flight. The following drag breakdown complies with this principle.

1. Zero-lift drag consists predominantly of friction drag – pressure drag is small in the design condition. The zero-lift profile drag coefficient of an airfoil section in incompressible flow is computed as the skin friction drag coefficient C_f of a flat plate multiplied by the form factor allowing for supervelocities and pressure drag,

$$c_{d_p} = 2\,\Phi_f\,C_f = 2\,(1 + r_\Phi\,t/c)\,C_f \tag{10.36}$$

Depending on the transition point and the location where maximum suction occurs, the magnitude of the shape factor r_Φ is between 2.5 and 3.5 for thickness ratios up to 0.15. In order to convert section drag to three-dimensional wing drag, wing sweep is incorporated as follows:

$$(\check{C}_{D_p})_w = 2\,\{1 + r_\Phi\,(\overline{t/c})_w\,(\cos \Lambda_w)^2\}\,C_f \tag{10.37}$$

The average thickness ratio $(\overline{t/c})_w$ equals the ratio of frontal area to planform area. A correction is made for camber, intended to minimize profile in the lifting condition. Since this term is typically proportional to the design lift coefficient squared, it is counted as an increment of the drag due to lift. Horizontal tail profile drag is assumed to be a constant fraction of the wing profile drag. The wing profile drag is thus multiplied by a factor

$$d_{w+h} \stackrel{\text{def}}{=} 1 + (\check{C}_{D_p} S)_h/(\check{C}_{D_p} S)_w \tag{10.38}$$

with a typical value between 1.2 and 1.3. Total wing plus tail profile drag thus becomes

$$\check{C}_{D_p} = 2\,d_{w+h}\,\{1 + r_\Phi\,(\overline{t/c})_w\,(\cos \Lambda_w)^2\}\,C_f \tag{10.39}$$

2. Drag due to lift is composed of vortex drag due to an elliptical lift distribution, a correction for non-ellipticity and lift-dependent profile drag, combined into

$$C_{D_L} = K_L \hat{C}_L^2 = \frac{\hat{C}_L^2}{\pi\,A_w\,\check{e}} \tag{10.40}$$

For a plane wing, the parameter $\check{e}$ is less than 1.0 but higher than the traditional Oswald factor which applies to an aircraft drag polar in design and off-design conditions. Flow separation is assumed to be avoided in the design condition by suitable airfoil shape and twist distribution.

3. Compressibility drag C_{D_c} can be experienced by all exposed aircraft components. However, only the wing contributes significantly to the initial drag rise – a typical drag curve for

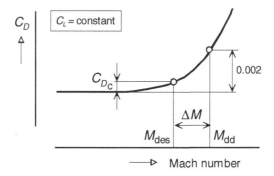

Figure 10.8   Variation of the drag coefficient with transonic Mach number

constant C_L is depicted in Figure 10.8. The NASA defines the drag divergence Mach number M_{dd} as the condition for $\partial C_D / \partial M = 0.10$. Other definitions base M_{dd} on a drag penalty, for example, $C_{D_c} = 0.0020$. The compressibility drag C_{D_c} in a fuel-efficient cruise over long ranges (LRC) is mostly less than ten drag counts; a higher drag penalty is accepted in high-speed cruising flight (HSC).

The drag divergence Mach number for a specified design Mach number is

$$M_{dd} = M_{des} + \Delta M \tag{10.41}$$

with a typical value of ΔM between 0 and 0.05. The total wing drag at M_{des} is obtained by adding the aforementioned contributions,

$$C_{D_w} = \check{C}_{D_p} + C_{D_L} + C_{D_c} \tag{10.42}$$

with C_{D_c} typically between 0.0005 and 0.0010.

10.6.3   Modified Wing Penalty Function

The wing penalty function (WPF) derived in Section 10.3 represents the objective function for wing design since its minimum value corresponds to minimum MTOW. The WPF for a subsonic airplane contains the wing weight parameter Φ_1 defined by Equation (10.12) for given thickness ratio $\overline{(t/c)}_w$ and sweep angle Λ_w. However, wing optimization for a specified transonic flight condition requires adaptation of $\overline{(t/c)}_w$ and Λ_w to the design Mach number and lift coefficient. In order to incorporate this design sensitivity Φ_1 is replaced by Φ_3 according to Equation (10.35). This leads to a modified version of Equation (10.14),

$$F_{wp} = \frac{\Phi_3 A_w \sqrt{A_w/\hat{C}_L}}{\overline{(t/c)}_w \cos^2 \Lambda_w} + \frac{\Phi_2}{\hat{C}_L} + F_{prop} \left(\frac{\check{C}_{D_p} + C_{D_c}}{\hat{C}_L} + \frac{\hat{C}_L}{\pi A \check{e}} \right) \tag{10.43}$$

This equation has two contributions containing $(\overline{t/c})_w$ and Λ_w:

$$\frac{\Phi_3 A_w \sqrt{A_w/\hat{C}_L}}{(\overline{t/c})_w \cos^2 \Lambda_w} \quad \text{and} \quad \check{C}_{D_p} = 2\{1 + r_\Phi\,(\overline{t/c})_w\,(\cos \Lambda_w)^2\}\,C_f \tag{10.44}$$

The first term is the primary wing structure weight which decreases when the thickness ratio increases. The second term representing fuel and power plant weight to balance wing thickness drag is reduced when the thickness ratio decreases. Their combination has a minimum value for

$$(\overline{t/c})_w (\cos \Lambda_w)^2 = A_w^{0.75} \hat{C}_L^{0.25} \sqrt{\frac{\Phi_3}{2\,d_{w+h}\,r_\Phi\,C_f\,F_{\text{prop}}}} \tag{10.45}$$

for which the two terms of Equation (10.44) are equal. This 'optimum thickness ratio' seems to be an intriguing result but in practice it is too high to have any practical significance for a transonic airliner. A more realistic design objective is to find the best value of $(\overline{t/c})_w \cos^2 \Lambda_w$ within the limit of acceptable compressibility drag in the cruise condition. The next paragraphs are drawn up to derive wing designs with a drag divergence Mach number at a specified margin above the design Mach number.

10.6.4 Thickness Ratio Limit

The thickness ratio of a transonic airfoil section has an upper limit determined by the ability to achieve a very limited amount of compressibility drag at the design Mach number. Aerodynamic wing design technology has a large effect on this limit. During the 1960s, long-range airliners and executive aircraft were designed with peaky airfoils which operated with a small pocket of supersonic flow above the wing nose. The more advanced supercritical airfoils developed between 1970 and 1990 are representative of the current state of the art. These are characterized by a relatively large leading-edge radius, reduced curvature over the middle region of the upper surface, and substantial camber of the aft surface; see Figure 10.9. This geometry results in a large region of supercritical flow on top closed by a weak shock wave and considerable aft loading in the design condition. The design limit of supercritical sections can be quantified by Korn's equation [20],

$$M_{dd} + t/c + 0.10\,c_l = \kappa \tag{10.46}$$

The aerodynamic technology factor κ for supercritical wing sections amounts to approximately 0.95. The following modified version of Korn's equation is based on empirical data of second generation supercritical sections [22]:

$$M_{dd} + t/c + 0.10\,c_l^{1.5} = M^\star \quad \text{with} \quad M^\star = 0.935 \tag{10.47}$$

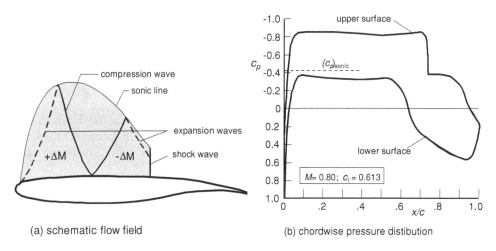

(a) schematic flow field   (b) chordwise pressure distibution

Figure 10.9 Flow properties for an 11% thick supercritical section near the design Mach number [22]

Equation (10.47) is adapted to three-dimensional airfoils by means of the simple sweep theory as follows:

$$M_{dd} \cos \Lambda_w + \frac{(\overline{t/c})_w}{\cos \Lambda_w} + 0.10 \left\{ \frac{1.1 \hat{C}_L}{(\cos \Lambda_w)^2} \right\}^{1.5} = M^\star \qquad (10.48)$$

The factor 1.1 accounts for the higher than average aerodynamic loading of the outboard wing and the sweep angle refers to the outboard mid-chord line. The upper thickness ratio limit follows from

$$(\overline{t/c})_w (\cos \Lambda_w)^2 = (\cos \Lambda_w)^3 (M^\star - M_{dd} \cos \Lambda_w) - 0.115 \hat{C}_L^{1.5} \qquad (10.49)$$

Figure 10.10 depicts this relationship and shows that for $M_{dd} \geq 0.75 M^\star$ a sweep angle can be identified for which the term $(\overline{t/c})_w (\cos \Lambda_w)^2$ has a maximum value,

$$\cos \Lambda_w = 0.75 \frac{M^\star}{M_{dd}} \qquad (10.50)$$

It is worth noting that this characteristic sweep angle is independent of the lift coefficient. For Mach 0.7 or less this condition means that a swept wing is not required to achieve the highest thickness ratio limit. If this condition is inserted in Equation (10.44), the dominant wing weight-related term has a minimum value. This criterion defines the condition of minimum wing penalty function. This is interpreted as a recommendation that for flying at M_{dd} the Mach number component normal to the mid-chord line should be equal to $0.75 M^\star$. Indeed, it is confirmed and appears statistically that sweep angles and maximum cruise Mach number of modern transport and executive aircraft are interrelated as follows: $M_{des} \cos \Lambda_w \approx 0.70$. The optimum sweep angle is thus primarily dependent on the design Mach number and on aerodynamic technology. An advanced aerodynamic wing design corresponds to a high $M^\star$ and allows a smaller sweepback angle compared to a less advanced wing. The best thickness

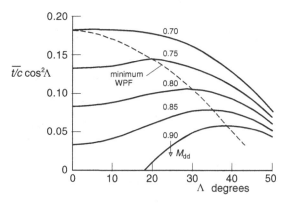

Figure 10.10 Upper limit of the thickness ratio related to the drag divergence Mach number, for $\hat{C}_L = 0.55$ and $M^\star = 0.935$

ratio is, however, definitely affected by the design lift coefficient. It is therefore rational to select the sweep angle first and then solve for the highest thickness ratio determined by Equation (10.48).

10.6.5 WPF Affected by Sweep Angle and Thickness Ratio

An interesting illustration of the effects of sweep and thickness ratio variation on the WPF is obtained by adapting the combination of Λ_w and $(\overline{t/c})_w$ to the design lift coefficient $\hat{C}_L$. The following observations are based on results for the optimization study in Chapter 8 depicted in Figure 10.11:

- In accordance with previous remarks, the sweep angle resulting in the lowest WPF – line I, $\Lambda_w = 31.2°$ – is independent of $\hat{C}_L$. Selecting a sweep angle a few degrees smaller should be favoured since this affects WPF only slightly whereas low-speed aerodynamic properties are improved. Another argument for less sweep is that it allows a higher aspect ratio in case of an active pitch-up limit.
- The partial optimum $\hat{C}_L$ for minimum F_{wp} is represented by curve II. Its analytical derivation is discussed in Section 10.7.
- The response surface of wing and tail L/D is a saddle surface. The partial optimum $\hat{C}_L$ differs little from the MTOW-optimum (curve II) and should be taken seriously. Sensitivity to sweep angle variation is low. Profile drag for zero sweep is low because the wing is very thin; it is low for high sweep angles due to the strong $(\cos \Lambda_w)^2$ effect.
- Reference [25] emphasizes the importance of achieving the highest possible cruise lift coefficient since this results in minimum wing area. The condition for maximum $\hat{C}_L$ is derived from Equation (10.49) by setting the partial derivative w.r.t. the sweep angle equal to zero, which yields

$$\cos \Lambda_w = \frac{3M^\star}{8\,M_{dd}} \left\{ 1 + \sqrt{1 - \frac{32\,(\overline{t/c})_w\,M_{dd}}{9\,(M^\star)^2}} \right\} \qquad (10.51)$$

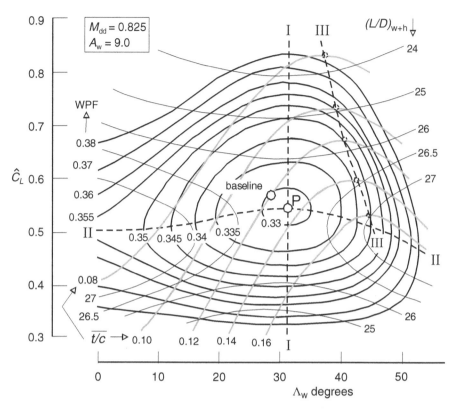

Figure 10.11 Wing penalty function and wing/tail aerodynamic efficiency affected by planform geometry and lift coefficient. Design: Figure 8.1

depicted by curve III in Figure 10.11. However, complying with this criterion requires considerably more sweepback and higher MTOW compared to Equation (10.50) defining minimum WPF.

These observations clearly illustrate that using a purely aerodynamic criterion as the objective function leads to sub-optimum solutions. The 30° sweepback angle of the baseline design is fully justified. The next section demonstrates that the choice of $(\overline{t/c})_w$ and $\hat{C}_L$ is closely related.

10.7 Lift Coefficient and Aspect Ratio

10.7.1 Partial Optima

Since for transonic aircraft the sweep angle and/or thickness ratio have to be treated as selection variables, the wing penalty function optimization must be reconsidered. Equations (10.43) and (10.49) are used to compute the WPF contours in Figure 10.12 which deviate significantly from those in Figure 10.5 for the turboprop freighter design. Figure 10.13 shows that for $A_w \geq 6$ the partial optimum $\hat{C}_L$ (curve I) is increasingly below the value defined for the case

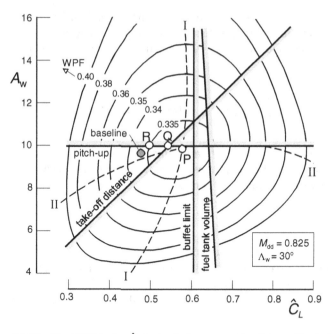

Figure 10.12 The WPF in the $\hat{C}_L - A_w$ design space. Conditions: Figure 10.11

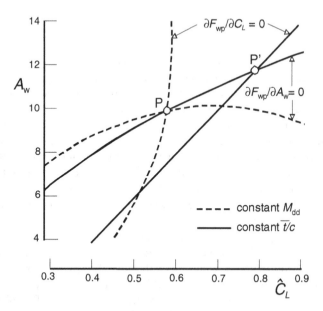

Figure 10.13 Optima for constant drag rise Mach number compared with those for constant thickness ratio

of constant thickness ratio, Equation (10.17). The partial optimum aspect ratio according to Equation (10.18) remains valid provided the wing weight factor Φ_1 is adapted for variation of the sweep angle and thickness ratio with Equation (10.35). The condition $\partial F_{\mathrm{wp}}/\partial A_{\mathrm{w}} = 0$ results in

$$A_{\mathrm{w}} = \hat{C}_L^{0.6} \left[\frac{2 F_{\mathrm{prop}}}{3\pi \,\breve{e}\, \Phi_3} \left\{ (\cos \Lambda_{\mathrm{w}})^3 (M^\star - M_{\mathrm{dd}} \cos \Lambda_{\mathrm{w}}) - 0.11\, \hat{C}_L^{1.5} \right\} \right]^{0.4} \qquad (10.52)$$

designated as curve II. This solution differs significantly from the subcritical case for $\hat{C}_L > 0.6$. The global optimum design P' for the transonic wing has a much lower wing loading and aspect ratio compared with the optimizer for constant thickness ratio.

10.7.2 Constraints

Design limitations such as those mentioned in Section 10.4 apply to subsonic as well as transonic aircraft. However, they have to be modified in order to account for sweep and thickness ratio variation. The span loading constraint indicated in Figure 10.12 is derived from the BFL limit. Attempts to derive an analytical solution for the constrained optimum (design Q) were unsuccessful. The wing box volume available for fuel is sensitive to thickness ratio variation. The volume limit happens to define a design lift coefficient which is almost independent of the aspect ratio. The corresponding thickness ratio may therefore be assumed to be constant and Equation (10.31) needs no modification.

Wing design for transonic flow raises some additional limitations which can be very restrictive but difficult to predict accurately.

- Deviations from the equilibrium in cruising flight can occur as a consequence of manoeuvres or a gust upset. The associated variations in incidence and Mach number may lead to shock waves, shock-boundary layer interactions and a turbulent wing wake impinging on the horizontal stabilizer. This phenomenon is felt in the aircraft as buffeting. Figure 10.14 shows a typical variation of the lift coefficient for initial buffet onset. Freedom from buffet is normally required during a manoeuvre load factor of at least 1.30 or a vertical gust velocity of 12.5 m/s from the trimmed condition. Since C_L for the buffet onset is difficult to predict during preliminary design, it is usually derived from wind tunnel and flight tests. For the study aircraft the buffet limitation indicated in Figure 10.12 is based on the assumption that with the current CFD technology a buffet onset limit of $C_L = 0.80$ can be realized for a 12% thick supercritical wing at the normal Mach 0.81 cruise speed.
- Another off-design requirement is related to handling characteristics at high speed. In response to an increase in incidence, the aircraft should tend to pitch nose-down. If the flow breaks down over the outer wing, this will move the centre of pressure forward and inboard, causing increased downwash at the stabilizer. The resulting pitch-up tendency will be stronger when the angle of sweep and the aspect ratio are larger. Therefore, a maximum aspect ratio depending on wing sweep and taper is imposed to avoid high speed pitch-up and critical aero-elastic effects. In accordance with current advanced civil aircraft, an aspect ratio $A_{\mathrm{w}} = 10$ is considered to form the pitch-up limitation for a $30°$ sweptback wing.

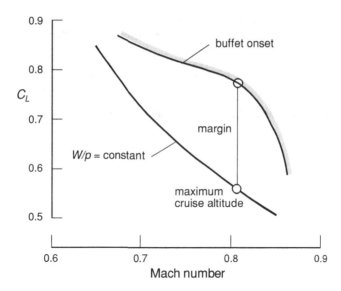

Figure 10.14 Buffet boundary compared with constant altitude cruise

The study design Q (Figure 10.12) has a constrained optimum $\hat{C}_L = 0.54$ and $A_w = 10$ – both figures are slightly higher than for the baseline design. The preferred modified design R with $\hat{C}_L = 0.50$ and $A_w = 10$ promises a useful reduction of the mission fuel at the cost of less than 1% MTOW penalty relative to the unconstrained optimum design P.

The analytical optima derived in this chapter are intended as guidelines for the initial selection of the wing geometry. For the present category of aircraft, explicit results such as Figure 10.12 are considered to be more useful than closed-form expressions for the various (partial and global) optima. Although computing iso-merit contours is more labour-intensive, visualization of the design sensitivity to excursions from the baseline and the global optimum is rewarding. The diagrams become even more useful when FOMs such as the propulsion function and fuel energy efficiency are shown explicitly.

10.7.3   Refining the Optimization

In this chapter we have treated the wing loading and aspect ratio as the primary selection variables. For the optimization of subsonic aircraft the wing was assumed to have zero sweep and constant thickness ratio. For transonic aircraft the sweep angle and thickness ratio were additional selection variables associated with obtaining a specified drag-divergence Mach number. Relatively simple relationships for the WPF according to Equations (10.14) and (10.43) enabled the use of explicit optimization. The two example designs appeared to have a feasible region in which a constrained optimizer could be allocated with modest penalties relative to the unconstrained one. Experienced designers may, however, argue that the situation in the real design world can be more complicated. This is illustrated hereafter for the cases where the load factor and the maximum lift coefficient for landing are treated as design parameters or constraints.

The ultimate load factor featuring in Equations (10.12) and (10.35) equals 1.5 times of the limit load factor, the greater of the manoeuvre load factor n_{man} and the gust load factor,

$$n_{gust} = 1 + \frac{0.5 K_g \, \rho_{sl} \, U_{DE} \, V_C \, S_w C_{L_\alpha}}{W_{MTO}} \tag{10.53}$$

For a gust load alleviation factor $K_g = 0.8$ the manoeuvre load factor is critical for

$$\frac{W_{MTO}}{S_w} \geq \frac{0.4 \, \rho_{sl} \, U_{DE} \, V_C \, C_{L_\alpha}}{n_{man} - 1} \tag{10.54}$$

If the wing loading does not comply with this condition, the gust load factor is dominant and the ultimate load factor is

$$n_{ult} = 1.5 + 0.6 \, \rho_{sl} \, U_{de} V_C \, C_{L_\alpha} \left(\frac{W_{MTO}}{S_w}\right)^{-1} \tag{10.55}$$

In the interest of simplicity it has been assumed that the manoeuvre load factor is dominant so that n_{ult} is not affected by wing design. This is justified only if Equation (10.54) is treated as an inequality constraint. In reality this equation merely represents a limit for the validity of the simplified objective function, and there is no valid argument to consider wing loadings not complying with Equation (10.54) as unfeasible. Instead, for these wing loadings the gust load factor should be treated as dominant. For a given aspect ratio and GW, the structure weight required to resist wing bending is then composed of a term proportional to $S_w^{1.5}$ and a term proportional to $S_w^{0.5}$, with a dominating first term.

The landing wing loading can be constrained by an upper limit of the approach speed to a lower value than that for minimum weight. As this limit is proportional to $C_{L_{max}}$ the refinement may be introduced to treat this coefficient as a design variable. This requires a design-sensitive relationship between high-lift system design parameters and wing weight. For a given system configuration, it is anticipated that increasing $C_{L_{max}}$ is possible to a limited extent by increasing flap chords and deflection angles. If this measure is expressed in terms of flap weight, the wing loading constraint can be relaxed. In practice this may be a laborious and complicated exercise.

At first sight it seems that the optimization problem has been oversimplified by using the same expression for the wing penalty function for all combinations of wing loading and aspect ratio. However, the design space can be subdivided into three regions: (1) for low wing loadings n_{ult} decreases when the wing loading increases; (2) for intermediate wing loadings $C_{L_{max}}$ and n_{ult} are constant; (3) for high wing loadings n_{ult} is constant and $C_{L_{max}}$ increases when the wing loading increases. Figure 10.15 depicts an example where it can be useful to treat a design criterion as a borderline between two regions of the design space where the objective function is computed differently. In Figure 10.15 (a) the stall speed constraint is active, making the unconstrained optimum design P unfeasible. The constrained optimum design Q applies to $C_{L_{max}} = 2.80$. The analysis can be refined by relaxing the inequality constraint and converting it into a borderline with the region where $C_{L_{max}}$ is variable. In Figure 10.15 (b) a new unconstrained optimum P corresponds to $C_{L_{max}} > 2.80$.

The wing loading limit for which the load factor is constant is an inequality constraint in Figure 10.15 (a). Reducing the wing loading below this limit renders the gust load dominant,

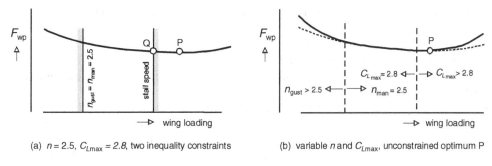

(a) $n = 2.5$, $C_{Lmax} = 2.8$, two inequality constraints   (b) variable n and C_{Lmax}, unconstrained optimum P

Figure 10.15   Treating a design limit as an inequality or as an equality constraint

the wing weight increases more progressively and the objective function becomes considerably more design-sensitive than for a constant load factor as in Figure 10.15 (b). Nevertheless, since the inequality constraint is not active, the optimum location is not affected. The relaxation is only worthwhile for an active constraint.

These observations justify the simplified approach taken in the present chapter; at the same time they warn against oversimplification. The moment that design variables are selected, it becomes necessary to find out whether the optimum is constrained or not. If, for instance, the wing loading appears to be highly constrained by a field length requirement, it is useful to compute the MTOW, the mission fuel and installed thrust penalties incurred. If these appear to be significant, a more sophisticated or powerful high-lift system should be considered. Although this principle applies to each active constraint, the experienced designer will anticipate whether the complication of dividing up the design space will be worthwhile. It is concluded that the first step of wing design optimization should be a design-sensitive analysis of active constraints forming the boundaries of the feasible design space. Unconstrained optimization forms the secondary activity.

10.8   Detailed Design

The optimum wing configuration will depend on the objective function which is considered to be dominant. Up to now, only wing structural weight, engine and fuel weight and MTOW were used as objectives for wing sizing. Although further details of the wing geometry will have some effects on weight sensitivity, many choices have to be based on practical considerations that cannot be expressed in a single unambiguous figure of merit – such as flying qualities, operational flexibility, safety and environmental impact. The aerodynamic design of a wing also requires due attention to off-design conditions, in particular, missions deviating from the design range, cruising flight at non-optimum speed and altitude and flight at low speeds with extended high-lift devices. The following overview is based on qualitative considerations.

10.8.1   Taper and Lift Distribution

A principle of aerodynamic wing design is that the lift distribution in the design condition should be nearly elliptical for minimum vortex drag. Although this can be achieved with an elliptical planform, (straight) tapered wings are favoured by airplane manufacturers for various reasons.

- Tapered wings are simpler and cheaper to manufacture.
- A highly tapered wing has a more triangular lift distribution, leading to reduced bending at the (highly loaded) wing root.
- For given planform area and span, the root section is thicker and the wing structure is lighter.
- At each lateral station, the structural layout can be better adapted to what is required to cope with the critical local loading case. This results in less non-ideal weight.
- Many sweptback wings have kinks in the leading and/or trailing edge which give them compound taper. A pronounced trailing edge kink is usually associated with an increased root chord length, creating space for retracting landing gears.

If the taper (ratio) is close to the value for minimum induced drag and the twist distribution is correct, the induced drag penalty due to straight taper is very small. A lower limit on the outboard wing taper ratio is imposed to avoid too high lift coefficients. The lift coefficient inboard of the kink is significantly lower which allows the root section to be thicker – useful for reducing wing structural weight.

10.8.2   Camber and Twist Distribution

Figure 10.16 illustrates that in the design (cruise) condition large regions of a supercritical wing have a nearly two-dimensional pressure distribution. The outboard wing is heavily loaded with a region of supersonic flow above the upper surface up to the crest with a plateau of $M \approx 1.20$. The highly cambered outboard wing is rear-loaded, resulting in high lift coefficients. A side effect of this lift distribution is a nose-down pitching moment that is to be balanced by a horizontal tail download, entailing a trim drag penalty. This disadvantage is reduced by twisting the inboard wing and applying negative camber near the wing/fuselage junction. The resulting pressure distribution causes a front loading at the root negating the outboard rear loading; see Figure 10.17. An increased leading-edge sweep of the inboard section will further

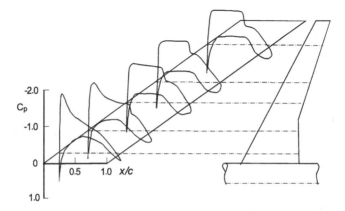

Figure 10.16   Pressure distribution of a supercritical wing in the design condition

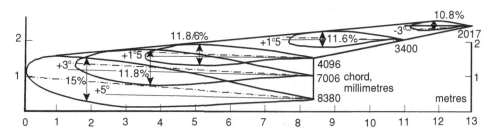

Figure 10.17   Geometry of the Airbus A 310 supercritical wing

reinforce this effect. The peaky pressure distribution is also effective for sweeping back the isobars, thereby minimizing the aerodynamic un-sweeping effect at the root. The variation of airfoil camber is combined with twist – especially the inboard wing – in order to optimize the lateral lift distribution and avoid too high outboard wing lift coefficients. Except for the tip region, the outboard wing in Figure 10.17 has little twist in order to approximate the two-dimensional pressure distribution as close as possible.[4] The net effect of these measures is a complicated wing geometry, in particular, inboard of the kink.

10.8.3   Forward Swept Wing (FSW)

Sweeping a wing provides an effective means of increasing the drag divergence Mach number, an effect that is obtained regardless of the direction of sweep. However, compared to the aft swept wing (ASW), the outboard section of a forward swept wing (FSW) has a reduced loading and the wing's centre of pressure is more inboard compared to a forward swept wing (FSW). Comparing an ASW and a FSW for equal planform area, span and lift, the FSW experiences a reduced bending moment at the structural wing root. Hence, if the FSW span is allowed to increase until the root bending moment equals that of the ASW, the reduced span loading reduces its induced drag. The FSW tends to stall first at the inboard wing and does not suffer from pitch up at high angles of attack, an inherent disadvantage of the ASW. It is also anticipated that forward sweep favours achievement of NLF since the inboard moving leading edge flow is not contaminated by the fuselage boundary layer.

When a FSW aircraft is manoeuvred at high dynamic pressure, wing bending leads to increased angle of attack and torsion of the outboard wing at the tips. The FSW tends to suffer from the aero-elastic phenomenon of structural divergence which cannot be solved with a conventional metallic structure without paying an excessive weight penalty. With composite wings, the skin laminates can be tailored to provide favourable deflections at dynamic pressures considerably higher than those previously obtained with conventional structures [42]. Advances in the use of composite materials have enabled the elimination of the divergence problem with little or no weight penalty. In spite of their favourable aerodynamic properties, only a few FSW aircraft have been designed, produced and flown. One notable example is the HFB 320 Hansa business jet [41], which was marketed (with little commercial success) in the early 1960s.

---

[4]Due to aero-elastic deformation, the jig shape built into the structure at the tip of a sweptback wing may differ up to several degrees from the desired 1-g loaded shape in flight.

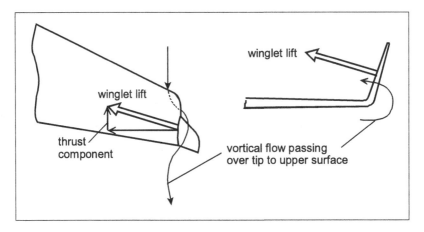

Figure 10.18   The winglet concept for drag reduction

10.8.4   Wing-Tip Devices

It has been known since many decennia that addition of wing tip-mounted lifting surfaces can diffuse tip vortex structures and reduce vortex drag. A favoured option is the use of winglets – relatively small vertical or outward canted lifting surfaces (Figure 10.18). They are immersed in the local vortex flow passing from the lower to the upper surface which has an inward velocity component above the wing tip. The winglet acts under an angle of incidence to this flow and experiences lift. Its component in the direction of flight – the figure indicates it as a thrust component – balances the winglet's vortex drag. The trailing vortex sheet is spread out in an upward and outward direction and the wing experiences reduced downwash at the tip. Dependent on the basic wing shape and winglet height, the resulting vortex drag reduction is between 10 and 15%. Wings designed with high outboard lift coefficients are good candidates for these wing tip devices. The vortex drag reduction is partly offset by the winglet's parasite drag and interference between the main wing, whereas the winglet flow may lead to formation of shock waves at the wing/winglet junction. The blended winglet aims at reducing these unfavourable effects [64]. Total drag reductions in cruising flight for several airliners are between 3.5 and 4.5% for winglet heights between 12 and 16% of the semi-span. This compares to a reduction of 5.5% for a 10% wing span increase by means of sheared tips.

Especially in the case of a constraint on the wing span, fitting winglets is a favoured option. In all other cases a fundamental issue is whether or not it is preferable to merely extend the wing tips instead of adding wing-tip devices. A parametric study [50] compared induced drag improvements achievable with wing tip extension and fitting of optimized winglets, assuming the same wing root bending moment for both cases. It was found that for almost every tip extension a better performing winglet could be found. Another investigation [55] showed more modest advantages. For aircraft with critical engine-out performance at high and hot airfields, a useful induced drag reduction is obtained during the take-off climb, allowing the plane to be taken off with increased weight. The question whether or not to apply winglets cannot be answered in general – each configuration must be examined for its net drag reduction, bending moment increase, weight penalty and manufacturing costs. Winglets designed as an integral wing component are widely accepted as beneficial, as add-on devices they can be detrimental. Aerodynamic tip devices are also introduced on new versions of an existing airliner if these can be installed with relatively minor structural modifications.

10.9 High Lift Devices

10.9.1 Aerodynamic Effects

Apart from powered systems such as blown and jet flaps, a high lift system is basically a variable geometry mechanical alteration of the (secondary) structure in front and aft of the primary box. Civil aircraft feature slotted flaps incorporated at the trailing edge of the inboard wing, many of them have leading edge flaps as well. Deflection of a trailing edge flap on an airfoil section increases its camber which increases the lift. If flap deflection does not extend the chord, the lift increment Δc_l is independent of the angle of attack α. Figure 10.19 (a) shows that, for small flap deflections, Δc_l is proportional to the flap angle δ_{fl}. The flap efficiency decreases at large deflections due to flow separation above the upper flap surface. For a plain flap (lacking slots), the separation begins at a relatively small flap angle, a single slotted flap is effective to higher deflections and double and triple slotted flaps can cope with deflections up to fifty degrees. Slotted flaps, in particular Fowler flaps, move backwards during extension. The increased chord length augments the lift increment proportional to α.

A wing with extended slotted flaps can be seen as a series of aerodynamically interacting airfoils. Each airfoil experiences lift-producing circulation, inducing upwash in the flow around the preceding airfoil and downwash at the following airfoil. The leading airfoil is deflected downwards to avoid separation due to the increased local upwash. The nose suction on the following airfoils acts to energize the trailing edge flow of the preceding one. These interactions postpone separation to higher angles of attack and deflection provided that the airfoils are carefully positioned relative to each other. Figure 10.19 (b) illustrates these effects

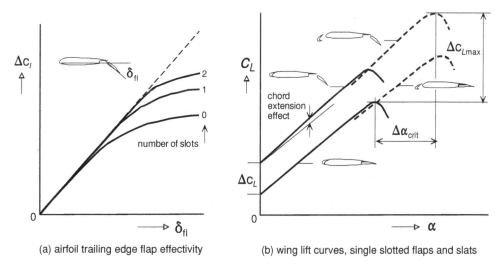

(a) airfoil trailing edge flap effectivity   (b) wing lift curves, single slotted flaps and slats

Figure 10.19 Effects on lift of extended high lift devices

for the lift curve of a three-dimensional wing. The achievable $\Delta C_{L_{\max}}$ of a wing with no leading edge devices is less than ΔC_L at low incidences whereas a slat extends the lift curve to a significantly higher α_{crit} and has little effect on lift at a given incidence. Multi-slotted part span Fowler flaps in combination with full span slats can increase the aircraft's $\Delta C_{L_{\max}}$ by up to 100% allowing the wing planform area to be selected without compromising high speed design. For take-off configurations, however, low drag is at least as important as high lift (Section 9.4). References [68, 71, 77] can be consulted for relevant further explanation.

10.9.2 Design Aspects

Application of slats on high speed aircraft with highly swept thin wings is almost unavoidable since it gives a high maximum lift in combination with a greater freedom to optimize the nose shape for high speed flight. A penalty for extending high lift devices is increased profile drag whereas part-span flaps increase vortex drag as well. Moreover, trailing edge flaps increase the wing's nose-down pitching moment. This effect must be compensated by a tail download at the cost of trim drag. Aircraft cruising at speeds below the Mach 0.70 to 0.75 region feature little sweepback and fairly thick airfoils. Even without slats they may achieve $C_{L_{\max}}$ up to 3.0 with deflected trailing edge devices. Several executive aircraft are optimized to cruise at high altitudes with a relatively low wing loading, and in combination with the large thrust for take-off their wings can be designed with no leading edge flaps. Avoiding the complication of a slat system simplifies the overall aircraft concept and operational properties, forming an invaluable cost saver. An aircraft featuring a modest $C_{L_{\max}}$ should therefore not be dismissed as technically inferior. In fact, apart from the application of advanced (composite) structural materials, there has not been any dramatic advancement in high lift system concepts of today's generation of aircraft compared to the historical trend up to 1984 depicted in Figure 10.20. The highest value in the landing configuration realized for civil transports amounts to $C_{L_{\max}} \approx 3.5 \cos \Lambda_{0.25}$ but many modern airliners do not need this. In association with the superior lifting properties of supercritical wings, current trends are towards mechanically simpler systems. These are mostly based on single slotted flaps and slats with increasing emphasis on reducing drag and aerodynamic noise emission by countering the effects of engineering excrescences such as

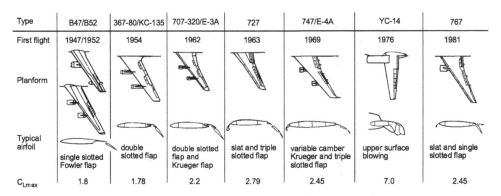

Type	B47/B52	367-80/KC-135	707-320/E-3A	727	747/E-4A	YC-14	767
First flight	1947/1952	1954	1962	1963	1969	1976	1981
Planform							
Typical airfoil	single slotted Fowler flap	double slotted flap	double slotted flap and Krueger flap	slat and triple slotted flap	variable camber Krueger and triple slotted flap	upper surface blowing	slat and single slotted flap
$C_{L_{\max}}$	1.8	1.78	2.2	2.79	2.45	7.0	2.45

Figure 10.20 Historical trends in Boeing transport aircraft high lift systems development. Copyright NATO STO - CSO

flap and slat tracks. A comprehensive overview of mechanical principles of high lift systems on modern commercial airliners is found in [27] and [82].

Bibliography

[1]   Küchemann, D., *The Aerodynamic Design of Aircraft*, Pergamon Press, Oxford, 1978.
[2]   Torenbeek, E., *Synthesis of Subsonic Airplane Design*, Appendix F, Delft University Press, 1981.
[3]   Shevell, R.S., *Fundamentals of Flight*, Chapter 11, Second Edition, Prentice-Hall, Englewood Cliffs, NJ, 1989.
[4]   Obert, E., *Aerodynamic Design of Transport Aircraft*, Part 5, Delft University of Technology, Faculty of Aerospace Engineering, IOS Press, Amsterdam, 2009.

Conventional and Supercritical Wing Design

[5]   McRae, D.M., "The Aerodynamic Development of the Wing of the A300B", *Aeronautical Journal*, Vol. 77, pp. 367–379, July 1973.
[6]   Whitcomb, R.T., "Review of NASA Supercritical Airfoils", ICAS Paper No. 74-10, August 1974.
[7]   Kyser, A.C., "An Elementary Analysis of the Effect of Sweep, Mach Number, and Lift Coefficient on Wing Structure Weight", NASA Technical Memorandum 74072, September 1977.
[8]   Hicks, R., and P.A. Henne, "Wing Design by Numerical Optimization", *Journal of Aircraft*, Vol. 15, pp. 407–412, 1978. Also: AIAA Paper No. 77-1247.
[9]   McGhee, R.J., and W.D. Beasly, "Low-Speed Aerodynamic Characteristics of a 13 Percent-Thick Medium-Speed Airfoil Designed for General Aviation Applications", NASA TP-1498, 1979.
[10] Henne, P.A., "Inverse Transonic Wing Design Method", *Journal of Aircraft*, Vol. 18, No. 2, pp. 121–127, February 1981. Also: AIAA Paper No. 77-1247. Also: AIAA Paper No. 80-0330.
[11] Chuprun, J., "The Evolution of Aircraft Wing Design", in Proceedings of the Symposium, Dayton, Ohio, March 18 and 19, 1980. Also: AIAA Paper No. 80-3031, 1980.
[12] Torenbeek, E., "Fundamentals of Conceptual Design Optimization of Subsonic Transport Aircraft", TU Delft, Department of Aerospace Engineering, Report LR-292, August 1980.
[13] Summerfield, M., "Transonic Aerodynamics", in *Progress in Astronautics and Aeronautics*, Volume 81, Edited by D. Nixon, Technical papers selected from the Transonic Perspective Symposium, Moffet Field, CA, February 1981.
[14] Large, E., "The Optimal Planform, Size and Mass of a Wing", *Aeronautical Journal*, pp. 103–110, March 1981.
[15] Aronson, M., "Optimum Configuration for a 10 Passenger Business Turbofan Jet Airplane", AIAA Paper No. 82-0365, 1982.
[16] Rettie, I.H., "Aerodynamic Design for Overall Vehicle Performance", AGARD Report R-712, Paper No. 6, May 1983.
[17] Kroo, I.M., "Design and Analysis of Optimally Loaded Lifting Systems", AIAA Paper No. 84-2507, 1984.
[18] Torenbeek, E., "On the Conceptual Design of Subsonic Transport Aircraft for Cruising Flight Optimized for Different Merit Functions", TU Delft, Department of Aerospace Engineering, Report LR-451, December 1985.
[19] Drela, M., and M.B. Giles, "ISES: A Two-Dimensional Viscous Aerodynamic Design and Analysis Code", AIAA Paper No. 87-0424, January 1987.
[20] Boppe, C.W., "CFD Drag Predictions for Aerodynamic Design", AGARD Advisory Report No. 256, Lisbon, 1988.
[21] Mabey, D.G., "Buffeting Criteria for a Systematic Series of Wings", *Journal of Aircraft*, Vol. 26, No. 6, pp. 576–589, 1989.
[22] Harris, C.D., "NASA Supercritical Airfoils", NASA TP 2969, March 1990.
[23] Desterac, D., and J, Reneaux, "Transport Aircraft Aerodynamic Improvement by Numerical Optimization", ICAS Paper No. 90-6.7.4, September 1990.
[24] Goldhammer, M.I., and F.W. Steinle, Jr., "Design and Validation of Advanced Transonic Wings Using CFD and Very High Reynolds Number Wind Tunnel Testing", ICAS Paper No. 90-2.6.2., September 1990.
[25] Mason, W., "Analytic Models for Technology Integration in Aircraft Design", AIAA Paper 90-3262, September 1990.

[26] Henne, P.A., and R.D. Gregg III, "New Airfoil Design Concept", *Journal of Aircraft*, Vol. 28, No. 5, pp. 300–311, May 1991.

[27] Jupp, J.A., "Wings for Airbus: Examples of Integrated Design", Material prepared for ECATA Course, Block 2, April 1992.

[28] Martínez-Val, R. and Pérez, E., "Optimum Cruise Lift Coefficient in Initial Design of Aircraft", *Journal of Aircraft*, Vol. 29, No. 4, pp. 712–714, July–August 1992.

[29] Van de Velden, A., "Aerodynamic Shape Optimization", AGARD R-803, AGARD-FDP-VK1 Special Course, April 1994.

[30] Streshinsky, J.R., and V.V. Ovcharenko, "Aerodynamic Design Transonic Wing Using CFD and Optimization Methods", ICAS Paper No. 94-2.1.4, September 1994.

[31] Torenbeek, E, "Optimum Wing Area, Aspect Ratio and Cruise Altitude for Long Range Transport Aircraft", TU Delft, Department of Aerospace Engineering, Report LR 775, October 1994.

[32] Hoffman, K.W., and P.F.H. Clignett, "Designing the Global Express", AIAA Paper No. 95-3953, September 1995.

[33] Vicini, A., and D. Quagliarella, "A Multiobjective Approach to Transonic Wing Design by Means of Genetic Algorithms", RTO/AVT Conference, Ottawa, Canada, Paper 22, October 1999.

[34] Kriszler, T., "A Conceptual Design Methodology to Predict the Wave Drag of a Transonic Wing", RTO/AVT Conference, Ottawa, Canada, Paper 27, October 1999.

[35] Schneider, G., F. van Dalen, T. Barth, H. Hörnlein, and M. Stettner, "Determining Wing Aspect Ratio of a Regional Aircraft with Multidisciplinary Optimisation", *CEAS Conference on Multidisciplinary Aircraft Design and Optimization*, Maternushaus, Köln, 25-26 June 2001.

[36] Dalhuijsen, J.L., and R. Slingerland, "Preliminary Wing Optimization for Very Large Transport Aircraft with Wingspan Constraints", AIAA Paper No. 2004-0699, January 2004.

[37] Jameson, A., "Efficient Aerodynamic Shape Optimization", 10th AIAA/ISSMO Multidisciplinary Analysis and Optimization Conference, Albany, New Jersey, Paper 2004-4369, August–September 2004.

[38] Scholz, D., and S. Ciornei, "Mach Number, Relative Thickness, Sweep and Lift Coefficient of the Wing: an Empirical Investigation of Parameters and Equations", *DGLR Jahrbuch 2005*, Paper 122, Friedrichshafen, September 2005.

[39] Bérard, A, and A.T. Isikveren, "Conceptual Design Prediction of the Buffet Envelope of Transport Aircraft", *Journal of Aircraft*, Vol. 46, No. 5, pp. 1593–1606, September–October 2009.

[40] Morris, A.M., C.B. Allen, and T.C.S. Rendall, "Aerodynamic Shape Optimization of a Modern Transport Wing Using Only Planform Variations", *Journal of Aerospace Engineering*, Proc. IMechE, Vol.223 Part G, pp. 843–851, 2009.

Forward Swept Wings

[41] Wocke, H., and L.W. Davis, "Sweptforward Wings for the HFB 320 Hansa", *Aircraft Engineering*, pp. 248–251, August 1964.

[42] Krone, N.J. Jr., "Divergence Elimination with Advanced Composites", AIAA Paper No. 75-1099, August 1975.

[43] Krone, N.J. Jr., "Forward Swept Wing Flight Demonstrator", AIAA Paper No. 80-1882, 1980.

[44] Spacht, G., "The Forward Swept Wing, A Unique Design Challenge", AIAA Paper No. 80-1885, 1980.

[45] Uhuad, G.C., T.M. Weeks, and R. Large, "Wind Tunnel Investigation of the Transonic Aerodynamic Characteristics of Forward Swept Wings", *Journal of Aircraft*, Vol. 20, No.3, pp. 195–202, March 1983.

[46] Smith, P.R., and A.J. Srokowski, "High Aspect Ratio Forward Sweep for Transport Aircraft", AIAA Paper No. 83-1832, 1983.

[47] Nangia, R.K., "Subsonic Investigations on Configurations with Forward- and Aft-Swept Wings of High Aspect Ratio", ICAS Paper No. 84-2.6.2, September 1984.

[48] Redeker, G., and G. Wichmann, "Forward Sweep: A Favourable Concept for a Laminar Flow Wing", *Journal of Aircraft*, Vol. 28, No. 2, pp. 97–103, February 1991.

Winglets, Sheared Tips and Crescent Moon-Shaped Wings

[49] Whitcomb, R.T., "A Design Approach and Selected High Speed Wind Tunnel Results at High Subsonic Speeds for Wing-Tip Mounted Winglets", NASA TN D-8260, July 1976.

[50] Heyson, H.H., G.D. Riebe, and C.L. Fulton, "Theoretical Parametric Study of the Relative Advantages of Winglets and Wing-Tip Extensions". NASA TP-1020, September 1977.

[51] Reynolds, P.T., W.M. Gersten, and C.G. Voorhees, "Gates Learjet Model 28/29, the First Longhorn Learjet", AIAA Paper No. 78-1445, 1978.

[52] Spillman, J.J., "The Use of Wing Tip Sails to Reduce Vortex Drag", *Aeronautical Journal*, pp. 387–395, September 1978.

[53] Gilkey, C.D., "Design and Wind Tunnel Results of Winglets on a DC-10 Wing", NASA CR-3119, April 1979.

[54] Van Dam, C.P., and B.J. Holmes, "Effect of Winglets on Performance and Handling Qualities of General Aviation Aircraft", AIAA Paper No. 80-1870, August 1980.

[55] Jones, R.T.+ and T.A. Lasinski, "Effect of Winglets on the Induced Drag of Ideal Wing Shapes", NASA TM 81230, 1980.

[56] Webber, G.W., and T. Dansby, "Wing-Tip Devices for Energy Conservation and Other Purposes", *Canadian Aeronautics and Space Journal*, Vol. 29, No. 2, p. 105, 1983.

[57] Weihs, D., and J. Ashenberg, "Minimum Induced Drag Characteristics of Wings with Curved Planform", *Journal of Aircraft*, Vol. 21, pp. 89–91, January 1984.

[58] Van Dam, C.P., "Induced-Drag Characteristics of Crescent-Moon-Shaped Wings," *Journal of Aircraft*, Vol. 24, No. 2, pp. 115–119, 1987.

[59] Burkett, C.W., "Reduction in Induced Drag by the Use of Aft Swept Wing Tips", *The Aeronautical Journal*, Vol. 93, pp. 400–405, December 1989.

[60] DeHaan M.A., "Induced Drag of Wings with Highly Swept and Tapered Wing Tips", AIAA Paper No. 90-3062, 1990.

[61] Van Dam, C.P., P.M.H.W. Vijgen, and B.J. Holmes, "Experimental Investigation on the Effect of Crescent Planform on Lift and Drag", *Journal of Aircraft*, Vol. 28, No. 11, pp. 713–720, November 1991. Also AIAA Paper No. 90-0300, January 1990.

[62] Rokhsaz, K., "A Brief Survey of Wing Tip Devices for Drag Reduction", SAWE Paper No. 93-2574, 1993.

[63] Kravchenko, S.A., "Wing Tip Lifting Surfaces: Aerodynamic Design and Comparative Analysis", AIAA Paper No. 95-3909, 1995.

[64] Faye, R., R. Laprete, and M. Winter, "Blended Winglets for Improved Airplane Performance", Boeing Commercial Airplane Compnay, AERO 16-31, 2007.

[65] Verstraten, J.G., and R. Slingerland, "Drag Characteristics for Optimally Span-Loaded Planar, Wingletted and C-Wings", *Journal of Aircraft*, Vol. 46, pp. 962–971, March 2009.

High Lift Systems

[66] Wimpenny, J.C., "The Design and Application of High Lift Devices", *Annals of the New York Academy of Sciences*, Vol. 154, Art. 2, pp. 245–1117, November 1968.

[67] McRae, D.M., "Aerodynamics of Mechanical High Lift Systems", AGARD Lecture Series 43, 1971.

[68] Smith, A.M.O., "High-Lift Aerodynamics", 37th Wright Brothers Lecture, *Journal of Aircraft*, Vol. 12, No. 6, pp. 501–530, June 1975. Also AIAA Paper No. 74-939, 1974.

[69] Liebeck, R.H., "Design of Airfoils for High Lift", AIAA Paper No. 80-3034, March 1980.

[70] Murillo, L.E., and J.H. McMasters, "A Method for Predicting Low-Speed Aerodynamic Characteristics of Transport Aircraft", *Journal of Aircraft*, Vol. 22, No. 3, pp. 168–174, March 1984.

[71] Butter, D.J., "Recent Progress on Development and Understanding of High Lift Systems", AGARD CP-365, Paper No. 1, May 1984.

[72] Dillner, B., F.W. May, and J.H. McMasters, "Aerodynamic Issues in the Design of High-Lift Systems for Transport Aircraft", AGARD-CP-365, Paper No. 9, May 1984.

[73] Brune, B.W., and J.H. McMasters, "Computational Aerodynamics Applied to High-Lift Systems", in *Progress in Aeronautics and Astronautics*, Vol. 125, AIAA, 1991.

[74] Valarezo, W.O., and V.D. Chin, "Maximum Lift Prediction for Multielement Wings", AIAA Paper No. 92-0401, January 1992.

[75] Valarezo, W.O., and R.J. McGhee, "Multielement Airfoil Performance Due to Reynolds and Mach Number Variations", *Journal of Aircraft*, Vol. 30, No. 5, pp. 689–694, September–October 1993.

[76] Flaig, A., and R. Hilbig, "High-Lift Design for Large Civil Aircraft", *High-Lift Systems Aerodynamics*, AGARD-CP-515, September 1993.

[77] Obert, E., "Forty Years of High-Lift R&D: An Aircraft Manufacturer's Experience", *High-Lift Systems Aerodynamics*, AGARD-CP-515, September 1993.

[78] Nield, B., "An Overview of the 777 High-Lift Aerodynamic Design", CEAS European Forum on High Lift and Separation Control", University of Bath, UK, March 29–31, 1995.

[79] Anonymous, "Increments in Aerofoil Lift Coefficient at Zero Angle of Attack and in Maximum Lift Coefficient due to Deployment of a Double-Slotted or Triple-Slotted Trailing-Edge Flap, with or without a Leading-Edge High-Lift Device, at Low Speeds", ESDU Item 94031, London, April 1995.

[80] Anonymous, "Maximum Lift of Wings with Leading-Edge Devices and Trailing-Edge Flaps Deployed", ESDU Item 92031, London, November 1995.

[81] Van Dam, C.P., and R.S. Pepper, "Design Methodology for Multi-Element High-Lift Systems on Subsonic Civil Transport Aircraft", NASA CR-202365, 1996.

[82] Rudolph, P.K.C., "High-Lift Systems on Commercial Subsonic Airliners", NASA CR-4746, September 1996.

Variable Camber

[83] Hilbig, R., and H. Wagner, "Variable Wing Camber for Civil Transport Aircraft", ICAS Paper No. 85-5.2.1., 1984.

[84] Szodruch, J., "The Influence of Wing Camber Variation on the Aerodynamics of Civil Transport Aircraft", AIAA Paper No. 85-0353, January 1985.

[85] Renken, J.H., "Mission Adaptive Wing Camber Control Systems for Transport Aircraft", AIAA Paper No. 85-5006, 1985.

[86] Szodruch, J., and R. Hilbig, "Variable Wing Camber for Transport Aircraft", *Progress in Aerospace Sciences*, Vol. 25, pp. 297–328, 1988.

[87] Greff, E., "Aerodynamic Design and Integration of a Variable Camber Wing for a New Generation Long/Medium Aircraft", ICAS Paper No. 88-2.2.3, September 1988.

[88] Greff, E., "The Development and Design Integration of a Variable Camber Wing for Long/Medium Range Aircraft", *The Aeronautical Journal*, Vol. 94, No. 939, pp. 301–312, 1990.

11

The Wing Structure and Its Weight

> The structural weight of wings on swept-wing aircraft is a subject that continues to be widely discussed and little understood except in terms of the results of a highly complex design process.
> —A.C. Kyser [26] (1977)

11.1 Introduction

The aerodynamic performance and weight of the wing are primary drivers of civil airplane optimization. Contrary to the fuselage body, wing design offers a wide range of possibilities to minimize aerodynamic drag and structure weight. However, the most effective measures to reduce drag tend to increase wing weight which makes a careful trade-off necessary. Although this observation applies to other airplane components as well, the wing is the most dominant of these and its weight prediction is therefore a classical subject of research. Several publications suggest that wing weight can be predicted within less than 1 or 2% of inaccuracy by applying a sophisticated computational programme based on finite elements methods. This may be true for the advanced stage of preliminary design, optimization of an airplane as a complete system does not allow a prediction which is both very accurate and design-sensitive. Even recent major airliner development programmes have proven that, in spite of 'very accurate weight prediction' and a radical weight reduction programme, the empty weight of a newly developed aircraft can be significantly higher than guaranteed by the design specification.

Wing weight prediction in conceptual design is actually more an art than a science, justifying the dedication of a complete chapter of this book. Since this deals exclusively with the wing structure, the index 'w' of geometric wing parameters – such as area, span, sweep angle, thickness, aspect and taper ratio – used in previous chapters have been left out for editorial clarity.

11.1.1 Statistics can be Useful

In the initial design stage, a Class I wing structure weight prediction is based on a statistical relationship between the wing weight and few primary influential characteristics (Section 8.1). For instance, if the maximum take-off weight (MTOW) is available, a collection of wing structure weight data for similar aircraft forms a useful point of departure. It is often found

Advanced Aircraft Design: Conceptual Design, Analysis and Optimization of Subsonic Civil Airplanes, First Edition. Egbert Torenbeek.
© 2013 by Egbert Torenbeek. Published 2013 by John Wiley & Sons, Ltd.

that wing weight as a fraction of the MTOW shows little variation with wing size and that, for a given class of aircraft,

$$W_w \propto W_{MTO} \tag{11.1}$$

A weight fraction of 0.115 is typical for jetliners and 0.10 for business jets, with a ballpark accuracy of about 10%. A design engineer may consider this heuristic estimation to be unsatisfactory since it ignores that wing weight depends on the load factor, geometry, type of high-lift devices, material properties and many other characteristics. Therefore, an alternative approach can be tried to find a correlation between the wing weight fraction and a few primary design parameters. Experience teaches that the wing span is the most influential parameter since it affects the bending load as well as the length of the primary structure whereas the area is less prominent because it determines mainly the weight of secondary structure. For instance, the following relationship for civil jet aircraft is based on trial and error:

$$W_w \propto b^2 (S\, W_{MZF}/W_{MTO})^{0.5} \tag{11.2}$$

where b and S denote the wing span[1] and the planform area, respectively, and the statistically determined factor of proportionality is 6.25 N m^{-3}. Even though this equation uses more detailed input, its standard deviation is 11%, similar to the previous equation.

A significant improvement is obtained by applying a trick. Since we have just proposed two expressions for the same quantity, multiplication and then drawing the square root yields

$$W_w \propto b (S\, W_{MZF} W_{MTO})^{0.25} \tag{11.3}$$

The factor of proportionality amounts to 0.86 N$^{0.5}$ m$^{-1.5}$ for jetliners and business jets with MTOW between 10 and 600 metric tonnes. Surprisingly, even though Equation (11.3) uses only four parameters, it has a standard deviation better than 6% and it is accurate for small as well as large aircraft. One explanation for this improved accuracy is a partial cancellation of positive and negative errors of the two previous statistical expressions. Moreover, most airplanes in the category considered operate in the same Mach number range leading to similarly optimized wing designs. Equation (11.3) is therefore sufficiently accurate to be used as a first step to start the iterative wing weight prediction method derived in this chapter.

11.1.2 Quasi-Analytical Weight Prediction

The sophistication of analysis methods used in advanced design (AD) increases when the design progresses through conceptual, preliminary and detail design stages. However, lack of time and detailed geometric data during the initial design stage often prohibits application of high fidelity CFD methods for computing aerodynamic loads and FEM for sizing the structure. Nevertheless, even in the initial conceptual phase the designer must use a validated weight prediction method which is design-sensitive and easy to apply – that is, it should not require

---

[1] Airplane wings designed with tip extensions such as winglets often have their span defined as the distance between the extended tips.

a large amount of (guessed) input data and expensive computations. As the computationally simple quasi-analytical approach proposed in this chapter complies with these requirements, it can be considered as a medium fidelity design tool. Application to several existing airplanes has shown that their wing weights are realistically predicted although a standard deviation significantly better than existing Class II methods is not claimed.

Attempts to make an accurate prediction of wing structure weight have been made through the years. During the 1950s, F.R. Shanley and M.E. Burt published methods based on analytical lift distributions, strength and stiffness considerations and systematic experiments [1, 9]. Their approaches paved the way to the (multiple) station analysis method enabling weight engineers to extend the method to components designed to resist different loading cases. A station analysis method is based on a computational procedure by which the amount of material required to resist the most significant loads is determined at a number of stations along the span. The material per unit span is integrated numerically to yield the basic (minimum) weight of the primary (box) structure. Penalties are made to allow for non-ideal structures and practical requirements unrelated to the aerodynamic loads, such as non-tapered and minimum gauge skins, joints, inspection hatches, etc. The significant weight contribution of secondary structures – leading edges, trailing edges, high-lift devices, flight control surfaces, and attachment structures for engines and landing gears – is usually obtained from semi-empirical analysis.

A Class II weight prediction used in the preliminary design of a modern aircraft wing is based on the station analysis approach and incorporates modifications required for structures with a specified lifetime requirement. However, the large amount of detailed data required for such a method discourages designers to apply them in conceptual stage. The present approach is based on first principles applying to wing box sizing and statistics for 1-g stress levels and secondary structure. The method was originally derived in [33] and its application resulted in several improvements and extensions. Primary structure weight is predicted with classical beam theory, using analytical integration along the span of material required to resist bending and shear. Allowances are made for stiffness required to cope with aero-elasticity and reduced stress levels for increasing lifetime. The summary in Section 11.9 describes a straightforward procedure for implementing the method. The user may choose between a basic method for a straight tapered wing and a more refined approach applying to a wing with kinked leading and trailing edges and a central carry-through structure. Another option is offered to select material properties different from Al-alloys and to account for winglets at the tips. The method should not be used without major modifications to unusual concepts such as boxplanes and the blended wing body (BWB) concept. In accordance with the SI system, weight components are stated in newtons or kN, as indicated by the derived equations. In order to minimize the possibility of confusion, some statistics-based equations are made more accessible to users of the Imperial System by scaling weights and geometry with reference values.

11.2   Methodology

11.2.1   Weight Breakdown and Structural Concept

The wing structure of conventional transport and business aircraft is composed of a large number of components classified in Figure 11.1. The primary structure is configured as a closed box beam featuring a multi-rib design concept with integral fuel tanks. Its main elements are upper and lower stiffened skin panels, a front spar, an aft spar, and ribs. The box

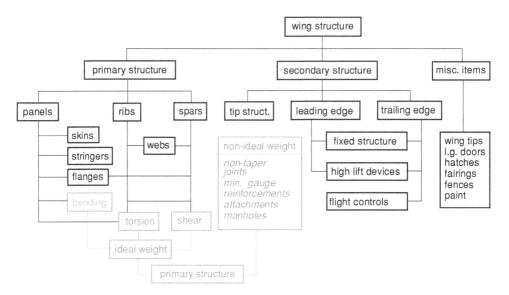

Figure 11.1   Subdivision of functional and hardware components of a civil aircraft wing structure

chord is typically between 45 and 50% of the wing chord for short- and long-range aircraft, respectively. The primary structure of most sweptback wings can be subdivided into inboard and outboard sections and a carry-through structure known as the centre section (CS). The dominant aerodynamic force is the lift on the exposed wing outboard of the wing/fuselage junction. Moreover, inertial forces act on the wing structure, its fuel contents and attached fixed masses. Since they oppose the wing bending due to lift, these forces entail a significant weight reduction which is known as inertia relief.

The box beam accommodates aerodynamic, inertia and ground loads acting on the wing and collects loads from secondary structures as well as wing-mounted items such as engines, nacelles and pylons. The load normal to the wing plane is transferred to the body structure by attachments at the structural root (SR, index sr) consisting of reinforced ribs where the CS is mated to the outboard sections. The in-plane load on the wing is a small fraction of the normal load which is ignored in most weight prediction methods – it is demonstrated that this is not always justified. The combination of all loads results in bending, shear and torsion of the box beam. Most of its weight is determined by the amount of material required to resist the critical loading conditions. Non-ideal weight increments are required to comply with principles of fail-safety, damage tolerance and manufacturing, as well as operational and maintenance considerations. Ground forces are exerted by landing gears through their attachment structure.

Secondary structure consists of components located to the fore of the front spar and behind the aft spar. Some components are fixed structures whereas high-lift devices and flight controls are movable. Depending mainly on the size and complexity of the movable devices and their mechanisms and supports, the summation of all secondary weight components forms a substantial fraction up to 35% of the wing weight. This weight cannot readily be predicted by means of first principles; hence, statistics have to be used in the conceptual design stage. Some of them can be viewed as non-ideal primary structures. Inspection hatches are classified as non-ideal box structure, flap support fairings are counted as components of flaps and controls. Most manufacturers do not classify a large wing/fuselage fairing as a component of the wing structure. Although the weight of miscellaneous items is not always considered as a separate category, it is mentioned here for completeness.

11.2.2 Basic Approach

The present weight prediction method is applicable to conventional primary structures predominantly made of stressed skin construction. Referring to the functional classification in Figure 11.1 and the terminology in Figure 11.2, the following basic principles are used for a generic weight prediction methodology:

- The method applies to conventional high aspect ratio cantilevered wings with a primary structure of predominantly Al-alloy. Guidelines are given to correct for the use of composite materials.
- The wing is treated as a statically determined rigid structure. A tentative allowance is given to ensure adequate stiffness resisting aero-elastic deformation.
- Loading cases considered are steady manoeuvring and atmospheric gusts, resulting in two weight predictions from which the critical case is selected. The theoretical minimum weight of the fully-stressed box structure is called the ideal weight.
- Wing bending is absorbed by the top skin compression panels and lower tension panels. Covers plates are stiffened by stringers and spar flanges which are 'buttered' with the skins

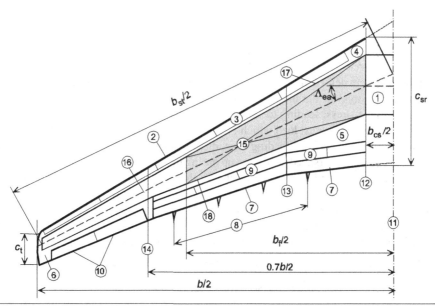

1 centre section 2 leading edge 3 slats 4 fixed leading edge 5 flap shroud 6 aileron shroud 7 trailing edge flaps 8 flap supports 9 lift dumpers / spoilers 10 ailerons 11 centreline 12 structural root 13 kink 14 reference chord 15 fuel tank 16 elastic axis 17 front spar 18 aft spar

Figure 11.2 Geometric definitions and nomenclature for a typical high-speed aircraft wing structure

to form an equivalent skin thickness (index es). Simple plate theory is used for the analysis of panel cross-section and weight.

- Shear loads are transmitted by stiffened spar webs using an equivalent skin thickness.
- Torsion is not taken into account explicitly. Instead, the box weight is corrected to ensure sufficient torsional stiffness.
- Rib weight is estimated by an empirical approach based on first principles.
- The emphasis is on accurate prediction of the stiffened panel loads at the wing root where the maximum bending moment occurs. The required material is obtained by analytical integration of the box weight per unit span.
- Non-ideal weight allows for local reinforcements near attachments introducing discrete forces and penalties due to minimum-gauge skin thickness, splices, joints, manholes, landing gear and engine attachments.
- Secondary structure weight is estimated for each category separately from statistical information and functional parameters.

11.2.3 Load Factors

The aerodynamic load considered explicitly in wing weight prediction is the normal load due to wing lift L causing bending, shear and torsion. The highest lift loads occur during manoeuvring and flight into a vertical gust. They are specified in relation to the aircraft gross weight W_G (GW) by the load factor,

$$n \overset{\text{def}}{=} \frac{L}{W_G} \tag{11.4}$$

The airframe must be designed to withstand limit and ultimate loads specified in the airworthiness requirements relevant for the aircraft category under consideration. The limit load must not lead to permanent deformation corresponding to the yield stress. Nowhere in the structure, may the material stress at the ultimate load – which is equal to 1.5 times the limit load – exceed the ultimate stress, leading to failure. Relatively small passenger aircraft with a low wing loading experience a gust load which is in excess of the manoeuvre load. Consequently, in order to determine the critical loading case, diagrams specifying gust and manoeuvre load factors versus equivalent airspeed are drawn up during the preliminary design phase for several loading conditions and airplane configurations. In the context of wing weight prediction we consider only loads caused by upward lift for the en-route configuration.

Manoeuvre loads (index man) are exerted during steady turns and pull up manoeuvres. Transport aircraft do not engage in extreme manoeuvres because of passenger comfort. FAR Chapter 25.337 specifies a positive manoeuvre limit load factor for speeds up to the maximum cruise speed

$$n_{\text{man}} = 2.50 \quad \text{for} \quad W_{\text{MTO}} \geq 50\,000 \tag{11.5}$$

or

$$n_{\text{man}} = 2.10 + \frac{24\,000}{W_{\text{MTO}} + 10\,000} \leq 3.8 \quad \text{for} \quad W_{\text{MTO}} < 50\,000 \tag{11.6}$$

with forces in lbf. For a given load factor, the normal wing load is highest when the airplane is flown at its MTOW. However, for a given GW, the zero fuel weight W_{ZF} (ZFW) depends on the payload (PL). Section 11.4 shows that the highest wing bending moment occurs when the ZFW is taken at its maximum value MZFW; in other words: the aircraft takes off with the maximum allowable payload.

Gust loads (index gust) are derived from the assumption that the aircraft flies through a sharp-edged upward gust with a derived (equivalent) gust speed U_{DE} increasing the angle of attack – downward gusts are usually not critical for most of the wing structure. The gust load factor is computed as follows:[2]

$$n_{\text{gust}} = 1 + \Delta n = 1 + K_{\text{g}} \frac{0.5 \, \rho_{\text{sl}} \, U_{\text{DE}} \, V_{\text{EAS}} \, S}{W_{\text{G}}} \frac{dC_L}{d\alpha} \tag{11.7}$$

where $\rho_{\text{sl}} = 1.275$ kg m^{-3} and EAS denotes equivalent airspeed. Since Equation (11.7) is derived for a quasi-steady gust load condition the empirical gust alleviation factor

$$K_{\text{g}} = \frac{0.88 \mu_{\text{g}}}{5.3 + \mu_{\text{g}}} \quad \text{with} \quad \mu_{\text{g}} \overset{\text{def}}{=} \frac{2 W_{\text{G}} \, b}{\rho \, g \, S^2} \left(\frac{dC_L}{d\alpha} \right)^{-1} \tag{11.8}$$

is introduced to correct for dynamic effects – the aircraft pitching and vertical motion and the time lag during which lift is building.

Many wing weight prediction methods assume *a priori* that manoeuvre loads are more critical than gust loads. This is not necessarily true for propeller aircraft and business jets with relatively low wing loading and maximum subsonic speed. Combining Equations (11.7) and (11.8) shows that the manoeuvre and gust load factors are equal for

$$\frac{W_{\text{G}}}{S} = \frac{dC_L}{d\alpha} \left(0.44 \frac{\rho_{\text{sl}} \, U_{\text{DE}} \, V_{\text{EAS}}}{n_{\text{man}} - 1} - 2.65 \frac{\rho \, g \, S}{b} \right) \tag{11.9}$$

The gust load factor exceeds the manoeuvre load factor for wing loadings below this value. This condition increases the sensitivity of wing weight to wing area variation.

Gust velocities featuring in discrete gust design criteria are specified in Figure 11.3 as linear functions of the altitude.[3] The highest gust velocity occurs at the design airspeed for maximum gust intensity V_B. Lower gust velocities apply to the maximum cruise speed V_C and the design diving speed V_D as specified by the aircraft's flight envelope. The highest gust load factor is determined by the highest value of $U_{DE} V_{EAS} dC_L / d\alpha$. Since all three parameters of this product are affected by the flight speed and altitude, the critical condition is not immediately obvious. A starting point is selecting the design cruise speed V_C @ 20 000 ft (6 100 m) altitude where the gust velocity amounts to 15 m/s as the critical condition. A more thorough approach requires drawing up the gust envelopes depicted in FAR/JAR Chapter 25.333 for

---

[2]The present simple method for computing gust load factors is intended for initial analysis. It is not as complete and accurate as the method used at large aircraft companies.
[3]Transport aircraft certified before 1990 comply with more stringent gust velocities.

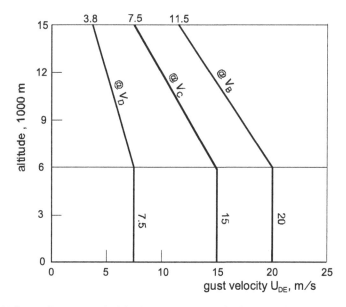

Figure 11.3   Derived gust velocities for discrete gust criteria according to FAR 25.341 (a)

several altitudes. The lift curve slope is required for computing the mass parameter and the gust load factor. It can be predicted with the DATCOM method [18] for straight-tapered airfoils,

$$\frac{dC_L}{d\alpha} = \frac{2\pi A}{\left\{(\tan^2 \Lambda_{0.50} + 1 - M^2)A^2 + 4\right\}^{1/2} + 2} \quad \text{rad}^{-1} \tag{11.10}$$

This equation does not account for transonic lift divergence. The maximum level flight speed of a transonic airliner will occur near the force break where Equation (11.10) underestimates the wing lift gradient by 5 to 10%. On the other hand, the lift on the complete aircraft is generally less than that of the exposed wing whereas the latter determines the wing load.

11.3   Basic Wing Box

The basic weight analysis is applied to a simplified wing geometry. In particular, the presence of the fuselage and the carry through structure are ignored, the wing is assumed to have no leading and trailing edge kinks and tip structures such as winglets and sheared tips are ignored. Hence, the wing is schematized into a straight-tapered ruled surface, an acceptable simplification when the major effects of wing geometry variation are required. Refinements are discussed in Section 11.8.

11.3.1   Bending due to Lift

Normal bending of the wing is caused by the upward aerodynamic force component normal to the wing plane caused by angle of attack increments. It can be shown that, within the flight envelope of a civil aircraft, this normal force is only slightly less than the lift. Hence, it is conservative to use the lift instead of the normal pressure force component for computing the

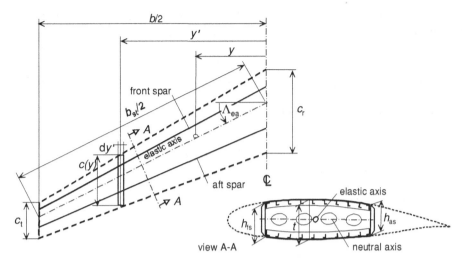

Figure 11.4 Nomenclature for computing the bending moment due to lift and the section efficiency factor

bending load. With reference to the notations in Figure 11.4, the bending moment due to lift (index BL) at a lateral wing station at distance y from the plane of symmetry is obtained from integration of the lift outboard of it,

$$M_{\text{BL}}(y) = \int_y^{b/2} \frac{y' - y}{\cos \Lambda_{\text{ea}}} \, \mathrm{d}L(y') \tag{11.11}$$

The sweep angle Λ_{ea} of the elastic axis (EA) is nearly equal to the mean sweep angle of a straight line midway between the front and the aft spars. The lift contribution $\mathrm{d}L(y')$ of each section acts at the centre of pressure (CP, index cp) which is assumed to coincide with the elastic axis. The lift distribution is defined by a generalized circulation function,

$$\gamma \stackrel{\text{def}}{=} \frac{c_l \, c(y) \, b}{C_L \, S} \tag{11.12}$$

The lift contribution at station y can be written as

$$\mathrm{d}L(y) = \frac{L}{2} \gamma \, \mathrm{d}\eta \quad \text{with} \quad \eta \stackrel{\text{def}}{=} \frac{y}{b/2} \tag{11.13}$$

and the bending moment due to lift is

$$M_{\text{BL}}(y) = \frac{L \, b_{\text{st}}}{4} \int_\eta^1 \gamma'(\eta' - \eta) \, \mathrm{d}\eta' \tag{11.14}$$

The structural span (index st) is defined as total length of the structural box measured along the elastic axis,

$$b_{st} = \frac{b}{\cos \Lambda_{ea}} \tag{11.15}$$

The bending moment at the wing root is

$$(M_{BL})_{y=0} = \frac{L}{4} b_{st} \int_0^1 \gamma \eta \, d\eta \tag{11.16}$$

and the resulting lift on a half wing acts at the CP with lateral location

$$\eta_{cp} = \int_0^1 \gamma \eta \, d\eta \tag{11.17}$$

The bending moment distribution along the span is

$$M_{BL}(y) = \frac{L}{4} \eta_{cp} b_{st} I_1(\eta) \tag{11.18}$$

with the bending moment function defined as

$$I_1(\eta) \stackrel{\text{def}}{=} \frac{M_{BL}(y)}{(M_{BL})_{y=0}} = \frac{\int_\eta^1 \gamma'(\eta' - \eta) \, d\eta'}{\int_0^1 \gamma \, \eta \, d\eta} \tag{11.19}$$

The lift distribution is affected by the following geometric properties and flight conditions.

- Taper moves the CP inboard, thereby reducing the bending moment at the root. For this reason, a more triangular lift distribution is applied than the elliptic distribution favoured for minimum drag.
- Aerodynamic washout decreases the effective angle of attack of the outboard wing and reduces the bending moment at the centreline proportional to the amount of washout.
- Compressibility affects the pressure distribution along the chord and the span. This effect is difficult – if not impossible – to quantify at the conceptual design stage. However, the compressibility effect on the lift curve slope is essential for computing the gust load.
- Flow interference between the wing and other airplane components can have a significant effect on total wing lift and its distribution.

Classical lifting line theory or vortex-lattice methods are suitable to compute the lift distribution for straight or swept wings, respectively. Panel methods are required for wing/body combinations when interference effects have to be modelled. This requires a more detailed geometry than is normally available in conceptual design. In order to illustrate the use of the bending moment function $I_1(\eta)$ the following elementary lift distributions are considered.

- If the two-dimensional lift coefficient is constant along the span, the circulation function equals the (relative) chord distribution,

$$\gamma = \{1 - \eta(1 - \lambda)\} \frac{2}{1 + \lambda} \qquad (11.20)$$

where $\lambda = c_t/c_r$ denotes the taper ratio. Substitution into Equation (11.17) yields the CP location

$$\eta_{cp} = \frac{1 + 2\lambda}{3(1 + \lambda)} \qquad (11.21)$$

and the lift distribution function according to Equation (11.19) becomes

$$I_1(\eta) = \{3(1 - \eta)^2 - (1 - \lambda)(2 - 3\eta + \eta^3)\} (1 + 2\lambda)^{-1} \qquad (11.22)$$

This function is plotted in Figure 11.5 for several taper ratios. It is found that these curves are accurately approximated as

$$I_1(\eta) = (1 - \eta)^{3 - 2\lambda + \lambda^2} \qquad (11.23)$$

- An elliptical lift distribution is defined as

$$\gamma = \frac{1}{\pi}\sqrt{1 - \eta^2} \qquad (11.24)$$

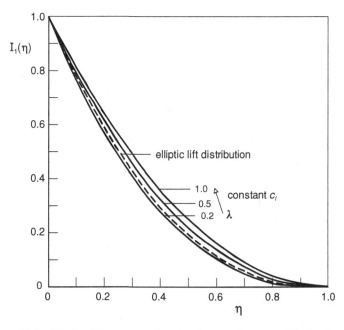

Figure 11.5   The bending moment function for two elementary lift distributions

with the CP location according to Equation (11.17),

$$\eta_{cp} = \frac{4}{3\pi} = 0.424 \qquad (11.25)$$

For this case, the lift distribution function according to Equation (11.19) is found from integration,

$$I_1(\eta) = (1 - \eta^2)^{0.5}(1 - 0.5\,\eta^2) - 1.5\,\eta\,(\cos\eta)^{-1} \qquad (11.26)$$

Figure 11.5 shows that this result is very similar to the previous case and it can be accurately approximated by

$$I_1(\eta) = (1 - \eta)^{2.39} \qquad (11.27)$$

This result is identical to Equation (11.23) for $\lambda = 0.376$, resulting in minimum induced drag for a straight-tapered planar wing.

The bending moment due to lift according to Equation (11.18) is used for computing the wing panel loads. For usual taper ratios, the function $I_1(\eta)$ appears to be relatively insensitive to the taper ratio and the lift distribution. Hence, the problem has been shifted to computing the CP location. A well-known approximation for a straight wing is taking the mean value of the two elementary cases considered above, resulting in

$$\eta_{cp} = \frac{2}{3\pi} + \frac{1 + 2\lambda}{6(1 + \lambda)} \qquad (11.28)$$

Sweepback can be accounted for by adding $\Delta\eta_{cp} = 0.02\sin\Lambda_{0.25}$ whereas the effective washout due to bending flexure shifts the CP inboard. The net effect on the CP location depends on the wing's flexibility – since it is difficult to compute in conceptual design, it has been ignored.

Another effect on the CP location is associated with the variation of lift distribution at load factors different from the 1-g flight condition. An aerodynamically optimized wing has a near-elliptical basic lift distribution in cruising flight by giving it an optimum angle of incidence (wash-out) variation along the span. At high angles of attack, the additional lift depending on the chord distribution is dominating. In order to account for this effect, the following expression is proposed for the effect of the load factor on the CP location:

$$\eta_{cp} = \frac{1}{3n}\left\{\frac{4}{\pi} + (n - 1)\frac{1 + 2\lambda}{1 + \lambda}\right\} \qquad (11.29)$$

When during a manoeuvre or a gust, the load factor is increased, the CP moves inboard of $\eta_{cp} = 0.4244$, provided $\lambda < 0.376$. For instance, a wing with taper ratio $\lambda = 0.2$ loaded to $n_{ULT} = 3.75$ enjoys 6% root bending moment reduction relative to an elliptic lift distribution.

11.3.2   Bending Material

The upper and lower stiffened skin panels including spar flanges (Figure 11.4) resist bending by normal loads leading to tensile and compressive stresses. The variation of stress along the span for resisting the bending due to lift (index BL) is

$$|\sigma(y)| = \frac{M_{\mathrm{BL}}(y)\,t(y)}{2I_z(y)} \tag{11.30}$$

where the neutral axis has been taken midway between the upper and lower panels. The bending moment of inertia I_z is obtained by replacing the panels by an equivalent skin with panel cross-section area $A(y)$ and equivalent thickness $\delta(y)$ at an effective mutual distance $\eta_t t(y)$, with

$$\eta_t \stackrel{\mathrm{def}}{=} \sqrt{\frac{2I_z(y)}{A(y)\,t^2(y)}} \tag{11.31}$$

This effectiveness factor depends on the wing section shape, the location of the front spar (index fs) and the aft spar (index as), and on panel geometry where the nomenclature (Figure 11.2) applies to the wing root. Methods for computing η_t in [1] and [22] require detailed input and apply to obsolete wing sections. Instead, the following first-order estimation is proposed instead:

$$\eta_t = \sqrt{\frac{1 + (h_{\mathrm{fs}}/t)^2 + (h_{\mathrm{as}}/t)^2}{3}} - t_{\mathrm{sk}}/t \tag{11.32}$$

where t_{sk}/t is the smeared skin thickness as a fraction of the section thickness. A typical value of this fraction is 0.025. If the section thickness at the spars is not known, the value of η_t may be assumed to be equal to 0.81 for subsonic and 0.84 for transonic (supercritical) sections. The required panel cross-section area follows from the two previous equations,

$$A(y) = \frac{M_{\mathrm{BL}}(y)}{\eta_t(y)\,|\sigma(y)|} \tag{11.33}$$

and the total weight of the upper and lower wing box covers is

$$W_{\mathrm{BL}} = 2\rho g \int_0^{b_{\mathrm{st}}/2} A(y)\,\mathrm{d}y \tag{11.34}$$

with ρ denoting the material's specific density. The maximum bending moment is exerted when the lift equals the ultimate load factor times the gross weight W_{G}. Combining Equations (11.18) and (11.33) yields the weight of material required to resist bending due to lift,

$$W_{\mathrm{BL}} = 0.5\, n_{\mathrm{ULT}}\, W_{\mathrm{G}}\, b_{\mathrm{st}}\, \frac{\eta_{\mathrm{cp}}}{\eta_t}\, \frac{b_{\mathrm{st}}}{2\,t_{\mathrm{r}}} \left\{ \frac{\rho g}{\sigma_{\mathrm{t}}(y=0)}\, (I_2)_{\mathrm{t}} + \frac{\rho g}{\sigma_{\mathrm{c}}(y=0)}\, (I_2)_{\mathrm{c}} \right\} \tag{11.35}$$

The indices t and c refer to tension and compression stress, respectively. The integral

$$I_2 = \int_0^1 I_1(\eta) \frac{(\sigma t)_{\eta=0}}{\sigma(\eta)t(\eta)} \, d\eta \qquad (11.36)$$

is developed differently for the lower (tension) and upper (compression) panels:

- The theoretical minimum weight of lower panels is obtained by assuming that the tensile stress is constant along the span. The allowed tensile stress is either the (fatigue) stress limit determined by the aircraft's specified lifetime, or the maximum tensile stress of the applied material. The basic approach assumes a constant thickness ratio along the span; hence, the variation of thickness taper $t(\eta)/t(\eta = 0)$ equals the planform taper. When using the lift distribution functions discussed previously it is found that $(I_2)_t = 0.36$ for most practical taper ratios. It is noteworthy that this value is 10% higher than $(I_2)_t = 1/3$ for a parabolic variation of material weight along the span.
- The allowable compressive stress in the upper panels is mostly based on the criterion that column buckling is to be avoided. The limit is assumed proportional to the square root of the local structural index defined as the panel load per unit width divided by the rib distance. For lightly loaded structures the tangent modulus is constant and for constant ratio of box chord to wing chord and constant rib pitch along the span we have

$$\sigma_c \propto \sqrt{\frac{M_{BL}(y)}{t(y)\,c(y)}} \qquad (11.37)$$

with the factor of proportionality depending on the stiffening configuration. The result for constant section thickness ratio is

$$(I_2)_c = \int_0^1 \sqrt{I_1(\eta)} \, d\eta \qquad (11.38)$$

Similar to the lower surface panels, the lift distributions considered do not lead to significant variation in this integral. A good average value is $(I_2)_c = 0.45$. For highly loaded structures the compressive stress is not allowed to exceed the maximum compressive stress. See Section 11.7 for more details.

Equation (11.35) shows that the weight of material required to resist bending is proportional to the cantilever ratio, broadly defined as the ratio of the structural semi-span to the maximum box thickness at the root,

$$R_{cant} = \frac{b_{st}}{2t_r} = \frac{b}{2t_r \cos \Lambda_{ea}} = \frac{A(1+\lambda)}{4(t/c)_r \cos \Lambda_{ea}} \qquad (11.39)$$

where $A = b^2/S$ denotes the aspect ratio. The last term on the right-hand side of Equation (11.39) features in several published wing weight prediction formulas, for instance, Equation (8.1). Introducing R_{cant} into Equation (11.35) leads to

$$W_{BL} = (I_2)_t \frac{\rho g}{\overline{\sigma_r}} n_{ULT} W_G R_{cant} \frac{\eta_{cp}}{\eta_t} b_{st} \qquad (11.40)$$

with the mean stress level at the root defined as

$$\frac{1}{\overline{\sigma_r}} = 0.5 \left\{ \frac{1}{\sigma_t} + \frac{1.25}{\sigma_c} \right\} \qquad (11.41)$$

Thanks to the relatively crude assumptions made so far, this result seems deceptively simple. However, its application requires knowledge of the distribution of fuel in the wing and allowable stress levels, subjects treated in Sections 11.4 and 11.7. It is worth noting that, for given material stresses, load factor and cantilever ratio, Equation (11.40) proves that the bending material weight fraction increases proportional to the wing span. This important consequence of the square cube law [15, 16] explains why, especially for large airliners, much effort must be spent in reducing wing weight.

11.3.3   Shear Material

The shear force due to wing lift is transferred by spar webs which are assumed to be of the full-depth type stiffened diagonal tension webs – spar flanges are counted as material resisting bending to which they contribute approximately 5%. Similar to the bending moment, the shear force due to lift (index SL) at any station is obtained from integration of the lift outboard of that station

$$F_{SL}(y) = \frac{n_{ULT} W_G}{2} I_3(\eta) \qquad (11.42)$$

where

$$I_3(\eta) \overset{def}{=} \int_{\eta}^{1} \gamma' \, d\eta' \qquad (11.43)$$

The required shear material is assumed to be divided equally over the shear webs. Its area $A(y)$ is equal to the shear force divided by the (mean) shear stress $\tau(y)$,

$$A_{SL}(y) = \frac{F_{SL}(y)}{\tau(y)} \qquad (11.44)$$

For constant shear stress along the span, the total weight of material required to transfer lift to the wing root is

$$W_{SL} = 0.5 \, n_{ULT} W_G \, b_{st} \frac{\rho g}{\tau} \int_0^1 I_3(\eta) \, d\eta \qquad (11.45)$$

The integral on the right-hand side equals the lateral coordinate of the CP. Similar to the compressive stress in the top cover panels, the allowable shear stress decreases in outboard direction. Its mean value is assumed equal to 50% of the mean centreline stress in the upper and lower cover panels. With an additional 20% for web stiffening, the shear material weight amounts to

$$W_{SL} = 1.20\, n_{ULT}\, W_G \eta_{cp}\, b_{st} \frac{\rho g}{\sigma} \tag{11.46}$$

11.3.4 In-Plane Loads and Torsion

In deriving the ideal basic box weight it was assumed that bending and shear loads due to upward lift determine the material required. In reality, coping with in-plane bending and shear causes a weight penalty that cannot be ignored.

- In addition to the normal load N, the wing experiences a tangential forward force which equals the leading edge suction force $L \sin \alpha$ minus the wing drag. Its magnitude can be considerably larger than expected. For instance, at $15°$ angle of attack the in-plane aerodynamic force amounts to roughly 20% of the lift. The resulting in-plane bending and shear loads entail shear stresses and increased tensile stresses behind and compressive stresses in front of the elastic axis.
- Straight wings experience a positive (nose-up) torsional moment which is caused by the location of the CP forward of the EA. Its magnitude is approximately equal to the bending moment due to lift divided by the aspect ratio. In addition, wing camber and twist cause a (generally negative) torsion which achieves its highest value at the design diving speed V_D.
- Sweepback shifts the CP backwards relative to the root. This causes a negative torsion of the inboard wing counteracting the positive torsional load mentioned for straight wings. The net effect is zero when the CP coincides with the shear centre at the structural root. This is the case for a sweepback angle of the order of $10°$. For the more usual sweep angles the exerted torsion requires the inboard rear spar to be beefed up. The magnitude of this effect is highly dependent on the structural topology.

The present method accounts for these effects by increasing the cover panel loads at the wing root by 5%.

11.3.5 Ribs

Ribs give shape to the wing section and are designed to fulfil a multitude of functions. General ribs support and stabilize the skin – in particular compression loaded panels – and transmit distributed aerodynamic pressure load from the upper and lower panels to the spars. They also stiffen the wing against torsion and withstand crushing loads due to wing bending. Special ribs collect concentrated loads from high-lift devices, control surfaces, landing gears and engine pods and pass them to the primary elements of the box. Several special ribs act as baffles to prevent the fuel surging around when the aircraft manoeuvres. The bulkheads of the structural root transfer the normal loads to the fuselage. The mass of a rib depends on the structural topology and, since only a few ribs can be considered as typical, an analytical approach to their weight prediction is precluded. In the CS of a large airliner, the mass of general ribs is roughly 10 kg m^{-2}, equivalent to a smeared thickness of 3.6 mm – the outboard rib mass is about half this value. However, special ribs carrying concentrated loads can be two to three times heavier.

Even if the mass of a typical rib could be predicted accurately, the number of ribs must be known to compute their total weight. Several authors have attempted to optimize the rib pitch with the aim of minimizing the weight of compression panels plus ribs. The outcome of this approach appears to be unrealistic since in practice rib pitches are selected on other than minimum weight considerations such as the location of flap and engine attachments. Although the rib pitch varies along the span, the average pitch of 0.75 m is not greatly different between inboard and outboard sections. The prediction method suggested here applies to the general rib weight, including the wing-to-fuselage interconnection bulkhead and fuel tank boundaries, but does not include penalties for support structures (Section 11.5). Basic assumptions are:

- The weight per unit of rib area is constant along the span.
- For constant rib pitch, the number of ribs is proportional to the span. Rib weight is then proportional to the product of the mean wing chord, the mean section thickness, and the span. In other words: rib weight is proportional to wing volume.
- If the rib pitch is proportional to the mean wing thickness the number of ribs per unit span is inversely proportional to it. Rib weight is then proportional to the mean rib chord times the span, that is, wing area.

If the actual rib pitch is determined partly by the first and partly by the second assumption, the general rib weight amounts to

$$W_{rib} = k_{rib}\, \rho\, g S \left(t_{ref} + \frac{t_r + t_t}{2} \right) \qquad (11.47)$$

including the wing-to-fuselage interconnection and fuel tank bulkheads. Factors to be calibrated are: $k_{rib} = 0.5 \times 10^{-3}$ and $t_{ref} = 1$ m.

11.4   Inertia Relief and Design Loads

The B-52 wing was much larger than the B-47 wing and thus had a high volume between the spars. In the B-52 this internal wing volume was used as a fuel tank. There were also large fuel tanks in the body. The use of wing fuel saved a lot of wing weight. The B-52 also had relatively small outboard wing fuel tank pods.

—G. Schairer, July/August 1989

The inertia forces acting on the wing structure, its contents and attached (engine) masses oppose the lift and lead to a relief of bending and shear loads. The large amount of fuel in the wing yields a significant load relief when a long-range airplane takes off; however, this is no longer the case at the end of the flight. Since the payload depends on the mission and fuel varies during the flight, critical loading conditions forming the design case for the wing structure must be identified. The subject of this section is to derive weight-critical loading conditions and corrections of the ideal box weight derived in the previous section for given gross weight and load factor.

11.4.1   Relief due to Fixed Masses

The resultant of the inertial forces on the aircraft and its contents in flight with a constant load factor equals the lift. In the (hypothetical) case of an all-wing aircraft with lift and inertial forces distributed identically along the span, the wing is not subjected to bending and shear. Although this span loading effect is less dominant for realistic designs, it forms an incentive for allotting as much useful load as possible to the wing (Chapter 5). The fuselage group and its contents of a conventional TAW configuration are concentrated inboard of the wing/fuselage connection. The lift required to carry their combined inertial force causes major bending and shear loads on the wing, whereas the inertial relief due to the wing group and its contents reduces the wing structural weight by a fraction of not more than 20%. The computation of inertia relief due to components of the empty weight is straightforward. Major relief stems from the wing structure and wing-mounted engines, the effects of wing-mounted landing gears and system components are usually negligible.

The mass of the wing structure, fuel system and local flight control systems is denoted as the wing group (index wg) mass. Similar to the lift, its weight is distributed between the root and the tip. Since wing structure weight per unit chord decreases in outboard direction, whereas the lift per unit chord is more or less constant, the resulting inertia relief is concentrated inboard of the CP. The relative bending moment reduction is some 20% less than the wing weight fraction but it is compensated by the inertia relief due to systems inside the wing. The result is expressed as a relief factor on the root bending moment,

$$(R_{in})_{wg} = \frac{y_{wg} W_{wg}}{y_{cp} W_{MTO}} \approx \frac{W_w}{W_{MTO}} \tag{11.48}$$

Application of this expression requires knowledge of the wing weight and its distribution which happens to be the outcome of the wing structure weight prediction. Consequently, computation of the inertia relief is essentially iterative. A practical approach to avoid iteration is to input a Class I wing weight estimation from a statistical method such as Equation (11.3). A second approximation is then usually not required.

A wing-mounted engine (index eng) causes a constant shear force and a linear bending moment reduction inboard of its attachment. It is found that the inertia relief factor due to one engine on each side is approximately

$$(R_{in})_{eng} = 3 \frac{\eta_{eng}^2}{\eta_{cp}} \frac{W_{eng}}{W_{MTO}} \tag{11.49}$$

where η_{eng} denotes the dimensionless lateral coordinate of the engine attachment (Figure 11.6) and W_{eng} is the installed weight of one engine including nacelle, thrust reverser and pylon. If two engines are mounted per wing half, their contributions must be added. Typical inertia relief factors are 0.03 and 0.08 for twin and quadruple wing-mounted turbofans, respectively.

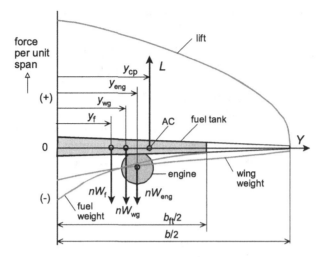

Figure 11.6   Forces contributing to inertia relief of wing bending due to lift

11.4.2   Weight-Critical UL and Design Weights

The effect of a filled fuel tank on the root bending moment is determined in a way similar to the bending moment due to lift. The inner tank bulkhead is assumed to coincide with the root station (Figure 11.7) and the upward moment due to lift and fuel (index LF) amounts to

$$(M_{\mathrm{LF}})_{y=0} = 0.25\, b_{\mathrm{st}}\, n_{\mathrm{ULT}}\, (\eta_{\mathrm{cp}}\, W_{\mathrm{G}} - \eta_{\mathrm{f}}\, W_{\mathrm{f}}) \tag{11.50}$$

or

$$(M_{\mathrm{LF}})_{y=0} = 0.25\, \eta_{\mathrm{cp}}\, b_{\mathrm{st}}\, n_{\mathrm{ULT}} \left\{ W_{\mathrm{ZF}} + \left(1 - \frac{\eta_{\mathrm{f}}}{\eta_{\mathrm{cp}}}\right) W_{\mathrm{f}} \right\} \tag{11.51}$$

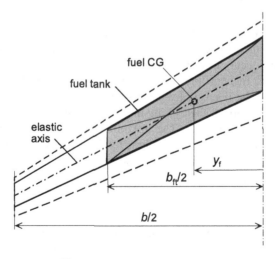

Figure 11.7   Fuel tank geometry

The lateral coordinate of the fuel CG in a filled straight tapered tank with taper ratio λ_{ft} is

$$\eta_{\mathrm{f}} = \frac{y_{\mathrm{f}}}{b/2} = \frac{1 + 2\lambda_{\mathrm{ft}} + 3\lambda_{\mathrm{ft}}^2}{4(1 + \lambda_{\mathrm{ft}} + \lambda_{\mathrm{ft}}^2)} \frac{b_{\mathrm{ft}}}{b} \tag{11.52}$$

For a completely filled full-span tank – the so-called 'wet wing' – the tank and wing spans and taper ratios are (nearly) equal and Equations (11.29) and (11.52) show that $\eta_{\mathrm{f}} < \eta_{\mathrm{cp}}$. For a given GW, reduction of the fuel load and its inertia relief is more than the reduction of the bending moment due to lift. The highest root bending moment therefore occurs when the aircraft has its maximum payload on board and takes off with MTOW,

$$(M_{\mathrm{LF}})_{y=0} = 0.25\, \eta_{\mathrm{cp}}\, b_{\mathrm{st}}\, n_{\mathrm{ULT}}\, W_{\mathrm{MTO}} \left\{ 1 - \frac{\eta_{\mathrm{f}}}{\eta_{\mathrm{cp}}} \left(1 - \frac{W_{\mathrm{MZF}}}{W_{\mathrm{MTO}}} \right) \right\} \tag{11.53}$$

The bending material reduction due to the inertia relief of fuel is derived from Equations (11.29), (11.52) and (11.53). If it is assumed that the variation along the span of bending material due to lift is proportional to the local chord and its reduction due to fuel relief is proportional to the square of the local chord, the solution is mathematically straightforward. However, the exact answer is cumbersome and the following inertia reduction factor for fuel is suggested if the wing and the wing box have the same taper ratios:

$$(R_{\mathrm{in}})_{\mathrm{f}} = 0.5\, \frac{b_{\mathrm{f}}}{b} \left(1 + \frac{3\lambda^2}{1 + 2\lambda} \right) \left(1 - \frac{W_{\mathrm{MZF}}}{W_{\mathrm{MTO}}} \right) \tag{11.54}$$

This equation should be applied so that b_{f}/b complies with the filled tank. It should also be noted that the amount of fuel corresponding to the MZFW is normally (much) less than the maximum fuel capacity. Hence, Equation (11.54) should not be applied to the wet wing since it does not represent the critical load condition. It is found that the bending material reduction due to fuel inertia relief is mostly less than 15% of the material required to withstand lift.

Further analysis is required if the inner fuel tank bulkhead is outboard from the wing root. For a given amount of fuel, this decreases the root bending moment up to the point that the fuel CG coincides with the wing CP. If the fuel tank CG is shifted outboard from the CP, the maximum root bending occurs when the tanks are empty and the wing structure weight must be computed for the zero-fuel condition. However, filled outboard tanks reduce wing bending in 1-g flight which is in the interest of reducing structural fatigue. Some turboprop aircraft have fuel tanks only outboard of their (wing-mounted) engines. A structural reserve fuel is then used to select the design condition. The airworthiness rules allow reduced maximum load factors; see FAR/JAR Chapter 25.343. From these results it is concluded that inertia relief due to fuel is important only for the manoeuvre-critical case.

11.5 Non-Ideal Weight

The ideal wing box weight is derived from static bending and shear loads and from allowable stress levels in every cross-section, sometimes referred to as a fully stressed structure. In a practical wing design, allowances are made for non-ideal features such as non-tapered skins,

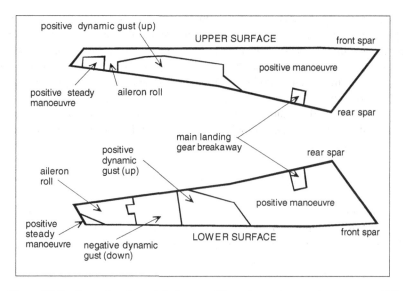

Figure 11.8 Example of panel design conditions for a primary wing structure [27]

partitioning into sub-assemblies, inspection hatches in the lower box cover plates, etc. Typical spar and panel surface design conditions for a large transport airplane in Figure 11.8 illustrate that several segments of the primary box are sized differently by manoeuvre and gust loads. Moreover, some structural elements are sized by aileron control and landing flap load cases, the landing impact load, dynamic loads, and torsional stiffness requirements to avoid critical aero-elastic effects.

Several publications refer to non-ideal weight penalties as 'non-optimum weight'. This is a misnomer because it suggests that non-ideal weight is not necessary and can be avoided if sufficient attention is paid to computational refinement. However, even if non-ideal weight can be minimized, it cannot altogether be eliminated. In terms of cost-effectiveness, one might even speak of 'optimum non-ideal weight'. Since most non-ideal weight penalties cannot be treated analytically, they must be derived from statistics and calibrated with data collected for manufactured aircraft. Several prediction methods compute the non-ideal weight by applying a correction factor to the ideal weight. This suggests that there exists a functional relationship between ideal and non-ideal weight terms. However, the design sensitivity is not properly represented in this way since most non-ideal weight is determined by design variables and sizing criteria conditions different from the ideal weight. The present alternative approach aims at avoiding these traps.

11.5.1 Non-Taper, Joints and Fasteners

Manufacturing constraints and cost considerations do not allow continuous adjustment to the theoretical optimum along the span of the skin/stringer configuration. For instance, minimum gauge sheet thickness (0.8 mm, typically) is applied in lightly loaded parts near the wing tips. Stepwise skin thickness adjustments can be made by subdividing a panel into sections joined by splices. Theoretically, there is an optimum number of steps for which the combined weight penalty due to non-tapered sheets and joints is minimized. Compared to constant thickness sheets, the application of linear sheet tapering and integrally machined panels reduces the non-taper weight penalty significantly. Penalties also occur where the wing is partitioned into sub-assemblies leading to major production joints and splices in cover panels. Depending on the wing size and its manufacturing concept, weight penalties add up to between 5 and 10% of the box weight. Since the major part of this non-ideal (index nid) weight is related to the surface area of the wing box, the following estimation can be made by applying a smeared thickness $\Delta\delta$ increment to panel skins and shear webs:

$$\Delta_{\mathrm{nid}} W = \delta_{\mathrm{nid}} \rho g \, \Sigma \, S_{\mathrm{box}} \approx 1.20 \, \delta_{\mathrm{nid}} \rho g \, S \tag{11.55}$$

with ΣS_{box} denoting the total projected area of box panels plus spar webs. For fabricated (built-up) structures based on high-strength fastener technology, a typical value is $\delta_{\mathrm{nid}} = 10^{-3}$ m. For bonded or integrally machined structures, this penalty can be reduced by 20% or 50%, respectively.

11.5.2   Fail Safety and Damage Tolerance

The ideal shear material weight has to be increased to allow for increased web thickness to cope with shear due to wing torsion and for provisions such as the need for web stiffeners and rib posts. Moreover, an essential requirement for civil aircraft states that the wing must be a fail safe and damage-tolerant structure. In particular, complete failure of a wing spar constitutes a potential hazard that must either be countered by an auxiliary spar or avoided altogether. The provision of crack stoppers is an effective measure to protect a shear web from complete failure, especially when it forms an integral structure. The combined effect of these provisions is estimated to be a 15% penalty on the basic shear material according to Equation (11.46).

$$\Delta_{\mathrm{nid}} W = 0.18 \, n_{\mathrm{ULT}} \, W_{\mathrm{G}} \eta_{\mathrm{cp}} \, b_{\mathrm{st}} \frac{\rho g}{\sigma} \tag{11.56}$$

11.5.3   Manholes and Access Hatches

Cutouts in structures invariably increase its weight because the structure adjacent to a cutout must be beefed up to redistribute the load. To allow inspection and maintenance of the structure and tank condition, the lower box cover features numerous manholes closed by inspection covers (index ic) which can be removed for access. Manholes cause weight penalties depending on the type of access. Inspection hatches in the outer wing are large relative to the chord and are stressed to reduce the weight penalty. Non-stressed hatches are applied mainly in the inboard and mid-wing sections and the carry-through structure. The weight penalty due to the reinforcement around manholes is estimated by assuming that: (a) the structural width of the lower box cover is effectively reduced by the cutout width; (b) the cover is reinforced by increasing its smeared thickness by the same amount as the cutout material; and (c) the inspection cover's smeared thickness is equal to the increased cover thickness. This leads effectively to a cover thickness increment along the structural span equal to the ratio of the cutout to the mean box width. If the mean box chord is assumed to be equal to 50% of the mean geometric chord, the ideal lower panel weight is corrected by a factor

$$R_{ic} = 1 + 2\frac{w_{ic}\, b_{st}}{S} \tag{11.57}$$

with a standard IC cutout width of 10 inch; $w_{ic} = 0.25$ m. Application of this correction yields a lower box cover weight penalty between 5% for large and 15% for small airliners. This result complies with the recommendation in [34] that the net weight penalty due to providing for a manhole with inspection hatch amounts to 1.5 times the cutout material weight.

11.5.4 Reinforcements, Attachments and Support Structure

Structural elements are required for mounting engines and landing gears to the wing box and connecting the wing to the fuselage.

- The installation of engines in pods under the wing leading edge requires pylon attachments, additional ribs, and reinforcements of spars, special ribs and covers. The following weight penalty is based on scanty statistical data for jet aircraft:

$$\Delta_{nid} W = 0.015(1 + 0.2 N_{eng}) W_{pp} \tag{11.58}$$

The number of wing-mounted engines is denoted N_{eng}, the power plant weight W_{ppp} includes the weight of engines, nacelles, thrust reversers and pylons (Section 8.3). An engine mounted to the outboard wing can have an appreciable effect on aero-elastic characteristics. Depending on the engine location relative to the elastic axis, this may lead to a weight reduction or a penalty.
- The landing impact load absorbed by wing-mounted landing gears requires provisions such as gear bulkheads, auxiliary spar structures transferring the impact loads to the fuselage, aft spar and rib reinforcements, and wheel well doors. The total non-optimum weight penalty is a fraction of the MLW, for example

$$\Delta_{nid} W = 0.006 W_{ML} \tag{11.59}$$

depending on the landing impact load which is affected by the impact load factor n_{land}. If the load factor and shock absorbing properties are known, the prediction can be refined to

$$\Delta_{nid} W = 0.0015\, n_{land} W_{ML} \tag{11.60}$$

This penalty is omitted if the gear legs are attached to the fuselage.
- Connecting the wing to the fuselage requires attachment structures, rib reinforcements and other hardware. For a continuous carry through structure this weight penalty amounts to

$$\Delta_{nid} W = 0.0003\, n_{ULT} W_{MTO} \tag{11.61}$$

11.5.5   Dynamic Over Swing

The wing of large jetliners is mostly designed by the manoeuvre load which is more critical than the gust load. Since during a manoeuvre the lift build-up is gradual, the dynamics will have little overall effect on the bending and shear loads. For the gust-critical case, however, dynamic effects may become important. As opposed to static gust load factors, the dynamic behaviour must be analyzed for flight through a standardized gust distribution, accounting for the degrees of freedom of the elastic structure [2]. Such an analysis needs knowledge of the complete structural design and yields the dynamic over-swing deflection for every station along the span, an analysis which is outside the scope of the present method. Instead of this, the gust load factor according to Equation (11.7) can be multiplied by a correction factor that should be obtained from past design experience and checked by a dynamic analysis downstream of the conceptual design. However, dynamic gust loads are usually concentrated on the outboard wing where extra stiffness is built in to cope with aero-elasticity. It is therefore unlikely that dynamic gust loads will entail a weight penalty in excess of the one to be treated hereafter.

11.5.6   Torsional Stiffness

In addition to static bending and shear, there may be additional considerations which require a beef up in the basic structure in specific areas, in particular the wing box in front of the ailerons. Some of these weight penalties are very design-specific, others are related to more general conditions. In principle, the need to avoid aero-elastic divergence when flying at a high dynamic pressure may cause a weight penalty depending on general design characteristics. When dealing with the wing lift distribution (Section 11.3), it was mentioned that there exists an interaction between the static air load distribution and wing distortion. Static deflections that tend to shift the lift inboard are found to be a function of the bending stiffness, those that tend to shift the lift outboard are a function of the torsional stiffness. The usual result is that there is a small inboard CP shift on sweptback wings and an even smaller outboard shift on straight wings. Other aero-elastic phenomena relevant to wing weight prediction are primarily affected by the sweepback angle and the dynamic pressure.

Figure 11.9 forms a qualitative indication of the speeds at which static aero-elastic divergence, bending-torsion flutter and aileron reversal occur dependent on the angle of sweep.

- For (nearly) straight wings, the divergence case is usually covered when the requirements for aileron reversal are met; the flutter case must be investigated separately.
- Whereas sweepback has a relieving effect in the flutter and divergence cases resulting from the outboard lift relief due to upward bending, aileron reversal usually becomes the critical design case. Some high-aspect ratio swept wings have been problematic with regard to aileron reversal. A solution has been found on the A310 by applying inboard high-speed ailerons in combination with spoilers.
- Divergence is a critical condition when designing a forward swept wing. Advanced composite construction is most adaptable to control the deflection characteristics by aero-elastic tailoring.

For a certain category of high speed airplanes, torsional stiffness requirements may become dominant and the extra structure required to safeguard against flutter may amount to as much as 20% of the wing weight. The location of outboard wing-mounted engines relative to the

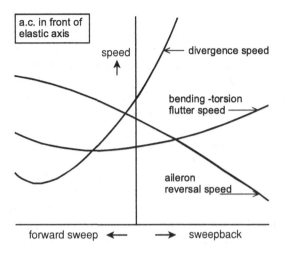

Figure 11.9   Influence of wing sweep on divergence, aileron reversal and flutter speed

wing's inertia axis has an appreciable effect on the flutter speed. When a high flight speed and a high wing aspect ratio are combined, weight penalties to avoid flutter are likely to be unavoidable. Several authors ([8], [14]) have proposed analytical criteria for the torsional strain energy in a linear torsion mode about the elastic axis required to avoid objectionable aero-elastic behaviour at high speed of the wing structure. These criteria have become obsolete with the introduction of supercritical wings, which have non-linear transonic aerodynamics. Nevertheless, achieving adequate torsional rigidity remains a valid principle for conceptual and preliminary design.

The basic effect of a torsional stiffness requirement on wing box panel weight is illustrated in Figure 11.10. Curve A indicates the optimum skin plus stringer equivalent thickness required to resist bending, curve B is just the skin thickness. The skin-to-stringer material ratio for optimum compression structures is between 40 : 60 and 50 : 50. Curve C depicts a possible

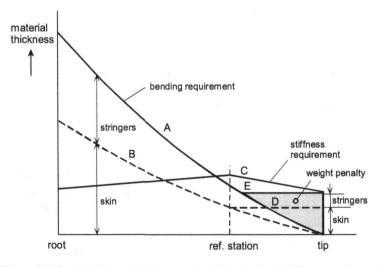

Figure 11.10   Material required to comply with a torsional stiffness requirement

variation along the span of skin thickness which provides the required torsional stiffness. The stiffness is defined in terms of the strain energy associated with a given torsional deflection. The reference station is usually at 70% semi-span outboard from the root. Since the amount of inboard wing skin material required to cope with bending exceeds the torsional requirement, the outboard skin thickness can be reduced from curve C to curve D. Adding stringers yields curve E, representing a mass distribution satisfying both bending and torsional requirements, with the hatched area defining the associated weight penalty. This penalty may be further reduced by increasing the stringer pitch in areas where the stiffness requirement dominates, resulting in modified skin-to-stringer material ratio to, for example, 55 : 45. Another option is increasing the rib pitch in combination with a skin thickness increment. Basically, this approach complies with design experience that a minimum weight penalty due to aero-elastic effects is obtained by a local increase in skin gauge over the outboard portion of the wing. Apart from skin thickening to comply with aero-elastic requirements, the outboard wing skin may have to be thickened to comply with a minimum gauge requirement, the transfer of aileron loads, dynamic gusts, etc. A detailed station analysis is required to find out which case is critical at any given lateral position.

The implementation of stiffness requirements is feasible only if the torsional stiffness variation along the span and details such as the location of the inertial and flexural axis are available. It was concluded from a literature survey that an empirical weight penalty for an Al-alloy box structure will have the following appearance:

$$\Delta_{\text{nid}} W = W_{\text{ref}} \frac{q_{\text{D}}}{q_{\text{ref}}} \left(\frac{b \cos \Lambda_0}{b_{\text{ref}}} \right)^3 \left(\frac{t}{c} \right)_{\text{ref}}^{-2} (1 - \sin \Lambda_{0.50})\{1 - (M_{\text{D}} \cos \Lambda_{0.50})^2\}^{-0.5} \quad (11.62)$$

with $q_{\text{ref}} = 30 \text{ kN m}^{-2}$ and $b_{\text{ref}} = 50 \text{ m}$. Leading edge and mid-chord sweep angles are denoted as Λ_0 and $\Lambda_{0.50}$, respectively, and the thickness ratio $(t/c)_{\text{ref}}$ refers to the reference wing chord just inside the aileron (Figure 11.2). The dynamic pressure q_{D} and the Mach number M_{D} at the design diving speed are obtained from the flight envelope. The calibration factor W_{ref} can be estimated from previous design information. If this is lacking, a typical $W_{\text{ref}} = 200 \text{ N}$ leads to a wing weight penalty of about 2% for regional jets up to 5% for wide-body airliners. This stiffness correction is most sensitive to the span and the thickness ratio.

11.6   Secondary Structures and Miscellaneous Items

> There is little evidence that specific weights of high-lift system components change noticeably with the size of the aircraft.
>
> —P.K.C. Rudolph in [36]

Even though secondary structure can contribute as much as 35% to the wing weight, published methods treat its prediction superficially due to the difficulty experienced in defining design criteria. Therefore, weight predictions are mostly based on statistical evidence using basic geometry and functional parameters as input. Since there is hardly any functional relation between secondary weight (components) and the airplane MTOW, statistical prediction methods may use the MTOW to account for the scale effect. On the other hand, it is not necessary to vary secondary structure weight (components) to account for minor MTOW variations during the design sizing process. This section presents equations for leading and trailing edge structure weight components and wing tip structures (Figure 11.1) based on first principles and statistical information.

Most presented equations are derived in [31] by regression analysis of some twenty civil aircraft in operation until the 1980s. Since they apply to basically Al-alloy structures it is recommended to calibrate component weights with information on recent aircraft, if available. Weight reduction factors obtainable with composite material applications are presented in Section 11.8. Most component weights are expressed in terms of specific weight Ω referred to the component's area in plan view – these areas are normalized by $S_{ref} = 10$ m^2. Since the basic secondary structures are lifting surfaces with a geometry somewhat similar to a wing section, a reference specific weight $\Omega_{ref} = 56$ N m^{-2} is defined equal to the hypothetical weight of a thin section built up from two Al-alloy skins with one millimetre thickness each.

11.6.1 Fixed Leading Edge

Fixed leading edge (index fle) structure (excluding the de-icing system) is the structure in front of the primary box remaining when all movable leading edge devices are removed. A typical fixed leading edge weighs between 4% and 7% of the complete wing. Leading edge surface panels are generally not designed to carry any part of the wing box bending and torsion loads. The primary design load is the aerodynamic surface pressure. The following expression for the specific weight has a standard deviation of about 10%:

$$\Omega_{fle} = 3.15 \, k_{fle} \Omega_{ref} \left(\frac{q_D}{q_{ref}} \right)^{0.25} \left(\frac{W_{MTO} \, b_{st}}{W_{ref} b_{ref}} \right)^{0.145} \tag{11.63}$$

with $q_{ref} = 30$ kN m^{-2}, $W_{ref} = 10^6$ N and $b_{ref} = 50$ m. The specific weight is referred to the leading edge plan view area with high-lift devices removed. Use $k_{fle} = 1.0$ if the leading edge carries no slats or Krueger flaps and $k_{fle} = 1.3$ to account for the strengthening required to support slats.

11.6.2 Leading Edge High-Lift Devices

Slats and Krueger flaps are the most frequently applied leading edge high-lift devices. Including the panel structure, actuator supports and tracks, their weight is referred to the plan view area in retracted position. Droop noses have been used in the past on a few civil airplanes. Compared to slats, droop noses are mechanically simple but aerodynamically less effective. Their specific weight is comparable to slats. Slats are primarily stressed to carry normal forces and their nose must protect the wing from rain erosion and hail damage. Closely spaced slat ribs are required to transfer the aerodynamic loads and there are heavy ribs where actuator supports and tracks are located. The tracks are usually made of high strength steel or titanium to react to the high bending loads on the extended slats. A complete slat system weighs typically 5% of the total wing weight.

The specific weight of slats is quite high as indicated by the following correlation which has a standard deviation of 14%:

$$\Omega_{\text{slat}} = 4.83\,\Omega_{\text{ref}} \left(\frac{S_{\text{slat}}}{S_{\text{ref}}} \right)^{0.183} \tag{11.64}$$

This weight is referred to the nested slat area and includes tracks and support but excludes slat actuation. Slats are not always fitted over the full span and their chords are usually between 15% and 20% of the wing chord. Krueger flaps are mostly used on the inboard section of a sweptback wing. Fixed geometry Krueger flaps are generally made of one piece of material. Their specific weight referred to the planform area in the retracted position amounts to roughly half the specific weight of a slat, but there is a large scatter due to the variety in the retraction mechanism. Variable camber increases the specific weight by about 60%.

11.6.3   Fixed Trailing Edge

The fixed trailing edge (index fte) is the structure aft of the rear spar remaining when trailing edge flaps and controls are removed. Spoilers and lift dumpers hinged and actuated from the rear spar are not included. However, moving shroud spoilers and lift dumpers are often highly integrated with the trailing edge structure and the distinction is difficult to make. In that case their combined weight has to be considered. Fixed trailing edge weight is of the order of 8% of the total wing weight. Trailing edge shrouds on large aircraft consists of skin panels, ribs supporting control surface hinges, intermediate supports and auxiliary beams. Part of the undercarriage support structure may also be integrated in this structure. The specific weight depends primarily on the overhang and on the configuration of flap supports. The specific weight based on the plan view area of trailing edge shrouds for single slotted flaps and ailerons is

$$\Omega_{\text{fte}} = 2.6\,\Omega_{\text{ref}} \left(\frac{W_{\text{MTO}}\, b_{\text{st}}}{W_{\text{ref}} b_{\text{ref}}} \right)^{0.0544} \tag{11.65}$$

with the reference weight and span used in Equation (11.63). This must be increased by 40 N m^{-2} for structures supporting single slotted Fowler flaps and double slotted flaps or by 100 N m^{-2} for supporting double-slotted Fowler and triple-slotted flaps. The trailing edge area excludes open shroud spoilers and depends to a large degree on the location of the rear spar, the presence of a trailing edge kink and the mechanical design of the flap support.

11.6.4   Trailing Edge Flaps

The aerodynamic effectiveness of high-lift devices at the trailing edge can be expressed as the lift coefficient increase at given angle of attack and deflection angle. It is affected mainly by the overall flap configuration and actuation, section and planform geometry. Flap loads are proportional to flap size, lift increment and dynamic pressure at the maximum extension speed. Since the complete system of trailing edge flaps weighs up to 20% of the total wing structure, any attempt to reduce flap weight and improve its prediction is worthwhile. A subdivision into subgroups – preferably for each flap segment – is essential to achieve the highest weight prediction accuracy. The following is a typical example for transport aircraft.

- Flap surfaces weight is determined by the air load and by the bending moment distribution, the latter being affected by the support structure lay-out. Another factor is the thickness/span ratio of a flap segment. The absolute thickness can be quite small for high speed wing flaps, in particular when they are retracted below the trailing edge shroud.
- Slotted flap supports and vane controls have a weight dependent on the type: fixed hinge, four-bar linkage or track systems. Their weight depends mainly on the chordwise extension and deflection angle in the fully extended position.
- Support fairings weight is proportional to wetted area.
- Shroud doors weight is proportional to projected area.
- Flap actuation is not included.

In the conceptual stage many flap system design criteria are still unknown. An attempt to develop a generalized weight prediction method that takes into account all considerations mentioned above is likely to result in a complex procedure requiring many input data, but with a mean prediction error no better than 20%. A more effective method is to use available information on a flap system of similar lay-out and size. The flap element weights are then scaled up or down to the actual size. If that is not feasible (due to lack of information), it can be assumed that the flap loads and specific weight are statistically related to the MTOW. The following specific weight of Al-alloy flaps is based on the plan view area of all flaps in nested position:

$$\Omega_{tef} = 1.7 \, k_{sup} \, k_{slot} \, \Omega_{ref} \{1 + (W_{MTO}/W_{ref})^{0.35}\} \qquad (11.66)$$

with $W_{ref} = 10^6$ N. The multiplication factors depend on the mechanical properties of the flap system:

- The factor k_{sup} represents the complexity of the primary flap motion support. It amounts to 1.0 for simple hinge external supports, 1.2 for link/track end supports, or 1.6 for Fowler flaps with hooked track external supports. An auxiliary trailing edge hinged flap increases k_{sup} by 0.2.
- The factor k_{slot} represents the number of slots and flap angle variability. It amounts to 1.0 for single-slotted flaps, 1.5 for double-slotted flaps with fixed nose vanes, and 2.0 for double-slotted flaps with articulating vanes.[4]

For wings equipped with different flap systems – for example, double-slotted inboard and single-slotted outboard – the specific weight and flap area are multiplied for each flap type separately. Trailing edge flap chords are between 25% and 35% of the local wing chord, their span is generally between 65% and 75% of the outboard wing span. If no details are available, the total nested area of uninterrupted flaps can be estimated at 20% of the wing planform area, the presence of inboard high speed ailerons brings this down to 16%. It is

---

[4]Triple-slotted Fowler flaps were used for the first versions of the Boeing 727, 737 and 747. They are mechanically complex and about 15% heavier than double-slotted systems.

noted that Equation (11.66) predicts a flap weight independent of segmentation. In reality, flap surface specific weight is reduced when more flap segments are used, while flap support and control weight will increase. As explained in [19], the number of flap segments is subject to optimization. However, an attempt to incorporate this into a practical weight prediction method is likely to fail.

11.6.5 Flight Control Devices

The weight of flight control devices includes the control structures, hinges and supports, and balance weights – it excludes actuators and controls. Similar to high-lift devices, the presented data on weight applies to Al-alloy structures. Application of composite materials reduces weight appreciably.

Aileron-specific weight depends primarily on the design diving speed, maximum deflection angles, size and thickness/chord ratio. However, this sensitivity is obscured by large variations associated with balancing and activation. For manually activated ailerons with a tab system, there is a substantial weight penalty due to mass balancing and aerodynamic balancing requiring large surfaces in front of the hinge. This may bring the total aileron weight up to 3% of the wing weight. Tabs and balancing masses are not required for triplex powered ailerons and their weight is about 1% of the wing weight, for power-boosted ailerons, this is typically 1.5%. Aileron design criteria and the method of balancing and activation depend on aileron size as reflected in the following empirical expression for total aileron weight:

$$\Omega_{ail} = 3.0 \, \Omega_{ref} \, k_{bal} \, (S_{ail}/S_{ref})^{0.044} \tag{11.67}$$

The area S_{ail} denotes the total planform area behind the hinge line of all ailerons in the neutral position which may vary between 3% and 5% of the wing area. The factor k_{bal} is 1.0 for unbalanced ailerons, 1.3 for aerodynamic-balanced and 1.54 for mass-balanced ailerons.

The spoiler group consists of spoilers, lift dumpers and air brakes fitted to the top surface of the fixed trailing edge structure. Two or more of their functions can be combined in one device. Spoiler weight, comprising control panels and the supporting bracket, is treated in similar fashion to aileron weight. The scanty available information suggests that

$$\Omega_{sp} = 2.2 \, \Omega_{ref} \, (S_{sp}/S_{ref})^{0.032} \tag{11.68}$$

Total spoiler and lift dumper plan view area S_{sp} is defined in the retracted controls position. For large jetliners this may amount to about 4% of the gross wing area.

11.6.6 Tip Structures

The complete tip assembly includes the structure, its attachments and provisions for tip lights. The great variety of tip shapes found on existing aircraft makes a statistical prediction of this weight component unavoidable. The following equation based on [31] is suggested for traditional tapered wings:

$$W_{tip} = 150 \left(\frac{W_{MTO}}{W_{ref}} \right)^{0.67} \tag{11.69}$$

with $W_{ref} = 10^6$ N. Its standard deviation of 33% is not objectionable for this small component. However, sheared wing tips and winglets are as heavily loaded aerodynamically as the wing itself and, depending on their size, their weight amounts to several percent of the wing weight. Since the structural layout of a tip extension is comparable to the fixed leading edge, Equation (11.63) is adapted to estimate the specific weight of a tip extension with length l_{tip},

$$\Omega_{tip} = 2.5 \, \Omega_{ref} \left(\frac{W_{MTO} \, l_{tip}}{W_{ref} l_{ref}} \right)^{0.145} \tag{11.70}$$

where l_{tip} is defined in Figure 11.12 and $l_{ref} = 5$ m. Unfortunately, statistical material available to check the accuracy of Equation (11.70) is not available. It is also emphasized that tip extensions increase the structural wing box span and therefore its bending load and weight. This effect is discussed in Section 11.8.

11.6.7 Miscellaneous Items

Miscellaneous items represent widely scattered weight components. Their summation depends on structural layout and on the manufacturer's empty weight breakdown. Typical items are paint, fuel tank sealant, undercarriage wheel well doors, jacking fittings, rivets, nuts and bolts, fences and flow spoilers, and wing tips. Fairings for wings without wheel wells are often relatively small, weighing less than 1% of the wing weight. Fairings for low wings housing undercarriage wheel wells can be quite large, weighing up to 5% of the wing structure. However, they are generally classified as a fuselage body component. If data is missing a total miscellaneous weight can be estimated by increasing the secondary wing weight by 10%.

11.7 Stress Levels in Aluminium Alloys

Structural material selection has a great impact on airliner design and is not just a matter of looking at the highest strength and stiffness values available. In addition to low empty weight, airplanes have service life requirements which will limit the freedom of choosing materials which can be used to perform the function. The allowable stresses for a given material are often determined by the proven ability to withstand minor damage in service without endangering structural integrity. The ground-air-ground (GAG) load cycle spectrum and gust loading experienced in 1-g flight cause the majority of the damage to transport aircraft structure. Variation in tension and stress concentrations lead to fatigue of structural material, reducing the operational lifetime and number of flight cycles. Many advanced designers initially express the allowable tensile stress of the lower box covers' surfaces in terms of a maximum 1-g stress in steady level flight. More refined analysis is based on the expected number of flights to be made during the aircraft's service life since this largely determines the allowable stress levels.

The present method requires knowledge of material properties for computing the material weight required to resist bending, shear and torsion, assuming that upward normal loads are the dominating design condition. They are combined in the term $\rho g / \overline{\sigma}$ at the wing root where maximum panel loads occur. Allowable stress levels depend largely on the type of material and, for compression panels, on the loading intensity, the structural configuration and fabrication technology. The present method is based on high-strength Al-alloys used in aircraft primary structures which are variants of the 2XXX- and the 7XXX-series with a specific weight $\rho g = 2\,796$ N m^{-3}. Typical yield tensile stresses are:

- 450 MPa for 2024-T3 (Dural) sheet
- 520 MPa for 7075-T6
- 550 MPa for 7150-T6, and
- 620 MPa for 7055-T7 materials.

Further developments of the 2XXX series alloys, such as 2324-T3 for plates and 2224-T3 for extrusions, have approximately 8% improved specific strength while durability and damage tolerance are also higher. These properties can be used to obtain a panel weight reduction, to design for an extended lifetime or both. Although the allowable stress levels mentioned hereafter are realistic, it is recommended to check them with structural design experts.

11.7.1   Lower Panels

Lower box covers are primarily sized to withstand application and relaxation of tensile stresses due to (static) upward bending. They are skin damage critical and, for this reason, the traditional material used is 2024-T3 which has good fatigue resistance. In order to cope with combined tension and shear loading, the ultimate tensile stress of 450 MPa for Dural is reduced by 5%. For an ultimate manoeuvre load factor of 3.75, this would yield a tensile stress in 1-g flight of 114 MPa – too high for most applications.

To prevent fatigue failure in wing structures of aircraft with a specified service life, an analysis must be performed dealing with frequency and magnitude of loads. Such a procedure results in an allowable stress dependent on the structural configuration and type of material [22]. The following recommendations are based on statistical data found in the literature.

- Short-range (regional) airliners must achieve a crack-free structure for more than 50 000 flights. For conventional fabricated and riveted skin-stringer panels in 2024-T3 alloy it is recommended to use a 1-g tensile stress no higher than 75 MPa. A stress between 80 and 85 MPa can be acceptable for adhesive-bonded structures which eliminate stress concentrations in rivet and bolt holes. Accordingly, for a gust-critical $n_{\text{ULT}} = 5.0$, the allowable tensile stress is between 375 and 425 MPa.
- Long-range airliners achieve a crack-free structure for about 25 000 flights. Integrally machined structures allow a 1-g stress of 110 MPa. For $n_{\text{ULT}} = 3.75$ this corresponds to an allowable tensile stress of about 410 MPa.

If service life and type of fabrication are not known, a conservative approach could be to select a maximum tensile stress between 375 MPa for a gust-critical and 410 MPa for a manoeuvre-critical wing structure.

11.7.2   Upper Panels

The upper box cover is designed to withstand primarily compression loads and is typically made of 7075-T6 alloy stiffened panels. Achievable compressive stress levels are based on the buckling behaviour of a rib-supported panel. Lightly loaded stiffened panels exhibit buckling failure in a flexural instability mode. For traditional skin-and-stringer combinations, a higher compressive stress is allowed with increasing loading intensity which is represented by the structural index (SI), defined as the panel end load $\mathcal{L}$ per unit width w divided by the rib pitch r (Figure 11.11). The compressive stress at failure is proportional to the square root of the structural index,

$$\sigma_c = k_F \sqrt{E_t \frac{\mathcal{L}}{wr}} \quad \text{for} \quad \sigma_c < 350 \, \text{MPa} \tag{11.71}$$

where E_t is the tangent modulus and k_F denotes Farrar's efficiency factor. The dotted line in Figure 11.11 represents Equation (11.71) for $k_F = 1.0$, with Young's modulus of 71 GPa substituted for E_t. For a given type of stiffeners, k_F has a maximum value depending on stringer size and spacing, rib spacing and other geometric details of the panel's cross-section. The obtainable efficiency for integrally stiffened panels with non-flanged stiffeners is no more than $k_F = 0.75$. Factors of 0.88 are obtained for Z-stringers, J-stringers and hat section stringers and efficiencies up to 1.15 are achievable for optimized Y-section stringers. Experiments indicate that, for stress levels greater than 350 MPa, the structure begins to lose its efficiency with increasing loading intensity. A yield stress of 440 MPa is considered the limit for avoiding skin buckling. A limit on the compressive stress of 420 MPa is recommended, taking into account a 5% reserve for combined compression and shear. This value applies to a transport aircraft with an MTOW higher than 50 tonnes, typically.

The curve recommended for weight prediction in Figure 11.11 is based on $k_F = 0.85$ in the elastic range and 420 MPa for highly loaded panels. Between 350 and 420 MPa, the curve is blended between these two values. Applying this curve to the wing root structure requires knowledge of the local SI. This is obtained from the bending moment according to Equation (11.18) and the effective thickness $\eta_t t_r$ as defined by Equation (11.32). Correcting the bending moment for inertia relief (Section 11.4) by the factor

$$R_{in} = 1 - \{(R_{in})_{wg} + (R_{in})_{eng} + (R_{in})_f\} \tag{11.72}$$

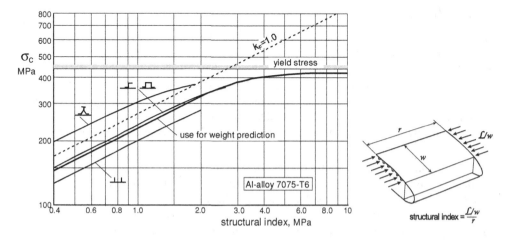

Figure 11.11 Allowable stress in compression structures

yields the structural index at the root,

$$\text{SI} = 0.5\,n_{\text{ULT}}\,W_{\text{MTO}}\,\frac{R_{\text{in}}\eta_{\text{cp}}b_{\text{st}}}{\eta_t t_r c_{\text{box}}r} \approx 0.2\,n_{\text{ULT}}\,W_{\text{MTO}}\,\frac{N_{\text{rib}}}{t_r c_{\text{box}}} \qquad (11.73)$$

where N_{rib} is the total number of general ribs. A more accurate computation requires knowledge of the structural topology – in particular, the distance between the front and aft spars and the rib pitch.

11.7.3 Shear Stress in Spar Webs

Similar to compression structures, the achievable shear stress in spar webs depends on the loading intensity and the type of material – 7075-T6 material can accept about 25% higher shear stress than 2024-T3. The loading intensity and the achievable shear stress decrease between the wing root and the tip. Accurate prediction of allowable stresses requires detailed information on the structural concept which is generally not available. Since shear web weight constitutes a relatively minor contribution, the conservative assumption has been that the mean shear stress $\bar{\tau}$ in the spar webs amounts to 50% of the average normal stress $\bar{\sigma}$ in the cover panels.

11.8 Refinements

Equation (11.35) was derived analytically for a straight-tapered wing with linearly decreasing wing thickness between the root and the tip, disregarding the presence of the centre section between the structural roots. In particular for high-speed wings, the prediction accuracy can be improved by adapting the structural box to account for the carry-through structure connecting the two wing halves, non-linear lofted wing sections and modifications such as winglets and sheared tips. In principle, these refinements result in a modified lateral lift distribution.

11.8.1 Tip Extensions

Sheared tips and winglets extend the trailing vortex sheet and shift the tip vortices in the direction of their own tips. This effectively increases the aerodynamic span and reduces induced drag. Blended winglets improve the flow at their attachment to the main wing and can be installed as retrofits as well as on clean sheet designs. Apart from their own structure weight, tip extensions have a significant effect on wing box bending load and weight. However, the author is not aware of a published validated method making allowance for this penalty. The following heuristic approach based on first principles is illustrated by Figure 11.12.

- By definition, the sheared wing tip (index tip) is an in-plane tip extension of the basic wing (index bw). Both surfaces are schematized as straight-tapered sections with aerodynamic lift distribution proportional to the local chord. Their lift is denoted L_{bw} and L_{tip}.
- A winglet is represented by rotating the sheared tip upwards around the basic wing tip over an angle Γ_{tip}. It is assumed that winglet lift and the CP location are unchanged during this rotation. This reduces the winglet's contribution to the total lift.

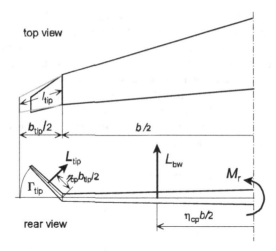

Figure 11.12   Sheared tip and winglet geometry

- In order to keep the basic wing and winglet lift together equal to the lift of the original wing, the basic lift is increased to compensate for the reduced winglet lift.

This approach is evolved in terms of a wing root bending moment increment which is translated into a correction factor on the cantilever ratio of the basic wing without tip extension:

$$R_{\text{tip}} = 1 + \frac{S_{\text{tip}}}{S} \left(\frac{1 - \eta_{\text{cp}}}{\eta_{\text{cp}}} \cos \Gamma_{tip} + \frac{b_{\text{tip}}}{b} \right) \tag{11.74}$$

with b_{tip} and Γ_{tip} denoting the span and the cant angle of the winglet. For a winglet area and span of 5% and 10% of the basic wing area and span, R_{tip} varies between 1.005 for vertical winglets and 1.08 for sheared tips, corresponding to a wing weight increment between 0.5% and 5%. More exotic load-carrying tip extensions are discussed in Section 10.8 – their effect on wing weight has to be investigated by means of a detailed structural layout including an investigation of aero-elastic effects.

11.8.2   Centre Section

The centre section (CS) depicted in Figure 11.2 is a prismatic structure with thickness t_{sr}, connected to the outboard wings by root joints. In order to resist a constant bending and torsion moment, the cover panels are stabilized by stiffeners, ribs and/or auxiliary spars. The impact on the present weight prediction method is summarized as follows:

- The cantilever ratio of the exposed wing is

$$R_{\text{cant}} = \frac{b - b_{\text{cs}}}{2 \, t_{\text{sr}} \cos \Lambda_{\text{ea}}} \tag{11.75}$$

- Since the CS is not exposed to the flow, the lift on a low-set wing acts only outboard of the structural root. For the same lift distribution, the CP of each half-wing moves to a lateral position $0.5\eta_{cp}(b - b_{cs})$ outboard from the structural root. Application of Equation (11.40) to the exposed wing demonstrates that the weight of its bending material is reduced by a factor $(1 - b_{cs}/b)^2$ relative to the wing with the same span but without the CS.
- The CS bending material resisting lift is derived by modifying Section 11.3. It is assumed that the bending moment and the stress levels are equal to the value at the structural root; hence, $I_1 = I_2 = 1$ are substituted in Equation (11.40). The local elastic axis is straight and the amount of bending material along the CS width is constant.

Application of these principles leads to a modified expression for the panel weight. Since the ratio σ_c/σ_t appears to have hardly any effect, it is treated as a constant. This yields the following correction factor on the box weight resisting bending due to lift:

$$R_{cs} = \left(1 - \frac{b_{cs}}{b}\right)\left\{1 + \frac{b_{cs}}{b}(2.45\cos\Lambda_{ea} - 1)\right\} \qquad (11.76)$$

For a straight wing, the CS leads to a typical cover weight penalty of 3%. For a 30° sweptback wing the un-sweeping effect of the CS reduces this penalty to almost zero.

Although the shear forces due to lift on both exposed wing halves are taken out at the wing root, shear webs are required in the central wing box to cope with torsion. Therefore, Equation (11.46) can be used without correction on the provision that the structural span includes the CS. Disregarding the weight penalty of joints at the wing roots, it is justified to assume that the same value of R_{cs} applies to shear material and to the wing box as a whole. In accordance with [26] it is concluded that – contrary to what is often suggested – there is no large penalty associated with the sweep break at the wing root, compared to a non-swept wing with the same structural span. The factor R_{cs} does not account for the small change of the inertia relief due to wing fuel if there is no fuel in the centre section. The factor $(R_{in})_f$ according to Equation (11.54) should therefore be applied to the case when there is no fuel in the wing outboard of the structural root. If the aircraft has a centre section tank[5] this is filled only for long-range flights when the outboard tanks are full – they are the first to be emptied. If (due to a failure) this is not possible, the aircraft is allowed to be manoeuvred to less than the limit load factor.

11.8.3 Compound Taper

Most published wing weight prediction methods contain the thickness ratio of the root section in the plane of symmetry, the aspect ratio and the taper ratio. Since an internationally accepted normalization of dimensions such as (gross) wing area and root chord does not exist, the definition of these parameters is not always unambiguous. This complicates calibration of the weight prediction method and may lead to confusion. In particular, the root chord at the centreline of a schematic swept wing is based on a somewhat arbitrary extension of the leading and trailing edges to the plane of symmetry. However, the (maximum) thickness of the

---

[5]Aircraft with a centre section tank have a zero wing fuel weight limitation (MZWFW) in addition to the maximum zero fuel weight (MZFW).

structural root is well defined since it equals the CS thickness – the same applies to the wing span. Consequently, the present method uses the cantilever ratio as defined by Equation (11.39) as the major shape parameter instead of using the root chord thickness ratio.

The variation of the chord and thickness along the span is non-linear for most high-speed airplane wings. In order to approximate an ideal planform, many wings feature one or two kinks in the trailing edge, some have a kinked leading edge as well. The inboard wing has a pronounced wash-out and its thickness tapers off more rapidly than the outboard wing taper. The reduced thickness of the outboard wing is in the interest of obtaining high local lift coefficients at the cost of increased panel loads. The weight penalty due to the reduced outboard thickness ratio and the presence of winglets or sheared tips is taken into account by modifying the cantilever as follows:

$$R'_{cant} = R_{tip} \frac{b - b_{cs}}{2\, t_{sr} \cos \Lambda_{ea}} \left\{ \frac{2}{3} + \frac{(t/c)_{sr}}{3\,(t/c)_{tk}} \right\} \tag{11.77}$$

The indices sr and tk denote the structural root and the thickness kink, respectively. Equation (11.74) is used for wings with winglets or sheared tips, $R_{tip} = 1$ is used for conventional wing tips. Compared to the basic wing with ruled surfaces, use of the modified cantilever ratio may lead to 10% of the bending material weight increment.

11.8.4   Exposed Wing Lift

In deriving the primary structure weight it was assumed that the wing lift equals the aircraft lift. In reality, the lift on the aircraft is made up from wing lift – this is the dominating term – and (upward or downward) lift contributions from other exposed components, in particular the fuselage and the horizontal tail. Depending on the CG location, the tail download required for manoeuvring can subtract an appreciable fraction from the total lift. Consequently, wing lift can be appreciably less or more than airplane lift.

Wing lift is generated by the exposed wing surface whereas lift carry-over on the fuselage is less than the (hypothetical) lift on the wing section covered by the fuselage. Lift losses must also be expected where engine nacelles are connected to the wing. If, in the design condition, the exposed wing lift is less than the complete aircraft lift, this may lead to a reduction of the bending and shear loads. This is taken into account by a pro-forma lift correction factor

$$R_L = \frac{\text{exposed wing lift}}{\text{aircraft lift}} \tag{11.78}$$

Exposed wing lift and its distribution along the span depend on the load case – manoeuvring or gust – and its computation can be time consuming. If we assume conservatively that the lift carried by the fuselage and – the tail download at the most forward CG location are compensating, we have $R_L \doteq 1.0$. This assumption may have to be revised for tail-first aircraft with a relatively large upward lift on the canard.

11.8.5   Advanced Materials

Composites archieve extremely high strength/weight ratios from cloths of carbon, aramid or glass fibres embedded within a thermosetting resin such as epoxy. They offer the best near-term

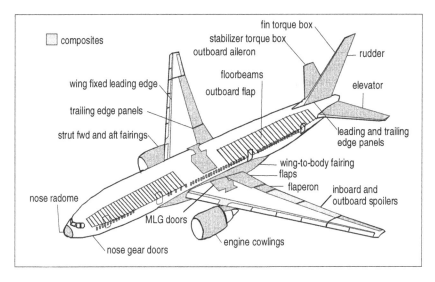

Figure 11.13 Composites application on the Boeing 777. Adapted from [4]

prospect for significantly reducing wing weight. The obtainable weight saving is, however, highly dependent on the application. For more than half a century, composites have been used for secondary structures such as trailing edges, flaps, control surfaces, fairings, and wheel well doors. Since the 1980s, composite tail units and significant parts of the wing structure have been made of CFC. The application of thermoplastics is increasing as well. The Airbus A320 and A330/340 have all-composite vertical and horizontal tails. Figure 11.13 illustrates a typical example for a modern airliner in which composites are applied in the empennage and in secondary structures. Several airliners have embraced a hybrid material application in the wing box by incorporating CFC for components such as ribs, the carry-through structure or the outboard wings. In the Boeing 787 and Airbus A350, composites have even become the dominant material in the complete wing structure.

Because of their remarkably high specific tensile strength and stiffness, CFCs offer the potential of weight savings up to 35%. Demonstrated secondary wing weight reductions collected in Table 11.1 show that actual weight savings are more modest. This applies in particular to spoilers and lift dumpers since CFC is not applied to their highly loaded supports. A practical wing structure will not completely be made of composites. Various studies have indicated that a CFC materials usage of approximately 80% is obtainable for wings, resulting in a weight reduction of 28%. Since the primary box of a modern wing features a mixed application of several materials, a simple correction factor on all-aluminium wing weight is not useful. A reliable prediction should be made by a multiple station analysis. This requires a more detailed knowledge of the structural design features than the elementary procedure exposed in the present prediction method.

Fibre composites are not the only advanced materials considered for wing structures. For instance, aluminium-lithium alloys have become a viable alternative because of the modulus increase and density decrease. Compared to standard Al-alloys, Al-Li has superior strength, stiffness and fatigue life. A density reduction of 8% to 10% has been achieved, although in

Table 11.1 Typical achievable secondary weight reductions with CFC relative to Al-alloy structures

Component	Weight reduction (%)
Fixed leading and trailing edge structures	20
Trailing edge flaps	15
Ailerons	25
Spoilers, lift dumpers and air brakes	10
Wing/fuselage fairing	20

many wing components Al-Li may not replace conventional materials because of their lower fracture toughness and higher cost. Another viable replacement for standard Al-alloys is Glare, a hybrid composite built up from layers of Al-alloy with glass fibre composite in between. The layers of glass fibres act as crack stoppers and this hybrid material has practically no lifetime restriction from fatigue. Glare is damage-tolerant and accepts increased tensile stress levels. For example, for fatigue-critical pressure cabin skins and lower wing panels, weight savings due to Glare application are comparable to CFC. Moreover, fabrication of Glare components is done with similar techniques and tools as with Al-alloys.

11.9 Application

The main body of this chapter is devoted to deriving the structure weight of a basically straight-tapered Al-alloy wing with constant section thickness ratios. This basic prediction method is also used as the starting point for the more complex wing of a high-speed aircraft. For this application, the basic weight is modified and augmented by several refinements accounting for the presence of a central wing box, compound taper, wing-tip extensions and application of composite material.

11.9.1 Basic Ideal Structure Weight

As indicated in Figure 11.1, the ideal weight (index id) of the primary wing structure represents the theoretical minimum amount of structural material required to resist bending, shear and torque loads. It is obtained by adding the bending material according to Equation (11.40) corrected by 5% for in-plane and torsion, shear material according to Equation (11.46) increased by 10% for damage tolerance and fail safety, and rib weight according to Equation (11.47).

The basic ideal structure weight is computed from the ideal box weight and rib weight,

$$W_{id} = (W_{id})_{box} + W_{rib} \qquad (11.79)$$

where

$$(W_{id})_{box} = 0.36\, n_{ULT} R_{in} W_G \eta_{cp} b_{st} \frac{\rho g}{\sigma_r} \left(1.05\, \frac{R_{cant}}{\eta_t} + 3.67 \right) \qquad (11.80)$$

and

$$W_{rib} = \rho g k_{rib} S \left(t_{ref} + \frac{t_r + t_t}{2} \right).$$ (11.81)

- The MTOW is substituted for the design gross weight W_G and the ultimate load factor n_{ULT} is determined according to the directives in Section 11.2. In order to decide which condition is critical, Equation (11.9) is consulted or the wing weight is be computed for ultimate maneuver as well as gust loads.
- The inertia relief due to wing, engine and fuel weight (Section 11.4) is

$$R_{in} = 1 - \{(R_{in})_w + (R_{in})_{eng} + (R_{in})_f\}$$ (11.82)

with the various terms defined by Equations (11.48), (11.49) and (11.54). The fuel weight equals MTOW – MZFW.
- The location of the centre of pressure, the section effectiveness factor η_t and the cantilever ratio R_{cant} are obtained from Equations (11.29), (11.32) and (11.39), respectively.
- The structural span b_{st} obtained from Equation (11.15) is measured along the elastic axis which can be assumed midway between the front and aft spars or defined at 40% of the chord.
- The specific weight of Al-alloy amounts to $\rho g = 2\,796$ N m^{-3}.
- The mean stress at the root section is obtained from

$$\frac{1}{\overline{\sigma}_r} = \frac{1}{2} \left(\frac{R_{ic}}{\sigma_t} + \frac{1.25}{\sigma_c} \right)$$ (11.83)

Allowable tensile and compressive stress levels comply with the recommendations of Section 11.7. The correction factor R_{ic} for manholes – this represents strictly a non-ideal weight penalty – is defined by Equation (11.57).
- Statistical rib weight coefficients are $k_{rib} = 0.005$ and $t_{ref} = 1.0$ m – calibration with in-house data is recommendable.

11.9.2   Refined Ideal Structure Weight

If sufficient details are available to make the refinements exposed in Section 11.8 the following modifications are applied:

- Sheared tips and winglets are categorized as tip extensions. Their structure weight is derived from Equation (11.70) and the effect on the wing box weight is taken into account by the correction factor R_{tip} on the cantilever ratio according to Equation (11.74).
- The presence of the centre section is taken into account by the correction factor R_{cs} according to Equation (11.76) to the ideal bending material weight.
- The effect of compound taper and varying section thickness ratio along the span is taken into account by the modified cantilever ratio R'_{cant} according to Equation (11.77). The inertia relief due to fuel is computed for fuel outboard of the centre section.

- If the exposed wing lift and its distribution are known, the pro forma correction factor R_L is applied to the ideal box weight according to Equation (11.78), $R_L = 1$ is assumed when details are lacking.
- Application of advanced materials to the wing box requires empirical modifications of the specific mass and/or the factor k_{rib} in Equation (11.79) and stress levels in Equation (11.83). This simple approach cannot be expected to give more than a first-order approximation.

Introduction of these refinements in Equation (11.79) yields

$$W_{id}' = (W_{id}')_{box} + W_{rib}' \qquad (11.84)$$

where

$$(W_{id}')_{box} = 0.36\, n_{ULT}\, R_{tip}\, R_{cs}\, R_L\, W_G \eta_{cp}\, b_{st} \frac{\rho g}{\bar{\sigma}} \left(1.05\, \frac{R_{cant}'}{\eta_t} + 3.67 \right) \qquad (11.85)$$

and

$$W_{rib}' = \rho g\, k_{rib}\, S \left(t_{ref} + \frac{t_r + t_t}{2} \right) \qquad (11.86)$$

11.9.3 Wing Structure Weight

The total wing weight is computed from Equation (11.79) or (11.84) by adding non-ideal weight penalties and secondary weight to the ideal box weight,

$$W_{wing} = W_{id} + \Sigma\, \Delta W_{nid} + 1.10\, \Sigma\, W_{sec} \qquad (11.87)$$

- Non-ideal penalties (index nid) are specified in Section 11.5, Equations (11.55) and (11.58) through (11.62). Dependent on the specific configuration and availability of data, some of these terms may be modified or calibrated with in-house data.
- Secondary weight (index sec) components are specified by Equations (11.63) through (11.68). Each component is computed in terms of its specific weight Ω multiplied by the item's planform area.
- The correction factor 1.10 for miscellaneous items should be improved if better information is available.

11.9.4 Accuracy

Different from many published methods, the present weight prediction uses a mixture of rational analysis based on first principles and statistical calibrations and formulas. It demonstrates the merits and pitfalls of a typical quasi-analytical approach intended to generate a result that is accurate in the absolute sense as well as sensitive to variations in the design characteristics. Most of the work to be done when using the method is input data collection of design weights, combinations of design speeds and Mach number and about thirty geometric properties. A worked example for the Boeing 747-100 has demonstrated that the method is easy to apply [33]. An prediction error of + 3% was found for the basic method and + 4% for the refined approach. The large reduction of the thickness ratio from 13.4% at the root to 8.0% at the outer wing increases the bending material considerably compared to a constant thickness ratio of 13.4%. The use of the modified cantilever ratio according to Equation (11.77) is therefore recommended provided the wing geometry is well defined.

The validation has been extended to three jetliners and a light business jet with maximum take-off weights between 5 and 350 tonnes. In all cases the prediction error was less than 4% and they were positive as well as negative. These results are encouraging in that the method appears to predict the size effect adequately even though it has not been calibrated for different airplane categories. The most influential parameter determining the prediction accuracy is the mean stress level at the wing root. In particular, it may be improved when sufficient data are available about the airplane's required lifetime in combination with a typical flight spectrum and the derivation points to several assumptions which can be revisited with the use of in-house data. In principle, application to non-conventional wing configurations is possible by extending it to a multiple station analysis with different load cases for each wing station. However, this may require a disproportional increase of detailed and complicated analysis which does not necessaryly comply with the character of the conceptual design process.

11.9.5 *Conclusion*

A comprehensive method has been developed to predict the structural wing weight of transport and business aircraft with the objective of obtaining a design-sensitive method which can be modified by the user. Accurate information on data such as tensile, compressive and shear stresses at the structural root, lift curve slope and location of the aerodynamic centre can be input to improve the presented approaches. The method has the following basic characteristics:

- The structure is broken down into functional components (Figure 11.1). Their weights are estimated in analysis based on first principles, where necessary augmented by statistics based expressions. Occasionally, missing information on some minor weight components has necessitated a heuristic approach.
- The bending and shear loads at the wing root are computed from the lift distribution and include inertia relief. The theoretical minimum material required to resist these loads is derived from analytical integration along the span. Effects of variation in thickness ratio and allowable stresses are taken into account.
- Semi-empirical methods are used for the weight of ribs, weight penalties due to non-ideal structural features, and for secondary weight of leading and trailing edge structures, including high-lift devices and flight controls.
- Considerations are given to select allowable stress levels in the wing box cover panels and spars, taking into account structural topology, expected lifetime, fail safe and damage tolerance structural design principles.
- The effect of inertia relief due to fixed masses and fuel is included and quantified by means of elementary equations.
- Refinements are suggested to account explicitly for the presence of tip extensions, carry-through structure, compound taper and advanced materials.

Advantages and limitations of the method have been clarified: many weight components are sensitive to computations and design decisions to be made downstream of conceptual design. It is therefore unrealistic to expect the method to have a standard deviation less than, say, 5% in the conceptual design phase, unless the wing is a derivative of an existing aircraft version. Although a standard deviation has not been established, application to several existing airplanes points to an acceptable accuracy.

Bibliography

[1] Shanley, F.R., *Weight-Strength Analysis of Aircraft Structures*, Second Edition, McGraw-Hill, NY, Reprinted by Dover Publications, 1960.

[2] Hoblit, F.M., *Gust Loads on Aircraft: Concepts and Applications*, AIAA Education Series, American Institute of Aeronautics and Astronautics, Inc., Washington, DC, 1988.

[3] Niu, C.Y., *Airframe Structures: Practical Design Information and Data*, Conmilit Press, Hong Kong, 1988.

[4] Niu, C.Y., *Composite Airframe Structures: Practical Design Information and Data*, Conmilit Press Ltd., Hong Kong, 1992.

Weight Prediction Methods

[5] Farrar, D.J., "The Design of Compression Structures for Minimum Weight", *Journal of the Royal Aeronautical Society*, pp. 1041–1052, November 1949.

[6] Carayette, J.F., "Aircraft Wing Weight Estimation", *Aircraft Engineering*, January 1950, pp. 8–11, and page 119, April 1950.

[7] Ripley, E.L., "A Method of Wing Weight Prediction", RAE Report Structures 109, May 1951.

[8] Broadbent, E.G., "Aeroelastic Problems in Connection with High-Speed Flight", *Journal of the Royal Aeronautical Society*, July 1956.

[9] Burt, M.E., "Structural Weight Estimation for Novel Configurations", Paper No. 11220, May 1961.

[10] Saelman, B., "Effect of Wing Geometry on Volume and Weight", *Journal of the Aeronautical Sciences*, Vol. 29, No. 11, November 1962, pp. 1390–1392. Further notes in Vol. 1, No. 5, pp. 305–306, September–October 1964.

[11] Crawford, R.F, and A.B. Burns, "Minimum Weight Potentials for Stiffened Plates and Shells", *AIAA Journal*, Vol. 1, No. 4, pp. 879–886, April 1963.

[12] Saelman, B., "Multitapered Wings", *AIAA Journal of Aircraft*, Vol. 2, No. 4, pp. 348–349, July–August 1965.

[13] Garrock, C.A., and J.T. Jackson, "Estimation of Wing Box Weight to Preclude Aeroelastic Instabilities", SAWE Paper No. 500, May 1966.

[14] Sanders, K.L., "A Review and Summary of Wing Torsional Stiffness Criteria for Pre-Design and Weight Estimations", SAWE Paper No. 1632, May 1967.

[15] Keith-Lucas, D., "Defeating the Square Cube Law", *Flight International*, Vol. 94, No. 3106, pp. 440–442, September 1968.

[16] Laser, "Design Probe: Another Look at the Square Cube Law", *Flight International*, Vol. 94, No. 3110, pp. 615–616, October 1968.

[17] Roland, H.R., "General Approach to Preliminary Design Weight Analysis and Structural Weight Prediction", Short Course in Modern Theory and Practice of Weight Optimization and Control for Advanced Aeronautical Systems, University of Tennessee, November 1968.

[18] Hoak, D.E., et al., "USAF Stability and Control DATCOM", 45433–0000 (revised), Flight Control Division, AFFDL, Wright Patterson AFB, Ohio, 1968.

[19] Sanders, K.L., "High-Lift Devices, a Weight and Performance Trade-off Methodology", SAWE Paper No. 761, 1969.

[20] Cate, D.M., "A Parametric Approach to Estimate Weights of Surface Control Systems of Combat and Transport Aircraft", SAWE Paper No. 812, May 1969.

[21] Torenbeek, E., "Prediction of Wing Group Weight for Preliminary Design", *Aircraft Engineering*, pp. 16–21, July 1971. (Summary in February 1972 issue, pp. 18–19).

[22] Schneider, W., "A Procedure for Calculating the Weight of Wing Structures with Increased Service Life", SAWE Paper No. 1021, May 1974.

[23] Hangartner, R., "Correlation of Fatigue Data for Aluminium Aircraft Wing and Tail Structures", National Aeronautical Establishment, Aeronautical Report LR-582, Ottawa, December 1974.

[24] Fritz, R.J., "Method for Determining the Maximum Allowable Stress for Preliminary Aircraft Wing Design", University of Witwatersrand, School of Mechanical Engineering Research Report No. 72, January 1977.

[25] Anderson, R.D., "Development of Weight Estimates for Lifting Surfaces with Active Controls", *SAWE Journal*, Vol. 36, No. 3, April 1977.

[26] Kyser, A.C., "An Elementary Analysis of the Effect of Sweep, Mach Number, and Lift Coefficient on Wing Structure Weight", NASA Technical Memorandum 74072, September 1977.

[27] Ramsey, H.D., and J.G. Lewolt, "Design Manoeuvre Loads for an Airplane with an Active Control System", 20th Structures, Structural Dynamics and Materials Conference, AIAA/ASME/ASCE/AHS, St. Louis, NASA CP 1795, pp. 456–464, 1979.

[28] Toll, T.A., "Parametric Study of Variation in Cargo-Airplane Performance Related to Progression from Current to Spanloader Designs", NASA Technical Paper No. 1625, April 1980.

[29] York, P., and R.W. Labell, "Aircraft Wing Weight Build-Up Methodology with Modification for Materials and Construction Techniques", NASA Contractor Report No. 166173, September 1980.

[30] Fairchild, M.P., "Structural Weight Comparison of a Joined Wing and a Conventional Wing", AIAA Paper No. 81–0366, January 1981.

[31] Schaeken, A.H.P.M., "A Detailed Wing Mass Estimation Method", MSc. thesis, Delft University of Technology, Dept. of Aerospace Engineering, June 1983.

[32] Udin, S.V., and W.J. Anderson, "Wing Mass Formula for Subsonic Aircraft", *Journal of Aircraft*, Vol. 29, No. 4, pp. 725–727, July–August 1992.

[33] Torenbeek, E., "Development and Application of a Comprehensive, Design-Sensitive Weight Prediction Method for Wing Structures of Transport Category Aircraft", Report LR 693, Department of Aerospace Engineering, Delft University of Technology, September 1992.

[34] Kelm, R., M. Läpple, and M. Grabietz, "Wing Primary Structure Weight Estimation of Transport Aircraft in the Pre-Development Phase of a New Aircraft", SAWE Paper No. 2283, May 1995.

[35] Macci, S.H., "Semi-Analytical Method for Predicting Wing Structural Mass", SAWE Paper No. 2285, May 1995.

[36] Rudolph, P.K.C., "High-Lift Systems on Commercial Subsonic Airliners", NASA CR-4746, September 1996.

12

Unified Cruise Performance

12.1 Introduction

The design of an airliner or a business aircraft inevitably confronts the analyst with one or several of the following questions:

1. What are the best cruise altitude and speed from the point of view of minimum fuel consumed per unit of time elapsed or per unit of distance travelled?
2. How to estimate the total amount of fuel and time required for a specified mission?
3. For a given mission range, how to determine the flight profile resulting in minimum fuel consumed or direct operating costs (DOC) incurred?

The first question can be treated as a point performance problem, with equilibrium of forces in steady level flight. Most educational texts are, however, focussed on flight at a low Mach number and treat the engines as if they were pure jets. Consequently, their answers are of little use to turbofan-powered aircraft flying at conditions where the compressibility of air affects the aerodynamic efficiency. The second of the above questions involves integration of point performances to obtain a flight path performance. Take-off gross weight (TOGW) can be an input for computing the fuel load or a result of the analysis in case the zero fuel weight (ZFW) is specified. Each regular flight consists of a number of segments – take-off, climb, cruise, descent, approach, and landing – and a diversion flight has to be analyzed as well. The computation is usually made by iterations on the TOGW. A rigorous solution of the third problem requires flight mechanics to be treated as a dynamic problem and the optimization requires application of the calculus of variations or optimal control theory [26]. Except for very short-range flights, dynamic optimization leads to marginal fuel savings and the problem is considered to be outside the context of advanced aircraft design.

12.1.1 Classical Solutions

The analysis and optimization of cruise performance have been treated in the extensive literature on flight mechanics [2, 3, 4, 18]. All sources are in agreement that, for specified initial

Advanced Aircraft Design: Conceptual Design, Analysis and Optimization of Subsonic Civil Airplanes, First Edition. Egbert Torenbeek.
© 2013 by Egbert Torenbeek. Published 2013 by John Wiley & Sons, Ltd.

altitude and speed, a cruise/climb flight program yields the longest range. Bréguet's equation modified for jet aircraft is widely accepted as a good representation of a long-range flight. There is, however, little consensus in derivations of the best cruise altitude and speed. The inexperienced design analyst of a transonic airplane who takes a close look at these solutions may become puzzled by remarkable inconsistencies; examples are given in [32]. The confusion is ascribed mainly to one or more of the following issues.

- Classical derivations of the maximum range of a jet aircraft are mostly based on subcritical flight conditions and they simplify the aerodynamic model by assuming that there is no compressibility drag. In reality, the optimum cruise condition of a jetliner is dominated by the fact that there is not a single unique drag polar in the transonic flight regime.
- The numerical model of engine characteristics is mostly oversimplified. In particular, the often made assumption that specific fuel consumption is independent of speed is invalid at high speeds.
- Many authors derive the optimum cruise condition by assuming a constraint on altitude, speed or engine rating. Though this may form a practical approach for optimizing an operational cruising flight of a given aircraft, the designer is also interested in achieving an unconstrained optimum initial design. This is achieved by matching the installed power or thrust to the airframe aerodynamics so that fuel efficiency closely approaches its highest achievable value (Chapter 8).

Classical fixed-wing aircraft performance analysis is based on a different set of equations for propeller and jet aircraft. This historical distinction has been made because the output of piston engines is specified as a shaft power whereas the performance of (now obsolete) straight jet engines is specified as a thrust. Piston engine shaft power and (straight) jet engine thrust were roughly independent of airspeed and the same could be said of specific fuel consumption (SFC). Different sets of optimum flight conditions for propeller and jet airplanes can be found in many textbooks on aircraft performance. However, the presently dominant categories of airplanes with turboprops and high bypass ratio turbofan engines require a different approach.

12.1.2   Unified Cruise Performance

The evolution of high-speed airliners and business jets has degraded the validity of the classical theory of cruise performance. Turboprop engines generate propeller shaft power as well as jet thrust and the available propulsive power and thrust vary considerably with forward speed. Dependent on (mainly) the bypass ratio, turbofan thrust decreases and TSFC increases with speed. This tendency will become more pronounced with the future introduction of open rotor and ultra-high bypass ratio engines. Arguably, the unification into a single category of gas turbine-powered propulsive devices has made the distinction between jet and propeller aircraft less relevant.

The theory of maximum specific range performance published in [25] forms the point of departure for this chapter which offers a unified approach covering the complete spectrum of gas turbine engine-powered subsonic and transonic aircraft. Overall efficiency rather than SFC is used to define the fuel flow and drag due to compressibility is taken into account. For editorial reasons, the original analysis is simplified by leaving out the usually minor effect of engine rating on propulsive efficiency. Several operational flight schedules are analyzed and compared for given initial and final gross weights. A closed-form solution for the mission fuel required to accomplish a flight over a given distance is also included. This analysis shows that a high accuracy of analytical performance optimization can be obtained with little extra complication.

No effort has been made to extend the present approach to low speed flight segments. Effects of wind on cruise performance [21], special flights such as extended range operations [30] and conditions for minimum direct operating costs (DOC) [33] are not treated. It is emphasized that the present chapter focusses on optimization of the cruising flight for a completely defined airplane with known aerodynamic and engine properties. The derivation of optimum cruise altitude and Mach number is, therefore, not applicable to optimize an airplane design, an example of analysis versus synthesis problem structure discussed in Section 1.7.

12.1.3   Specific Range and the Range Parameter

The range R is the distance that an airplane can cruise with a given amount of fuel (index f). It is computed by integrating the specific air range (SAR),

$$\frac{V}{F} = \frac{\mathrm{d}R}{\mathrm{d}W_f} = \frac{\mathrm{d}R/\mathrm{d}t}{\dot{m}_f\, g} \tag{12.1}$$

The term $F = \dot{m}_f g$ is the fuel weight flow rate. Engine characteristics are introduced in the form of the overall efficiency of propulsion as defined in Section 3.2, Equation (3.10),

$$\eta_o = \frac{\text{thrust power developed by the engine}}{\text{rate of fuel energy added to the engine}} = \frac{TV}{\dot{m}_f\, H} \tag{12.2}$$

where H denotes the calorific value of the fuel. For a given airplane GW, the SAR is written in terms of the range parameter,

$$\mathcal{P} \stackrel{\text{def}}{=} \frac{W}{H/g}\frac{V}{F} = \eta_o\frac{W/\delta}{T/\delta} \tag{12.3}$$

where δ denotes the relative ambient pressure (Appendix B). The range parameter is maximized for quasi-steady level flight with equilibrium conditions

$$W/\delta = L/\delta = \frac{1}{2}\gamma p_{sl}M^2 SC_L \quad \text{and} \quad T/\delta = D/\delta = \frac{1}{2}\gamma p_{sl}M^2 SC_D \tag{12.4}$$

Different from previous chapters, the wing area S is defined without the index w for wing. The range parameter and specific range in steady level flight are

$$\mathcal{P} = \eta_o\frac{C_L}{C_D} \quad \rightarrow \quad \frac{V}{F} = \mathcal{P}\frac{H/g}{W} \tag{12.5}$$

For jet aircraft, the specific range is usually written in terms of the thrust-specific fuel consumption (TSFC),

$$\frac{V}{F} = \frac{V}{C_T W} \frac{C_L}{C_D} = \frac{a_{sl} M L/D}{W C_T/\sqrt{\theta}} \tag{12.6}$$

with symbols defined in Section 3.2 and in Appendix B. Translated into the classical equation for propeller aircraft, the use of the power-specific fuel consumption (PSFC) is

$$\frac{V}{F} = \frac{\eta_{prop}}{C_P W} \frac{C_L}{C_D} \tag{12.7}$$

The relation between overall efficiency and flight conditions is based on knowledge of the fuel flow required to produce a given amount of installed thrust. This accounts for engine and propeller installation effects such as intake and exhaust losses, bleed air and power off-takes, and propeller slipstream wash.

The objective of cruise point optimization is to maximize the product of overall propulsive and aerodynamic efficiencies. Both are mutually unrelated functions of cruise speed and altitude. For transonic Mach numbers, their combined optimum does not coincide with the maxima of either η_o or C_L/C_D. Moreover, η_o does not always attain an extremal value in the operational flight regime. In order to introduce the reader to the optimization of the specific range for high speed flight, the following sections deal with C_L/C_D, MC_L/C_D and $\mathcal{P}$ separately. The conditions for maximizing the range parameter are explained and applied to derive the maximum cruise range.

12.2   Maximum Aerodynamic Efficiency

The aerodynamic performance of a transonic airplane is defined by a set of drag polars depending on the Mach number (Section 4.4). An example of aerodynamic efficiency curves derived from these polars is depicted on Figure 12.1. Since several curves are intersecting, this representation is not entirely unambiguous. The preferred rendering of the same data in Figure 12.2 shows contours connecting points with the same aerodynamic efficiency which enable graphical identification of conditions for minimum drag [17]. Below the critical Mach number, C_D at constant C_L decreases slightly with Mach number and C_L/C_D increases gradually until the critical Mach number is exceeded and the drag begins to increase. Point A identifies the global minimum drag condition with $M = 0.675$, $C_L = 0.515$ and $C_L/C_D = 18.4$. For a given GW this defines a unique combination of pressure altitude and flight Mach number, The C_L/C_D contours become closed in the drag rise where the highest L/D at given Mach number is identified by a vertical tangent to the contour. Due to the deteriorating aerodynamic efficiency at transonic Mach numbers, the conventional definition of the minimum drag speed has become irrelevant. Figures 12.1 and 12.2 illustrate the essential message that in the transonic flight regime the minimum drag condition is defined by a different value of C_L for each Mach number.

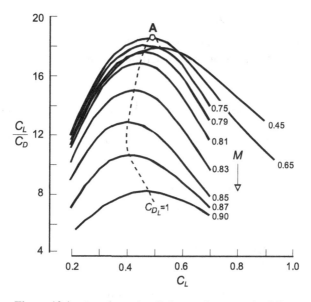

Figure 12.1 Aerodynamic efficiency of a transonic airliner

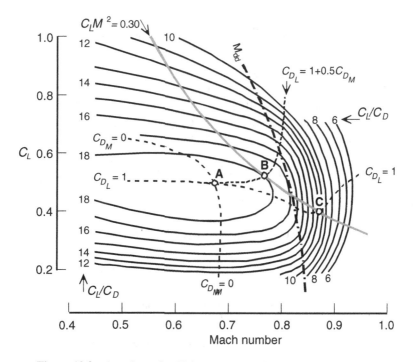

Figure 12.2 Aerodynamic efficiency contours in the C_L versus M plane

12.2.1   Logarithmic Drag Derivatives

The previous section has shown qualitatively that lift and drag coefficients at high-subsonic and transonic Mach speeds are not uniquely determined by the angle of attack but also by the Mach number,

$$C_L = C_L(\alpha, M) \quad \text{and} \quad C_D = C_D(\alpha, M) \quad \rightarrow \quad C_D = C_D(C_L, M) \qquad (12.8)$$

Analytical conditions for the unconstrained maximum C_L/C_D are readily obtained by means of logarithmic differentiation as follows:

$$d\log(C_L/C_D) = d\log C_L - d\log C_D = 0 \qquad (12.9)$$

with

$$d\log C_D = \frac{\partial \log C_D}{\partial \log C_L} d\log C_L + \frac{\partial \log C_D}{\partial \log M} d\log M \qquad (12.10)$$

Logarithmic derivatives are written in short-hand notation,

$$C_{D_{\mathcal{L}}} \stackrel{\text{def}}{=} \frac{\partial \log C_D}{\partial \log C_L} = \frac{C_L}{C_D} \frac{\partial C_D}{\partial C_L} \qquad (\text{constant } M) \qquad (12.11)$$

$$C_{D_M} \stackrel{\text{def}}{=} \frac{\partial \log C_D}{\partial \log M} = \frac{M}{C_D} \frac{\partial C_D}{\partial M} \qquad (\text{constant } C_L) \qquad (12.12)$$

Dimensionless log-derivatives are interpreted as a percentage change in a dependent variable divided by a given percentage change of the independent variable. The derivative $C_{D_{\mathcal{L}}}$ should not be confused with the drag due to lift coefficient C_{D_L} (Section 4.4). Numerical values of $C_{D_{\mathcal{L}}}$ and C_{D_M} are between zero and two, although C_{D_M} can be larger in the drag rise. In terms of log-derivatives, Equation (12.10) reads as follows:

$$d\log C_D = C_{D_{\mathcal{L}}} d\log C_L + C_{D_M} d\log M \qquad (12.13)$$

By substitution of this result, Equation (12.9) becomes

$$(1 - C_{D_{\mathcal{L}}}) d\log C_L - C_{D_M} d\log M = 0 \qquad (12.14)$$

Partial optima for C_L and M are found by setting $d\log M = 0$ and $d\log C_L = 0$,

$$C_{D_{\mathcal{L}}} = 1 \quad \rightarrow \quad \frac{\partial C_D}{\partial C_L} = \frac{C_D}{C_L} \quad (\text{constant } M) \qquad (12.15)$$

$$C_{D_M} = 0 \quad \rightarrow \quad \frac{\partial C_D}{\partial M} = 0 \quad (\text{constant } C_L) \qquad (12.16)$$

For a given Mach number, the condition $C_{D_{\mathcal{L}}} = 1$ defines the minimum drag condition which is found in Figure 12.1 graphically by drawing horizontal tangents to the C_L/C_D curves or in

Figure 12.2 by drawing vertical tangents. The condition $C_{D_M} = 0$ forms the lower boundary of the Mach number region where compressibility degrades the aerodynamic efficiency. The partial optima intersect in the global optimizer point A defined by unique values of the lift coefficient and the Mach number. This defines the most fuel-efficient cruise condition only if the overall efficiency of the engines has a stationary value in the same point. In general, this is not the case for jet aircraft although a (future) propulsion systems such as the open rotor engine may achieve its maximum overall efficiency at a subsonic Mach number.

12.2.2 Interpretation of Log-Derivatives

Dependent on the aircraft design stage, drag polars of different appearance may be used (Section 4.4). For the purpose of initial performance analysis, the drag polar is often approximated by a two-term approximation,

$$C_D = C_{D_0} + K_L C_L^2, \quad \text{with} \quad K_L \overset{\text{def}}{=} \mathrm{d}C_D / \mathrm{d}(C_L^2) \tag{12.17}$$

For this case the log-derivative C_{D_L} is

$$C_{D_L} = \frac{2 K_L C_L^2}{C_{D_0} + K_L C_L^2} = 2 \frac{\text{drag due to lift}}{\text{total drag}} \tag{12.18}$$

Conditions for minimum drag are

$$C_{L_{\mathrm{MD}}} = \sqrt{\frac{C_{D_0}}{K_L}} \quad \text{and} \quad C_{D_{\mathrm{MD}}} = 2 C_{D_0} \quad \rightarrow \quad \left(\frac{C_L}{C_D}\right)_{\mathrm{MD}} = \frac{1}{2\sqrt{K_L C_{D_0}}} \tag{12.19}$$

This is the classical result that the aerodynamic efficiency is maximized for equal zero-lift drag and drag due to lift. The relationship between C_{D_L} and C_L is

$$C_{D_L} = \frac{2(C_L/C_{L_{\mathrm{MD}}})^2}{1 + (C_L/C_{L_{\mathrm{MD}}})^2} \quad \text{or} \quad \frac{C_L}{C_{L_{\mathrm{MD}}}} = \sqrt{\frac{C_{D_L}}{2 - C_{D_L}}} \tag{12.20}$$

The aerodynamic efficiency in relation to its maximum value amounts to

$$\frac{C_L/C_D}{(C_L/C_D)_{\mathrm{MD}}} = \frac{2(C_L/C_{L_{\mathrm{MD}}})}{1 + (C_L/C_{L_{\mathrm{MD}}})^2} = \sqrt{C_{D_L}(2 - C_{D_L})} \tag{12.21}$$

which proves that $0 \le C_{D_L} < 2$. The relationships derived above are depicted in a generalized form in Figure 12.3.

In principle, a two-term drag polar can be used at low as well as high Mach numbers. In the latter case, Mach number-dependent coefficients have to be used,

$$C_D = C_{D_0}(M) + K_L(M) C_L^2 \tag{12.22}$$

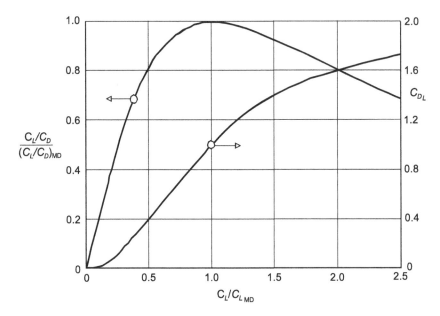

Figure 12.3   Generalized aerodynamic efficiency and the log-derivative C_{D_C} for a two-term drag polar

leading to maximum aerodynamic efficiency dependent on the Mach-number. However, a two-term approximation may be unsatisfactory at transonic Mach numbers. The accuracy is improved when the polar is approximated by a parabola which is symmetrical about a positive C_L^* axis (Section 4.4),

$$C_D = C_D^*(M) + K_L^*(M)\{ C_L - C_L^*(M)\}^2 \tag{12.23}$$

This three-term approximation requires suitably selected values of $C_D^*(M), C_L^*(M)$ and $K_L^*(M)$. Figure 4.9 (a) shows that, for a given polar, C_L^* and C_D^* are determined from the horizontal tangent and $(C_L/C_D)_{MD}$ is found from the tangent from the origin to the drag polar. It can be shown that

$$K_L^*(M) = \left[4(C_L/C_D)_{MD}\{C_D^*(C_L/C_D)_{MD} - C_L^*\}\right]^{-1} \tag{12.24}$$

The best curve fit near the design condition is found by trying different values of C_L^*. It is then noted that $K_L^*(M)$ is larger than $K_L(M)$ in Equation (12.22). Expressing C_D^*, C_L^* and K_L^* as polynomials in M yields a complete set of drag data in a format suitable for numerical optimization of the optimum flight condition.

12.2.3   Altitude Constraint

For an aircraft flying at a specified pressure altitude, the equilibrium $L = W$ dictates that the minimum drag condition complies with

$$C_L M^2 = \frac{W}{\frac{1}{2}\gamma p\, S} = \text{constant} \quad \rightarrow \quad \text{d}\log C_L + 2\,\text{d}\log M = 0 \tag{12.25}$$

and substitution into Equation (12.14) yields

$$C_{D_{\mathcal{L}}} = 1 + 0.5 C_{D_M} \tag{12.26}$$

Visualized in Figure 12.2, this condition is identical to Equation (12.15) below the critical Mach number where $C_{D_M} \approx 0$. The occurrence of high compressibility drag leads to $C_{D_M} > 0$ and a constrained minimum drag condition $C_{D_{\mathcal{L}}} > 1$; hence, $C_L > C_{L_{\mathrm{MD}}}$. As an example, Figure 12.2 depicts the altitude constraint $C_L M^2 = 0.30$ which has a point of tangency to a C_L / C_D contour in point B, defining the intersection with Equation (12.26). If we had derived this condition by intersecting the same altitude curve with $C_{D_{\mathcal{L}}} = 1$ the erroneous result would be point C with a much higher Mach number and lower C_L than point B.

12.3   The Parameter *ML/D*

Classical cruise performance optimization of jet aircraft assumes that the TSFC is independent of the flight speed. Overall efficiency is then proportional to Mach number and, according to Equation (12.6), the objective function is $M C_L / C_D$, or $M L / D$. In terms of log-derivatives its maximum value is determined from

$$\mathrm{d} \log M + \mathrm{d} \log C_L - \mathrm{d} \log C_D = 0 \tag{12.27}$$

After expansion of $\mathrm{d} \log C_D$ in terms of the lift coefficient and Mach number, Equation (12.13) is used to find that

$$(1 - C_{D_{\mathcal{L}}}) \mathrm{d} \log C_L + (1 - C_{D_M}) \mathrm{d} \log M = 0 \tag{12.28}$$

where the log-derivatives $C_{D_{\mathcal{L}}}$ and C_{D_M} are defined by Equations (12.11) and (12.12), respectively. The parameter $M L / D$ can be interpreted as the airframe's contribution to specific range, an acceptable simplification when aircraft with similar engines are compared within a small range of Mach numbers. However, $M L / D$ should not be used for comparing range performances of aircraft cruising at widely different speeds.

12.3.1   Subsonic Flight Mach Number

The traditional approach to the analysis of optimum cruising flight is based on the assumption that there is no drag due to compressibility ($C_{D_M} = 0$). For this case, meaningful results are obtained only by imposing a constraint on some operational parameter limiting the flight speed to a subcritical Mach number.

- For given Mach number, the optimizer according to Equation (12.20) is $C_{D_{\mathcal{L}}} = 1$. This condition defines the altitude where $C_L = C_{L_{\mathrm{MD}}}$.
- For given altitude, the product $C_L M^2$ is prescribed. Equation (12.28) yields $C_{D_{\mathcal{L}}} = 0.5$ which is identical to maximizing $\sqrt{C_L}/C_D$. For a two-term parabolic polar, Equation (12.20) yields $C_L = C_{L_{\mathrm{MD}}} / \sqrt{3}$, speed $3^{1/4} V_{\mathrm{MD}}$ and $C_D = \frac{4}{3} C_{D_0}$. This is a classical solution for the optimum cruise speed of a jet aircraft.

• For a constant engine rating – this is (nearly) equivalent to a constraint on the corrected thrust T/δ – the equation for horizontal equilibrium yields

$$C_D M^2 = \frac{2\,T/\delta}{\gamma p_{sl} S} \qquad (12.29)$$

Hence, the term $C_D M^2$ is constant, or $d\log C_D + 2\,d\log M = 0$. Substitution into Equation (12.28) yields $C_{D_\mathcal{L}} = 2/3$. This condition is identical to maximizing C_L^2/C_D^3 for which $C_L = C_{L_{MD}}/\sqrt{2}$, $C_D = 1.5\,C_{D_0}$ and $V = 2^{1/4} V_{MD}$, another classical solution.

These conditions are incompatible and there exists no unconstrained optimizer.

12.3.2   Transonic Flight Mach Number

The 'fingerprint' contours of constant MC_L/C_D in Figure 12.4 are obtained by multiplying C_L/C_D contours (Figure 12.2) by the Mach number. Partial optima are defined as follows:

• The optimum lift coefficient for given Mach number $C_{D_\mathcal{L}} = 1$ (curve I) is identical to that for maximum C_L/C_D.
• The optimum Mach number for given lift coefficient obtained from Equation (12.28) is in the drag rise: $C_{D_M} = 1$ (curve II).

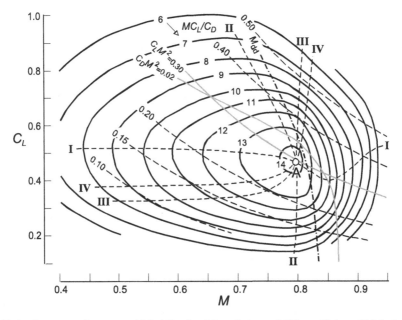

Figure 12.4   Contours of constant MC_L/C_D. Partial optimizers: I lift coefficient; II Mach number. Constrained optima: III specified altitude; IV specified thrust

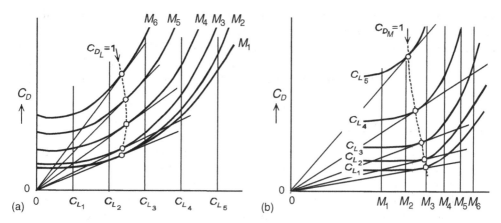

Figure 12.5 Graphical construction of the partial optima for ML/D. (a) Optimum lift coefficient for given Mach number. (b) Optimum Mach number for given lift coefficient

- The unconstrained maximizer of ML/D id defined by the intersection of curves I and II in point A. In the example this flight condition is located just below the drag divergence Mach number, but it is clearly in the drag rise since the aerodynamic efficiency has decreased to 96% of $(C_L/C_D)_{MD}$.

Although the unconstrained optimum is illustrated by means of a contour plot, its allocation does not require the drawing up of these explicit contours. A direct solution of the solution is provided by the graphical construction using two representations of drag curves illustrated in Figure 12.5. The partial optimizers $C_{D_L} = 1$ and $C_{D_M} = 1$ are transferred as curves I and II, respectively, to the C_L versus M plane (Figure 12.4). Their intersection (point A) defines the global optimizer. A global optimizer may not be achievable due to one of the following equality constraints.

- For an equality constraint on the pressure altitude, the optimizer follows from substitution of $d \log C_L + 2 \, d \log M = 0$ into Equation (12.28) resulting in $2C_{D_L} = 1 + C_{D_M}$. Depicted as curve III, this condition is identical to the one found for subsonic Mach speed in the previous paragraph. Curve III bends sharply upwards for $M > 0.70$ and intersects curves I and II in Point A. Its intersection with the altitude constraint defines the constrained optimum very close to point A.
- For an equality constraint on the corrected thrust, the optimizer follows from substitution of $d \log C_D + 2 \, d \log M = 0$ into Equation (12.28) resulting in $C_{D_L} = 2 + C_{D_M}$. Depicted as curve IV, this condition intersects curves I, II and III in point A. The intersection of curve IV with the thrust constraint defines the constrained optimum above point A.

If the equality constraints on altitude and thrust were treated as inequality constraints allowing to fly at reduced speed and/or altitude, the global optimum flight condition point A would become the best cruise condition for the present example.

Curves I and II are readily obtained without the use of explicit L/D contours. This is shown in Figure 12.5 where tangents from the origin identify the condition $\partial C_D / \partial C_L = C_D / C_L$ and $\partial C_D / \partial M = C_D / M$. After transfer to Figure 12.4 the intersecting of these partial optima yields the global unconstrained optimizer. This demonstrates that an unconstrained optimum flight condition for maximum ML/D exists only when compressibility effects are present. *Hence, the global optimizer is located in the drag rise.* For the example aircraft, the criterion $\partial C_D/\partial M = C_D/M = 0.033$ defines a Mach number below the drag divergence Mach number with $\partial C_D / \partial M = 0.10$. A more accurate criterion for the maximum specific range will be derived below.

12.4   The Range Parameter

The range parameter $\mathcal{P}$ defined by Equation (12.5) is a primary characteristic contributing to the energy efficiency of an airliner. It is obtained from multiplication of the overall efficiency and the aerodynamic efficiency. As explained in Chapter 3, the overall efficiency η_0 of a gas turbine-based propulsion system is a function of the Mach number and the corrected thrust T/δ representing the engine rating. Figure 12.6 is an example derived from engine performance information such as Figure 3.4. It shows T/δ made dimensionless with the take-off thrust T_{TO}. Since in steady level flight

$$T/\delta = D/\delta = 0.5\,\gamma\,p_{sl}M^2C_D S \tag{12.30}$$

the corrected thrust is a function of C_D and M. Consequently, *the range parameter is fully determined by C_L and M and the range parameter contours on Figure 12.7 are independent of the plane's gross weight.*

12.4.1   Unconstrained Optima

General conditions for the maximum specific range have been derived in [25]. Since Mach number variation has a far more significant effect than engine rating, the results have been simplified by considering only the log-derivative of the overall efficiency with respect to the Mach number,

$$\eta_{\mathcal{M}} \overset{\text{def}}{=} \frac{d\log \eta_0}{d\log M} = \frac{M}{\eta_0}\frac{d\eta_0}{dM} \tag{12.31}$$

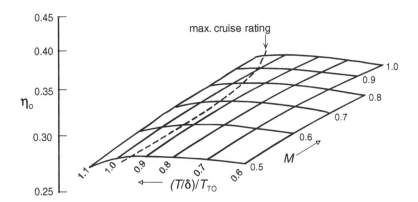

Figure 12.6   Overall efficiency of a high bypass ratio turbofan on a logarithmic scale

Normally, this parameter has a value between zero for constant overall efficiency and one and for constant TSFC. Conditions for achieving the maximum range parameter are derived by logarithmic differentiation,

$$\mathrm{d}\log \mathcal{P} = \mathrm{d}\log \eta_o \, C_L/C_D = \mathrm{d}\log \eta_o + \mathrm{d}\log C_L - \mathrm{d}\log C_D = 0 \qquad (12.32)$$

which can be expanded in terms of partial derivatives as follows:

$$(1 - C_{D_\mathcal{L}})\,\mathrm{d}\log C_L + (\eta_M - C_{D_M})\,\mathrm{d}\log M = 0 \qquad (12.33)$$

This leads to the following partial optima for C_L and M:

$$C_{D_\mathcal{L}} = 1 \quad \text{or} \quad \frac{\partial\,C_D}{\partial\,C_L} = \frac{C_D}{C_L} \quad (\text{constant } M) \qquad (12.34)$$

$$C_{D_M} = \eta_M \quad \text{or} \quad \frac{\partial\,C_D}{\partial\,M} = \eta_M \frac{C_D}{M} \quad (\text{constant } C_L) \qquad (12.35)$$

The unconstrained global optimizer (point A) is found by combining these conditions. At subsonic and transonic Mach numbers, the overall efficiency of a turbofan engine mostly increases monotonically with Mach number; hence, Equation (12.35) indicates that the unconstrained optimum is in the drag rise. In other words: *the best cruise Mach number of a jet airplane*

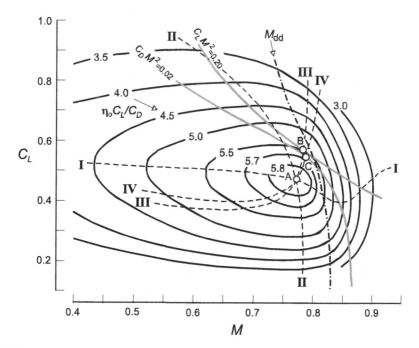

Figure 12.7 Range parameter of a turbofan powered airliner. Partial optima: I lift coefficient; II Mach number. Constrained optima: III specified altitude; IV specified thrust

is in the drag rise. Partial optima for C_L and M are indicated in Figure 12.7 as curves I and II, respectively. Although the general shape of this diagram is similar to Figure 12.4 there are some noticeable differences. In particular, since $\eta_M < 1$, the best cruise condition (point A) is located at a lower Mach number than for the case of maximum ML/D – the difference is approximately $\Delta M = 0.025$ – and the range parameter is less sensitive to flying at an off-optimum flight Mach number.

12.4.2 Constrained Optima

For an equality constraint on the altitude, the optimum is found by inserting a constant $C_L M^2$ into Equation (12.33) which yields

$$C_{D_c} = 1 + 0.5\,(C_{D_M} - \eta_M) \tag{12.36}$$

In Figure 12.7 this condition is represented by curve III, which intersects the specified value of $C_L M^2$ in point B. The optimizer with an equality constraint on the corrected thrust is found by inserting a constant $C_D M^2$ which yields

$$C_{D_c} = \frac{2 + C_{D_M}}{2 + \eta_M} \tag{12.37}$$

This condition is represented by curve IV which intersects the specified value of $C_D M^2$ in point C. These results can be simplified for aircraft having a single parabolic drag polar cruising in subcritical flow ($C_{D_M} = 0$) which yields the optimum conditions of Table 12.1. For idealized jet propulsion ($\eta_M = 1$) they are in accordance with classical performance analysis.

12.4.3 Interpretation of η_M

The log-derivative η_M represents the percentage change in overall engine efficiency as a fraction of the same percentage change in Mach number. In order to make it a practical concept, this parameter deserves further explanation based on its relation to the corrected TSFC,

$$\eta_M = 1 - \frac{M}{C_T/\sqrt{\theta}} \frac{d(C_T/\sqrt{\theta})}{dM} \tag{12.38}$$

The overall efficiency can be expressed as the product of the combustion efficiency η_{cb}, thermal efficiency η_{th} and propulsive efficiency η_{prop},

$$\eta_o = \eta_{cb}\eta_{th}\eta_{prop} \tag{12.39}$$

Table 12.1 Constrained optima for maximum subsonic specific range.

Parameter	Altitude constraint	Thrust constraint
C_{D_c}	$1 - 0.5\,\eta_M$	$(1 + 0.5\,\eta_M)^{-1}$
$C_L/C_{L_{MD}}$	$\{(2 - \eta_M)/(2 + \eta_M)\}^{1/2}$	$(1 + \eta_M)^{-0.5}$
V/V_{MD}	$\{(2 + \eta_M)/(2 - \eta_M)\}^{1/4}$	$\{(1 + \eta_M)/(1 + 0.5\,\eta_M)\}^{1/2}$
$(L/D)/(L/D)_{MD}$	$(1 - 0.25\,\eta_M^2)^{1/2}$	$(1 + \eta_M)^{1/2}/(1 + 0.5\,\eta_M)$

Logarithmic differentiation yields

$$\eta_M = \frac{d \log \eta_{cb}}{d \log M} + \frac{d \log \eta_{th}}{d \log M} + \frac{d \log \eta_{prop}}{d \log M} \qquad (12.40)$$

The first of these terms is close to zero, the second has a small positive value due to the ram effect on the engine pressure ratio. The third term is dominant – it can be associated with the Froude equation (3.15). This approach can be further evolved resulting in $\eta_M = 1 - \text{factor} \times \eta_{prop}$ where the factor depends on the specific thrust [32]. Since $\eta_{prop} = 0$ for $M = 0$, η_M approaches 1.0 for $M \downarrow 0$. For subsonic Mach numbers η_M varies typically between 0.8 for low bypass turbofans, 0.5 for high bypass turbofans and 0.3 for very high bypass ratios. The propulsive efficiency of a turboprop system at cruise conditions varies little with subsonic speed; hence, $\eta_M \approx 0$.

Application to cruise performance optimization requires input of a realistic value of η_M. The following approaches are suggested.

- If engine performance data is available, the corrected TSFC versus Mach number is approximated by a linear relationship,

$$C_T / \sqrt{\theta} = C_0 (1 + C_M M) \quad \rightarrow \quad \eta_M = (1 + C_M M)^{-1} \qquad (12.41)$$

Figure 12.8 shows an example for a typical high bypass turbofan. In the Mach number range of interest, the variation of η_M with Mach number is relatively small so that for optimization purposes η_M can be considered as constant.

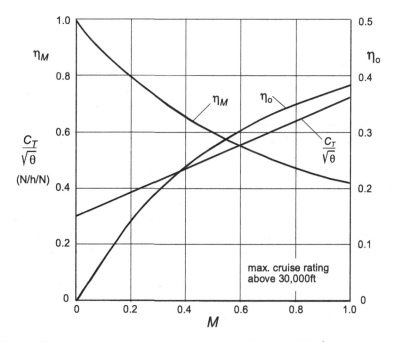

Figure 12.8   Properties of a high bypass turbofan with $C_0 = 0.3 \text{ h}^{-1}$ and $C_M = 1.4$

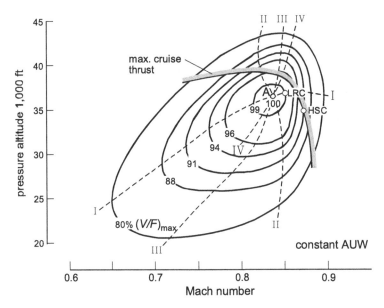

Figure 12.9 Specific range of a jet airliner with given gross weight. Partial optima: I constant Mach number, II constant lift coefficient. Constrained optima: III altitude constraint, IV thrust constraint

- If only the engine bypass ratio or the specific thrust is available, the corrected TSFC can be approximated as an exponential function of the Mach number: $C_T/\sqrt{\theta} \propto M^n \to \eta_M = 1 - n$. Statistical information in [24] or Figure 10 of [32] can be used to select the parameter n.

12.4.4 Optimum Cruise Condition

The present analysis has proven that the range parameter is a unique function of the lift coefficient and the Mach number (Figure 12.7). Although the translation of airframe aerodynamic properties and engine fuel consumption into the range parameter requires some computational effort, the result forms a valuable starting point for cruise performance analysis and optimization. On the other hand, cruise performance analysis requires the specific range to be determined from Equation (12.6) using the condition of vertical equilibrium to compute the lift coefficient from the pressure altitude. Figure 12.9 shows an example of specific range versus pressure altitude and Mach number. When preparing this diagram for a low GW, it is found that the maximum specific range (point A) is not constrained by the available thrust. For high weights the engine rating limit or the buffet margin required (Figure 10.14) or the maximum cabin pressure may drive the flight condition to a Mach number and/or altitude below the unconstrained optimum.

Since the GW varies during cruising, it seems necessary to prepare different versions of Figure 12.9 for each weight. This can be avoided by selecting W/p instead of the pressure altitude along the vertical axis and plotting VW/F instead of V/F contours. An engine rating constraint is inserted as an upper limit on T/p and the buffet margin becomes a single curve. The general appearance of such a diagram is similar to Figure 12.9 and only one figure is required to cover different weights. A choice of the initial and final operating conditions – for example, for long-range cruising (LRC) or high speed cruising (HSC) – is necessary for carrying out the integration of specific range, the subject of the following text.

12.5   Range in Cruising Flight

Different from the point performance problem treated in the previous sections, the distance attainable in cruising flight is an integrated performance. In order to maximize the range, an optimal control law has to be be derived for the independent variables. Several cruise techniques leading to closed-form solutions have been reported in the literature; an overview is given in [22]. Solutions for the case of specified initial conditions are accurate on the provision that the flight condition does not vary considerably during the flight. However, the optimum initial cruise condition and maximum range are inaccurate if propeller efficiency or engine TSFC are assumed to be invariable with speed and altitude. Computed optimum cruise conditions may even be unrealistic when compressibility effects are ignored. The unified cruise performance analysis exposed in this section is applied to derive analytical solutions for the cruise range corresponding to several control laws. It applies to airplanes powered by any category of gas turbine engines flying at low as well as high airspeeds.

12.5.1   Bréguet Range Equation

The widely used expression ascribed to Louis Bréguet (1880–1955) for the range obtained in steady level cruising flight was originally derived for propeller aircraft with piston engines flying at constant angle of attack. If we denote the engine's SFC based on shaft horse power (SHP) by C_P, propeller efficiency by η_{pr}, the initial GW by W_i and the final GW by W_e, Bréguet's classical equation reads as follows:

$$R_{\mathrm{Br}} = \frac{\eta_{\mathrm{pr}}}{C_P} \frac{C_L}{C_D} \ln \frac{W_i}{W_e} \qquad (12.42)$$

This expression can be used for turboprop engines as well, provided SFC is based on equivalent engine power (Section 3.3). When jet engines were introduced during the 1940s, it was noted that Bréguet's equation can be modified so that it applies to the range of jet aircraft flying with constant speed and angle of attack,

$$R = \frac{V}{C_T} \frac{C_L}{C_D} \ln \frac{W_i}{W_e} \qquad (12.43)$$

Equations (12.42) and (12.43) can be replaced by a single one. The range in quasi-steady cruising flight is obtained from integration of the specific range according to Equation (12.1),

$$R = \int_{W_e}^{W_i} \frac{V}{F} \, \mathrm{d}W \qquad (12.44)$$

If V/F were constant during the flight the solution would simply be

$$R = \frac{V}{F} (W_i - W_e) = \frac{V}{F} W_f \qquad (12.45)$$

where W_f denotes the weight of fuel burnt. In reality, the specific range increases during the flight due the decreasing GW. This is taken into account by substitution of Equation (12.5) for the specific range $\mathcal{P}$,

$$R = \frac{H}{g} \int_{W_e}^{W_i} \eta_0 \frac{C_L}{C_D} \frac{\mathrm{d}W}{W} = \frac{H}{g} \int_{W_e}^{W_i} \mathcal{P} \frac{\mathrm{d}W}{W} \qquad (12.46)$$

If the angle of attack and the flight speed are controlled so that $\mathcal{P}$ is constant during the flight, the integral can be solved analytically,

$$R = \frac{H}{g} \mathcal{P} \ln \frac{W_i}{W_e} \qquad (12.47)$$

Substitution of the overall efficiency according to Equations (3.12) and (3.28) yields the Bréguet range equation for jet and propeller aircraft, respectively. Equation (12.47) represents therefore a unified expression for the range of aircraft with various types of propulsion systems when flying with a constant range parameter.

The theoretical maximum distance is covered when $\mathcal{P}$ stays constant at its maximum value. For transonic jet aircraft this means that C_L and M are constant. Since this is not always feasible, alternative schedules are investigated using the following notations for frequently occurring combinations of terms:

$$\zeta \stackrel{\text{def}}{=} W_e/W_i = 1 - W_f/W_i \quad \text{and} \quad y_i \stackrel{\text{def}}{=} C_{L_i}/C_{L_{MD}} \qquad (12.48)$$

12.5.2   Continuous Cruise/Climb

Aircraft cruising at constant angle of attack and Mach number fly at constant L/D and increase their altitude so that W/p is constant as well. In the (isothermal) stratosphere the aircraft flies with constant engine rating, and hence constant corrected thrust and overall efficiency. This flight schedule is known as the continuous cruise/climb (index ccc) technique[1] for which integration of Equation (12.46) yields

$$R_{ccc} = \frac{H}{g} \mathcal{P} \ln(1 - \zeta)^{-1} \qquad (12.49)$$

If the initial conditions are such that $\mathcal{P}$ has its maximum value this equation represents the longest range theoretically achievable in quasi-steady flight. Although Equations (12.49) and (12.47) are identical, the original Bréguet equation was derived for airplanes with constant propeller efficiency cruising at constant altitude. The continuous cruise/climb technique may, however, not form a practical control law since the continuously varying altitude is considered objectionable from the air traffic control (ATC) point of view. Several alternative schedules are available resulting in slightly shorter ranges than the Bréguet range.

---

[1] Some extra thrust is required to climb to higher altitude and strictly speaking, there is no equilibrium between thrust and drag. However, having completed the cruising flight, the aircraft descends back to the initial altitude with reduced thrust, thus regaining the range loss during the climb.

12.5.3   Horizontal Cruise, Constant Speed

Different from the continuous cruise/climb, cruising horizontally at constant speed (index hccs) leads to a decreasing angle of attack and lift coefficient. The engines must be gradually throttled back so that the thrust matches the decreasing drag due to lift. This procedure is readily accomplished with modern engine control systems and it is beneficial from the operational point of view. Integration of $\mathcal{P}$ for the case of a two-term parabolic drag polar [20] results in

$$R_{\text{hccs}} = 2\frac{H}{g}(\eta_0)_i \left(\frac{C_L}{C_D}\right)_{\text{MD}} \{\arctan y_i - \arctan y_i (1 - \zeta)\} \qquad (12.50)$$

The difference between the two angles in the parentheses can be evolved by means of goniometric equations into a single one. After introduction of the range parameter for the initial cruise condition one finds

$$R_{\text{hccs}} = \frac{H}{g}\mathcal{P}_i (y_i^{-1} + y_i) \arctan \left\{\frac{\zeta}{y_i^{-1} + y_i (1 - \zeta)}\right\} \qquad (12.51)$$

Since for optimum cruising usually $C_{L_i} \leq C_{L_{\text{MD}}}$ (or $y_i \leq 1.0$) the angle defined by the arctan is small so that it can be approximated by its argument, resulting in

$$R_{\text{hccs}} = \frac{H}{g}\mathcal{P}_i \zeta \left(1 - \zeta\frac{y_i^2}{1 + y_i^2}\right)^{-1} \qquad (12.52)$$

This equation overestimates the exact solution by less than 1%. If the initial cruise altitude is sufficiently high to obtain $y_i \approx 1$ the constant altitude/speed schedule predicts a slightly shorter range than the continuous cruise/climb range. If, however, the initial cruise altitude is constrained by the available engine thrust, Equation (12.52) may predict a significant fuel penalty for a long range flight. As soon as the engine rating falls below the optimum cruise rating, the overall efficiency begins to deteriorate. The usual operational practice is therefore to approximate the continuous cruise/climb by executing several horizontal cruise segments with intermediate climb segments to increased flight levels. This is known as the stepped cruise/climb.

12.5.4   Horizontal Cruise, Constant Lift Coefficient

The constant incidence schedule has traditionally been used to compute the range of jet aircraft. In subsonic flow C_L and C_D are constant, speed is reduced steadily and the engines are throttled back so that T/W is constant as well. The following expression is found for constant η_M by integration of the specific range:

$$R_{\text{hccl}} = 2\frac{H}{g}\mathcal{P}_i \frac{1 - (1 - \zeta)^{\eta_M/2}}{\eta_M} \qquad (12.53)$$

This equation applies to any value of $\eta_\mathcal{M}$ between zero and one. If the TSFC of a jet engine is assumed independent of speed, $\eta_\mathcal{M} = 1$ and Equation (12.53) complies with the classical square root range equation,

$$R_{\text{hccl}} = 2\frac{H}{g}\mathcal{P}_i(1 - \sqrt{1 - \zeta})$$ (12.54)

The decreasing flight speed reduces the overall efficiency of turbofan engines causing a range loss at subcritical Mach numbers. If the flight is initiated in the drag rise, speed reduction improves the aerodynamic efficiency and the range parameter may actually stay nearly constant. From the operational point of view, the continuously reducing flight speed is objectionable. The overall efficiency of a turboprop system is nearly invariable in high-speed flight and Equation (12.53) applies to propeller aircraft for $\eta_\mathcal{M} \downarrow 0$. It is worth noting that this outcome appears to be numerically identical to the Bréguet range.

12.6 Cruise Procedures and Mission Fuel

The continuous cruise/climb is the only schedule based on constant overall propulsive and aerodynamic efficiencies. In the absence of an active constraint on altitude or engine rating, the longest cruising range with a given amount of fuel is thus obtained when the range parameter $\mathcal{P}$ has its unconstrained maximum value throughout the flight (Section 12.4). For subsonic as well as transonic Mach numbers this schedule results in the generalized Bréguet range Equation (12.47). Other cruise techniques lead to a few percentage range penalty, dependent on range.

12.6.1 Subsonic Flight

The case of zero compressibility drag is important mainly for propeller aircraft although jet aircraft may be forced to fly at reduced speed and/or altitude; for example, when engine failure has occurred [30]. Steady level flight is a practical schedule approximating the cruise/climb range closely. For this case the mean value of the range parameter should be maximized instead of its initial value. In Equation (12.51) this is manifest in the ratio $y_i = C_{L_i}/C_{L_{\text{MD}}}$. Its modification Equation (12.52) defines the optimum initial altitude for given speed as

$$y_i = \frac{1}{\sqrt{1 - \zeta}}$$ (12.55)

whereas the best initial speed at given altitude is

$$y_i = \sqrt{\frac{2 - \eta_\mathcal{M}}{(2 + \eta_\mathcal{M})(1 - \zeta)}}$$ (12.56)

These optima are identical for (idealized) propeller aircraft with $\eta_\mathcal{M} = 0$, defining the initial cruise speed to be lower or the altitude to be slightly above the condition for minimum drag ($y_i > 1$). For jet propulsion with $\eta_\mathcal{M} > 0$, the partial optima are incompatible – this is not surprising in view of the findings in Section 12.3. In this case, cruise performance is improved by flying fast and high until an engine thrust limit is met, forming the ultimate constraint. The best range performance is obtained for an initial lift coefficient determined by

$$y_i = \frac{1}{\sqrt{(1 + \eta_M)(1 - \zeta)}} \tag{12.57}$$

Contrary to what is usually stated in the literature, this equation shows that for long-range flight the optimal initial lift coefficient is quite close to the minimum drag condition. For example, a fuel fraction of 0.30 for a turbofan powered airplane with $\eta_M = 0.6$ yields $y_i = 0.945$. The initial flight speed is then only 3% above the speed for minimum drag.

12.6.2 Transonic Flight

The best cruising flight of a transonic jet airplane is executed in the drag rise. Figure 12.10 shows an enlarged sector of Figure 12.7 near the maximum range condition (point A) which yields the unconstrained maximum range in a continuous cruise/climb,

$$R_{max} = (H/g)\mathcal{P}_{max} \ln(1 - \zeta)^{-1} \tag{12.58}$$

Although in the present example the cruise thrust limit allows the aircraft to fly continuously at this combination of C_L and Mach number, the required cruise/climb control law is, in general, not selected in operational practice. With reference to Figure 12.9, long-range cruising (LRC) is executed in a flight condition where the specific range is 98 or 99% of its maximum value, provided the higher cruise speed makes the fuel penalty acceptable in view of the time saved. A larger fuel penalty may be acceptable for short haul flights and the aircraft may be flown

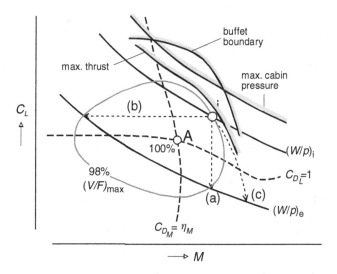

Figure 12.10   Flight schedules for high speed cruising

at a lower altitude with the engines operating at their maximum cruise rating, known as high speed cruising (HSC).[2]

As an example we assume that in the initial flight condition – point 'i' in Figure 12.10 – the engines are running at the recommended cruise rating and the altitude is above the minimum drag condition. Several flight schedules can be considered with ranges slightly below the unconstrained maximum.

(a) During a horizontal constant speed cruise, the GW decreases from the initial value W_i to the final value W_e, the ambient pressure is constant at p_i. The gradually decreasing C_L allows the range parameter to remain close to the partial optimum $C_{D_L} = 1$ and the range is longer than for a continuous cruise/climb in point i.

(b) During a horizontal constant C_L cruise, the Mach number decreases and the specific range varies close to the partial optimum $C_{D_M} = \eta_M$. In contrast with subsonic flight, this ensures good range performance. However, a significant speed reduction during the flight is unlikely to be preferred in commercial operation.

(c) A horizontal constant engine rating cruise penalizes the range significantly because the drag rise is deeply penetrated. For a reduced (but constant) rating with point i moved to the left and/or downwards, good range performance is combined with the drawback of increasing speed.

(d) In a stepped cruise/climb (not indicated in Figure 12.10) the lift coefficient initially follows schedule (a) in horizontal flight at constant speed. As soon as the gross weight reduction enables a 4 000 ft altitude increment, the plane climbs to the altitude for minimum drag and starts another steady level cruise segment. Dependent on the number of climb steps, the complete flight closely approximates $C_{D_L} = 1$.

The question remains, at what altitude and speed the cruising flight should be commenced. For given Mach number, the specific range is maximum for $C_{D_L} = 1$ but for a level constant speed cruise C_L decreases during the flight and the optimum initial C_L is higher. Moreover, the altitude for maximum range may initially not be attainable due to an engine rating or a buffet margin constraint as indicated. The preferred initial cruise condition in Figure 12.10 is then at the intersection of the 98% or 99% maximum V/F contour with the critical constraint.

12.6.3 Cruise Fuel

The present analysis has resulted in expressions for the range in cruising flight applying to several Mach number regimes and flight schedules. The horizontal constant speed schedule is mostly used for short to medium range flights and the stepped cruise/climb is preferred for long distance flights. Equation (12.52) for horizontal cruising at constant Mach number is therefore considered as representative of operational practice. Although it is derived for the case of a two-term parabolic drag polar this does not invalidate the result, provided compressibility drag

---

[2] Weather conditions may force the pilot to reduce the speed below the theoretical best value in order to improve ride comfort and to increase the buffet margin.

is taken into account. The following versatile expression applies to several flight schedules and has the advantage that it provides the fuel fraction in analytical closed form:

$$\frac{W_f}{W_i} = \left(\frac{H/g}{R} \mathcal{P}_i + 0.5 \, f_{cc} \right)^{-1} \tag{12.59}$$

where $\mathcal{P} = \eta_o C_L / C_D$. Equation (12.52) is approximated by selecting the cruise control factor f_{cc} as follows:

$$f_{cc} = 1 + \frac{R}{6 \, \mathcal{P}_i \, H/g} \quad \text{continuous cruise/climb @ constant } C_L \text{ and } M,$$

$$f_{cc} = 1 - \frac{R}{6 \, \mathcal{P}_i \, H/g} \quad \text{horizontal cruise @ constant } M.$$

Using the first of these factors closely approximates the cruise fuel (index crf) based on Bréguet's range equation. The following simple approximation is based on the stepped cruise/climb with $f_{cc} = 1$:

$$\frac{W_{crf}}{W_i} = \left\{ \frac{H/g}{R} \mathcal{P}_i + 0.5 \right\}^{-1} \tag{12.60}$$

with $H/g = 4\,350$ km (2 700 sm). It is worth noting that this result would also be found from Equation (12.45) provided V/F specific range is computed for the mean GW during the cruising flight, $W = W_i - 0.5 \, W_f$.

The following recommendation for the initial cruise condition is based upon the previous analysis, Section 12.4 in particular.

- Subsonic propeller aircraft: the optimum speed is V_{MD} at the altitude where η_o achieves its maximum value, on the provision that there is no engine power constraint.
- Subsonic jets: the fuel-optimum is on the engine cruise rating (corrected thrust) constraint; see Table 12.1.
- High speed propeller and open rotor aircraft: the fuel-optimum is located halfway between the Mach numbers for maximum aerodynamic efficiency and maximum overall efficiency. The best (pressure) altitude is determined by the condition $C_{D_\mathcal{L}} = 1$.
- Transonic jet aircraft: The unconstrained maximum V/F should be determined first, as well as the 98 or 99% contour (Figure 12.10). Its intersection at high speed with the recommended engine cruise rating defines a good initial cruise condition.

12.6.4 Mission Fuel

The profile of a typical commercial flight is shown in Figure 12.11. Total fuel is subdivided into block fuel and reserve fuel. Block fuel is equal to mission fuel plus fuel for taxying out.[3]

---

[3]Except for multi-mission operations, reserve fuel may be used for taxying after landing.

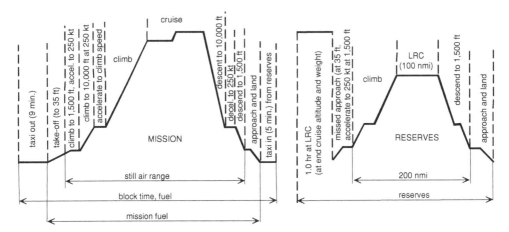

Figure 12.11 Flight profile for an international flight, with components for the mission and the reserve flight

The mission begins at take-off brake release and ends at standstill after landing. Mission fuel is broken down into fuel for take-off, acceleration and climb to the initial cruise altitude and speed, cruising flight, descent and deceleration for approach and landing at the destination. Accurate calculation of the total fuel load is possible when detailed data are available in the form of design weights, drag polar curves, engine thrust and fuel flow diagrams or tables, including corrections for bleed air and power take-off for driving on-board systems.

In the conceptual stage it may be acceptable to derive the mission fuel weight from the cruise fuel with estimated corrections to allow for the non-cruising segments. The dominant factor to be used as input for such a method is the range factor $\mathcal{P}$ for the initial cruise condition. The analysis of range performance in Section 12.6 has resulted in expressions for the range in cruising flight according to several techniques. These were used to derive Equation (12.60) for for the cruise fuel fraction. A procedure for mission fuel prediction recommended in [32] is to determine first the amount of fuel required for a (hypothetical) cruising flight over the required mission range R_{mis} assuming the MTOW to be the initial weight. An amount of fuel is then added referred to as lost fuel $(W_f)_{lost}$. This is not the fuel used in climb but rather the increase in fuel required to take off, climb and acceleration to the cruise altitude and speed compared to cruising the distance covered in climb [24]. The lost fuel is obtained from a simplified energy balance, assuming constant GW. In terms of the energy height at the initial cruise condition,

$$(h_e)_i = h_i + \frac{V_i^2}{2g} \tag{12.61}$$

the lost fuel fraction is determined from the following expression [32]:

$$\frac{(W_f)_{lost}}{W_{TO}} = (1.1 + 0.5\,\eta_M)\,\frac{(h_e)_i}{(\eta_0)_i\,H/g} \tag{12.62}$$

The term $1.1 + 0.5\,\eta_M$ accounts for the factors that are suboptimal in climbing flight: lower speed, altitude and engine efficiency. It shows that the higher the bypass ratio – hence, the lower η_M – the smaller the lost fuel fraction. The fuel consumed during descent, approach and landing is conservatively assumed equal to the fuel used during a cruising flight over the same distance. In other words: lost fuel for this flight segment is ignored. The lost fuel may be converted into a lost range

$$R_{\text{lost}} = (1.1 + 0.5\,\eta_M)(C_L/C_D)_{\text{i}}\,(h_e)_{\text{i}} \qquad (12.63)$$

representing the difference between the actual mission range and the distance travelled in cruising flight with the same amount of fuel. The mission fuel fraction is found by adding the lost range to the mission range,

$$\frac{(W_{\text{f}})_{\text{mis}}}{W_{\text{TO}}} = \frac{R_{\text{mis}} + R_{\text{lost}}}{\mathcal{P}_{\text{i}}\,H/g + 0.5\,(R_{\text{mis}} + R_{\text{lost}})} \qquad (12.64)$$

12.6.5 Reserve Fuel

The following reserve fuel requirements are specified in the FAR.141.645 regulation for international flights.

1. After completing the mission, flying 10% of the mission time.
2. After that, to fly to the most distant alternate airport.
3. Thereafter, to fly 30 minutes at holding speed 1 500 ft above the airport.

Reserve fuel requirements for US domestic operations are specified in FAR 121.639 as follows:

1. After completing the mission, to fly to the most distant alternate airport.
2. Thereafter, to fly for 45 minutes at normal cruising fuel consumption.

There are important differences between different nations and operations in specifications of standard reserve fuel. The Association of European Airlines (AEA) specifies a 200 nm (370 km) diversion flight for short and medium range aircraft or a 250 nm (463 km) diversion for long range aircraft, 30 minutes holding at 1 500 ft (457 m) altitude, and 5% of the mission fuel for contingency reserve [28]. Reserves are less for domestic flights in the USA, for example 130 nm (241 km) diversion and 30 minutes holding at 1 500 ft (457 m) altitude. Long-range international flights may require an extension of the cruise sector, for example one hour at LRC flying. For business aircraft, the total reserves are frequently defined as an extension of the cruising flight by 45 minutes.

Statistical values of the total reserve fuel are between 4.5 and 5% MTOW. A more accurate computation of reserve fuel in the preliminary design stage allows the user to insert the reserve policy specified in the TLRs. This results in the selection of several of the following allowances:

1. A diversion distance equal to R_{div} flown at the MLW. This fuel weight fraction can be expressed as a mission range increment of $r_{\text{div}} R_{\text{div}}$ times the ratio $W_{\text{ML}}/W_{\text{MTO}}$ where r_{div} accounts for the fuel penalty due to an aborted landing and the reduced specific range due to flying at reduced speed and altitude, partly in low-speed configuration.
2. An allowance for a holding period t_{hold} flown at the MLW. This reserve fuel weight fraction is considered as a range increment $r_{\text{hold}} V_{\text{hold}} t_{\text{hold}}$ times the ratio $W_{\text{ML}}/W_{\text{MTO}}$ where r_{hold} accounts for the reduced specific endurance during holding relative to cruising.
3. Contingency fuel is accounted as a fraction – for example, 0.05 or 0.10 – of the mission fuel weight.
4. An extension time Δt_{cr} of the cruising flight, equivalent to a mission range increment of $V_{\text{cr}} \Delta t_{\text{cr}}$.

12.7 Reflection

12.7.1 Summary of Results

Comprehensive treatment of high speed cruise optimization requires the availability of drag polars for different Mach numbers. It is recommended to convert these to contours of constant C_L/C_D values as depicted in Figure 12.2. Drag polars at transonic Mach numbers can be quite different from those at subsonic speeds and the present theory does not assume them to be two-term parabolic approximations. Global unconstrained as well as constrained optimizers are derived from contour plots for $M C_L/C_D$ or $\eta_o C_L/C_D$, as summarized below.

- The lift coefficient and the Mach number are treated as independent control variables affecting the drag coefficient. Different from the classical solutions for subsonic airplanes, the aerodynamic efficiency C_L/C_D has a global unconstrained maximum for a unique combination of C_L and M referred to as $(C_L)_{\text{MD}}$ and M_{MD}, respectively. This optimizer is defined by the intersection of partial optima with respect to the two control variables.
- The parameter $M L/D$ attains a global unconstrained optimizer well into the drag rise at a lift coefficient not much different from $(C_L)_{\text{MD}}$. However, the optimum Mach number is defined by the partial optimum $d\,C_D/d\,M = C_D/M$ which is readily constructed graphically; see Figure 12.5.
- The range parameter $\mathcal{P} = \eta_o L/D$ is a unique function of C_L and M and can be visualized in a similar contour plot as $M L/D$; see Figure 12.7. The unconstrained global optimizer is defined by the partial optimum $d \log C_D / d \log M = \mathrm{dlog}\,\eta_o/\mathrm{dlog}\,M = \eta_{\text{M}}$. It is located at a Mach number between M_{MD} and the Mach number for maximum $M L/D$. The parameter $\eta_{\mathcal{M}}$ is determined mainly by the engine's propulsive efficiency and is readily obtained from engine performance information. Propeller and open rotor engines at high subsonic speeds have $\eta_{\mathcal{M}} \approx 0$. High-bypass ratio turbofans have $\eta_{\mathcal{M}} \approx 0.5$ at transonic flight speeds.
- The range in cruising flight is derived for three flight control schedules: continuous cruise/climb, level flight at constant Mach number and level flight at constant lift coefficient.

An example for high-speed cruising shows that, different from subsonic cruising, these schedules lead to a few percent shorter range compared to the Bréguet range equation. Level flight at constant altitude is recommendable if the attainable cruise altitude for the initial gross weight is sufficiently high to cruise the minimum drag condition.

- A constraint on the cruise altitude imposed by the available engine thrust may cause a significant range loss if the aircraft takes off with maximum take-off weight. A stepped cruise/climb is recommendable for long range flights.
- A simple closed-form equation containing a parameter dependent on the flight control schedule is derived for computing the cruise fuel weight. Analytical approximations for mission and reserve fuel are presented which are useful for the quasi-analytical optimization procedure treated in Chapters 8 to 10.

Although the mission fuel depends to a large extent on the cruise conditions, there is limited freedom to deviate from the global optimum with a small fuel penalty. A penalty of 1 or 2% is accepted on a long-range cruising flight if the increased speed results in a worthwhile time saving.

12.7.2 The Design Connection

The unconstrained global optimizer for the cruise altitude and Mach number derived in this chapter applies to conditions which lead to maximum range for given fuel or minimum fuel for a specified range. A constraint on the (initial) cruise altitude may cause a significant range loss or fuel penalty and the designer may be tempted to increase the installed engine power or thrust. However, it was found in Chapter 8 that the altitude for minimum MTOW is below the altitude for minimum cruise fuel. This may be the explanation why existing aircraft are equipped with engines that do not allow them to climb to the most fuel-efficient cruise altitude after taking off with MTOW. Chapter 9 showed that installed engine thrust may have to be increased in excess of the value for minimum MTOW in order to comply with low speed requirements or with the requirement of reducing mission fuel burn-off. An increased installed thrust also helps to fly at a higher Mach number in the interest of economy.

Selection of the cruise Mach number for optimum aerodynamic wing design is more critical than selecting optimum cruise conditions for a given (wing) design. Figure 12.7 illustrates that the global optimum Mach number is about 0.05 below the drag-divergence value Mach number. It can be argued that for this example the wing is too thin or has too much sweep, resulting in a heavier wing than necessary. This may lead to the decision to reduce the design cruise Mach number. The greatest uncertainty is that airliners are operated by different operators on widely different missions. The design mission with maximum payload or maximum range is therefore not often flown and the most economical operation should be defined with lower than the structural payload over shorter distances than the maximum range.

The question may thus arise in the advanced aircraft designer's mind: *Which figure of merit should I select to optimize my design?* The answer cannot be left to the individual designer. The well-tempered airplane is the result of system engineering in which the designer/analyst plays a modest but invaluable part on the instrument of advanced optimization.

Bibliography

[1] Miele, A., *Flight Mechanics: Theory of Flight Paths*, Addison-Wesley, Reading, MA, 1962.

[2] Hale, F.J., *Introduction to Aircraft Performance, Selection and Design*, John Wiley & Sons, New York, NJ, 1984.

[3] Mair, W.A. and Birdsall, D.L., *Aircraft Performance*, Cambridge Aerospace Series 5, Cambridge University Press, Cambridge, 1992.

[4] Vinh, N.X., *Flight Mechanics of High-Performance Aircraft*, Cambridge Aerospace Series 4, Cambridge University Press, Cambridge, 1993.

[5] Ojha, S.K., *Flight Performance of Aircraft*, Chapters 8 and 9, American Institute of Aeronautics and Astronautics, Education Series, Washington, DC, 1995.

[6] Padilla, C.E., *Optimizing Jet Transport Effeiciency, Performance, Operations & Economics*, McGraw-Hill, New York, 1996.

[7] Eshelby, M., *Aircraft Performance: Theory and Practice*, Chapter 4, Arnold Publishing, London, 2000.

[8] Filippone, A., *Flight Performance of Fixed and Rotary Wing Aircraft*, Chapter 9, Elsevier Aerospace Engineering Series, Oxford, 2006.

[9] Roux, E., *Avions Civils à Reaction: Plan 3 Vues et Données Charactéristiques,* Éditions Élodie Roux, ISBNN 978-2-9529380-2-0, 2007.

[10] Torenbeek, E., and H. Wittenberg, *Flight Physics: Essentials of Aeronautical Disciplines and Technology, with Historical Notes*, Chapters 6 and 9, Springer Verlag, Heidelberg, 2009.

Optimum Cruise Performance and Prediction of Range

[11] Jonas, J., "Jet Airplane Range Considerations", *Journal of the Aeronautical Sciences*, Vol. 14, No. 2, pp. 124–128, February 1946.

[12] Ashkenas, I.L., "Range Performance of Turbojet Airplanes", *Journal of the Aeronautical Sciences*, Vol. 15, No. 2, pp. 97–101, February 1948.

[13] Edwards, A.D., "Performance Estimation of Civil Jet Aircraft", *Aircraft Engineering*, pp. 95–99, April 1950.

[14] Pearson, H., "The Estimation of Range of Jet-Propelled Aircraft", *The Aeronautical Quarterly*, Vol. II, pp. 167–182, November 1950.

[15] Schairer, G.S. and Olason, M.L., "Some Performance Considerations of a Jet Transport Airplane", First Turbine Powered Air Transportation Meeting, Institute of the Aeronautical Sciences, Seattle, Washington, USA, August 9–11, 1954.

[16] Backhaus, G., "Grundbeziehungen für den Entwurf Optimaler Verkehrsflugzeugen", *Jahrbuch der WGLR*, pp. 201–213, 1958.

[17] Ward, K.E., "Contour Plotting for the Transonic Flight-Test Drag Polar", *Journal of Aircraft*, Vol. 5, No. 2, pp. 191–192, March–April 1968.

[18] Page, R.K., "Range and Radius-of-Action Performance Prediction for Transport and Combat Aircraft", AGARD Lecture Series No. 56, Paper No. 1, March 1973.

[19] ESDU, "Approximate Methods for Estimation of Cruise Range and Endurance: Aeroplanes with Turbo-jet and Turbo-fan Engines", Data Sheet No. 73019, October 1973.

[20] Peckham, D.H., "Range Performance in Cruising Flight," Royal Aircraft Establishment Technical Report 73164, March 1974.

[21] Hale, F.J. and Steiger, A.R., "Effects of Wind on Aircraft Performance", *Journal of Aircraft*, Vol. 16, No. 6, pp. 382–387, June 1979.

[22] ESDU, "Introduction to Estimation of Range and Endurance", Data Item No. 73018, Performance, Vol. 6, October 1980.

[23] Bert, C.W., "Prediction of Range and Endurance of Jet Aircraft at Constant Altitude", *Journal of Aircraft*, Vol. 18, No. 10, pp. 890–892, October 1981.

[24] ESDU, "Approximate Methods for Estimation of Cruise Range and Endurance: Aeroplanes with Turbojet and Turbofan Engines", Data Item No. 73019, Revised May 1982.

[25] Torenbeek, E. and Wittenberg, H., "Generalized Maximum Specific Range Performance", *Journal of Aircraft*, Vol. 20, No. 7, pp. 617–622, July 1983.

[26] Sachs, G., and T. Christodoulou, "Reducing Fuel Consumption of Subsonic Aircraft by Optimal Cyclic Cruise", *Journal of Aircraft*, Volume 24, No. 9, pp. 616–622, September 1987.

[27] Menon, P.K.A., "Study of Aircraft Cruise", *Journal of Guidance, Control, and Dynamics*, Vol. 12, No. 5, pp. 631–639, September–October 1989.

[28] AEA, "Definitions and Input for Range and Direct Operating Cost Calculation", Appendix;1, G(T) 5656, 1990.

[29] Martínez-Val, R. and Pérez, E., "Optimum Cruise Lift Coefficient in Initial Design of Jet Aircraft", *Journal of Aircraft*, Vol. 29, No. 4, pp. 712–714, July–August 1992.

[30] Martínez-Val, R. and Pérez, E., "Extended Range Operations of Two and Three Turbofan Engined Airplanes", *Journal of Aircraft*, Vol. 30, No. 3, pp. 382–386, May–June 1993.

[31] Miller, L.E., "Optimal Cruise Performance", *Journal of Aircraft*, Vol. 30, No. 3, pp. 403–405, May–June 1993.

[32] Torenbeek, E., "Cruise Performance and Range Prediction Reconsidered", *Progress in Aerospace Sciences*, Vol. 33, No. 5/6, pp. 285–321, 1997.

[33] Isikveren, A.T., "Identifying Economically Optimal Flight Techniques of Transport Aircraft", *Journal of Aircraft*, Vol. 39, No. 4, pp. 528–544, July–August 2002.

[34] Cavcar, A., and M. Cavcar, "Approximate Solutions for Range at Constant Altitude: Constant High Speed Subsonic Flight of Transport Aircraft", *Aerospace Science and Technology*, Vol. 8, No. 6, pp. 557–567, 2004.

[35] Pargett, D.M., and M.D. Ardema, "Flight Path Optimization at Constant Altitude", *Journal of Guidance, Control, and Dynamics*, Vol. 30, No. 4, pp. 1197–1201, July–August 2007.

[36] Rivas, D., and A. Valenzuela, "Compressibility Effects on Maximum Range Cruise at Constant Altitude", *Journal of Guidance, Control, and Dynamics*, Vol. 32, No. 5, pp. 1654–1658, September-October 2009.

Appendix A

Volumes, Surface and Wetted Areas

The surface area of a fully defined airplane component can be computed by dividing its external surface into a sufficiently large number of nearly-flat panels and adding their areas. Alternatively, a double-curved streamline body (section) can be decomposed into a number of (thin) disks with easy-to-compute ring shaped external areas. Quicker – but less accurate – analytical approximations are useful for conceptual design. This appendix presents formulas for the volume and surface area of airplane major components derived in papers mentioned in the bibliography.

A.1 Wing

The straight-tapered wing of a subsonic airplane can be considered as a linear lofted three-dimensional airfoil. Each half-wing is treated as a truncated pyramid with parallel end planes in the form of airfoil sections. Its volume is related to the planform area S_w by

$$Q_w = \frac{k_Q \, \overline{(t/c)}_w}{\sqrt{1 + \lambda_w}} \, S_w \sqrt{\frac{S_w}{A_w}} \tag{A.1}$$

with the aspect ratio defined as $A_w = b_w^2 / S_w$. The mean thickness ratio $\overline{t/c}$ is the ratio of frontal to planform area and the taper ratio λ is the ratio of tip chord to root chord. The volume factor of a prismatic wing (section) is $k_Q = 1$ – a typical value for a tapered wing is $k_Q = 0.95$. If the surface area is approximated as twice the planform area plus half the frontal area it is related to the volume as follows:

$$(S_w)_{surf} = \{2 + 0.5(\overline{t/c})_w\} \left\{ \frac{Q_w \sqrt{A_w(1 + \lambda_w)}}{k_Q \, t/c} \right\}^{2/3} \tag{A.2}$$

Advanced Aircraft Design: Conceptual Design, Analysis and Optimization of Subsonic Civil Airplanes, First Edition. Egbert Torenbeek.
© 2013 by Egbert Torenbeek. Published 2013 by John Wiley & Sons, Ltd.

It is worth noting that the sweep angle does not appear in these equations. However, the wing of many high-speed airplanes features a kink in the leading and/or the trailing edge. Such wings are composed of distinct inboard and outboard sections. Equations (A.1) and (A.2) are then used to compute the properties of the combined inboard and outboard sections, treating both combinations as a low-aspect ratio wing. Even a blended wing body (BWB) configuration can be treated in this way.

A.2 Fuselage

The fuselage of passenger aircraft with a pressure cabin is approximated as a slender body of revolution with a cylindrical mid section and streamlined nose and tail sections. A fuselage with overall length l_{fus} and further geometry depicted in Figure A.1 has a frontal area according to

$$(S_{\text{fus}})_{\text{front}} = (\pi/4)\, w_{\text{fus}} h_{\text{fus}} = (\pi/4)\, \overline{d}_{\text{fus}}^{2} \tag{A.3}$$

Its volume equals

$$Q_{\text{fus}} = (S_{\text{fus}})_{\text{front}}\,(l_{\text{fus}} - 2\,\overline{d}_{\text{fus}}) = (\pi/4)\, \overline{d}_{\text{fus}}^{2}\, l_{\text{fus}}(1 - 2/\lambda_{\text{fus}}) \tag{A.4}$$

where the fineness ratio is defined as $\lambda_{\text{fus}} = l_{\text{fus}}/\overline{d}_{\text{fus}}$. Surface area and volume are interrelated as follows:

$$(S_{\text{fus}})_{\text{surf}} = 2\,(2\pi\lambda_{\text{fus}})^{1/3}\, Q_{\text{fus}}^{2/3}(1 + 1/\lambda_{\text{fus}}^{2}) \approx \pi\,\overline{d}_{\text{fus}}(l_{\text{fus}} - 1.3\,d_{\text{fus}}) \tag{A.5}$$

and the ratio of frontal to surface area is approximated as

$$\frac{(S_{\text{fus}})_{\text{front}}}{(S_{\text{fus}})_{\text{surf}}} = \frac{1}{4(\lambda_{\text{fus}} - 1.30)} \tag{A.6}$$

This equation can be used for streamlined fuselage bodies with or without a cylindrical mid-section.

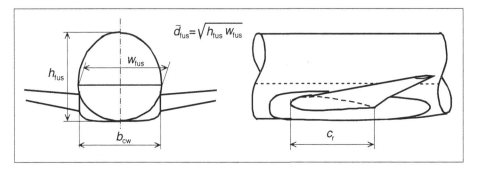

Figure A.1   Fuselage cylindrical mid section and fairing geometry of a low-wing jetliner

A.3 Tail Surfaces

The geometry and location of tail surfaces become available downstream of the airplane sizing and balancing process. However, the design analyst may prefer to make a first estimation of tail surface areas using only wing and fuselage dimensions as input. The following equations were derived with inspiration from [2] and Section 6.3. They predict the empennage area of post-1970 jetliners with a standard deviation of less than 10%. Referring to the geometry in Figure A.1 the horizontal tail volume is obtained from

$$S_h l_h = 0.2\,(S_w \bar{c}_w + 2\,w_{\text{fus}}^2 l_{\text{fus}}) \frac{A_w + 2}{A_w - 2} \qquad \text{(A.7)}$$

The term $(A_w + 2)/(A_w - 2)$ accounts for the effect of wing aspect ratio on the downwash at the tail and $\bar{c}_w = S_w/b_w$ denotes the mean aerodynamic wing chord. The expectation that the reduced downwash at a high-set horizontal tail allows a smaller area is not confirmed by statistics. This may be ascribed to the tail size required to cope with 'deep stall'. The vertical tail volume is approximated as

$$S_v l_v = 0.03\,(S_w b_w + 10\,h_{\text{fus}}^2\,l_{\text{fus}}) \qquad \text{(A.8)}$$

A.4 Engine Nacelles and Pylons

The external surface area of a turbofan nacelle depends primarily on the maximum engine airflow which is determined by the SLS take-off thrust T_{TO} and the specific thrust; that is, the thrust per unit airflow rate. The specific thrust is closely related to the bypass ratio B (BPR). The expression derived in [6] for the nacelle's surface area uses the engine thrust and the BPR as the basic parameters,

$$S_{\text{nac}} = S_{\text{ref}}\,(1 + B)^{0.2}\,(T_{\text{TO}}/T_{\text{ref}})^{0.8} \qquad \text{(A.9)}$$

with $S_{\text{ref}} = 25$ m^2 and $T_{\text{ref}} = 100$ kN. For take-off thrusts between 50 and 250 kN per engine the surface area of N_{eng} nacelles is linearized as follows:

$$S_{\text{nac}} = 0.8\,S_{\text{ref}} N_{\text{eng}}(1 + B)^{0.2}\,(T_{\text{TO}}/T_{\text{ref}} + 0.25) \qquad \text{(A.10)}$$

The surface area of engine pylons which amounts to typically 20% of the nacelle area has to be added.

A.5 Airframe Wetted Area

The analytical methods of this appendix are used to compute gross volumes and surface areas of isolated major airplane components – their accuracy may be improved by calibration with data for a design under development. The wetted area of the airframe is the net surface area obtained after merging of major components and addition of aerodynamic fairings. The resulting airframe surface area is less than the sum of all component gross areas and the combination of the fuselage and the wing needs particular attention. Figure A.1 illustrates a typical fairing which causes a surface area penalty dependent on local wing and body dimensions according to the following considerations:

- High-speed airliners with a low wing mostly have a sizable fairing covering the wing-to-fuselage intersection. The fairing increases the body volume and surface area by a few percent whereas the wing's wetted area is (considerably) reduced. Some business aircraft have the wing passing below the pressure cabin resulting in a rather large fairing increasing the body cross-sectional and surface area over a long stretch.
- The wetted area of a high wing is not much less than its gross surface area since only the lower surface of the carry-through structure is covered by the fuselage. A small fairing is usually sufficient.
- The net (exposed) area of a mid wing is considerably smaller than its gross surface area. The fuselage area covered by the intersecting wing is small and a fairing has a negligible effect on the wetted area.

Instead of computing exposed areas of al major items, adding them and correcting for fairings, a simple estimation for the airframe (less nacelles) with a near-circular fuselage cross-section is obtained from

$$S_{wet} = \pi \overline{d}_{fus}(l_{fus} - 1.3\,\overline{d}_{fus}) + (S_w)_{net}\{2 + 0.5\,(\overline{t/c})_w\} + k_{fair}\,b_{cw}c_r + 2\,(S_h + S_v) \quad (A.11)$$

The first term on the r.h.s. is the gross surface area of the fuselage, the second is the surface area of the exposed wing outboard of the fuselage or the fairing. The net area of a shoulder wing is equal to its gross area reduced by the (small) lower skin surface covered by the fuselage. The third term of Equation (A.11) corrects for the area penalty due to the fairing (Figure A.1) which is proportional to the centre section span b_{cs} and the wing root chord c_r. Typical values of k_{fair} are 0 for a mid wing, 0.2 for a high wing and 0.3 for a low wing. The total airframe wetted area may have to be corrected for miscellaneous items such as winglets, ventral fins and fairings covering landing gear bays, flap tracks and horizontal/vertical tail intersections.

Bibliography

[1] Schmidt, A.H., "A Simplified Method for Estimating the Wetted Area of Aircraft Fuselages", SAWE Paper No. 308, 1962.
[2] Morris, J., and D.M. Ashford, "Fuselage Configuration Studies", SAE Paper No. 670370, April 1967.
[3] Caddell, W.E., "On the Use of Aircraft Density in Preliminary Design", SAWE Paper No. 813, May 1969.
[4] Torenbeek, E., "The Computation of Characteristic Areas and Volumes of Major Aircraft Components in Project Design", TU Delft, Department of Aerospace Engineering, Memorandum M-188, January 1973.
[5] Bullis, R.T., "Geometric Analysis", SAWE Paper No. 1025, May 1974. Also: *SAWE Journal*, pp. 19–29, April/May 1975.
[6] Torenbeek, E., and G.H. Berenschot, "De Berekening van het Omspoeld Gondeloppervlak van Enkel- en Dubbel-stroomstraalmotoren voor Civiele Vliegtuigen" [Computation of the Wetted Area of Turbojet and Turbofan Engines for Civil Aircraft], Memorandum M-445, TU Delft, Department of Aerospace Engineering, January 1983.

Appendix B

International Standard Atmosphere

The International Standardization Organization (ISO) has defined a standard atmosphere based on the model established by the ICAO in their Document 7488/2, Second Edition, 1964. The data in Table B.1 define the state variables at sea level (SL) and the gravitational acceleration at 45° northern latitude on which the potential altitude is based. The (absolute) temperature varies linearly in the troposphere between SL and 11 km with a gradient of −6.5°C/km. The stratospheric layer between 11 and 20 km is isothermal.

Instead of absolute quantities, atmospheric state variables are often quoted as fractions of their SL values as follows:

- relative temperature: $\theta = T/T_{\mathrm{sl}}$
- relative pressure: $\delta = p/p_{\mathrm{sl}}$
- relative density: $\sigma = \rho/\rho_{\mathrm{sl}}$

Table B.1 Basic properties of the International Standard Atmosphere (ISA)

Standard values at SL	
Pressure	$p = 1.013250 \times 10^5$ Pa (760 mm Hg)
Temperature	$T = 15°$C (288.15 K)
Density	$\rho = 1.2250$ kg/m^3
Speed of sound	$a = 340.29$ m/s
Absolute viscosity coefficient	$\mu = 1.7894 \times 10^{-5}$ kg/(ms)
Gravitational acceleration	$g = 9.80665$ m/s^2
Other standard values	
Molar weight of air	$\hat{M} = 28.9644$ kg/kmol
Gas constant of air	$R = 287.05287$ (J/kg)/K
Ratio of specific heats	$\gamma = c_p/c_v = 1.4$
Tropopause	$h_{\mathrm{pot}} = 11\,000$ m
Troposphere	$T = -56.5°$C (216.65 K)

Advanced Aircraft Design: Conceptual Design, Analysis and Optimization of Subsonic Civil Airplanes, First Edition. Egbert Torenbeek.
© 2013 by Egbert Torenbeek. Published 2013 by John Wiley & Sons, Ltd.

Table B.2   Properties of the International Standard Atmosphere (ISA) between sea level and 20 km
geopotential altitude

Alt. (m)	T (K)	θ	a (m/s)	p (Pa)	δ	ρ (kg/m^3)	σ	μ/μ_{sl}
				Standard values at altitude				
0	288.15	1	340.29	101,325	1	1.2250	1	1
500	284.90	0.9887	338.37	95,461	0.9421	1.1673	0.9529	0.9912
1,000	281.65	0.9774	336.43	89,874	0.8870	1.1117	0.9075	0.9823
1,500	278.40	0.9662	334.49	84,556	0.8345	1.0581	0.8638	0.9735
2,000	275.15	0.9549	332.53	79,495	0.7846	1.0065	0.8216	0.9645
2,500	271.90	0.9436	330.56	74,682	0.7371	0.9569	0.7811	0.9556
3,000	268.65	0.9306	328.58	70,108	0.6919	0.9091	0.7421	0.9465
3,500	265.40	0.9210	326.58	65,764	0.6490	0.8632	0.7055	0.9375
4,000	262.15	0.9098	324.58	61,640	0.6083	0.8191	0.6686	0.9283
4,500	258.90	0.8985	322.56	57,728	0.5697	0.7768	0.6341	0.9191
5,000	255.65	0.8872	320.53	54,020	0.5331	0.7361	0.6009	0.9099
5,500	252.40	0.8759	318.48	50,506	0.4985	0.6971	0.5691	0.9006
6,000	249.15	0.8647	316.43	47,181	0.4656	0.6597	0.5385	0.8911
6,500	245.90	0.8534	314.36	44,034	0.4346	0.6238	0.5092	0.8818
7,000	242.65	0.8421	312.27	41,060	0.4052	0.5895	0.4812	0.8724
7,500	239.40	0.8308	310.17	38,251	0.3775	0.5566	0.4544	0.8628
8,000	236.15	0.8195	308.06	35,599	0.3513	0.5252	0.4287	0.8532
8,500	232.90	0.8083	305.93	33,099	0.3267	0.4951	0.4042	0.8436
9,000	229.65	0.7970	303.79	30,742	0.3040	0.4663	0.3807	0.8339
9,500	226.40	0.7857	301.63	28,523	0.2815	0.4389	0.3583	0.8241
10,000	223.15	0.7744	299.46	26,436	0.2609	0.4127	0.3369	0.8143
10,500	219.90	0.7631	297.27	24,474	0.2415	0.3877	0.3165	0.8044
11,000	216.65	0.7519	295.07	22,632	0.2234	0.3639	0.2971	0.7944
12,000	216.65	0.7519	295.07	19,330	0.1908	0.3108	0.2537	0.7944
13,000	216.65	0.7519	295.07	16,510	0.1629	0.2655	0.2167	0.7944
14,000	216.65	0.,7519	295.07	14,101	0.1392	0.2268	0.1851	0.7944
15,000	216.65	0.7519	295.07	12,044	0.1189	0.1937	0.1581	0.7944
16,000	216.65	0.7519	295.07	10,287	0.1015	0.1654	0.1350	0.7944
17,000	216.65	0.7519	295.07	8,786	0.0867	0.1413	0.1153	0.7944
18,000	216.65	0.7519	295.07	7,505	0.0741	0.1207	0.0985	0.7944
19,000	216.65	0.7519	295.07	6,410	0.0633	0.1031	0.0842	0.7944
20,000	216.65	0.7519	295.07	5,475	0.0540	0.0880	0.0718	0.7944

ISA properties for potential altitudes up to 20 km are given in Table B.2. The dynamic viscosity
is computed with Sutherland's Equation

$$\frac{\mu}{\mu_{sl}} = \left(\frac{T}{T_{sl}}\right)^{3/2} \frac{T_{sl} + T_S}{T + T_S}$$

with the Sutherland constant $T_S = 110$ K.

Appendix C

Abbreviations

AC	aerodynamic centre
ACARE	Advisory Council for Aeronautics in Europe
AD	advanced design
ADS	automated design synthesis
ADSE	Aircraft Design and Systems Engineering
AEA	Association of European Airlines
AGARD	Advisory Group for Aeronautical Research and Development
AI	artificial intelligence
AIAA	American Institute of Aeronautics and Astronautics
ANOPP	Aircraft Noise Prediction Program
APU	auxiliary power unit
ASD	accelerate-stop distance
ATC	Air Traffic Control
ATM	air traffic management
ATR	air traffic control
AUW	all-up weight
AWA	all-wing aircraft
BFL	balanced field length
BPR	by-pass ratio
BWB	blended wing body
CAEP	Committee on Aviation Environmental Protection
CFC	carbon fibre composite
CFD	computational fluid dynamics
CG	centre of gravity
CP	centre of pressure
CRDF	contra-rotating ducted fans
CRPF	contra-rotating propfan
CS	centre section
CTO	continued take off

CTP	conventional turboprop
DAR	Design of Aircraft and Rotorcraft
DDTF	direct driven turbofan
DOC	direct operating costs
DPL	design payload
DSM	design structure matrix
DWB	discrete wing and body
EA	elastic axis
EAS	equivalent airspeed
EASA	European Aeronautical Safety Administration
ECU	environmental control unit
EEF	energy efficiency
EFM	engine figure of merit
EIS	entry into service
EOAC	engine out altitude capability
EPNdB	equivalent perceived noise level dB
ESDU	Engineering Sciences Data Unit
ESHP	equivalent shaft horsepower
FAA	Federal Aviation Administration
FAR	Federal Aviation Regulations
FEM	finite element method
FM	flight manual
FOM	figure of merit
FPR	fan pressure ratio
GP	geometric programming
GTF	geared turbofan
GW	gross weight
HFW	hybrid flying wing
HLFC	hybrid laminar flow control
HP	high pressure
HSC	high-speed cruise
HSTP	high speed turboprop
ICAC	initial cruise altitude capability
ICAO	International Civil Aviation Organization
ICAS	International Council of Aeronautical Societies
ICT	information and communication technology
IFSD	in-flight shutdown
IP	intermediate pressure
IPCC	Intergovernmental Panel on Climate Change
IPD	integrated product development
ISA	ICAO Standard Atmosphere
IWB	integrated wing body
JAR	Joint Aviation Regulations
JWA	joined wing aircraft
KBE	knowledge based engineering
LCC	life cycle costs

LEMAC   leading edge of the MAC
LFC   laminar flow control
LFL   landing field length
LH2   liquid hydrogen
LOF   lifting off
LP   low pressure
LRC   long-range cruise
MAC   mean aerodynamic chord
MDO   multidisciplinary design optimization
MEW   manufacturers empty weight
MLW   maximum landing weight
MOO   multi-objective optimization
MPL   maximum payload
MTOW   maximum take-off weight
MVO   multivariate optimization
MZFW   maximum zero fuel weight
NACA   National Advisory Committee for Aeronautics
NASA   National Aeronautics and Space Administration
NBAA   National Business Aviation Association
NLF   natural laminar flow
OEF   overall engine efficiency
OEI   one engine inoperative
OEW   operating empty weight
OI   operator's item
OPR   overall pressure ratio
PAD   payload accommodation density
PDM   product data model
PEF   propulsive efficiency
PFE   payload fuel efficiency
PSFC   power-specific fuel consumption
ROC   rate of climb
RTO   rejected take-off
RTOD   regular take-off distance
R&T   research and technology
SAR   specific air range
SAS   stability augmentation system
SAWE   Society of Allied Weight Engineers
SBW   strut-braced wing
SDL   span distributed loading
SEAD   Systems Engineering and Aircraft Design
SFC   specific fuel consumption
SHP   shaft horsepower
SI   structural index
SL   sea level
SLP   sequential linear programming
SLPL   space limited payload

SPL	structural payload
SQP	sequential quadratic programming
SR	structural root
SSPF	single stage propfans
TBA	twin-body aircraft
TEF	thermal efficiency
TET	turbine entry temperature
TLR	thrust lapse rate
TLR	top level requirement
TOC	top of climb
TOD	top of descent
TOFL	take-off field length
TOGW	take-off gross weight
TSFC	thrust-specific fuel consumption
TTHP	total thrust horsepower
UDF	unducted fan
UL	useful load
VLJ	very light jet
WAT	wing and tail
WPF	wing penalty function

Index

80 m box, 176, 186

accelerate-stop distance, 268
active constraint, 11, 201, 205
active point, 208
advanced design, 3, 4, 6, 31, 81, 206, 213, 320
aerodynamic centre, 165, 180
aerodynamic efficiency, 32, 43, 45, 47, 83, 102,
    125, 131, 136, 239, 256, 281, 286, 287,
    366, 388
aerodynamic limit, 16, 165, 168
aft spar, 322
aileron, 348
air brake, 348
air traffic control, 380
air traffic management, 24
airborne distance, 273
airborne systems, 24
airfoil, 82
airframe noise, 77
airport compatibility, 35
airworthiness
    certification, 7
    code, 12
    requirement, 32, 39
all engines operating, 268
all-wing aircraft, 11, 24, 108, 121, 128, 129, 162
alternative fuel, 76
ambient conditions, 262
ambient pressure, 233
analytical optimization, 5
approach climb, 266
approach speed, 35, 262
artificial intelligence, 212

artificial stabilization, 145
aspect ratio, 90, 127, 128, 135, 158, 182, 218,
    221, 272, 284, 333, 393
automated design synthesis, 19
automated optimization, 197
average deceleration, 274

balanced design, 167, 168
balanced field length, 17, 268
baseline design, 10, 13, 17
bi-level optimization, 217
biofuel, 76
biplane, 91, 115, 159–161, 179
    theory, 91
blended wing body, 11, 45, 55, 124, 128, 144,
    159, 162, 321, 394
blended winglet, 312
block fuel, 385
boundary domain, 199
boundary layer, 85
box beam, 321
boxplane, 94, 115, 161, 162, 179, 321
    transonic, 161, 162, 173
Bréguet range equation, 43, 379, 385
branched tips, 161
Brayton cycle, 60
buffet
    boundary, 276
    margin, 378
buffeting, 306
business jet, 34
bypass engine, 2
bypass ratio, 15, 21, 159, 244, 395
    ultra-high, 23

C-wing, 162, 163
calculus of variations, 200
calorific value, 43, 51, 63, 365
canard configuration, 11, 93, 159, 162, 170
cantilever ratio, 332, 353
cantilevered wing, 179, 323
carpet plot, 21
centre of gravity, 12, 16, 36, 100, 143, 158, 165
centre of lift, 90
centre of pressure, 236, 311, 327
centre section, 322, 353
    tank, 354
certification category, 36
climb gradient, 262
climb rating, 262
climb-out performance, 32
combustion efficiency, 64, 376
community noise, 149
competition evaluation, 7
component build-up technique, 99
composite material, 24
compound taper, 310, 354
compressibility drag, 49, 98, 107, 241, 299, 301, 364
compressive stress, 332, 350
computational fluid dynamics, 12, 37, 85
computational system, 19
computer assisted design, 13, 197
concept
    definition, 10
    finding, 4, 7, 198
    sizing, 213
conceptual design, 4, 8, 9, 84, 198
concurrent engineering, 5
configuration, 10
configuration freeze, 9, 13
conjugate gradient algorithm, 209
constant speed propeller, 69
constrained optimization, 200
constraint, 21, 25, 200
continued take-off, 266, 268
continuous cruise/climb, 380
continuous optimization, 199
contour plot, 203
contra-rotating fans, 71
control parameter, 220
control vector, 201, 208
core engine, 61, 62, 74
corrected
    lapse rate, 262

performance, 65
rotor speed, 65
thrust, 136, 138, 374
cost function, 201
critical Mach number, 87, 98, 285, 366
cruise
    altitude, 25, 26, 232, 233, 256
    condition, 32, 244
    Mach number, 221
    speed, 8, 262
cruise control factor, 385

decision speed, 268
dependent design variable, 368
dependent variable, 199, 291
derivative design, 36
design
    condition, 26, 232, 299
    constraint, 8, 233
    definition, 11
    diving speed, 334
    efficiency, 104
    lift coefficient, 89, 285
    mission, 14, 42
    parameter, 197
    payload, 40
    range, 40, 45
    sensitivity, 37, 222
    space, 21, 25, 199, 202, 218, 285
    structure matrix, 216
    synthesis, 31
    technology, 1
    validation, 11
    variable, 19, 47
    weight, 14, 32, 36, 39, 45
design optimization, 4, 11
    analytical, 4
    automated, 5
    multidisciplinary, 5
design selection chart, 263, 297
design variable
    dependent, 25
    independent, 14, 25
detail design, 4, 9, 13, 31, 37
deterministic method, 212
development process, 8
direct lift control, 181
direct operating costs, 218, 276, 365
direct search, 210
discrete-variable optimization, 200

discrete wing and body, 128, 129
diversion distance, 388
downwash angle, 86
downwash gradient, 167
drag
   area, 88, 103
   bucket, 97
   build-up technique, 84
   coefficient, 82
   count, 83, 131, 300
   creep, 98
   divergence, 241, 300
   due to lift, 47, 87, 95, 131, 240, 287, 299
   inflation, 85
   parameter, 130, 139
   polar, 47, 91, 95–97, 100, 105, 240
   rise, 138, 299, 366
drag area, 130, 240, 241, 287
drag-divergence Mach number, 99
durability, 36
dynamic pressure, 82, 87, 131
dynamic programming, 200
dynamic viscosity, 398

elastic axis, 327
empennage, 158
empty weight, 13
enabling technology, 7, 22, 31
end plate, 161
energy efficiency, 32, 51, 159, 230, 233, 252, 374
energy height, 386
energy transfer efficiency, 64
engine
   cycle, 15
   failure, 18, 268
   figure of merit, 276
   nacelle, 1
   noise, 71, 77
   rating, 65
   rubberizing, 15, 26
   selection, 7
engine-airframe matching, 138
engine-out altitude capability, 262
environmental issues, 32, 35
equality constraint, 134, 201, 204
equivalent
   power, 291
   range, 245, 277, 288, 296
   shaft power, 68

skin friction, 102, 107
skin thickness, 324
exhaust emissions, 51
experimental design, 211
expert system, 212

fairing, 323
fairing geometry, 394
fan, 61
fatigue, 349
feasible design, 201
feasible region, 21, 201, 205
figure of merit, 19, 198, 200, 214, 233, 251, 254, 263, 283, 290
fineness ratio, 88, 394
finite element method, 12, 37, 295
fixed
   leading edge, 345
   trailing edge, 346
   weight, 42
flat plate analogy, 88, 103
flat rating, 267
flight
   envelope, 145, 325
   manual, 268
   profile, 17
flight control
   device, 348
   system, 12, 24
flow mixer, 62
fly-by-wire, 3
flying wing, 121, 159
   controversy, 123, 124, 127
   hybrid, 146, 163, 175
foreplane, 126, 162, 170
form drag, 85, 87, 106, 160
form factor, 88, 103
forward swept wing, 311
freighter, 44
friction drag, 106, 128, 299
front spar, 322
Froude equation, 65, 377
fuel
   energy efficiency, 307
   load, 38
   tank capacity, 39, 286
   weight, 32, 41
   weight flow, 62, 65
fuel tank
   volume, 107

fully stressed structure, 338
functional group, 37
functional sensitivity, 286
fuselage cross-section, 10
future projects, 6

geared turbofan, 23, 51, 71, 76, 277
genetic algorithm, 212
geometric programming, 207
global model, 208
global optimizer, 204
gradient search, 208
gross weight, 37, 82, 107
gross wing area, 83
ground load limit, 165
gust
    alleviation factor, 325
    load, 236, 325
    speed, 325

harmonic range, 40, 45, 52, 253, 254
heuristic method, 212
high speed cruising, 384
holding period, 388
horizontal tail, 2, 16, 34
    volume coefficient, 166
hybrid laminar flow control, 111
hydrogen fuelled engine, 24

ideal weight, 323, 357
inactive constraint, 205
inboard profile, 10
independent variable, 199, 203, 286, 368
induced drag, 24, 86, 87, 114, 128
inequality constraint, 201, 204
inertia relief, 236, 289, 322, 336, 351, 358
inherent stability, 167, 168
inoperative engine, 18
installed thrust, 240, 244, 251, 255
installed thrust lapse, 263
integer optimization, 199
integrated
    configuration, 11, 24, 135, 140, 149
    product development, 198, 215
wing body, 128, 129, 141, 146
interference drag, 100, 161

joined wing, 115, 159, 161, 162, 177, 178
joined-wing aircraft, 177

knowledge-based engineering, 21
Korn's equation, 301
Krueger flap, 345

Lagrange multiplier, 206
Lagrangian function, 206
laminar boundary layer, 108
laminar flow control, 24, 104, 110, 293
laminar flow technology, 15
landing
    climb, 267
    distance, 273
    field length, 35, 262, 273
    gear, 16
    weight, 38
leading edge, 34, 165
    flap, 313
    suction, 85
lift
    coefficient, 82
    curve slope, 167, 326
    dumper, 348, 356
    gradient, 286
lifting off, 266
lifting system, 160
limit load, 324
    factor, 236, 308, 354
liquid hydrogen, 77, 186
load and balance, 16
load factor, 324
loading cases, 323
loading diagram, 165
local
    minimizer, 203
    model, 207
    optimizer, 132, 202, 208
logarithmic derivative, 368
long-coupled canard, 170
long-range cruising, 379, 383
lost fuel, 44, 288, 386
lost range, 44, 46, 245, 387

manhole, 340
manoeuvre load, 324
manoeuvring, 324
manufacturer's empty weight, 234
manufacturing capabilities, 7
market analysis, 7
mass engineering, 37

maximum
  cruise rating, 244
  landing weight, 39
  payload, 253
  range, 40, 296
  take-off weight, 230, 290, 319
  zero fuel weight, 39
mean aerodynamic chord, 165
merit function, 201
minimum drag, 96
minimum drag speed, 366
minimum unstick speed, 168
miscellaneous items, 323, 349
missed approach, 266
mission analysis, 17
mission fuel, 17, 26, 38, 39, 43, 45, 113, 245,
  288, 292, 365, 386
mission range, 44
modified Oswald factor, 106
momentum equation, 60
monoplane, 90, 160, 161
multi-body configuration, 141
multidisciplinary optimization, 114
multiplane, 160
multiple station method, 231, 232, 289
Munk, M., 91
Munk stagger theorem, 173

narrow body aircraft, 33
natural laminar flow, 24, 90, 108, 163,
  173
noise footprint, 77
noise shielding, 51
non-ideal weight, 310, 324
nonplanar wing, 11, 24
normal force, 85

object-oriented engineering, 21
objective function, 25, 27, 129, 198, 200, 203,
  222, 232, 233, 283, 285
objective space, 218
obstacle height, 268
open rotor engine, 23, 51, 70, 72, 74,
  140, 159, 364, 369
operating costs, 2, 12, 32
operating empty weight, 37, 230
operational research, 7
operator's items, 37
optimal control, 199, 379

optimization
  algorithm, 197
  multi-objective, 198, 214, 218, 223, 276, 285
  multidisciplinary, 190, 213, 215
  multilevel, 211, 217
  multivariate, 206, 207
  problem structure, 25
optimizer, 200
Oswald factor, 47, 96, 105, 114, 299
overall efficiency, 15, 32, 43, 45, 49, 63, 65, 102,
  245, 276, 277, 291, 364, 369, 374

parametric survey, 5, 21, 128, 134, 198, 207
parasite drag, 87, 125, 128, 130, 140, 162, 172,
  180, 184, 240, 287
Pareto front, 218
partial optimum, 203
payload, 38, 325
  accommodation density, 44
  fuel efficiency, 51
  versus range diagram, 40
peaky airfoil, 301
penalty function, 211
pitch up, 145
planar wing, 160
planform area, 285, 320
power loading, 221
power plant, 12
Prandtl, L., 91
Prandtl Plane, 174
pre-assigned parameter, 199, 221, 233, 291
pre-conceptual study, 7, 221
pre-flight fuel, 39
preliminary design, 4, 9, 11, 18, 84, 97, 214, 231,
  283, 299
pressure cabin, 14, 150
pressure drag, 85, 299
primary structure, 158, 178, 321
product data model, 215
product development, 8
profile drag, 87, 287
project go-ahead, 9, 13
propeller
  diameter, 244
  efficiency, 291, 379
propfan, 23, 70, 72
propulsion
  function, 246, 276, 289, 307
  weight penalty, 243, 245, 252, 276, 289, 294

propulsive efficiency, 59, 64, 74, 376
pure flying wing, 130, 138

radiative forcing, 54
ram drag, 60
range, 40
range parameter, 45, 46, 365, 374, 382
reduced lift coefficient, 132
regional jet aircraft, 33
regional propeller aircraft, 33
rejected take-off, 268
relative
    density, 397
    pressure, 65, 397
    temperature, 65, 397
relative ambient pressure, 242
relaxed stability, 169
reliability, 36
reserve fuel, 17, 38, 39, 45, 250, 385
response surface, 202, 203, 207, 209, 230
resulting aerodynamic force, 85
reverse thrust, 273
rhomboidal wing, 163
rib weight, 324
riblets, 108
rigid structure, 323
rotation speed, 269
rubberizing, 66, 240, 244, 262
ruddervon, 147
    drag, 143

saddle point, 203
safety speed, 265
scale effect, 47
scissor plot, 168, 170
screen height, 273
sea level, 48, 397
second segment climb, 267
secondary structure, 322, 324
selection diagram, 283
selection variable, 21, 31, 37, 199, 202, 220, 232,
    233, 283, 285, 286
sensitivity information, 208
sequential linear programming, 210
sequential quadratic programming, 210
shear load, 324
shear stress, 352
sheared wing tip, 104, 312, 326, 349, 352
shock wave, 86
shroudless propeller, 67

side constraint, 201
simple sweep theory, 302
simplex algorithm, 210
single-level optimization, 214
single-objective optimization, 201, 206
skin friction drag, 47, 85, 98, 160
slat, 345
slotted flap, 313
span distributed loading, 182
span efficiency factor, 90, 92, 100, 115, 161, 175,
    179
span loader, 147, 162
span loading, 114, 127, 272, 286, 296, 306, 336
spanwise camber, 115
specific
    fuel consumption, 15, 48, 59, 62, 364, 366
    range, 40, 43, 46, 49, 51, 138, 379
    thrust, 60, 65, 70, 71, 74, 267, 395
spillage drag, 100
spoiler, 348, 356
square cube law, 42, 148, 236, 237, 333
stability augmentation system, 169
stability margin, 100, 167, 169
stagger theorem, 92
stalling speed, 265
standard atmosphere, 66, 262, 397, 398
static margin, 180
station analysis method, 321
steepest descent, 209
stepped cruise/climb, 381, 384
straight jet engine, 59, 60, 364
straight-taper, 284
structural
    configuration, 16
    divergence, 311
    efficiency, 52
    index, 332, 351
    material, 32
    payload, 39
    root, 322
    span, 328
structure
    damage tolerant, 340
    fail safe, 340
strut-braced wing, 11, 141, 163, 284
subsonic aircraft, 284
supercritical
    airfoil, 301
    flow, 284
    technology, 15

sweep angle, 88, 221, 298
sweepback angle, 284
synthesis program, 5, 197, 234, 253
system decomposition, 215
system engineering, 12, 197, 216

T-tail, 34, 169
tail volume coefficient, 168
tailless aircraft, 121, 141, 162
take-off
    distance, 268
    field length, 32, 35, 262, 268, 273
    run, 269
    thrust, 14, 15, 66, 262
    weight, 37, 39, 127
tandem wing, 162
tangential force, 85
taper ratio, 221, 236, 329, 393
technology assessment, 22
tensile stress, 332
thermal efficiency, 59, 64, 376
thermodynamic efficiency, 64
thickness ratio, 88, 140, 221, 236, 298, 301,
    393
three-surface aircraft, 11, 162, 172, 176
thrust, 60, 61
    gross, 60
    ideal, 60
    net, 60
    propeller, 48
    standard net, 60
thrust lapse, 244, 263
    parameter, 66
    rate, 15, 66, 67, 262, 289
thrust loading, 221
tip extension, 352
tool development, 6
top level requirements, 6, 7, 31, 35, 205,
    262
top of climb, 262
torsion, 324
torsional stiffness, 324
total fuel, 39
total thrust horsepower, 67
transonic aircraft, 284
transonic biplane, 115
transport capability, 40
trim drag, 100, 169, 310, 314
triplane, 94
tropopause, 263

tube and wing, 108, 125, 157, 232, 282
tumbling, 145
turbofan engine, 34, 61, 375
turboprop engine, 33, 67
turbulent boundary layer, 108
twin-fuselage aircraft, 11, 163, 182
type specification, 12, 37

ultimate load, 324
    factor, 236, 237, 308, 331, 358
ultimate range, 40
ultimate stress, 324
ultra-high bypass ratio, 364
unconstrained optimization, 261
unducted fan, 3, 23, 73
unfeasible region, 205
unity equation, 41, 42, 45
useful load, 37, 102, 126, 128, 165, 230, 249,
    282
useful volume, 128

V-tail, 162
V-wing, 161
variable weight, 42
vertical gap, 160
vertical gust, 324
vertical tail, 16
viscous drag, 85
volume ratio, 129, 132, 134
volumetric efficiency, 130
volumetric payload, 39
vortex drag, 86, 106, 160, 161, 173, 299,
    312

wake vortex, 149
wave drag, 48, 86, 99, 107, 114, 180
weight
    and balance, 10
    empty, 32
    engineering, 36
    fraction, 41
    growth, 13
    limit, 39
    reduction, 37
    sensitivity, 32
    statement, 36
well-tempered design, 276, 389
wetted area, 47, 71, 107
wide body aircraft, 33

wing
  area, 221
  bending, 323
  box, 158
  loading, 14, 221, 232, 233, 250, 256, 285
  planform, 10
  position, 10
  span, 86, 135, 140, 221, 222, 286, 320
wing and tail, 162, 165
wing penalty function, 290, 300

winglet, 24, 48, 104, 148, 161, 162, 172, 180,
  312, 321, 326, 349, 352

X-wing, 162

yield stress, 324

zero fuel weight, 38, 325
zero-lift drag, 87, 95, 101, 130, 240, 299

Printed in the United States
By Bookmasters